Springer Handbook of Nanotechnology

Bharat Bhushan (Ed.)

3rd revised and extended edition

Springer Handbook of Nanotechnology

Since 2004 and with the 2nd edition in 2006, the Springer Handbook of Nanotechnology has established itself as the definitive reference in the nanoscience and nanotechnology area. It integrates the knowledge from nanofabrication, nanodevices, nanomechanics, nanotribology, materials science, and reliability engineering in just one volume. Beside the presentation of nanostructures, micro/nanofabrication, and micro/nanodevices, special emphasis is on scanning probe microscopy, nanotribology and nanomechanics, molecularly thick films, industrial applications and microdevice reliability, and on social aspects. In its 3rd edition, the book grew from 8 to 9 parts now including a part with chapters on biomimetics. More information is added to such fields as bionanotechnology, nanorobotics, and (bio) MEMS/NEMS, bio/nanotribology and bio/nanomechanics. The book is organized by an experienced editor with a universal knowledge and written by an international team of over 145 distinguished experts. It addresses mechanical and electrical engineers, materials scientists, physicists and chemists who work either in the nano area or in a field that is or will be influenced by this new key technology.

"The strong point is its focus on many of the practical aspects of nanotechnology... Anyone working in or learning about the field of nanotechnology would find this an excellent working handbook."

IEEE Electrical Insulation Magazine

"Outstandingly succeeds in its aim... It really is a magnificent volume and every scientific library and nanotechnology group should have a copy."

Materials World

"The integrity and authoritativeness... is guaranteed by an experienced editor and an international team of authors which have well summarized in their chapters information on fundamentals and applications."

Polymer News

List of Abbreviations

1 Introduction to Nanotechnology

Part A Nanostructures, Micro-/Nanofabrication and Materials

2 Nanomaterials Synthesis and Applications: Molecule-Based Devices
3 Introduction to Carbon Nanotubes
4 Nanowires
5 Template-Based Synthesis of Nanorod or Nanowire Arrays
6 Templated Self-Assembly of Particles
7 Three-Dimensional Nanostructure Fabrication by Focused Ion Beam Chemical Vapor Deposition
8 Introduction to Micro-/Nanofabrication
9 Nanoimprint Lithography-Patterning of Resists Using Molding
10 Stamping Techniques for Micro- and Nanofabrication
11 Material Aspects of Micro- and Nanoelectromechanical Systems

Part B MEMS/NEMS and BioMEMS/NEMS

12 MEMS/NEMS Devices and Applications
13 Next-Generation DNA Hybridization and Self-Assembly Nanofabrication Devices
14 Single-Walled Carbon Nanotube Sensor Concepts
15 Nanomechanical Cantilever Array Sensors
16 Biological Molecules in Therapeutic Nanodevices
17 G-Protein Coupled Receptors: Progress in Surface Display and Biosensor Technology
18 Microfluidic Devices and Their Applications to Lab-on-a-Chip
19 Centrifuge-Based Fluidic Platforms
20 Micro-/Nanodroplets in Microfluidic Devices

Part C Scanning-Probe Microscopy
21 Scanning Probe Microscopy-Principle of Operation, Instrumentation,and Probes
22 General and Special Probes in Scanning Microscopies
23 Noncontact Atomic Force Microscopy and Related Topics
24 Low-Temperature Scanning Probe Microscopy
25 Higher Harmonics and Time-Varying Forces in Dynamic Force Microscopy
26 Dynamic Modes of Atomic Force Microscopy
27 Molecular Recognition Force Microscopy: From Molecular Bonds to Complex Energy Landscapes

Part D Bio-/Nanotribology and Bio-/ Nanomechanics
28 Nanotribology, Nanomechanics, and Materials Characterization
29 Surface Forces and Nanorheology of Molecularly Thin Films
30 Friction and Wear on the Atomic Scale
31 Computer Simulations of Nanometer-Scale Indentation and Friction
32 Force Measurements with Optical Tweezers
33 Scale Effect in Mechanical Properties and Tribology
34 Structural, Nanomechanical, and Nanotribological Characterization of Human Hair Using Atomic Force Microscopy and Nanoindentation
35 Cellular Nanomechanics
36 Optical Cell Manipulation
37 Mechanical Properties of Nanostructures

Part E Molecularly Thick Films for Lubrication
38 Nanotribology of Ultrathin and Hard Amorphous Carbon Films
39 Self-Assembled Monolayers for Nanotribology and Surface Protection
40 Nanoscale Boundary Lubrication Studies

Part F Biomimetics
41 Multifunctional Plant Surfaces and Smart Materials
42 Lotus Effect: Surfaces with Roughness-Induced Superhydrophobicity, Self-Cleaning, and Low Adhesion
43 Biological and Biologically Inspired Attachment Systems
44 Gecko Feet: Natural Hairy Attachment Systems for Smart Adhesion

Part G Industrial Applications
45 The *Millipede*-A Nanotechnology-Based AFM Data-Storage System
46 Nanorobotics

Part H Micro-/Nanodevice Reliability
47 MEMS/NEMS and BioMEMS/BioNEMS: Materials, Devices, and Biomimetics
48 Friction and Wear in Micro- and Nanomachines
49 Failure Mechanisms in MEMS/NEMS Devices
50 Mechanical Properties of Micromachined Structures
51 High-Volume Manufacturing and Field Stability of MEMS Products
52 Packaging and Reliability Issues in Micro-/ Nanosystems

Part I Technological Convergence and Governing Nanotechnology
53 Governing Nanotechnology: Social, Ethical and Human Issues

Subject Index

使 用 说 明

1.《纳米技术手册》原版为一册，分为A～I部分。考虑到使用方便以及内容一致，影印版分为7册：第1册—Part A，第2册—Part B，第3册—Part C，第4册—Part D，第5册—Part E，第6册—Part F, 第7册—Part G、H、I。

2.各册在页脚重新编排页码，该页码对应中文目录。保留了原书页眉及页码，其页码对应原书目录及主题索引。

3.各册均给出完整7册书的章目录。

4.作者及其联系方式、缩略语表各册均完整呈现。

5.主题索引安排在第7册。

6.目录等采用中英文对照形式给出，方便读者快速浏览。

材料科学与工程图书工作室

联系电话　0451-86412421

　　　　　0451-86414559

邮　　箱　yh_bj@yahoo.com.cn

　　　　　xuyaying81823@gmail.com

　　　　　zhxh6414559@yahoo.com.cn

Springer 手册精选系列

纳米技术手册

纳米技术的应用

【第7册】

Springer Handbook of Nanotechnology

〔美〕Bharat Bhushan　主编

（第三版影印版）

黑版贸审字 08-2013-001号

图书在版编目（CIP）数据

纳米技术手册 : 第3版. 7, 纳米技术的应用 =Handbook of Nanotechnology. 7, Application of Nanotechnology : 英文 / (美) 布尚 (Bhushan,B.) 主编. — 影印本. — 哈尔滨 : 哈尔滨工业大学出版社,2013.1
（Springer手册精选系列）
ISBN 978-7-5603-3953-5

Ⅰ. ①纳… Ⅱ. ①布… Ⅲ. ①纳米技术 – 手册 – 英文 Ⅳ. ①TB303-62

中国版本图书馆CIP数据核字(2013)第004227号

责任编辑 杨　桦　许雅莹　张秀华
出版发行 哈尔滨工业大学出版社
社　　址 哈尔滨市南岗区复华四道街10号　邮编 150006
传　　真 0451-86414749
网　　址 http://hitpress.hit.edu.cn
印　　刷 哈尔滨市石桥印务有限公司
开　　本 787mm × 960mm　1/16　印张 23.25
版　　次 2013年1月第1版　2013年1月第1次印刷
书　　号 ISBN 978-7-5603-3953-5
定　　价 68.00元

Foreword by Neal Lane

In a January 2000 speech at the California Institute of Technology, former President W.J. Clinton talked about the exciting promise of *nanotechnology* and the importance of expanding research in nanoscale science and engineering and, more broadly, in the physical sciences. Later that month, he announced in his State of the Union Address an ambitious US$ 497 million federal, multiagency national nanotechnology initiative (NNI) in the fiscal year 2001 budget; and he made the NNI a top science and technology priority within a budget that emphasized increased investment in US scientific research. With strong bipartisan support in Congress, most of this request was appropriated, and the NNI was born. Often, federal budget initiatives only last a year or so. It is most encouraging that the NNI has remained a high priority of the G.W. Bush Administration and Congress, reflecting enormous progress in the field and continued strong interest and support by industry.

Nanotechnology is the ability to manipulate individual atoms and molecules to produce nanostructured materials and submicron objects that have applications in the real world. Nanotechnology involves the production and application of physical, chemical and biological systems at scales ranging from individual atoms or molecules to about 100 nm, as well as the integration of the resulting nanostructures into larger systems. Nanotechnology is likely to have a profound impact on our economy and society in the early 21st century, perhaps comparable to that of information technology or cellular and molecular biology. Science and engineering research in nanotechnology promises breakthroughs in areas such as materials and manufacturing, electronics, medicine and healthcare, energy and the environment, biotechnology, information technology and national security. Clinical trials are already underway for nanomaterials that offer the promise of cures for certain cancers. It is widely felt that nanotechnology will be the next industrial revolution.

Nanometer-scale features are built up from their elemental constituents. Micro- and nanosystems components are fabricated using batch-processing techniques that are compatible with integrated circuits and range in size from micro- to nanometers. Micro- and nanosystems include micro/nanoelectro-mechanical systems (MEMS/NEMS), micromechatronics, optoelectronics, microfluidics and systems integration. These systems can sense, control, and activate on the micro/nanoscale and can function individually or in arrays to generate effects on the macroscale. Due to the enabling nature of these systems and the significant impact they can have on both the commercial and defense applications, industry as well as the federal government have taken special interest in seeing growth nurtured in this field. Micro- and nanosystems are the next logical step in the *silicon revolution*.

Prof. Neal Lane
Malcolm Gillis University Professor,
Department of Physics and Astronomy,
Senior Fellow,
James A. Baker III Institute for Public Policy
Rice University
Houston, Texas

Served in the Clinton Administration as Assistant to the President for Science and Technology and Director of the White House Office of Science and Technology Policy (1998–2001) and, prior to that, as Director of the National Science Foundation (1993–1998). While at the White House, he was a key figure in the creation of the NNI.

The discovery of novel materials, processes, and phenomena at the nanoscale and the development of new experimental and theoretical techniques for research provide fresh opportunities for the development of innovative nanosystems and nanostructured materials. There is an increasing need for a multidisciplinary, systems-oriented approach to manufacturing micro/nanodevices which function reliably. This can only be achieved through the cross-fertilization of ideas from different disciplines and the systematic flow of information and people among research groups.

Nanotechnology is a broad, highly interdisciplinary, and still evolving field. Covering even the most important aspects of nanotechnology in a single book that reaches readers ranging from students to active researchers in academia and industry is an enormous challenge. To prepare such a wide-ranging book on nanotechnology, Prof. Bhushan has harnessed his own knowledge and experience, gained in several industries and universities, and has assembled internationally recognized authorities from four continents to write chapters covering a wide array of nanotechnology topics, including the latest advances. The authors come from both academia and industry. The topics include major advances in many fields where nanoscale science and engineering is being pursued and illustrate how the field of nanotechnology has continued to emerge and blossom. Given the accelerating pace of discovery and applications in nanotechnology, it is a challenge to capture it all in one volume. As in earlier editions, professor Bhushan does an admirable job.

Professor Bharat Bhushan's comprehensive book is intended to serve both as a textbook for university courses as well as a reference for researchers. The first and second editions were timely additions to the literature on nanotechnology and stimulated further interest in this important new field, while serving as invaluable resources to members of the international scientific and industrial community. The increasing demand for up-to-date information on this fast moving field led to this third edition. It is increasingly important that scientists and engineers, whatever their specialty, have a solid grounding in the fundamentals and potential applications of nanotechnology. This third edition addresses that need by giving particular attention to the widening audience of readers. It also includes a discussion of the social, ethical and political issues that tend to surround any emerging technology.

The editor and his team are to be warmly congratulated for bringing together this exclusive, timely, and useful nanotechnology handbook.

Foreword by James R. Heath

Nanotechnology has become an increasingly popular buzzword over the past five years or so, a trend that has been fueled by a global set of publicly funded nanotechnology initiatives. Even as researchers have been struggling to demonstrate some of the most fundamental and simple aspects of this field, the term nanotechnology has entered into the public consciousness through articles in the popular press and popular fiction. As a consequence, the expectations of the public are high for nanotechnology, even while the actual public definition of nanotechnology remains a bit fuzzy.

Why shouldn't those expectations be high? The late 1990s witnessed a major information technology (IT) revolution and a minor biotechnology revolution. The IT revolution impacted virtually every aspect of life in the western world. I am sitting on an airplane at 30 000 feet at the moment, working on my laptop, as are about half of the other passengers on this plane. The plane itself is riddled with computational and communications equipment. As soon as we land, many of us will pull out cell phones, others will check e-mail via wireless modem, some will do both. This picture would be the same if I was landing in Los Angeles, Beijing, or Capetown. I will probably never actually print this text, but will instead submit it electronically. All of this was unthinkable a dozen years ago. It is therefore no wonder that the public expects marvelous things to happen quickly. However, the science that laid the groundwork for the IT revolution dates back 60 years or more, with its origins in fundamental solid-state physics.

By contrast, the biotech revolution was relatively minor and, at least to date, not particularly effective. The major diseases that plagued mankind a quarter century ago are still here. In some third-world countries, the average lifespan of individuals has actually decreased from where it was a full century ago. While the costs of electronics technologies have plummeted, health care costs have continued to rise. The biotech revolution may have a profound impact, but the task at hand is substantially more difficult than what was required for the IT revolution. In effect, the IT revolution was based on the advanced engineering of two-dimensional digital circuits constructed from relatively simple components – extended solids. The biotech revolution is really dependent upon the ability to reverse engineer three-dimensional analog systems constructed from quite complex components – proteins. Given that the basic science behind biotech is substantially younger than the science that has supported IT, it is perhaps not surprising that the biotech revolution has not really been a proper revolution yet, and it likely needs at least another decade or so to come into fruition.

Where does nanotechnology fit into this picture? In many ways, nanotechnology depends upon the ability to engineer two- and three-dimensional systems constructed from complex components such as macromolecules, biomolecules, nanostructured solids, etc. Furthermore, in terms of patents, publications, and other metrics that can be used to gauge the birth and evolution of a field, nanotech lags some 15–20 years behind biotech. Thus, now is the time that the fundamental science behind nanotechnology is being explored and developed. Nevertheless, progress with that science is moving forward at a dramatic pace. If the scientific community can keep up this pace and if the public sector will continue to support this science, then it is possible, and even perhaps likely, that in 20 years we may be speaking of the nanotech revolution.

The first edition of Springer Handbook of Nanotechnology was timely to assemble chapters in the broad field of nanotechnology. Given the fact that the second edition was in press one year after the publication of the first edition in April 2004, it is clear that the handbook has shown to be a valuable reference for experienced researchers as well as for a novice in the field. The third edition has one Part added and an expanded scope should have a wider appeal.

Prof. James R. Heath

Department of Chemistry
California Institute of Technology
Pasadena, California

Worked in the group of Nobel Laureate Richard E. Smalley at Rice University (1984–88) and co-invented Fullerene molecules which led to a revolution in Chemistry including the realization of nanotubes. The work on Fullerene molecules was cited for the 1996 Nobel Prize in Chemistry. Later he joined the University of California at Los Angeles (1994–2002), and co-founded and served as a Scientific Director of The California Nanosystems Institute.

Preface to the 3rd Edition

On December 29, 1959 at the California Institute of Technology, Nobel Laureate Richard P. Feynman gave at talk at the Annual meeting of the American Physical Society that has become one of the 20th century classic science lectures, titled *There's Plenty of Room at the Bottom*. He presented a technological vision of extreme miniaturization in 1959, several years before the word *chip* became part of the lexicon. He talked about the problem of manipulating and controlling things on a small scale. Extrapolating from known physical laws, Feynman envisioned a technology using the ultimate toolbox of nature, building nanoobjects atom by atom or molecule by molecule. Since the 1980s, many inventions and discoveries in fabrication of nanoobjects have been testament to his vision. In recognition of this reality, National Science and Technology Council (NSTC) of the White House created the Interagency Working Group on Nanoscience, Engineering and Technology (IWGN) in 1998. In a January 2000 speech at the same institute, former President W.J. Clinton talked about the exciting promise of *nanotechnology* and the importance of expanding research in nanoscale science and technology, more broadly. Later that month, he announced in his State of the Union Address an ambitious US$ 497 million federal, multi-agency national nanotechnology initiative (NNI) in the fiscal year 2001 budget, and made the NNI a top science and technology priority. The objective of this initiative was to form a broad-based coalition in which the academe, the private sector, and local, state, and federal governments work together to push the envelop of nanoscience and nanoengineering to reap nanotechnology's potential social and economic benefits.

The funding in the US has continued to increase. In January 2003, the US senate introduced a bill to establish a National Nanotechnology Program. On December 3, 2003, President George W. Bush signed into law the 21st Century Nanotechnology Research and Development Act. The legislation put into law programs and activities supported by the National Nanotechnology Initiative. The bill gave nanotechnology a permanent home in the federal government and authorized US$ 3.7 billion to be spent in the four year period beginning in October 2005, for nanotechnology initiatives at five federal agencies. The funds would provide grants to researchers, coordinate R&D across five federal agencies (National Science Foundation (NSF), Department of Energy (DOE), NASA, National Institute of Standards and Technology (NIST), and Environmental Protection Agency (EPA)), establish interdisciplinary research centers, and accelerate technology transfer into the private sector. In addition, Department of Defense (DOD), Homeland Security, Agriculture and Justice as well as the National Institutes of Health (NIH) also fund large R&D activities. They currently account for more than one-third of the federal budget for nanotechnology.

European Union (EU) made nanosciences and nanotechnologies a priority in Sixth Framework Program (FP6) in 2002 for a period of 2003–2006. They had dedicated small funds in FP4 and FP5 before. FP6 was tailored to help better structure European research and to cope with the strategic objectives set out in Lisbon in 2000. Japan identified nanotechnology as one of its main research priorities in 2001. The funding levels increases sharply from US$ 400 million in 2001 to around US$ 950 million in 2004. In 2003, South Korea embarked upon a ten-year program with around US$ 2 billion of public funding, and Taiwan has committed around US$ 600 million of public funding over six years. Singapore and China are also investing on a large scale. Russia is well funded as well.

Nanotechnology literally means any technology done on a nanoscale that has applications in the real world. Nanotechnology encompasses production and application of physical, chemical and biological systems at scales, ranging from individual atoms or molecules to submicron dimensions, as well as the integration of the resulting nanostructures into larger systems. Nanotechnology is likely to have a profound impact on our economy and society in the early 21st century, comparable to that of semiconductor technology, information technology, or cellular and molecular biology. Science and technology research in nanotechnology promises breakthroughs in areas such as materials and manufacturing, nanoelectronics, medicine and healthcare, energy, biotechnology, information technology and national security. It is widely felt that nanotechnology will be the next industrial revolution.

There is an increasing need for a multidisciplinary, system-oriented approach to design and manufacturing of micro/nanodevices which function reliably. This can only be achieved through the cross-fertilization of ideas from different disciplines and the systematic flow of information and people among research groups. Reliability is a critical technology for many micro- and nanosystems and nanostructured materials. A broad based handbook was needed, and the first edition of Springer Handbook of Nanotechnology was published in April 2004. It presented an overview of nanomaterial synthesis, micro/nanofabrication, micro- and nanocomponents and systems, scanning probe microscopy, reliability issues (including nanotribology and nanomechanics) for nanotechnology, and industrial applications. When the handbook went for sale in Europe, it was sold out in ten days. Reviews on the handbook were very flattering.

Given the explosive growth in nanoscience and nanotechnology, the publisher and the editor decided to develop a second edition after merely six months of publication of the first edition. The second edition (2007) came out in December 2006. The publisher and the editor again decided to develop a third edition after six month of publication of the second edition. This edition of the handbook integrates the knowledge from nanostructures, fabrication, materials science, devices, and reliability point of view. It covers various industrial applications. It also addresses social, ethical, and political issues. Given the significant interest in biomedical applications, and biomimetics a number of additional chapters in this arena have been added. The third edition consists of 53 chapters (new 10, revised 28, and as is 15). The chapters have been written by 139 internationally recognized experts in the field, from academia, national research labs, and industry, and from all over the world.

This handbook is intended for three types of readers: graduate students of nanotechnology, researchers in academia and industry who are active or intend to become active in this field, and practicing engineers and scientists who have encountered a problem and hope to solve it as expeditiously as possible. The handbook should serve as an excellent text for one or two semester graduate courses in nanotechnology in mechanical engineering, materials science, applied physics, or applied chemistry.

We embarked on the development of third edition in June 2007, and we worked very hard to get all the chapters to the publisher in a record time of about 12 months. I wish to sincerely thank the authors for offering to write comprehensive chapters on a tight schedule. This is generally an added responsibility in the hectic work schedules of researchers today. I depended on a large number of reviewers who provided critical reviews. I would like to thank Dr. Phillip J. Bond, Chief of Staff and Under Secretary for Technology, US Department of Commerce, Washington, D.C. for suggestions for chapters as well as authors in the handbook. Last but not the least, I would like to thank my secretary Caterina Runyon-Spears for various administrative duties and her tireless efforts are highly appreciated.

I hope that this handbook will stimulate further interest in this important new field, and the readers of this handbook will find it useful.

February 2010

Bharat Bhushan
Editor

Preface to the 2nd Edition

On 29 December 1959 at the California Institute of Technology, Nobel Laureate Richard P. Feynman gave at talk at the Annual meeting of the American Physical Society that has become one of the 20th century classic science lectures, titled "There's Plenty of Room at the Bottom." He presented a technological vision of extreme miniaturization in 1959, several years before the word "chip" became part of the lexicon. He talked about the problem of manipulating and controlling things on a small scale. Extrapolating from known physical laws, Feynman envisioned a technology using the ultimate toolbox of nature, building nanoobjects atom by atom or molecule by molecule. Since the 1980s, many inventions and discoveries in the fabrication of nanoobjects have been a testament to his vision. In recognition of this reality, the National Science and Technology Council (NSTC) of the White House created the Interagency Working Group on Nanoscience, Engineering and Technology (IWGN) in 1998. In a January 2000 speech at the same institute, former President W. J. Clinton talked about the exciting promise of "nanotechnology" and the importance of expanding research in nanoscale science and, more broadly, technology. Later that month, he announced in his State of the Union Address an ambitious $497 million federal, multiagency national nanotechnology initiative (NNI) in the fiscal year 2001 budget, and made the NNI a top science and technology priority. The objective of this initiative was to form a broad-based coalition in which the academe, the private sector, and local, state, and federal governments work together to push the envelope of nanoscience and nanoengineering to reap nanotechnology's potential social and economic benefits.

The funding in the U.S. has continued to increase. In January 2003, the U. S. senate introduced a bill to establish a National Nanotechnology Program. On 3 December 2003, President George W. Bush signed into law the 21st Century Nanotechnology Research and Development Act. The legislation put into law programs and activities supported by the National Nanotechnology Initiative. The bill gave nanotechnology a permanent home in the federal government and authorized $3.7 billion to be spent in the four year period beginning in October 2005, for nanotechnology initiatives at five federal agencies. The funds would provide grants to researchers, coordinate R&D across five federal agencies (National Science Foundation (NSF), Department of Energy (DOE), NASA, National Institute of Standards and Technology (NIST), and Environmental Protection Agency (EPA)), establish interdisciplinary research centers, and accelerate technology transfer into the private sector. In addition, Department of Defense (DOD), Homeland Security, Agriculture and Justice as well as the National Institutes of Health (NIH) would also fund large R&D activities. They currently account for more than one-third of the federal budget for nanotechnology.

The European Union made nanosciences and nanotechnologies a priority in the Sixth Framework Program (FP6) in 2002 for the period of 2003-2006. They had dedicated small funds in FP4 and FP5 before. FP6 was tailored to help better structure European research and to cope with the strategic objectives set out in Lisbon in 2000. Japan identified nanotechnology as one of its main research priorities in 2001. The funding levels increased sharply from $400 million in 2001 to around $950 million in 2004. In 2003, South Korea embarked upon a ten-year program with around $2 billion of public funding, and Taiwan has committed around $600 million of public funding over six years. Singapore and China are also investing on a large scale. Russia is well funded as well.

Nanotechnology literally means any technology done on a nanoscale that has applications in the real world. Nanotechnology encompasses production and application of physical, chemical and biological systems at scales, ranging from individual atoms or molecules to submicron dimensions, as well as the integration of the resulting nanostructures into larger systems. Nanotechnology is likely to have a profound impact on our economy and society in the early 21st century, comparable to that of semiconductor technology, information technology, or cellular and molecular biology. Science and technology research in nanotechnology promises breakthroughs in areas such as materials and manufacturing, nanoelectronics, medicine and healthcare, energy, biotechnology, information technology and national security. It is widely felt that nanotechnology will be the next industrial revolution.

There is an increasing need for a multidisciplinary, system-oriented approach to design and manufacturing of micro/nanodevices that function reliably. This can only be achieved through the cross-fertilization of ideas from different disciplines and the systematic flow of information and people among research groups. Reliability is a critical technology for many micro- and nanosystems and nanostructured materials. A broad-based handbook was needed, and thus the first edition of Springer Handbook of Nanotechnology was published in April 2004. It presented an overview of nanomaterial synthesis, micro/nanofabrication, micro- and nanocomponents and systems, scanning probe microscopy, reliability issues (including nanotribology and nanomechanics) for nanotechnology, and industrial applications. When the handbook went for sale in Europe, it sold out in ten days. Reviews on the handbook were very flattering.

Given the explosive growth in nanoscience and nanotechnology, the publisher and the editor decided to develop a second edition merely six months after publication of the first edition. This edition of the handbook integrates the knowledge from the nanostructure, fabrication, materials science, devices, and reliability point of view. It covers various industrial applications. It also addresses social, ethical, and political issues. Given the significant interest in biomedical applications, a number of chapters in this arena have been added. The second edition consists of 59 chapters (new: 23; revised: 27; unchanged: 9). The chapters have been written by 154 internationally recognized experts in the field, from academia, national research labs, and industry.

This book is intended for three types of readers: graduate students of nanotechnology, researchers in academia and industry who are active or intend to become active in this field, and practicing engineers and scientists who have encountered a problem and hope to solve it as expeditiously as possible. The handbook should serve as an excellent text for one or two semester graduate courses in nanotechnology in mechanical engineering, materials science, applied physics, or applied chemistry.

We embarked on the development of the second edition in October 2004, and we worked very hard to get all the chapters to the publisher in a record time of about 7 months. I wish to sincerely thank the authors for offering to write comprehensive chapters on a tight schedule. This is generally an added responsibility to the hectic work schedules of researchers today. I depended on a large number of reviewers who provided critical reviews. I would like to thank Dr. Phillip J. Bond, Chief of Staff and Under Secretary for Technology, US Department of Commerce, Washington, D.C. for chapter suggestions as well as authors in the handbook. I would also like to thank my colleague, Dr. Zhenhua Tao, whose efforts during the preparation of this handbook were very useful. Last but not the least, I would like to thank my secretary Caterina Runyon-Spears for various administrative duties; her tireless efforts are highly appreciated.

I hope that this handbook will stimulate further interest in this important new field, and the readers of this handbook will find it useful.

May 2005

Bharat Bhushan
Editor

Preface to the 1st Edition

On December 29, 1959 at the California Institute of Technology, Nobel Laureate Richard P. Feynman gave a talk at the Annual meeting of the American Physical Society that has become one classic science lecture of the 20th century, titled "There's Plenty of Room at the Bottom." He presented a technological vision of extreme miniaturization in 1959, several years before the word "chip" became part of the lexicon. He talked about the problem of manipulating and controlling things on a small scale. Extrapolating from known physical laws, Feynman envisioned a technology using the ultimate toolbox of nature, building nanoobjects atom by atom or molecule by molecule. Since the 1980s, many inventions and discoveries in fabrication of nanoobjects have been a testament to his vision. In recognition of this reality, in a January 2000 speech at the same institute, former President W. J. Clinton talked about the exciting promise of "nanotechnology" and the importance of expanding research in nanoscale science and engineering. Later that month, he announced in his State of the Union Address an ambitious $497 million federal, multi-agency national nanotechnology initiative (NNI) in the fiscal year 2001 budget, and made the NNI a top science and technology priority. Nanotechnology literally means any technology done on a nanoscale that has applications in the real world. Nanotechnology encompasses production and application of physical, chemical and biological systems at size scales, ranging from individual atoms or molecules to submicron dimensions as well as the integration of the resulting nanostructures into larger systems. Nanofabrication methods include the manipulation or self-assembly of individual atoms, molecules, or molecular structures to produce nanostructured materials and sub-micron devices. Micro- and nanosystems components are fabricated using top-down lithographic and nonlithographic fabrication techniques. Nanotechnology will have a profound impact on our economy and society in the early 21st century, comparable to that of semiconductor technology, information technology, or advances in cellular and molecular biology. The research and development in nanotechnology will lead to potential breakthroughs in areas such as materials and manufacturing, nanoelectronics, medicine and healthcare, energy, biotechnology, information technology and national security. It is widely felt that nanotechnology will lead to the next industrial revolution.

Reliability is a critical technology for many micro- and nanosystems and nanostructured materials. No book exists on this emerging field. A broad based handbook is needed. The purpose of this handbook is to present an overview of nanomaterial synthesis, micro/nanofabrication, micro- and nanocomponents and systems, reliability issues (including nanotribology and nanomechanics) for nanotechnology, and industrial applications. The chapters have been written by internationally recognized experts in the field, from academia, national research labs and industry from all over the world.

The handbook integrates knowledge from the fabrication, mechanics, materials science and reliability points of view. This book is intended for three types of readers: graduate students of nanotechnology, researchers in academia and industry who are active or intend to become active in this field, and practicing engineers and scientists who have encountered a problem and hope to solve it as expeditiously as possible. The handbook should serve as an excellent text for one or two semester graduate courses in nanotechnology in mechanical engineering, materials science, applied physics, or applied chemistry.

We embarked on this project in February 2002, and we worked very hard to get all the chapters to the publisher in a record time of about 1 year. I wish to sincerely thank the authors for offering to write comprehensive chapters on a tight schedule. This is generally an added responsibility in the hectic work schedules of researchers today. I depended on a large number of reviewers who provided critical reviews. I would like to thank Dr. Phillip J. Bond, Chief of Staff and Under Secretary for Technology, US Department of Commerce, Washington, D.C. for suggestions for chapters as well as authors in the handbook. I would also like to thank my colleague, Dr. Huiwen Liu, whose efforts during the preparation of this handbook were very useful.

I hope that this handbook will stimulate further interest in this important new field, and the readers of this handbook will find it useful.

September 2003

Bharat Bhushan
Editor

Editors Vita

Dr. Bharat Bhushan received an M.S. in mechanical engineering from the Massachusetts Institute of Technology in 1971, an M.S. in mechanics and a Ph.D. in mechanical engineering from the University of Colorado at Boulder in 1973 and 1976, respectively, an MBA from Rensselaer Polytechnic Institute at Troy, NY in 1980, Doctor Technicae from the University of Trondheim at Trondheim, Norway in 1990, a Doctor of Technical Sciences from the Warsaw University of Technology at Warsaw, Poland in 1996, and Doctor Honouris Causa from the National Academy of Sciences at Gomel, Belarus in 2000. He is a registered professional engineer. He is presently an Ohio Eminent Scholar and The Howard D. Winbigler Professor in the College of Engineering, and the Director of the Nanoprobe Laboratory for Bio- and Nanotechnology and Biomimetics (NLB^2) at the Ohio State University, Columbus, Ohio. His research interests include fundamental studies with a focus on scanning probe techniques in the interdisciplinary areas of bio/nanotribology, bio/nanomechanics and bio/nanomaterials characterization, and applications to bio/nanotechnology and biomimetics. He is an internationally recognized expert of bio/nanotribology and bio/nanomechanics using scanning probe microscopy, and is one of the most prolific authors. He is considered by some a pioneer of the tribology and mechanics of magnetic storage devices. He has authored 6 scientific books, more than 90 handbook chapters, more than 700 scientific papers (*h* factor – 45+; ISI Highly Cited in Materials Science, since 2007), and more than 60 technical reports, edited more than 45 books, and holds 17 US and foreign patents. He is co-editor of Springer NanoScience and Technology Series and co-editor of Microsystem Technologies. He has given more than 400 invited presentations on six continents and more than 140 keynote/plenary addresses at major international conferences.

Dr. Bhushan is an accomplished organizer. He organized the first symposium on Tribology and Mechanics of Magnetic Storage Systems in 1984 and the first international symposium on Advances in Information Storage Systems in 1990, both of which are now held annually. He is the founder of an ASME Information Storage and Processing Systems Division founded in 1993 and served as the founding chair during 1993–1998. His biography has been listed in over two dozen Who's Who books including Who's Who in the World and has received more than two dozen awards for his contributions to science and technology from professional societies, industry, and US government agencies. He is also the recipient of various international fellowships including the Alexander von Humboldt Research Prize for Senior Scientists, Max Planck Foundation Research Award for Outstanding Foreign Scientists, and the Fulbright Senior Scholar Award. He is a foreign member of the International Academy of Engineering (Russia), Byelorussian Academy of Engineering and Technology and the Academy of Triboengineering of Ukraine, an honorary member of the Society of Tribologists of Belarus, a fellow of ASME, IEEE, STLE, and the New York Academy of Sciences, and a member of ASEE, Sigma Xi and Tau Beta Pi.

Dr. Bhushan has previously worked for the R&D Division of Mechanical Technology Inc., Latham, NY; the Technology Services Division of SKF Industries Inc., King of Prussia, PA; the General Products Division Laboratory of IBM Corporation, Tucson, AZ; and the Almaden Research Center of IBM Corporation, San Jose, CA. He has held visiting professor appointments at University of California at Berkeley, University of Cambridge, UK, Technical University Vienna, Austria, University of Paris, Orsay, ETH Zurich and EPFL Lausanne.

List of Authors

Chong H. Ahn
University of Cincinnati
Department of Electrical
and Computer Engineering
Cincinnati, OH 45221, USA
e-mail: *chong.ahn@uc.edu*

Boris Anczykowski
nanoAnalytics GmbH
Münster, Germany
e-mail: *anczykowski@nanoanalytics.com*

W. Robert Ashurst
Auburn University
Department of Chemical Engineering
Auburn, AL 36849, USA
e-mail: *ashurst@auburn.edu*

Massood Z. Atashbar
Western Michigan University
Department of Electrical
and Computer Engineering
Kalamazoo, MI 49008-5329, USA
e-mail: *massood.atashbar@wmich.edu*

Wolfgang Bacsa
University of Toulouse III (Paul Sabatier)
Laboratoire de Physique des Solides (LPST),
UMR 5477 CNRS
Toulouse, France
e-mail: *bacsa@ramansco.ups-tlse.fr; bacsa@lpst.ups-tlse.fr*

Kelly Bailey
University of Adelaide
CSIRO Human Nutrition
Adelaide SA 5005, Australia
e-mail: *kelly.bailey@csiro.au*

William Sims Bainbridge
National Science Foundation
Division of Information, Science and Engineering
Arlington, VA, USA
e-mail: *wsbainbridge@yahoo.com*

Antonio Baldi
Institut de Microelectronica de Barcelona (IMB)
Centro National Microelectrónica (CNM-CSIC)
Barcelona, Spain
e-mail: *antoni.baldi@cnm.es*

Wilhelm Barthlott
University of Bonn
Nees Institute for Biodiversity of Plants
Meckenheimer Allee 170
53115 Bonn, Germany
e-mail: *barthlott@uni-bonn.de*

Roland Bennewitz
INM – Leibniz Institute for New Materials
66123 Saarbrücken, Germany
e-mail: *roland.bennewitz@inm-gmbh.de*

Bharat Bhushan
Ohio State University
Nanoprobe Laboratory for Bio- and
Nanotechnology and Biomimetics (NLB^2)
201 W. 19th Avenue
Columbus, OH 43210-1142, USA
e-mail: *bhushan.2@osu.edu*

Gerd K. Binnig
Definiens AG
Trappentreustr. 1
80339 Munich, Germany
e-mail: *gbinnig@definiens.com*

Marcie R. Black
Bandgap Engineering Inc.
1344 Main St.
Waltham, MA 02451, USA
e-mail: *marcie@alum.mit.edu; marcie@bandgap.com*

Donald W. Brenner
Department of Materials Science and Engineering
Raleigh, NC, USA
e-mail: *brenner@ncsu.edu*

Jean-Marc Broto
Institut National des Sciences Appliquées of Toulouse
Laboratoire National des Champs Magnétiques Pulsés (LNCMP)
Toulouse, France
e-mail: *broto@lncmp.fr*

Guozhong Cao
University of Washington
Dept. of Materials Science and Engineering
302M Roberts Hall
Seattle, WA 98195-2120, USA
e-mail: *gzcao@u.washington.edu*

Edin (I-Chen) Chen
National Central University
Institute of Materials Science and Engineering
Department of Mechanical Engineering
Chung-Li, 320, Taiwan
e-mail: *ichen@ncu.edu.tw*

Yu-Ting Cheng
National Chiao Tung University
Department of Electronics Engineering
& Institute of Electronics
1001, Ta-Hsueh Rd.
Hsinchu, 300, Taiwan, R.O.C.
e-mail: *ytcheng@mail.nctu.edu.tw*

Giovanni Cherubini
IBM Zurich Research Laboratory
Tape Technologies
8803 Rüschlikon, Switzerland
e-mail: *cbi@zurich.ibm.com*

Mu Chiao
Department of Mechanical Engineering
6250 Applied Science Lane
Vancouver, BC V6T 1Z4, Canada
e-mail: *muchiao@mech.ubc.ca*

Jin-Woo Choi
Louisiana State University
Department of Electrical and Computer Engineering
Baton Rouge, LA 70803, USA
e-mail: *choi@ece.lsu.edu*

Tamara H. Cooper
University of Adelaide
CSIRO Human Nutrition
Adelaide SA 5005, Australia
e-mail: *tamara.cooper@csiro.au*

Alex D. Corwin
GE Global Research
1 Research Circle
Niskayuna, NY 12309, USA
e-mail: *corwin@ge.com*

Maarten P. de Boer
Carnegie Mellon University
Department of Mechanical Engineering
5000 Forbes Avenue
Pittsburgh, PA 15213, USA
e-mail: *mpdebo@andrew.cmu.edu*

Dietrich Dehlinger
Lawrence Livermore National Laboratory
Engineering
Livermore, CA 94551, USA
e-mail: *dehlinger1@llnl.gov*

Frank W. DelRio
National Institute of Standards and Technology
100 Bureau Drive, Stop 8520
Gaithersburg, MD 20899-8520, USA
e-mail: *frank.delrio@nist.gov*

Michel Despont
IBM Zurich Research Laboratory
Micro- and Nanofabrication
8803 Rüschlikon, Switzerland
e-mail: *dpt@zurich.ibm.com*

Lixin Dong
Michigan State University
Electrical and Computer Engineering
2120 Engineering Building
East Lansing, MI 48824-1226, USA
e-mail: *ldong@egr.msu.edu*

Gene Dresselhaus
Massachusetts Institute of Technology
Francis Bitter Magnet Laboratory
Cambridge, MA 02139, USA
e-mail: *gene@mgm.mit.edu*

Mildred S. Dresselhaus
Massachusetts Institute of Technology
Department of Electrical Engineering
and Computer Science
Department of Physics
Cambridge, MA, USA
e-mail: *millie@mgm.mit.edu*

Urs T. Dürig
IBM Zurich Research Laboratory
Micro-/Nanofabrication
8803 Rüschlikon, Switzerland
e-mail: *drg@zurich.ibm.com*

Andreas Ebner
Johannes Kepler University Linz
Institute for Biophysics
Altenberger Str. 69
4040 Linz, Austria
e-mail: *andreas.ebner@jku.at*

Evangelos Eleftheriou
IBM Zurich Research Laboratory
8803 Rüschlikon, Switzerland
e-mail: *ele@zurich.ibm.com*

Emmanuel Flahaut
Université Paul Sabatier
CIRIMAT, Centre Interuniversitaire de Recherche
et d'Ingénierie des Matériaux, UMR 5085 CNRS
118 Route de Narbonne
31062 Toulouse, France
e-mail: *flahaut@chimie.ups-tlse.fr*

Anatol Fritsch
University of Leipzig
Institute of Experimental Physics I
Division of Soft Matter Physics
Linnéstr. 5
04103 Leipzig, Germany
e-mail: *anatol.fritsch@uni-leipzig.de*

Harald Fuchs
Universität Münster
Physikalisches Institut
Münster, Germany
e-mail: *fuchsh@uni-muenster.de*

Christoph Gerber
University of Basel
Institute of Physics
National Competence Center for Research
in Nanoscale Science (NCCR) Basel
Klingelbergstr. 82
4056 Basel, Switzerland
e-mail: *christoph.gerber@unibas.ch*

Franz J. Giessibl
Universität Regensburg
Institute of Experimental and Applied Physics
Universitätsstr. 31
93053 Regensburg, Germany
e-mail: *franz.giessibl@physik.uni-regensburg.de*

Enrico Gnecco
University of Basel
National Center of Competence in Research
Department of Physics
Klingelbergstr. 82
4056 Basel, Switzerland
e-mail: *enrico.gnecco@unibas.ch*

Stanislav N. Gorb
Max Planck Institut für Metallforschung
Evolutionary Biomaterials Group
Heisenbergstr. 3
70569 Stuttgart, Germany
e-mail: *s.gorb@mf.mpg.de*

Hermann Gruber
University of Linz
Institute of Biophysics
Altenberger Str. 69
4040 Linz, Austria
e-mail: *hermann.gruber@jku.at*

Jason Hafner
Rice University
Department of Physics and Astronomy
Houston, TX 77251, USA
e-mail: *hafner@rice.edu*

Judith A. Harrison
U.S. Naval Academy
Chemistry Department
572 Holloway Road
Annapolis, MD 21402-5026, USA
e-mail: *jah@usna.edu*

Martin Hegner
CRANN – The Naughton Institute
Trinity College, University of Dublin
School of Physics
Dublin, 2, Ireland
e-mail: *martin.hegner@tcd.ie*

Thomas Helbling
ETH Zurich
Micro and Nanosystems
Department of Mechanical
and Process Engineering
8092 Zurich, Switzerland
e-mail: *thomas.helbling@micro.mavt.ethz.ch*

Michael J. Heller
University of California San Diego
Department of Bioengineering
Dept. of Electrical and Computer Engineering
La Jolla, CA, USA
e-mail: *mjheller@ucsd.edu*

Seong-Jun Heo
Lam Research Corp.
4650 Cushing Parkway
Fremont, CA 94538, USA
e-mail: *seongjun.heo@lamrc.com*

Christofer Hierold
ETH Zurich
Micro and Nanosystems
Department of Mechanical
and Process Engineering
8092 Zurich, Switzerland
e-mail: *christofer.hierold@micro.mavt.ethz.ch*

Peter Hinterdorfer
University of Linz
Institute for Biophysics
Altenberger Str. 69
4040 Linz, Austria
e-mail: *peter.hinterdorfer@jku.at*

Dalibor Hodko
Nanogen, Inc.
10498 Pacific Center Court
San Diego, CA 92121, USA
e-mail: *dhodko@nanogen.com*

Hendrik Hölscher
Forschungszentrum Karlsruhe
Institute of Microstructure Technology
Linnéstr. 5
76021 Karlsruhe, Germany
e-mail: *hendrik.hoelscher@imt.fzk.de*

Hirotaka Hosoi
Hokkaido University
Creative Research Initiative Sousei
Kita 21, Nishi 10, Kita-ku
Sapporo, Japan
e-mail: *hosoi@cris.hokudai.ac.jp*

Katrin Hübner
Staatliche Fachoberschule Neu-Ulm
89231 Neu-Ulm, Germany
e-mail: *katrin.huebner1@web.de*

Douglas L. Irving
North Carolina State University
Materials Science and Engineering
Raleigh, NC 27695-7907, USA
e-mail: *doug_irving@ncsu.edu*

Jacob N. Israelachvili
University of California
Department of Chemical Engineering
and Materials Department
Santa Barbara, CA 93106-5080, USA
e-mail: *jacob@engineering.ucsb.edu*

Guangyao Jia
University of California, Irvine
Department of Mechanical
and Aerospace Engineering
Irvine, CA, USA
e-mail: *gjia@uci.edu*

Sungho Jin
University of California, San Diego
Department of Mechanical
and Aerospace Engineering
9500 Gilman Drive
La Jolla, CA 92093-0411, USA
e-mail: *jin@ucsd.edu*

Anne Jourdain
Interuniversity Microelectronics Center (IMEC)
Leuven, Belgium
e-mail: *jourdain@imec.be*

Yong Chae Jung
Samsung Electronics C., Ltd.
Senior Engineer Process Development Team
San #16 Banwol-Dong, Hwasung-City
Gyeonggi-Do 445-701, Korea
e-mail: *yc423.jung@samsung.com*

Harold Kahn
Case Western Reserve University
Department of Materials Science and Engineering
Cleveland, OH , USA
e-mail: *kahn@cwru.edu*

Roger Kamm
Massachusetts Institute of Technology
Department of Biological Engineering
77 Massachusetts Avenue
Cambridge, MA 02139, USA
e-mail: *rdkamm@mit.edu*

Ruti Kapon
Weizmann Institute of Science
Department of Biological Chemistry
Rehovot 76100, Israel
e-mail: *ruti.kapon@weizmann.ac.il*

Josef Käs
University of Leipzig
Institute of Experimental Physics I
Division of Soft Matter Physics
Linnéstr. 5
04103 Leipzig, Germany
e-mail: *jkaes@physik.uni-leipzig.de*

Horacio Kido
University of California at Irvine
Mechanical and Aerospace Engineering
Irvine, CA, USA
e-mail: *hkido@uci.edu*

Tobias Kießling
University of Leipzig
Institute of Experimental Physics I
Division of Soft Matter Physics
Linnéstr. 5
04103 Leipzig, Germany
e-mail: *Tobias.Kiessling@uni-leipzig.de*

Jitae Kim
University of California at Irvine
Department of Mechanical
and Aerospace Engineering
Irvine, CA, USA
e-mail: *jitaekim@uci.edu*

Jongbaeg Kim
Yonsei University
School of Mechanical Engineering
1st Engineering Bldg.
Seoul, 120-749, South Korea
e-mail: *kimjb@yonsei.ac.kr*

Nahui Kim
Samsung Advanced Institute of Technology
Research and Development
Seoul, South Korea
e-mail: *nahui.kim@samsung.com*

Kerstin Koch
Rhine-Waal University of Applied Science
Department of Life Science, Biology
and Nanobiotechnology
Landwehr 4
47533 Kleve, Germany
e-mail: *kerstin.koch@hochschule.rhein-waal.de*

Jing Kong
Massachusetts Institute of Technology
Department of Electrical Engineering
and Computer Science
Cambridge, MA, USA
e-mail: *jingkong@mit.edu*

Tobias Kraus
Leibniz-Institut für Neue Materialien gGmbH
Campus D2 2
66123 Saarbrücken, Germany
e-mail: *tobias.kraus@inm-gmbh.de*

Anders Kristensen
Technical University of Denmark
DTU Nanotech
2800 Kongens Lyngby, Denmark
e-mail: *anders.kristensen@nanotech.dtu.dk*

Ratnesh Lal
University of Chicago
Center for Nanomedicine
5841 S Maryland Av
Chicago, IL 60637, USA
e-mail: *rlal@uchicago.edu*

Jan Lammerding
Harvard Medical School
Brigham and Women's Hospital
65 Landsdowne St
Cambridge, MA 02139, USA
e-mail: *jlammerding@rics.bwh.harvard.edu*

Hans Peter Lang
University of Basel
Institute of Physics, National Competence Center for Research in Nanoscale Science (NCCR) Basel
Klingelbergstr. 82
4056 Basel, Switzerland
e-mail: *hans-peter.lang@unibas.ch*

Carmen LaTorre
Owens Corning Science and Technology
Roofing and Asphalt
2790 Columbus Road
Granville, OH 43023, USA
e-mail: *carmen.latorre@owenscorning.com*

Christophe Laurent
Université Paul Sabatier
CIRIMAT UMR 5085 CNRS
118 Route de Narbonne
31062 Toulouse, France
e-mail: *laurent@chimie.ups-tlse.fr*

Abraham P. Lee
University of California Irvine
Department of Biomedical Engineering
Department of Mechanical
and Aerospace Engineering
Irvine, CA 92697, USA
e-mail: *aplee@uci.edu*

Stephen C. Lee
Ohio State University
Biomedical Engineering Center
Columbus, OH 43210, USA
e-mail: *lee@bme.ohio-state.edu*

Wayne R. Leifert
Adelaide Business Centre
CSIRO Human Nutrition
Adelaide SA 5000, Australia
e-mail: *wayne.leifert@csiro.au*

Liwei Lin
UC Berkeley
Mechanical Engineering Department
5126 Etcheverry
Berkeley, CA 94720-1740, USA
e-mail: *lwlin@me.berkeley.edu*

Yu-Ming Lin
IBM T.J. Watson Research Center
Nanometer Scale Science & Technology
1101 Kitchawan Road
Yorktown Heigths, NY 10598, USA
e-mail: *yming@us.ibm.com*

Marc J. Madou
University of California Irvine
Department of Mechanical and Aerospace
and Biomedical Engineering
Irvine, CA, USA
e-mail: *mmadou@uci.edu*

Othmar Marti
Ulm University
Institute of Experimental Physics
Albert-Einstein-Allee 11
89069 Ulm, Germany
e-mail: *othmar.marti@uni-ulm.de*

Jack Martin
66 Summer Street
Foxborough, MA 02035, USA
e-mail: *jack.martin@alumni.tufts.edu*

Shinji Matsui
University of Hyogo
Laboratory of Advanced Science
and Technology for Industry
Hyogo, Japan
e-mail: *matsui@lasti.u-hyogo.ac.jp*

Mehran Mehregany
Case Western Reserve University
Department of Electrical Engineering
and Computer Science
Cleveland, OH 44106, USA
e-mail: *mxm31@cwru.edu*

Etienne Menard
Semprius, Inc.
4915 Prospectus Dr.
Durham, NC 27713, USA
e-mail: *etienne.menard@semprius.com*

Ernst Meyer
University of Basel
Institute of Physics
Basel, Switzerland
e-mail: *ernst.meyer@unibas.ch*

Robert Modliñski
Baolab Microsystems
Terrassa 08220, Spain
e-mail: *rmodlinski@gmx.com*

Mohammad Mofrad
University of California, Berkeley
Department of Bioengineering
Berkeley, CA 94720, USA
e-mail: *mofrad@berkeley.edu*

Marc Monthioux
CEMES - UPR A-8011 CNRS
Carbones et Matériaux Carbonés,
Carbons and Carbon-Containing Materials
29 Rue Jeanne Marvig
31055 Toulouse 4, France
e-mail: *monthiou@cemes.fr*

Markus Morgenstern
RWTH Aachen University
II. Institute of Physics B and JARA-FIT
52056 Aachen, Germany
e-mail: *mmorgens@physik.rwth-aachen.de*

Seizo Morita
Osaka University
Department of Electronic Engineering
Suita-City
Osaka, Japan
e-mail: *smorita@ele.eng.osaka-u.ac.jp*

Koichi Mukasa
Hokkaido University
Nanoelectronics Laboratory
Sapporo, Japan
e-mail: *mukasa@nano.eng.hokudai.ac.jp*

Bradley J. Nelson
Swiss Federal Institute of Technology (ETH)
Institute of Robotics and Intelligent Systems
8092 Zurich, Switzerland
e-mail: *bnelson@ethz.ch*

Michael Nosonovsky
University of Wisconsin-Milwaukee
Department of Mechanical Engineering
3200 N. Cramer St.
Milwaukee, WI 53211, USA
e-mail: *nosonovs@uwm.edu*

Hiroshi Onishi
Kanagawa Academy of Science and Technology
Surface Chemistry Laboratory
Kanagawa, Japan
e-mail: *oni@net.ksp.or.jp*

Alain Peigney
Centre Inter-universitaire de Recherche
sur l'Industrialisation des Matériaux (CIRIMAT)
Toulouse 4, France
e-mail: *peigney@chimie.ups-tlse.fr*

Oliver Pfeiffer
Individual Computing GmbH
Ingelsteinweg 2d
4143 Dornach, Switzerland
e-mail: *oliver.pfeiffer@gmail.com*

Haralampos Pozidis
IBM Zurich Research Laboratory
Storage Technologies
Rüschlikon, Switzerland
e-mail: *hap@zurich.ibm.com*

Robert Puers
Katholieke Universiteit Leuven
ESAT/MICAS
Leuven, Belgium
e-mail: *bob.puers@esat.kuleuven.ac.be*

Calvin F. Quate
Stanford University
Edward L. Ginzton Laboratory
450 Via Palou
Stanford, CA 94305-4088, USA
e-mail: *quate@stanford.edu*

Oded Rabin
University of Maryland
Department of Materials Science and Engineering
College Park, MD, USA
e-mail: *oded@umd.edu*

Françisco M. Raymo
University of Miami
Department of Chemistry
1301 Memorial Drive
Coral Gables, FL 33146-0431, USA
e-mail: *fraymo@miami.edu*

Manitra Razafinimanana
University of Toulouse III (Paul Sabatier)
Centre de Physique des Plasmas
et leurs Applications (CPPAT)
Toulouse, France
e-mail: *razafinimanana@cpat.ups-tlse.fr*

Ziv Reich
Weizmann Institute of Science Ha'Nesi Ha'Rishon
Department of Biological Chemistry
Rehovot 76100, Israel
e-mail: *ziv.reich@weizmann.ac.il*

John A. Rogers
University of Illinois
Department of Materials Science and Engineering
Urbana, IL, USA
e-mail: *jrogers@uiuc.edu*

Cosmin Roman
ETH Zurich
Micro and Nanosystems Department of Mechanical and Process Engineering
8092 Zurich, Switzerland
e-mail: *cosmin.roman@micro.mavt.ethz.ch*

Marina Ruths
University of Massachusetts Lowell
Department of Chemistry
1 University Avenue
Lowell, MA 01854, USA
e-mail: *marina_ruths@uml.edu*

Ozgur Sahin
The Rowland Institute at Harvard
100 Edwin H. Land Blvd
Cambridge, MA 02142, USA
e-mail: *sahin@rowland.harvard.edu*

Akira Sasahara
Japan Advanced Institute
of Science and Technology
School of Materials Science
1-1 Asahidai
923-1292 Nomi, Japan
e-mail: *sasahara@jaist.ac.jp*

Helmut Schift
Paul Scherrer Institute
Laboratory for Micro- and Nanotechnology
5232 Villigen PSI, Switzerland
e-mail: *helmut.schift@psi.ch*

André Schirmeisen
University of Münster
Institute of Physics
Wilhelm-Klemm-Str. 10
48149 Münster, Germany
e-mail: *schirmeisen@uni-muenster.de*

Christian Schulze
Beiersdorf AG
Research & Development
Unnastr. 48
20245 Hamburg, Germany
e-mail: *christian.schulze@beiersdorf.com;*
christian.schulze@uni-leipzig.de

Alexander Schwarz
University of Hamburg
Institute of Applied Physics
Jungiusstr. 11
20355 Hamburg, Germany
e-mail: *aschwarz@physnet.uni-hamburg.de*

Udo D. Schwarz
Yale University
Department of Mechanical Engineering
15 Prospect Street
New Haven, CT 06520-8284, USA
e-mail: *udo.schwarz@yale.edu*

Philippe Serp
Ecole Nationale Supérieure d'Ingénieurs
en Arts Chimiques et Technologiques
Laboratoire de Chimie de Coordination (LCC)
118 Route de Narbonne
31077 Toulouse, France
e-mail: *philippe.serp@ensiacet.fr*

Huamei (Mary) Shang
GE Healthcare
4855 W. Electric Ave.
Milwaukee, WI 53219, USA
e-mail: *huamei.shang@ge.com*

Susan B. Sinnott
University of Florida
Department of Materials Science and Engineering
154 Rhines Hall
Gainesville, FL 32611-6400, USA
e-mail: *ssinn@mse.ufl.edu*

Anisoara Socoliuc
SPECS Zurich GmbH
Technoparkstr. 1
8005 Zurich, Switzerland
e-mail: *socoliuc@nanonis.com*

Olav Solgaard
Stanford University
E.L. Ginzton Laboratory
450 Via Palou
Stanford, CA 94305-4088, USA
e-mail: *solgaard@stanford.edu*

Dan Strehle
University of Leipzig
Institute of Experimental Physics I
Division of Soft Matter Physics
Linnéstr. 5
04103 Leipzig, Germany
e-mail: *dan.strehle@uni-leipzig.de*

Carsten Stüber
University of Leipzig
Institute of Experimental Physics I
Division of Soft Matter Physics
Linnéstr. 5
04103 Leipzig, Germany
e-mail: *stueber@rz.uni-leipzig.de*

Yu-Chuan Su
ESS 210
Department of Engineering and System Science 101
Kuang-Fu Road
Hsinchu, 30013, Taiwan
e-mail: *ycsu@ess.nthu.edu.tw*

Kazuhisa Sueoka
Graduate School of Information Science
and Technology
Hokkaido University
Nanoelectronics Laboratory
Kita-14, Nishi-9, Kita-ku
060-0814 Sapporo, Japan
e-mail: *sueoka@nano.isthokudai.ac.jp*

Yasuhiro Sugawara
Osaka University
Department of Applied Physics
Yamada-Oka 2-1, Suita
565-0871 Osaka, Japan
e-mail: *sugawara@ap.eng.osaka-u.ac.jp*

Benjamin Sullivan
TearLab Corp.
11025 Roselle Street
San Diego, CA 92121, USA
e-mail: *bdsulliv@TearLab.com*

Paul Swanson
Nexogen, Inc.
Engineering
8360 C Camino Santa Fe
San Diego, CA 92121, USA
e-mail: *pswanson@nexogentech.com*

Yung-Chieh Tan
Washington University School of Medicine
Department of Medicine
Division of Dermatology
660 S. Euclid Ave.
St. Louis, MO 63110, USA
e-mail: *ytanster@gmail.com*

Shia-Yen Teh
University of California at Irvine
Biomedical Engineering Department
3120 Natural Sciences II
Irvine, CA 92697-2715, USA
e-mail: *steh@uci.edu*

W. Merlijn van Spengen
Leiden University
Kamerlingh Onnes Laboratory
Niels Bohrweg 2
Leiden, CA 2333, The Netherlands
e-mail: *spengen@physics.leidenuniv.nl*

Peter Vettiger
University of Neuchâtel
SAMLAB
Jaquet-Droz 1
2002 Neuchâtel, Switzerland
e-mail: *peter.vettiger@unine.ch*

Franziska Wetzel
University of Leipzig
Institute of Experimental Physics I
Division of Soft Matter Physics
Linnéstr. 5
04103 Leipzig, Germany
e-mail: *franziska.wetzel@uni-leipzig.de*

Heiko Wolf
IBM Research GmbH
Zurich Research Laboratory
Säumerstr. 4
8803 Rüschlikon, Switzerland
e-mail: *hwo@zurich.ibm.com*

Darrin J. Young
Case Western Reserve University
Department of EECS, Glennan 510
10900 Euclid Avenue
Cleveland, OH 44106, USA
e-mail: *djy@po.cwru.edu*

Babak Ziaie
Purdue University
Birck Nanotechnology Center
1205 W. State St.
West Lafayette, IN 47907-2035, USA
e-mail: *bziaie@purdue.edu*

Christian A. Zorman
Case Western Reserve University
Department of Electrical Engineering
and Computer Science
10900 Euclid Avenue
Cleveland, OH 44106, USA
e-mail: *caz@case.edu*

Jim V. Zoval
Saddleback College
Department of Math and Science
28000 Marguerite Parkway
Mission Viejo, CA 92692, USA
e-mail: *jzoval@saddleback.edu*

Acknowledgements

H.48 Friction and Wear in Micro- and Nanomachines

by Maarten P. de Boer, Alex D. Corwin, Frank W. DelRio, W. Robert Ashurst

The nanotractor data reported in this chapter were measured by A.D.C. and M.P.d.B. at Sandia National Laboratories and the work was supported by Laboratory Directed Research and Development (LDRD) and Campaign 6 programs. Certain commercial equipment, instruments or materials are identified in this report in order to specify the experimental procedure adequately. Such identification is not intended to imply recommendation or endorsement by the National Institute of Standards and Technology, nor is it intended to imply that the materials or equipment identified are necessarily the best available for the purpose.

目 录

缩略语

Part G 工业应用

45. 千足虫：基于原子力显微镜数据存储系统的纳米技术 …… 3
45.1 千足虫的概念 …… 5
45.2 热力学原子力显微镜的数据存储 …… 6
45.3 阵列设计、技术和制造 …… 8
45.4 阵列的特性 …… 9
45.5 三端悬臂的设计 …… 11
45.6 x，y，z 介质微型扫描仪 …… 12
45.7 首次32×32阵列芯片写/读结果 …… 15
45.8 聚合物介质 …… 16
45.9 读取信道模型 …… 23
45.10 系统特性 …… 26
45.11 结 论 …… 31
参考文献 …… 32
46. 纳米机器人 …… 35
46.1 纳米机器人概述 …… 36
46.2 纳米量级的驱动 …… 37
46.3 纳米机器人的操作系统 …… 39
46.4 纳米机器人的装配 …… 44
46.5 应 用 …… 53
参考文献 …… 56

Part H 微/纳米器件的可靠性

47. 微/纳机电系统和生物微/纳机电系统：材料、器件、仿生技术 …… 65
47.1 微/纳机电系统基础 …… 66
47.2 硅及相关材料的纳米摩擦学和纳米力学研究 …… 85
47.3 微/纳机电系统的润滑研究 …… 93
47.4 基于硅和多晶表面生物分子的摩擦学研究以及微纳米级颗粒治疗与诊断 …… 100

47.5 带有粗糙度诱导超疏水、自清洁、低附着的表面 …… 110
47.6 组件级研究 …… 119
47.7 结 论 …… 130
47.A 微米-纳米制造技术 …… 131
参考文献 …… 135
48. 微米-纳米机械中的摩擦和磨损 …… 143
48.1 从单粗糙到多粗糙摩擦 …… 145
48.2 纳米反应器概述 …… 149
48.3 结束语 …… 157
参考文献 …… 158
49. 微/纳机电系统中的失效机制 …… 163
49.1 失效模式和失效机制 …… 164
49.2 静摩擦和充电相关的失效机制 …… 165
49.3 蔓延、疲劳、磨损及封装相关的失效 …… 171
49.4 结 论 …… 181
参考文献 …… 181
50. 微机械结构的机械特性 …… 185
50.1 薄膜基板的机械性能检测 …… 185
50.2 测量机械性能的微机械结构 …… 187
50.3 机械性能的检测系统 …… 197
参考文献 …… 201
51. 微机械产品的大批量生产和市场稳定性 …… 205
51.1 背 景 …… 206
51.2 制造策略 …… 208
51.3 强大的制造业 …… 210
51.4 稳定的市场特征 …… 227
参考文献 …… 230
52. 微/纳米系统的封装和可靠性问题 …… 237
52.1 微机电系统封装简介 …… 237
52.2 密封和真空封装及应用 …… 243
52.3 散热问题和可靠性封装 …… 253
52.4 未来趋势和总结 …… 260
参考文献 …… 261

Part I 技术的融合和纳米技术的管理

53. 纳米技术的管理：社会的、伦理的和人权的问题 …… 269
53.1 社会科学背景 …… 269
53.2 人类对纳米技术的影响 …… 273
53.3 纳米技术的调节 …… 276
53.4 纳米技术的文化背景 …… 278
53.5 结 论 …… 281
参考文献 …… 282

主题索引 …… 287

Contents

List of Abbreviations

Part G Industrial Applications

45 **The *Millipede* – A Nanotechnology-Based AFM Data-Storage System**
Gerd K. Binnig, Giovanni Cherubini, Michel Despont, Urs T. Dürig, Evangelos Eleftheriou, Haralampos Pozidis, Peter Vettiger 1601
45.1 The Millipede Concept 1603
45.2 Thermomechanical AFM Data Storage 1604
45.3 Array Design, Technology, and Fabrication 1606
45.4 Array Characterization 1607
45.5 Three-Terminal Cantilever Design 1609
45.6 *x,y,z* Medium Microscanner 1610
45.7 First Write/Read Results with the 32×32 Array Chip 1613
45.8 Polymer Medium 1614
45.9 Read Channel Model 1621
45.10 System Aspects 1624
45.11 Conclusions 1629
References 1630

46 **Nanorobotics**
Bradley J. Nelson, Lixin Dong 1633
46.1 Overview of Nanorobotics 1634
46.2 Actuation at Nanoscales 1635
46.3 Nanorobotic Manipulation Systems 1637
46.4 Nanorobotic Assembly 1642
46.5 Applications 1651
References 1654

Part H Micro-/Nanodevice Reliability

47 **MEMS/NEMS and BioMEMS/BioNEMS: Materials, Devices, and Biomimetics**
Bharat Bhushan 1663
47.1 MEMS/NEMS Basics 1664
47.2 Nanotribology and Nanomechanics Studies of Silicon and Related Materials 1683

47.3 Lubrication Studies for MEMS/NEMS 1691
47.4 Nanotribological Studies of Biological Molecules on Silicon-Based and Polymer Surfaces and Submicron Particles for Therapeutics and Diagnostics 1698
47.5 Surfaces with Roughness-Induced Superhydrophobicity, Self-Cleaning, and Low Adhesion 1708
47.6 Component-Level Studies 1717
47.7 Conclusions 1728
47.A Micro-Nanofabrication Techniques 1729
References 1733

48 **Friction and Wear in Micro- and Nanomachines**
Maarten P. de Boer, Alex D. Corwin, Frank W. DelRio, W. Robert Ashurst 1741
48.1 From Single- to Multiple-Asperity Friction 1743
48.2 Nanotractor Device Description 1747
48.3 Concluding Remarks 1755
References 1756

49 **Failure Mechanisms in MEMS/NEMS Devices**
W. Merlijn van Spengen, Robert Modliński, Robert Puers, Anne Jourdain 1761
49.1 Failure Modes and Failure Mechanisms 1762
49.2 Stiction and Charge-Related Failure Mechanisms 1763
49.3 Creep, Fatigue, Wear, and Packaging-Related Failures 1769
49.4 Conclusions 1779
References 1779

50 **Mechanical Properties of Micromachined Structures**
Harold Kahn 1783
50.1 Measuring Mechanical Properties of Films on Substrates 1783
50.2 Micromachined Structures for Measuring Mechanical Properties 1785
50.3 Measurements of Mechanical Properties 1795
References 1799

51 **High-Volume Manufacturing and Field Stability of MEMS Products**
Jack Martin 1803
51.1 Background 1804
51.2 Manufacturing Strategy 1806
51.3 Robust Manufacturing 1808
51.4 Stable Field Performance 1825
References 1828

52 **Packaging and Reliability Issues in Micro-/Nanosystems**
Yu-Chuan Su, Jongbaeg Kim, Yu-Ting Cheng, Mu Chiao, Liwei Lin 1835
52.1 Introduction MEMS Packaging 1835

52.2 Hermetic and Vacuum Packaging and Applications 1841
52.3 Thermal Issues and Packaging Reliability 1851
52.4 Future Trends and Summary 1858
References 1859

Part I Technological Convergence and Governing Nanotechnology

53 **Governing Nanotechnology: Social, Ethical and Human Issues**
William Sims Bainbridge 1867
53.1 Social Science Background 1867
53.2 Human Impacts of Nanotechnology 1871
53.3 Regulating Nanotechnology 1874
53.4 The Cultural Context for Nanotechnology 1876
53.5 Conclusions 1879
References 1880

Subject Index 1919

List of Abbreviations

μCP	microcontact printing
1-D	one-dimensional
18-MEA	18-methyl eicosanoic acid
2-D	two-dimensional
2-DEG	two-dimensional electron gas
3-APTES	3-aminopropyltriethoxysilane
3-D	three-dimensional

A

a-BSA	anti-bovine serum albumin
a-C	amorphous carbon
A/D	analog-to-digital
AA	amino acid
AAM	anodized alumina membrane
ABP	actin binding protein
AC	alternating-current
AC	amorphous carbon
ACF	autocorrelation function
ADC	analog-to-digital converter
ADXL	analog devices accelerometer
AFAM	atomic force acoustic microscopy
AFM	atomic force microscope
AFM	atomic force microscopy
AKD	alkylketene dimer
ALD	atomic layer deposition
AM	amplitude modulation
AMU	atomic mass unit
AOD	acoustooptical deflector
AOM	acoustooptical modulator
AP	alkaline phosphatase
APB	actin binding protein
APCVD	atmospheric-pressure chemical vapor deposition
APDMES	aminopropyldimethylethoxysilane
APTES	aminopropyltriethoxysilane
ASIC	application-specific integrated circuit
ASR	analyte-specific reagent
ATP	adenosine triphosphate

B

BAP	barometric absolute pressure
BAPDMA	behenyl amidopropyl dimethylamine glutamate
bcc	body-centered cubic
BCH	brucite-type cobalt hydroxide
BCS	Bardeen–Cooper–Schrieffer
BD	blu-ray disc
BDCS	biphenyldimethylchlorosilane
BE	boundary element
BFP	biomembrane force probe
BGA	ball grid array
BHF	buffered HF
BHPET	1,1'-(3,6,9,12,15-pentaoxapentadecane-1,15-diyl)bis(3-hydroxyethyl-1H-imidazolium-1-yl) di[bis(trifluoromethanesulfonyl)imide]
BHPT	1,1'-(pentane-1,5-diyl)bis(3-hydroxyethyl-1H-imidazolium-1-yl) di[bis(trifluoromethanesulfonyl)imide]
BiCMOS	bipolar CMOS
bioMEMS	biomedical microelectromechanical system
bioNEMS	biomedical nanoelectromechanical system
BMIM	1-butyl-3-methylimidazolium
BP	bit pitch
BPAG1	bullous pemphigoid antigen 1
BPT	biphenyl-4-thiol
BPTC	cross-linked BPT
BSA	bovine serum albumin
BST	barium strontium titanate
BTMAC	behentrimonium chloride

C

CA	constant amplitude
CA	contact angle
CAD	computer-aided design
CAH	contact angle hysteresis
cAMP	cyclic adenosine monophosphate
CAS	Crk-associated substrate
CBA	cantilever beam array
CBD	chemical bath deposition
CCD	charge-coupled device
CCVD	catalytic chemical vapor deposition
CD	compact disc
CD	critical dimension
CDR	complementarity determining region
CDW	charge density wave
CE	capillary electrophoresis
CE	constant excitation
CEW	continuous electrowetting
CG	controlled geometry
CHO	Chinese hamster ovary
CIC	cantilever in cantilever
CMC	cell membrane complex
CMC	critical micelle concentration
CMOS	complementary metal–oxide–semiconductor
CMP	chemical mechanical polishing

CNF	carbon nanofiber
CNFET	carbon nanotube field-effect transistor
CNT	carbon nanotube
COC	cyclic olefin copolymer
COF	chip-on-flex
COF	coefficient of friction
COG	cost of goods
CoO	cost of ownership
COS	CV-1 in origin with SV40
CP	circularly permuted
CPU	central processing unit
CRP	C-reactive protein
CSK	cytoskeleton
CSM	continuous stiffness measurement
CTE	coefficient of thermal expansion
Cu-TBBP	Cu-tetra-3,5 di-tertiary-butyl-phenyl porphyrin
CVD	chemical vapor deposition

D

DBR	distributed Bragg reflector
DC-PECVD	direct-current plasma-enhanced CVD
DC	direct-current
DDT	dichlorodiphenyltrichloroethane
DEP	dielectrophoresis
DFB	distributed feedback
DFM	dynamic force microscopy
DFS	dynamic force spectroscopy
DGU	density gradient ultracentrifugation
DI FESP	digital instrument force modulation etched Si probe
DI TESP	digital instrument tapping mode etched Si probe
DI	digital instrument
DI	deionized
DIMP	diisopropylmethylphosphonate
DIP	dual inline packaging
DIPS	industrial postpackaging
DLC	diamondlike carbon
DLP	digital light processing
DLVO	Derjaguin–Landau–Verwey–Overbeek
DMD	deformable mirror display
DMD	digital mirror device
DMDM	1,3-dimethylol-5,5-dimethyl
DMMP	dimethylmethylphosphonate
DMSO	dimethyl sulfoxide
DMT	Derjaguin–Muller–Toporov
DNA	deoxyribonucleic acid
DNT	2,4-dinitrotoluene
DOD	Department of Defense
DOE	Department of Energy
DOE	diffractive optical element
DOF	degree of freedom
DOPC	1,2-dioleoyl-sn-glycero-3-phosphocholine
DOS	density of states
DP	decylphosphonate
DPN	dip-pen nanolithography
DRAM	dynamic random-access memory
DRIE	deep reactive ion etching
ds	double-stranded
DSC	differential scanning calorimetry
DSP	digital signal processor
DTR	discrete track recording
DTSSP	3,3'-dithio-bis(sulfosuccinimidylproprionate)
DUV	deep-ultraviolet
DVD	digital versatile disc
DWNT	double-walled CNT

E

EAM	embedded atom method
EB	electron beam
EBD	electron beam deposition
EBID	electron-beam-induced deposition
EBL	electron-beam lithography
ECM	extracellular matrix
ECR-CVD	electron cyclotron resonance chemical vapor deposition
ED	electron diffraction
EDC	1-ethyl-3-(3-diamethylaminopropyl) carbodiimide
EDL	electrostatic double layer
EDP	ethylene diamine pyrochatechol
EDTA	ethylenediamine tetraacetic acid
EDX	energy-dispersive x-ray
EELS	electron energy loss spectra
EFM	electric field gradient microscopy
EFM	electrostatic force microscopy
EHD	elastohydrodynamic
EO	electroosmosis
EOF	electroosmotic flow
EOS	electrical overstress
EPA	Environmental Protection Agency
EPB	electrical parking brake
ESD	electrostatic discharge
ESEM	environmental scanning electron microscope
EU	European Union
EUV	extreme ultraviolet
EW	electrowetting
EWOD	electrowetting on dielectric

F

F-actin	filamentous actin
FA	focal adhesion
FAA	formaldehyde–acetic acid–ethanol
FACS	fluorescence-activated cell sorting

FAK focal adhesion kinase
FBS fetal bovine serum
FC flip-chip
FCA filtered cathodic arc
fcc face-centered cubic
FCP force calibration plot
FCS fluorescence correlation spectroscopy
FD finite difference
FDA Food and Drug Administration
FE finite element
FEM finite element method
FEM finite element modeling
FESEM field emission SEM
FESP force modulation etched Si probe
FET field-effect transistor
FFM friction force microscope
FFM friction force microscopy
FIB-CVD focused ion beam chemical vapor deposition
FIB focused ion beam
FIM field ion microscope
FIP feline coronavirus
FKT Frenkel–Kontorova–Tomlinson
FM frequency modulation
FMEA failure-mode effect analysis
FP6 Sixth Framework Program
FP fluorescence polarization
FPR *N*-formyl peptide receptor
FS force spectroscopy
FTIR Fourier-transform infrared
FV force–volume

G

GABA γ-aminobutyric acid
GDP guanosine diphosphate
GF gauge factor
GFP green fluorescent protein
GMR giant magnetoresistive
GOD glucose oxidase
GPCR G-protein coupled receptor
GPS global positioning system
GSED gaseous secondary-electron detector
GTP guanosine triphosphate
GW Greenwood and Williamson

H

HAR high aspect ratio
HARMEMS high-aspect-ratio MEMS
HARPSS high-aspect-ratio combined poly- and single-crystal silicon
HBM human body model
hcp hexagonal close-packed
HDD hard-disk drive
HDT hexadecanethiol
HDTV high-definition television
HEK human embryonic kidney 293
HEL hot embossing lithography
HEXSIL hexagonal honeycomb polysilicon
HF hydrofluoric
HMDS hexamethyldisilazane
HNA hydrofluoric-nitric-acetic
HOMO highest occupied molecular orbital
HOP highly oriented pyrolytic
HOPG highly oriented pyrolytic graphite
HOT holographic optical tweezer
HP hot-pressing
HPI hexagonally packed intermediate
HRTEM high-resolution transmission electron microscope
HSA human serum albumin
HtBDC hexa-*tert*-butyl-decacyclene
HTCS high-temperature superconductivity
HTS high throughput screening
HUVEC human umbilical venous endothelial cell

I

IBD ion beam deposition
IC integrated circuit
ICA independent component analysis
ICAM-1 intercellular adhesion molecules 1
ICAM-2 intercellular adhesion molecules 2
ICT information and communication technology
IDA interdigitated array
IF intermediate filament
IF intermediate-frequency
IFN interferon
IgG immunoglobulin G
IKVAV isoleucine–lysine–valine–alanine–valine
IL ionic liquid
IMAC immobilized metal ion affinity chromatography
IMEC Interuniversity MicroElectronics Center
IR infrared
ISE indentation size effect
ITO indium tin oxide
ITRS International Technology Roadmap for Semiconductors
IWGN Interagency Working Group on Nanoscience, Engineering, and Technology

J

JC jump-to-contact
JFIL jet-and-flash imprint lithography
JKR Johnson–Kendall–Roberts

K

KASH	Klarsicht, ANC-1, Syne Homology
KPFM	Kelvin probe force microscopy

L

LA	lauric acid
LAR	low aspect ratio
LB	Langmuir–Blodgett
LBL	layer-by-layer
LCC	leadless chip carrier
LCD	liquid-crystal display
LCoS	liquid crystal on silicon
LCP	liquid-crystal polymer
LDL	low-density lipoprotein
LDOS	local density of states
LED	light-emitting diode
LFA-1	leukocyte function-associated antigen-1
LFM	lateral force microscope
LFM	lateral force microscopy
LIGA	Lithographie Galvanoformung Abformung
LJ	Lennard-Jones
LMD	laser microdissection
LMPC	laser microdissection and pressure catapulting
LN	liquid-nitrogen
LoD	limit-of-detection
LOR	lift-off resist
LPC	laser pressure catapulting
LPCVD	low-pressure chemical vapor deposition
LSC	laser scanning cytometry
LSN	low-stress silicon nitride
LT-SFM	low-temperature scanning force microscope
LT-SPM	low-temperature scanning probe microscopy
LT-STM	low-temperature scanning tunneling microscope
LT	low-temperature
LTM	laser tracking microrheology
LTO	low-temperature oxide
LTRS	laser tweezers Raman spectroscopy
LUMO	lowest unoccupied molecular orbital
LVDT	linear variable differential transformer

M

MALDI	matrix assisted laser desorption ionization
MAP	manifold absolute pressure
MAPK	mitogen-activated protein kinase
MAPL	molecular assembly patterning by lift-off
MBE	molecular-beam epitaxy
MC	microcantilever
MC	microcapillary
MCM	multi-chip module
MD	molecular dynamics
ME	metal-evaporated
MEMS	microelectromechanical system
MExFM	magnetic exchange force microscopy
MFM	magnetic field microscopy
MFM	magnetic force microscope
MFM	magnetic force microscopy
MHD	magnetohydrodynamic
MIM	metal–insulator–metal
MIMIC	micromolding in capillaries
MLE	maximum likelihood estimator
MOCVD	metalorganic chemical vapor deposition
MOEMS	microoptoelectromechanical system
MOS	metal–oxide–semiconductor
MOSFET	metal–oxide–semiconductor field-effect transistor
MP	metal particle
MPTMS	mercaptopropyltrimethoxysilane
MRFM	magnetic resonance force microscopy
MRFM	molecular recognition force microscopy
MRI	magnetic resonance imaging
MRP	molecular recognition phase
MscL	mechanosensitive channel of large conductance
MST	microsystem technology
MT	microtubule
mTAS	micro total analysis system
MTTF	mean time to failure
MUMP	multiuser MEMS process
MVD	molecular vapor deposition
MWCNT	multiwall carbon nanotube
MWNT	multiwall nanotube
MYD/BHW	Muller–Yushchenko–Derjaguin/Burgess–Hughes–White

N

NA	numerical aperture
NADIS	nanoscale dispensing
NASA	National Aeronautics and Space Administration
NC-AFM	noncontact atomic force microscopy
NEMS	nanoelectromechanical system
NGL	next-generation lithography
NHS	*N*-hydroxysuccinimidyl
NIH	National Institute of Health
NIL	nanoimprint lithography
NIST	National Institute of Standards and Technology
NMP	no-moving-part
NMR	nuclear magnetic resonance
NMR	nuclear mass resonance
NNI	National Nanotechnology Initiative

NOEMS nanooptoelectromechanical system
NP nanoparticle
NP nanoprobe
NSF National Science Foundation
NSOM near-field scanning optical microscopy
NSTC National Science and Technology Council
NTA nitrilotriacetate
nTP nanotransfer printing

O

ODA octadecylamine
ODDMS *n*-octadecyldimethyl(dimethylamino)silane
ODMS *n*-octyldimethyl(dimethylamino)silane
ODP octadecylphosphonate
ODTS octadecyltrichlorosilane
OLED organic light-emitting device
OM optical microscope
OMVPE organometallic vapor-phase epitaxy
OS optical stretcher
OT optical tweezers
OTRS optical tweezers Raman spectroscopy
OTS octadecyltrichlorosilane
oxLDL oxidized low-density lipoprotein

P

P–V peak-to-valley
PAA poly(acrylic acid)
PAA porous anodic alumina
PAH poly(allylamine hydrochloride)
PAPP *p*-aminophenyl phosphate
Pax paxillin
PBC periodic boundary condition
PBS phosphate-buffered saline
PC polycarbonate
PCB printed circuit board
PCL polycaprolactone
PCR polymerase chain reaction
PDA personal digital assistant
PDMS polydimethylsiloxane
PDP 2-pyridyldithiopropionyl
PDP pyridyldithiopropionate
PE polyethylene
PECVD plasma-enhanced chemical vapor deposition
PEEK polyetheretherketone
PEG polyethylene glycol
PEI polyethyleneimine
PEN polyethylene naphthalate
PES photoemission spectroscopy
PES position error signal
PET poly(ethyleneterephthalate)
PETN pentaerythritol tetranitrate
PFDA perfluorodecanoic acid
PFDP perfluorodecylphosphonate
PFDTES perfluorodecyltriethoxysilane
PFM photonic force microscope
PFOS perfluorooctanesulfonate
PFPE perfluoropolyether
PFTS perfluorodecyltricholorosilane
PhC photonic crystal
PI3K phosphatidylinositol-3-kinase
PI polyisoprene
PID proportional–integral–differential
PKA protein kinase
PKC protein kinase C
PKI protein kinase inhibitor
PL photolithography
PLC phospholipase C
PLD pulsed laser deposition
PMAA poly(methacrylic acid)
PML promyelocytic leukemia
PMMA poly(methyl methacrylate)
POCT point-of-care testing
POM polyoxy-methylene
PP polypropylene
PPD *p*-phenylenediamine
PPMA poly(propyl methacrylate)
PPy polypyrrole
PS-PDMS poly(styrene-b-dimethylsiloxane)
PS/clay polystyrene/nanoclay composite
PS polystyrene
PSA prostate-specific antigen
PSD position-sensitive detector
PSD position-sensitive diode
PSD power-spectral density
PSG phosphosilicate glass
PSGL-1 P-selectin glycoprotein ligand-1
PTFE polytetrafluoroethylene
PUA polyurethane acrylate
PUR polyurethane
PVA polyvinyl alcohol
PVD physical vapor deposition
PVDC polyvinylidene chloride
PVDF polyvinyledene fluoride
PVS polyvinylsiloxane
PWR plasmon-waveguide resonance
PZT lead zirconate titanate

Q

QB quantum box
QCM quartz crystal microbalance
QFN quad flat no-lead
QPD quadrant photodiode
QWR quantum wire

R

RBC red blood cell
RCA Radio Corporation of America
RF radiofrequency
RFID radiofrequency identification
RGD arginine–glycine–aspartic
RH relative humidity
RHEED reflection high-energy electron diffraction
RICM reflection interference contrast microscopy
RIE reactive-ion etching
RKKY Ruderman–Kittel–Kasuya–Yoshida
RMS root mean square
RNA ribonucleic acid
ROS reactive oxygen species
RPC reverse phase column
RPM revolutions per minute
RSA random sequential adsorption
RT room temperature
RTP rapid thermal processing

S

SAE specific adhesion energy
SAM scanning acoustic microscopy
SAM self-assembled monolayer
SARS-CoV syndrome associated coronavirus
SATI self-assembly, transfer, and integration
SATP (*S*-acetylthio)propionate
SAW surface acoustic wave
SB Schottky barrier
SCFv single-chain fragment variable
SCM scanning capacitance microscopy
SCPM scanning chemical potential microscopy
SCREAM single-crystal reactive etching and metallization
SDA scratch drive actuator
SEcM scanning electrochemical microscopy
SEFM scanning electrostatic force microscopy
SEM scanning electron microscope
SEM scanning electron microscopy
SFA surface forces apparatus
SFAM scanning force acoustic microscopy
SFD shear flow detachment
SFIL step and flash imprint lithography
SFM scanning force microscope
SFM scanning force microscopy
SGS small-gap semiconducting
SICM scanning ion conductance microscopy
SIM scanning ion microscope
SIP single inline package
SKPM scanning Kelvin probe microscopy
SL soft lithography
SLIGA sacrificial LIGA
SLL sacrificial layer lithography
SLM spatial light modulator
SMA shape memory alloy
SMM scanning magnetic microscopy
SNOM scanning near field optical microscopy
SNP single nucleotide polymorphisms
SNR signal-to-noise ratio
SOG spin-on-glass
SOI silicon-on-insulator
SOIC small outline integrated circuit
SoS silicon-on-sapphire
SP-STM spin-polarized STM
SPM scanning probe microscope
SPM scanning probe microscopy
SPR surface plasmon resonance
sPROM structurally programmable microfluidic system
SPS spark plasma sintering
SRAM static random access memory
SRC sampling rate converter
SSIL step-and-stamp imprint lithography
SSRM scanning spreading resistance microscopy
STED stimulated emission depletion
SThM scanning thermal microscope
STM scanning tunneling microscope
STM scanning tunneling microscopy
STORM statistical optical reconstruction microscopy
STP standard temperature and pressure
STS scanning tunneling spectroscopy
SUN Sad1p/UNC-84
SWCNT single-wall carbon nanotube
SWCNT single-walled carbon nanotube
SWNT single wall nanotube
SWNT single-wall nanotube

T

TA tilt angle
TASA template-assisted self-assembly
TCM tetracysteine motif
TCNQ tetracyanoquinodimethane
TCP tricresyl phosphate
TEM transmission electron microscope
TEM transmission electron microscopy
TESP tapping mode etched silicon probe
TGA thermogravimetric analysis
TI Texas Instruments
TIRF total internal reflection fluorescence
TIRM total internal reflection microscopy
TLP transmission-line pulse
TM tapping mode
TMAH tetramethyl ammonium hydroxide
TMR tetramethylrhodamine
TMS tetramethylsilane

TMS	trimethylsilyl
TNT	trinitrotoluene
TP	track pitch
TPE-FCCS	two-photon excitation fluorescence cross-correlation spectroscopy
TPI	threads per inch
TPMS	tire pressure monitoring system
TR	torsional resonance
TREC	topography and recognition
TRIM	transport of ions in matter
TSDC	thermally stimulated depolarization current
TTF	tetrathiafulvalene
TV	television

U

UAA	unnatural AA
UHV	ultrahigh vacuum
ULSI	ultralarge-scale integration
UML	unified modeling language
UNCD	ultrananocrystalline diamond
UV	ultraviolet
UVA	ultraviolet A

V

VBS	vinculin binding site
VCO	voltage-controlled oscillator
VCSEL	vertical-cavity surface-emitting laser
vdW	van der Waals
VHH	variable heavy–heavy
VLSI	very large-scale integration
VOC	volatile organic compound
VPE	vapor-phase epitaxy
VSC	vehicle stability control

X

XPS	x-ray photon spectroscopy
XRD	x-ray powder diffraction

Y

YFP	yellow fluorescent protein

Z

Z-DOL	perfluoropolyether

Part G

Industria

Part G Industrial Applications

45 The Millipede – A Nanotechnology-Based AFM Data-Storage System

Gerd K. Binnig, Munich, Germany
Giovanni Cherubini, Rüschlikon, Switzerland
Michel Despont, Rüschlikon, Switzerland
Urs T. Dürig, Rüschlikon, Switzerland
Evangelos Eleftheriou, Rüschlikon, Switzerland
Haralampos Pozidis, Rüschlikon, Switzerland
Peter Vettiger, Neuchâtel, Switzerland

46 Nanorobotics

Bradley J. Nelson, Zurich, Switzerland
Lixin Dong, East Lansing, USA

45. The *Millipede* – A Nanotechnology-Based AFM Data-Storage System

Gerd K. Binnig, Giovanni Cherubini, Michel Despont, Urs T. Dürig, Evangelos Eleftheriou, Haralampos Pozidis, Peter Vettiger

The *millipede* concept presented in this chapter is a new approach to storing data at high speed and ultrahigh density. The interesting part is that millipede stores digital information in a completely different way from magnetic hard disks, optical disks, and transistor-based memory chips. The ultimate locality is provided by a tip, and high data rates are a result of massive parallel operation of such tips. As storage medium, polymer films are being considered, although the use of other media, in particular magnetic materials, has not been ruled out. The current effort is focused on demonstrating the millipede concept with areal densities higher than 1 Tb/inch^2 and parallel operation of very large two-dimensional (2-D) (up to 64 × 64) atomic force microscopy (AFM) cantilever arrays with integrated tips and write/read/erase functionality. The fabrication and integration of such a large number of mechanical devices (cantilever beams) will lead to what we envision as the very large-scale integration (VLSI) age of micro- and nanomechanics.

In this chapter, the millipede concept for a microelectromechanical systems (MEMS)-based storage device is described in detail. In particular, various aspects pertaining to AFM thermomechanical read/write/erase functions, 2-D array fabrication and characteristics, *x,y,z* microscanner design, polymer media properties, read channel modeling, servo control and synchronization, as well as modulation coding techniques suitable for probe-based data-storage devices are discussed.

45.1 **The Millipede Concept** 1603
45.2 **Thermomechanical AFM Data Storage** ... 1604
45.3 **Array Design, Technology, and Fabrication** 1606
45.4 **Array Characterization** 1607
45.5 **Three-Terminal Cantilever Design** 1609
45.6 ***x,y,z* Medium Microscanner** 1610
45.7 **First Write/Read Results with the 32×32 Array Chip** 1613
45.8 **Polymer Medium** 1614
45.8.1 Writing Mechanism 1614
45.8.2 Erasing Mechanism 1618
45.8.3 Overwriting Mechanism 1619
45.9 **Read Channel Model** 1621
45.10 **System Aspects** 1624
45.10.1 PES Generation for the Servo Loop 1624
45.10.2 Timing Recovery 1627
45.10.3 Considerations on Capacity and Data Rate 1628
45.11 **Conclusions** 1629
References 1630

In the 21st century, the nanometer will probably play a role similar to that played by the micrometer in the 20th century. The nanometer scale will presumably also pervade the field of data storage, although there is so far no obvious way in conventional magnetic, optical or transistor-based storage to achieve the nanometer scale in all three dimensions. After decades of spectacular progress, those mature technologies have entered the home stretch; imposing physical limitations loom before them.

One promising method involves the use of patterned magnetic media, for which the ideal write/read concept still needs to be demonstrated. The biggest challenge, however, is to pattern the magnetic disk in a cost-effective way. If such approaches are successful, the basis for large-capacity storage in the 21st century might still be magnetism.

Other proposals call for totally different media and techniques such as local probes, near-field optics, magnetic superresolution or holographic methods. In general, when an existing technology is about to reach its limits in the course of its evolution and alternatives are emerging in parallel, two things usually happen: First, the existing and well-established technology is explored further and every effort is made to push its limits to the utmost in order to get the maximum return on the considerable investments made. Then, when all possibilities for improvement have been exhausted, the technology may still survive for certain niche applications, but the emerging technology will take over, opening new perspectives and new directions.

In many fields today we are witnessing the transition of structures from the micrometer to the nanometer scale, a dimension that nature has long been using to build the finest devices with a high degree of local functionality. Many of the techniques we use today are not suitable for the nanometer age; some will require minor or major modifications, while others will be partially or entirely replaced. It is certainly difficult to predict which techniques will fall into which category. In key information technology areas, such as nanoelectronics and data storage, it is not yet obvious which technologies and materials will be used in the future.

In any case, for an emerging technology to be seriously considered as a candidate to replace an existing technology that is approaching its inherent limits, it must provide a long-term perspective. For instance, the silicon microelectronics and storage industries are huge and have exacted correspondingly enormous investments, which makes them long-term oriented by nature. The consequence for storage is that any novel technology with higher areal storage density than today's magnetic recording [45.1, 2] should have long-term potential for further scaling, preferably down to the nanometer or even atomic scale.

The only available tool known today that is simple and yet provides this very long-term perspective is the nanometer-sharp tip. The simple tip is a very reliable tool that concentrates on one functionality: the ultimate local confinement of interaction. Techniques that use nanometer-sharp tips for imaging and investigating the structure of materials down to the atomic scale, such as the atomic force microscope (AFM) and the scanning tunneling microscope (STM) [45.3, 4], are suitable for the development of ultrahigh-density storage devices.

In the early 1990s, Mamin and Rugar at the IBM Almaden Research Center pioneered the capability of using an AFM tip for writing and read-back of topographic features for data storage. In one of their schemes [45.5], writing and reading were demonstrated with a single AFM tip in contact with a rotating polycarbonate substrate. The writing was performed thermomechanically by heating the tip. In this way, storage densities of up to 30 Gb/inch^2 were achieved, constituting a significant advance over the densities of that time. Later refinements included increasing read-back speeds to attain a data rate of 10 Mb/s [45.6] and the implementation of track servoing [45.7].

When using single tips in AFM or STM operation for storage, one has to deal with their fundamental data-rate limitations. At present, the mechanical resonance frequencies of the AFM cantilevers limit the data rates of a single cantilever to a few Mb/s for AFM data storage [45.8, 9]. The feedback speed and low tunneling currents limit STM-based storage approaches to even lower data rates.

Currently a single AFM operates at best on the microsecond time scale. Conventional magnetic storage, however, operates at best on the nanosecond time scale, making it clear that AFM data rates have to be improved by at least three orders of magnitude to be competitive with current and future magnetic recording. One solution for substantially increasing the data rates achievable by tip-based storage devices is to employ microelectromechanical system (MEMS) arrays of cantilevers operating in parallel, with each cantilever performing write/read/erase operations in an individual storage field. It is our conviction that very large-scale integrated (VLSI) micro/nanomechanics will greatly complement future micro- and nanoelectronics (integrated or hybrid) and may generate hitherto inconceivable applications of VLSI MEMS.

Various efforts are underway to develop MEMS-based storage devices. For example, a MEMS-actuated magnetic-probe-based storage device capable of storing 2 Gb of data on 2 cm^2 of die area and whose fabrication is compatible with a standard integrated-circuit manufacturing process is described by *Carley* et al. [45.10]. In their device, a magnetic storage medium is positioned in the xy-plane, and writing is achieved magnetically by using an array of probe tips, each tip being actuated in the z-direction. Another concept is the atomic-resolution storage described by *Gibson* et al. [45.11], who employ electron field emitters to change the state of a phase-change medium in a bitwise fashion from polycrystalline to amorphous or vice versa. Reading is done with lower currents by detecting either backscattered electrons or changes in the semiconductor properties in the media.

The *millipede* concept presented in this chapter is a new approach for storing data at high speed and ultrahigh density. The interesting part is that millipede stores digital information in a completely different way from magnetic hard disks, optical disks, and transistor-based memory chips. The ultimate locality is provided by a tip, and high data rates are a result of massive parallel operation of such tips. As storage medium, polymer films are being considered, although the use of other media, in particular magnetic materials, has not been ruled out. Our current effort is focused on demonstrating the millipede concept with areal densities higher than 1 Tb/inch^2 and parallel operation of very large 2-D (up to 64×64) AFM cantilever arrays with integrated tips and write/read/erase functionality. The fabrication and integration of such a large number of mechanical devices (cantilever beams) will lead to what we envision as the VLSI age of micro- and nanomechanics.

In this chapter, the millipede concept for a MEMS-based storage device is described in detail. In particular, various aspects pertaining to AFM thermomechanical read/write/erase functions, 2-D array fabrication and characteristics, x, y, z microscanner design, polymer media properties, read channel modeling, servo control and synchronization, as well as modulation coding techniques suitable for probe-based data-storage devices are discussed.

45.1 The Millipede Concept

A 2-D AFM cantilever array storage technique [45.12–15], internally called the millipede at the Zurich Research lab, is illustrated in Fig. 45.1. Information is stored as sequences of indentations and no indentations written on nanometer-thick polymer films using an array of AFM cantilevers. The presence and absence of indentations will also be referred to as logical marks. Each cantilever performs write/read/erase operations over an individual storage field with an area on the order of $100 \times 100\,\mu m^2$. Write/read operations depend on mechanical parallel x, y-scanning of either the entire cantilever array chip or the storage medium. Tip–medium contact is maintained and controlled globally, i. e., not on an individual cantilever basis, by using feedback control in the z-direction for the entire chip, which greatly simplifies the system. This basic concept of the entire chip approach/leveling was tested and demonstrated for the first time by parallel imaging with a 5×5 array chip [45.16, 17]. These parallel imaging results have shown that all 25 cantilever tips approached the substrate within less than 1 μm of z-actuation, which indicates that overall chip tip-apex height control to within 500 nm is feasible. The stringent requirement for tip-apex uniformity over the entire chip is determined by the uniform force required to reduce tip and medium wear due to large force variations resulting from large tip-height nonuniformities [45.7]. Moreover, as the entire array is tracked without individual lateral cantilever positioning, thermal expansion of the array chip has to be small or well controlled. Thermal expansion considerations are a strong argument for a 2-D instead of a one-dimensional (1-D) array arrangement.

Efficient parallel operations of large 2-D arrays can be achieved by a row/column time-multiplexed addressing scheme similar to that implemented in dynamic random-access memories (DRAMs). In the case of the millipede, the multiplexing scheme is used to address the array column by column with full parallel write/read operation within one column [45.18]. In particular, read-back signal samples are obtained by applying an electrical read pulse to the cantilevers in a column of the array, low-pass filtering the cantilever response signals, and finally sampling the filter output signals. This process is repeated sequentially until all columns of the array have been addressed, and then restarted from the first column. The time between two pulses applied to the cantilevers of the same column corresponds to the time it takes for a cantilever to move from one logical mark position to the next.

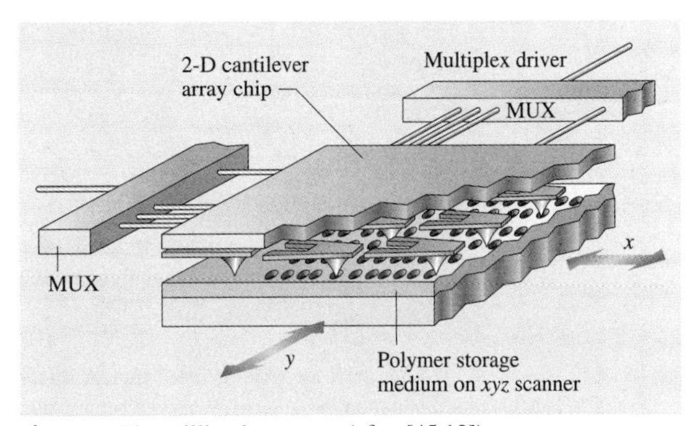

Fig. 45.1 The millipede concept (after [45.12])

An alternative approach is to access all or a subset of the cantilevers simultaneously without resorting to a row/column multiplexing scheme. Clearly, the latter scheme yields higher data rates, whereas the former leads to lower implementation complexity of the channel electronics.

45.2 Thermomechanical AFM Data Storage

In recent years, AFM thermomechanical recording in polymer storage media has undergone extensive modifications, mainly with respect to the integration of sensors and heaters designed to enhance simplicity and to increase data rate and storage density. Thermomechanical writing in polycarbonate films and optical read-back was first investigated and demonstrated by *Mamin* and *Rugar* [45.5]. Using heater cantilevers, thermomechanical recording at 30 Gb/inch^2 storage density and data rates of a few Mb/s for reading and 100 kb/s for writing have been demonstrated [45.5, 6, 19]. Although the storage density of 30 Gb/inch^2 obtained originally was not overwhelming, the results were encouraging enough to consider using polymer films to achieve density improvements. The current millipede storage approach is based on a new thermomechanical write/read process in nanometer-thick polymer films.

Thermomechanical writing is achieved by applying a local force through the cantilever tip to the polymer layer and simultaneously softening the polymer layer by local heating. Initially, the heat transfer from the tip to the polymer through the small contact area is very poor, but it improves as the contact area increases. This means that the tip must be heated to a relatively high temperature of about 400 °C to initiate softening. Once softening has been initiated, the tip is pressed into the polymer, and hence the indentation size is increased. Rough estimates [45.20, 21] indicate that at the beginning of the writing process only about 0.2% of the heating power is used in the very small contact zone (10–40 nm^2) to soften the polymer locally, whereas about 80% is lost through the cantilever legs to the chip body and about 20% is radiated from the heater platform through the air gap to the medium/substrate. After softening has started and the contact area has increased, the heating power available for generating the indentations increases at least ten times to reach 2% or more of the total heating power.

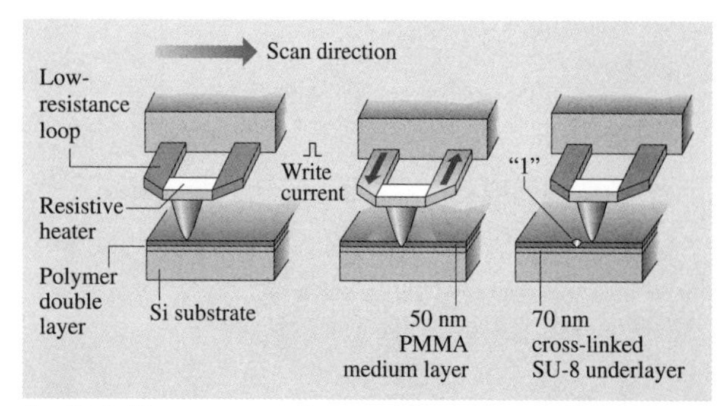

Fig. 45.2 New storage medium used for writing small bits. A thin, writable PMMA layer is deposited on a Si substrate separated by a cross-linked film of epoxy photoresist (after [45.15])

With this highly nonlinear heat-transfer mechanism it is very difficult to achieve small tip penetration and hence small bit sizes, as well as to control and reproduce the thermomechanical writing process. This situation can be improved if the thermal conductivity of the substrate is increased and if the depth of tip penetration is limited. We have explored the use of very thin polymer layers deposited on Si substrates to improve these characteristics [45.22, 23], as illustrated in Fig. 45.2. The hard Si substrate prevents the tip from penetrating farther than the film thickness, and it enables more rapid transport of heat away from the heated region, as Si is a much better conductor of heat than the polymer. Using coated Si substrates with a 50 nm film of polymethylmethacrylate (PMMA), we have achieved indentation sizes of between 10 and 50 nm. However, increased tip wear has occurred, probably caused by contact between the Si tip and the Si substrate during writing. Therefore, a 70 nm layer of cross-linked photoresist (SU-8) was introduced between the Si substrate and the PMMA film to act as a softer penetration stop that avoids tip wear but remains thermally stable.

Using this layered storage medium, indentations 40 nm in diameter have been written, as shown in Fig. 45.3. These experiments were performed using a 1 μm-thick, 70 μm-long, two-legged Si cantilever [45.19]. The cantilever legs are made highly conducting by high-dose ion implantation, whereas the heater region remains low doped. Electrical pulses 2 μs in duration were applied to the cantilever with a period of 50 μs. Figure 45.3a demonstrates that 40 nm bits can be written with 120 nm pitch or, as shown in Fig. 45.3b, very close to each other without merging, implying a potential areal density of 400 Gb/inch^2. Fig-

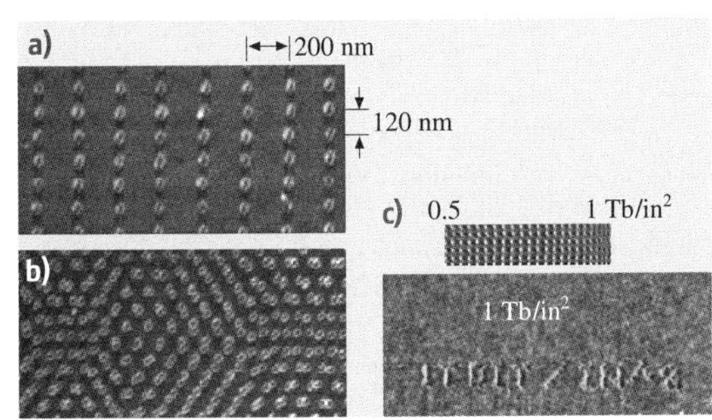

Fig. 45.3a–c Series of 40 nm indentations formed in a uniform array with (**a**) 120 nm pitch and (**b**) variable pitch (≥ 40 nm), resulting in areal densities of up to 400 Gb/inch^2. Images obtained with a thermal read-back technique. (**c**) Ultrahigh-density bit writing with areal densities approaching 1 Tb/inch^2. The scale is the same for all three images (from [45.15], © 2002 IEEE)

ure 45.3c shows results from a single-lever experiment, where indentations are spaced as closely as 25 nm apart, resulting in areal densities of up to 1 Tb/inch^2, although with a somewhat degraded write/read quality.

Imaging and reading are performed by a new thermomechanical sensing concept [45.24]. To read the written information, the heater cantilever originally used for writing is given the additional function of a thermal read-back sensor by exploiting its temperature-dependent resistance. In general, the resistance increases nonlinearly with heating power/temperature from room temperature to a peak value at 500–700 °C. The peak temperature is determined by the doping concentration of the heater platform, which ranges from 1×10^{17} to 2×10^{18} atoms/cm^3. Above the peak temperature, the resistance drops as the number of intrinsic carriers increases due to thermal excitation [45.25]. For sensing, the resistor is operated at about 350 °C, a temperature that is not high enough to soften the polymer as in the case of writing. The principle of thermal sensing is based on the fact that the thermal conductance between the heater platform and the storage substrate changes according to the distance between them. The medium between the heater platform and the storage substrate, in our case air, transports heat from the cantilever to the substrate. When the distance between cantilever and storage substrate is reduced as the tip moves into an indentation, the heat transport through the air becomes more efficient. As a result, the evolution of the heater temperature differs in response to a pulse applied to the cantilever. In particular, the maximum value achieved by the heater temperature is higher in the absence of an indentation. As the value of the variable resistance depends on the temperature of the cantilever, the maximum value achieved by the resistance will be lower as the tip moves into an indentation. Therefore, during the read process, the cantilever resistance reaches different values depending on whether the tip moves into an indentation (logical bit "1") or over a region without an indentation (logical bit "0"). Figure 45.4 illustrates this concept.

Under typical operating conditions, the sensitivity of thermomechanical sensing is even greater than that of piezoresistive-strain sensing, which is not surprising because thermal effects in semiconductors are stronger than strain effects. The good sensitivity is demonstrated by the images of the 40 nm-sized indentations in Fig. 45.3, which were obtained using the described thermal sensing technique.

The thermomechanical cantilever sensor, which transforms temperature into an electrical signal that carries information, is the electrical equivalent, to a first degree of approximation, of a variable resistance. A detection circuit must, therefore, sense a voltage that depends on the value of the cantilever resistance to decide whether a "1" or a "0" is written. The relative variation of thermal resistance is on the order of 10^{-5}/nm. Hence a written "1" typically produces a relative change of the cantilever thermal resistance $\Delta R^{\Theta}/R^{\Theta}$ of about 10^{-4} to 5×10^{-4}. Note that the relative change of the cantilever electrical resistance is of the same order of magnitude. Thus, one of the most critical issues in detecting the presence or absence of an indentation is the high resolution required to extract the

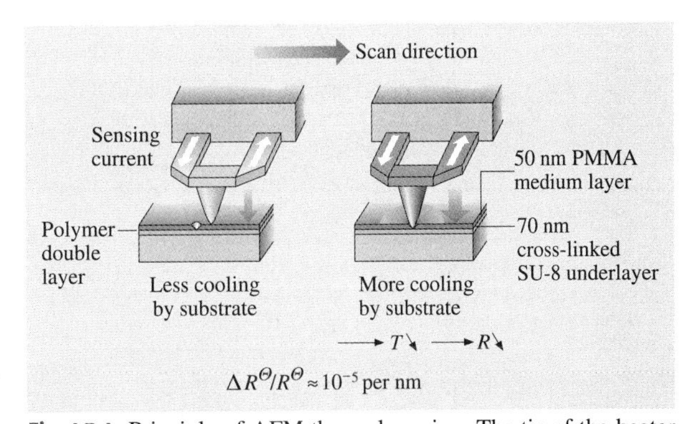

Fig. 45.4 Principle of AFM thermal sensing. The tip of the heater cantilever is continuously heated by a DC power supply while the cantilever is being scanned and the heater resistivity measured (after [45.15])

signal that contains the information about the logical bit being "1" or "0." The signal carrying the information can be regarded as a small signal superimposed on a very large offset signal. The large offset problem can be mitigated by subtracting a suitable reference signal [45.12, 26, 27].

More recently, single-probe experimental results have been obtained in which large data sets were recorded at 641 Gbit/inch^2 and read back with raw bit-error rates better than 10^{-4}, measured using the methodology of the magnetic-recording industry [45.28]. Although there are still aspects of thermomechanical recording that need further scrutiny before it can be employed in useful storage devices, it is the first scanning-probe-based method for which a thorough ultrahigh areal density demonstration has been performed.

45.3 Array Design, Technology, and Fabrication

Encouraged by the results of the 5×5 cantilever array [45.16, 17], a 32×32 array chip was designed and fabricated [45.18]. With the findings from the fabrication and operation of the 5×5 array and the very dense thermomechanical writing/reading in thin polymers with single cantilevers, some important changes of the chip functionality and fabrication processes were made. The major differences are (1) surface micromachining to form cantilevers at the wafer surface, (2) all-silicon cantilevers, (3) thermal instead of piezoresistive sensing, and (4) first- and second-level wiring with an insulating layer for a multiplexed row/column addressing scheme.

As the heater platform functions as a read/write element and no individual cantilever actuation is required, the basic array cantilever cell becomes a simple two-terminal device addressed by multiplexed x, y-wiring, as shown in Fig. 45.5. The cell area and x, y-cantilever pitch are 92×92 μm^2, which results in a total array size of less than 3×3 mm^2 for the 1024 cantilevers. The cantilevers are fabricated entirely of silicon for good thermal and mechanical stability. They consist of a heater platform with the tip on top, legs acting as soft mechanical springs, and electrical connections to the heater. They are highly doped to minimize interconnect resistance and to replace the metal wiring on the cantilever in order to eliminate electromigration and parasitic z-actuation of the cantilever due to a bimorph effect. The resistive ratio between the heater and the silicon interconnect sections should be as high as possible; currently the resistance of the highly doped interconnections is ≈ 400 Ω and that of the heater platform is 5 kΩ (at 3 V reading bias).

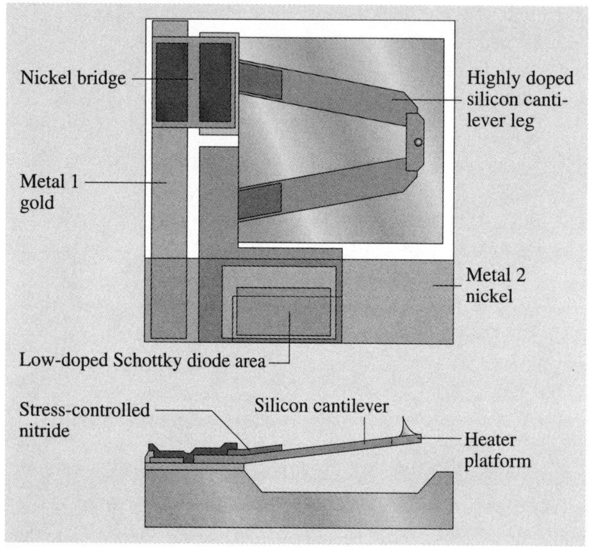

Fig. 45.5 Layout and cross-section of one cantilever cell (after [45.29])

The cantilever mass has to be minimized to obtain soft, high-resonance-frequency cantilevers. Soft cantilevers are required for a low loading force in order to eliminate or reduce tip and medium wear, whereas a high resonance frequency allows high-speed scanning. In addition, sufficiently wide cantilever legs are required for a small thermal time constant, which is partly determined by cooling via the cantilever legs [45.19]. These design considerations led to an array cantilever with 50 μm-long, 10 μm-wide, 0.5 μm-thick legs, and a 5 μm-wide, 10 μm-long, 0.5 μm-thick heater platform. Such a cantilever has a stiffness of ≈ 1 N/m and a resonance frequency of ≈ 200 kHz. The heater time constant is a few μs, which should allow a multiplexing rate of up to 100 kHz.

The tip height should be as small as possible because the heater platform sensitivity depends strongly on the platform-to-medium distance. This contradicts the requirement of a large gap between the chip surface and the storage medium to ensure that only the tips, and not the chip surface, make contact with the medium. Instead of making the tips longer, we purposely bent the cantilevers a few micrometers out of the chip plane by depositing a stress-controlled plasma-enhanced chemical vapor deposition (PECVD) silicon nitride layer at the base of the cantilever (Fig. 45.5). This bending as well as the tip height must be well controlled in order to maintain an equal loading force for all cantilevers of an array.

Cantilevers are released from the crystalline Si substrate by surface micromachining using either plasma or wet chemical etching to form a cavity underneath the cantilever. Compared with a bulk-micromachined through-wafer cantilever-release process, as was done for our 5×5 array [45.16, 17], the surface-micromachining technique allows even higher array density and yields better mechanical chip stability and heat sinking. As mentioned above, the entire array is tracked without individual lateral cantilever positioning, therefore thermal expansion of the array chip has to be small or well controlled. For a $3\times3\,\mathrm{mm}^2$ silicon array area and 10 nm tip-position accuracy, the temperature difference between array chip and medium substrate has to be controlled to $\approx 1\,°\mathrm{C}$. This is ensured by four temperature sensors in the corners of the array and heater elements on each side of the array.

The photograph in Fig. 45.6 shows a fabricated chip with the 32×32 array located in the center ($3\times3\,\mathrm{mm}^2$) and the electrical wiring interconnecting the array with the bonding pads at the chip periphery. Figure 45.7 shows the 32×32 array section of the chip with the independent approach/heat sensors in the four corners and the heaters on each side of the array, as well as zoomed scanning electron micrographs (SEMs) of an array section, a single cantilever, and a tip apex. The tip height is 1.7 μm and the apex radius is smaller than 20 nm, which is achieved by oxidation sharpening [45.30]. The cantilevers are connected to the column and row address lines using integrated Schottky diodes in series with the cantilevers. The diode is operated in reverse bias (high resistance) if the cantilever is not addressed, thereby greatly reducing cross-talk between cantilevers. More details about the array fabrication are given in [45.29, 31].

Fig. 45.6 Photograph of fabricated chip ($14\times7\,\mathrm{mm}^2$). The 32×32 cantilever array is located at the center with bond pads distributed on either side (from [45.29], © 1999 IEEE)

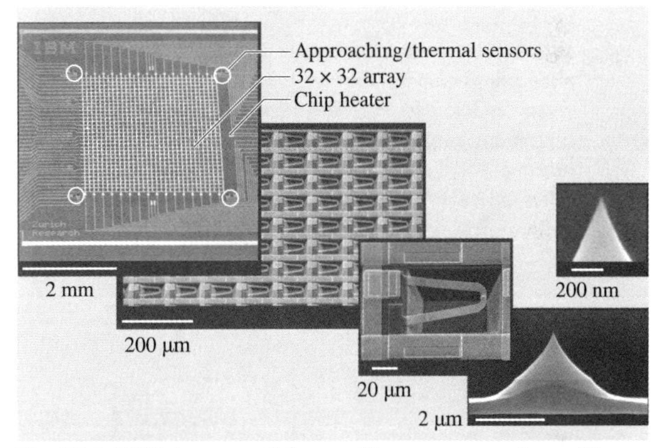

Fig. 45.7 SEM images of the cantilever array section with approaching and thermal sensors in the corners, array and single cantilever details, and tip apex (after [45.18], © 2000 International Business Machines Corporation)

45.4 Array Characterization

The array's independent cantilevers, which are located in the four corners of the array and used for approaching and leveling the chip and storage medium, serve to initially characterize the interconnected array cantilevers. Additional cantilever test structures are distributed over the wafer; they are equivalent to but independent of the array cantilevers. Figure 45.8 shows an I–V curve of such a cantilever; note the nonlinearity of the resistance. In the low-power part of the curve, the resistance increases as a function of heating power, whereas in the high-power regime it decreases.

In the low-power, low-temperature regime, silicon mobility is affected by phonon scattering, which depends on temperature, whereas at higher power, the intrinsic temperature of the semiconductor is reached, which results in a resistivity drop owing to

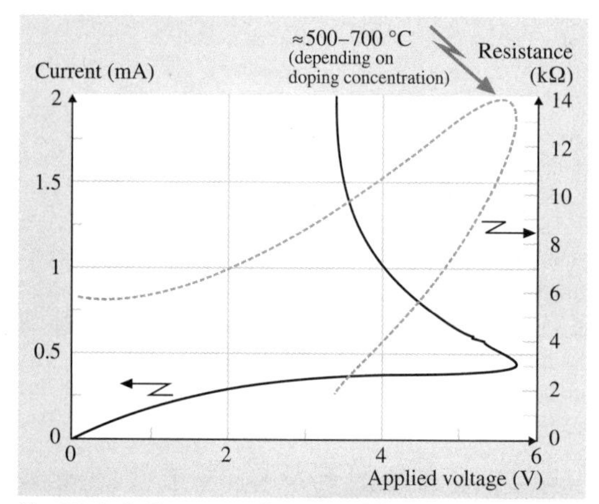

Fig. 45.8 I–V curve of one cantilever. The curve is nonlinear owing to the heating of the platform as the power and temperature are increased. For doping concentrations between 1×10^{17} and 2×10^{18} atoms/cm^3, the maximum temperature varies between 500 °C and 700 °C (after [45.31])

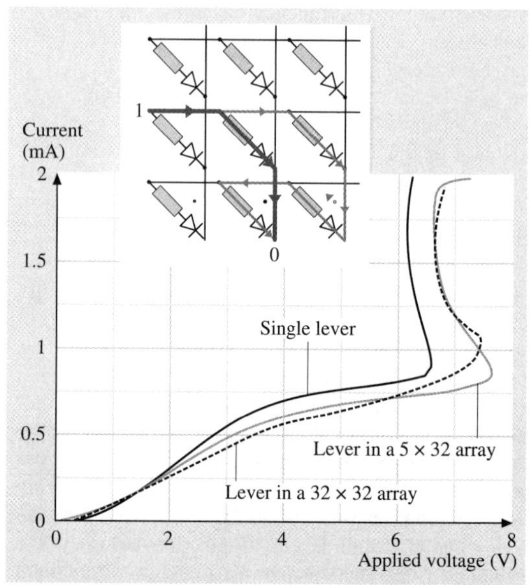

Fig. 45.9 Comparison of the I–V curve of an independent cantilever (*solid black line*) with the current response when addressing a cantilever in a 5×5 (*solid brown line*) or a 32×32 (*dashed line*) array with a Schottky diode serially connected to the cantilever. Little change is observed in the I–V curve between the different cases. *Inset*: sketch of the direct path (*bold line*) and a parasitic path (*thin line*) in a cantilever–diode array. In the parasitic path there is always one diode in reverse bias that reduces the parasitic current (after [45.31])

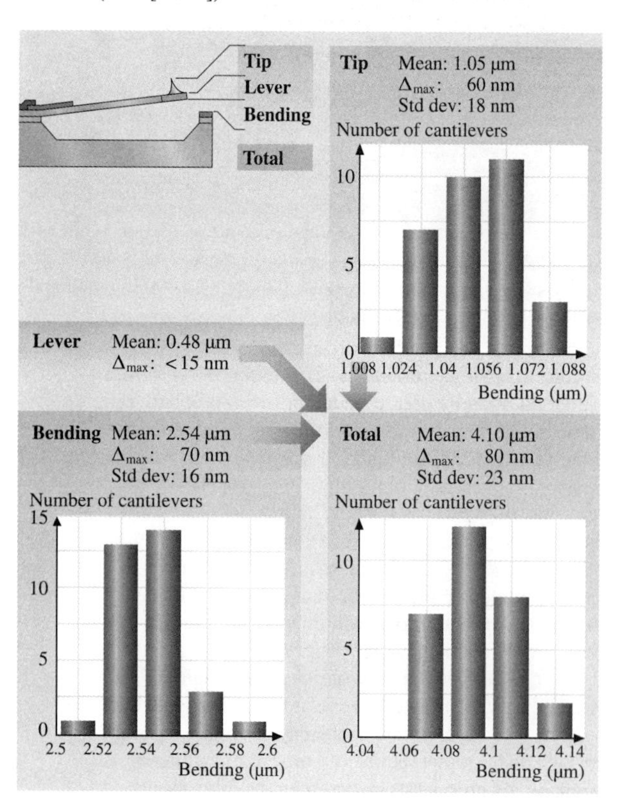

the increasing number of carriers [45.25]. Depending on the heater-platform doping concentration of 1×10^{17}–2×10^{18} atoms/cm^3, our calculations estimate a resistance maximum at a temperature of 500–700 °C, respectively.

The cantilevers within the array are electrically isolated from one another by integrated Schottky diodes. As every parasitic path in the array to the cantilever addressed contains a reverse-biased diode, the cross-talk current is drastically reduced, as shown in Fig. 45.9. Thus, the current response of an addressed cantilever in an array is nearly independent of the size of the array, as demonstrated by the I–V curves in Fig. 45.9. Hence, the power applied to address a cantilever is not shunted

Fig. 45.10 Tip-apex height uniformity across one cantilever row of the array with individual contributions from the tip height and cantilever bending (after [45.18], © 2000 International Business Machines Corporation) ◀

by other cantilevers, and the reading sensitivity is not degraded – not even for very large arrays (32 × 32). The introduction of the electrical isolation using integrated Schottky diodes turned out to be crucial for successful operation of interconnected cantilever arrays with a simple time-multiplexed addressing scheme.

The tip-apex height uniformity within an array is very important because it determines the force of each cantilever while in contact with the medium and hence influences write/read performance, as well as medium and tip wear. Wear investigations suggest that a tip-apex height uniformity across the chip of less than 500 nm is required [45.7], with the exact value depending on the spring constant of the cantilever. In the case of the millipede, the tip-apex height is determined by the tip height and the cantilever bending. Figure 45.10 shows the tip-apex height uniformity of one row of the array (32 tips) due to tip height and cantilever bending. It demonstrates that our uniformity is on the order of 100 nm, thus meeting the requirements.

45.5 Three-Terminal Cantilever Design

Figure 45.11 shows a scanning electron microscopy (SEM) image of the bottom of the write/read cantilever probe used in the high-areal-density experiments [45.28]. Unlike the two-terminal cantilevers described in Sect. 45.3, the cantilevers of this new generation employ a novel three-terminal design in which there are separate resistive heaters for reading and writing, and a capacitive platform for enhanced electrostatic force. The use of separate read and write heaters allows the read heater, which is located at a certain distance from the tip, to be operated at a higher temperature for better sensitivity, without incurring the problem of heating the tip to temperatures that could degrade previously recorded indentations in the polymer medium.

The main body of the cantilever consists of Si with a phosphorous doping concentration of $10^{20}\,\mathrm{cm}^{-3}$, and has a total length of 65 μm and a thickness of 0.5 μm. Thinned hinge regions (0.25 μm thick) near the points where the cantilever legs attach to the rigid support structure reduce the spring constant of the lever to a value of $\approx 0.05\,\mathrm{N/m}$. The two resistive heaters are regions of lower doping ($5 \times 10^{17}\,\mathrm{cm}^{-3}$). The write heater, from which the sharp tip protrudes, is 5 μm long and 2 μm wide. To write indentations into the polymer medium, a voltage pulse is applied across the appropriate cantilever terminals to heat both the write heater and the tip. At the same time, a voltage pulse is also applied to the substrate of the polymer medium to create a simultaneous force on the tip against the polymer film. The magnitudes of the heat and force pulses and their relative timing can be adjusted independently. The large area of the electrostatic platform allows forces as high as 1 μN to be generated at voltages of < 20 V. The read operation exploits the high-temperature sensitivity of the resistivity of doped semiconductors. The size of the

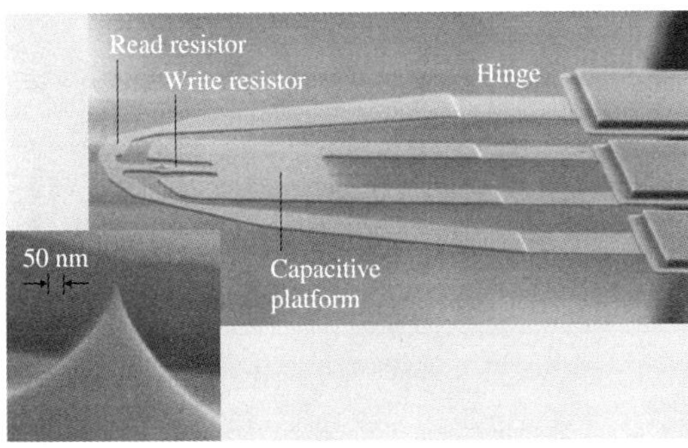

Fig. 45.11 Three-terminal integrated probe. *Inset*: enlarged view of the tip, which is located on the write heater (after [45.28], © IEEE 2004)

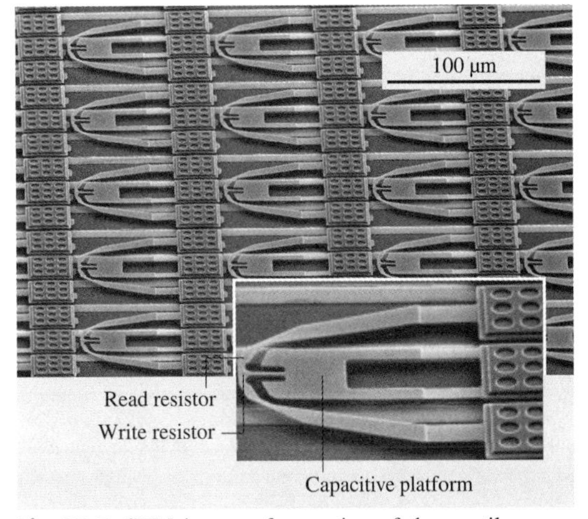

Fig. 45.12 SEM image of a section of the cantilever array transferred and interconnected onto its carrier wafer. *Inset*: close-up of a three-terminal integrated cantilever (after [45.32], © IOP Publishing Ltd. 2004)

read heater is $3\times4.5\,\mu m^2$ and it is heated by a constant applied voltage.

The tip sharpness is crucial for the creation of well-controlled indentations of sufficiently small diameter to achieve the densities reported here. Tip radii of $\approx 10\,nm$ are typical for the probes used in the new design. The tip is formed from the same Si material as the rest of the structure.

These concepts led to a second generation of array chips consisting of up to 4096 cantilevers with integrated tips, sensors, and actuators [45.33]. Figure 45.12 shows a section of this new array chip.

45.6 *x,y,z* Medium Microscanner

A key issue for the millipede concept is the need for a low-cost, miniaturized scanner with x, y, z motion capabilities and a lateral scanning range on the order of $100\,\mu m$. Multiple-probe systems arranged as 1-D or 2-D arrays [45.18] must also be able to control, by means of tilt capabilities, the parallelism between the probe array and the sample [45.18, 35].

We have developed a microscanner with these properties based on electromagnetic actuation. It consists of a mobile platform supported by springs and containing integrated planar coils positioned over a set of miniature permanent magnets [45.36]. A suitable arrangement of the coils and magnets allows us, by electrically addressing the various coils, to apply magnetically induced forces to the platform and drive it in the x-, y-, z-, and tilt directions. Our first silicon/copper-based version of this device has proved the validity of the concept [45.37], and variations of it have since been used elsewhere [45.38]. However, the undamped copper spring system gave rise to excessive cross-talk and ringing when driven in an open loop, and its layout limited the compactness of the overall device.

We investigate a modified microscanner that uses flexible rubber posts as a spring system and a copper-epoxy-based mobile platform (Fig. 45.13). The platform is made of a thick, epoxy-based SU-8 resist [45.39], in which the copper coils are embedded. The posts are made of polydimethylsiloxane (PDMS, Sylgard 184 silicon elastomer, Dow Corning, Midland, MI) and are fastened at the corners of the platform and at the ground plate, providing an optimally compact device by sharing the space below the platform with the magnets. The shape of the posts allows their lateral and longitudinal stiffness to be adjusted, and the dissipative rubber-like properties of PDMS provide damping to avoid platform ringing and to suppress nonlinearities.

Figure 45.14 shows the layout of the platform, which is scaled laterally, so that the long segments of the *racetrack* coils used for in-plane actuation coincide with commercially available $24\,mm^2$ SmCo magnets. The thickness of the device is determined by that of the magnets (1 mm), the clearance between magnet and platform ($500\,\mu m$), and the thickness of the platform itself, which is $250\,\mu m$ and determined mainly by the aspect ratio achievable in SU-8 resist during the ex-

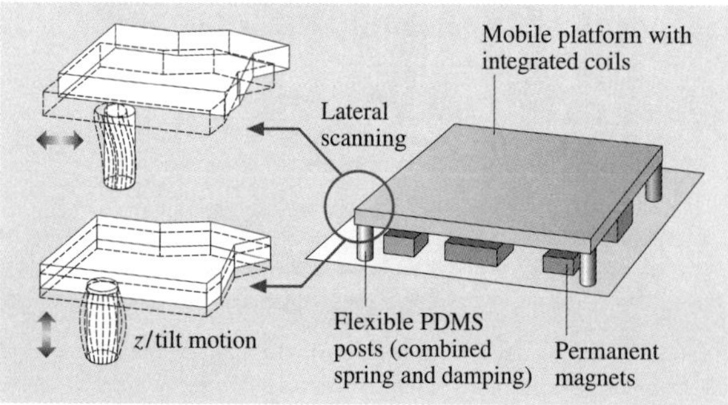

Fig. 45.13 Microscanner concept using a mobile platform and flexible posts (after [45.34])

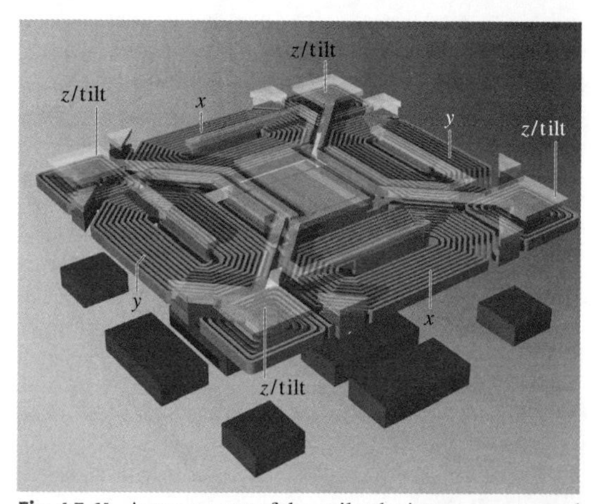

Fig. 45.14 Arrangement of the coils, the interconnects, and the permanent magnets, as well as the various motions addressed by the corresponding coils (after [45.15], © IEEE 2002)

posure of the coil plating mold. The resulting device volume is $\approx 15 \times 15 \times 1.6\,\text{mm}^3$.

The SmCo magnets produce a measured magnetic field intensity of $\approx 0.14\,\text{T}$ at the mid-thickness of the coils. The effective coil length is 320 mm, yielding an expected force $F_{x,y}$ of 45 μN per mA of drive current.

The principal design issue of the spring system is the ratio of its stiffnesses for in-plane and out-of-plane motion. Whereas for many scanning probe applications the required z-axis range need not be much larger than a few microns, it is necessary to ensure that the z-axis retraction of the platform due to the shortening of the posts as they take on an S-shape at large in-plane deflections can be compensated for at acceptable z-coil current levels. Various PDMS post shapes have been investigated to optimize and trade off the various requirements. Satisfactory performance was found for simple O-shapes [45.34].

The fabrication of the scanner (Fig. 45.15) starts on a silicon wafer with a seed layer and a lithographically patterned 200 μm-thick SU-8 layer, in which copper is electroplated to form the coils (Fig. 45.15a). The coils typically have 20 turns, with a pitch of 100 μm and a spacing of 20 μm. Special care was taken in the resist processing and platform design to achieve the necessary aspect ratio and to overcome adhesion and stress problems of SU-8. A second SU-8 layer, which serves as an insulator, is patterned with via holes, and another seed layer is then deposited (Fig. 45.15b). Next, an interconnect level is formed using a Novolac-type resist mask and a second copper-electroplating step (Fig. 45.15c). After stripping the resist, the silicon wafer is dissolved by a sequence of wet and dry etching, and the exposed seed layers are sputtered away to prevent shorts (Fig. 45.15d).

The motion of the scanner was characterized using a microvision strobe technique [45.40]. The results presented below are based on O-type PDMS ports. Frequency response curves for in-plane motion (Fig. 45.16) show broad peaks (characteristic of a large degree of damping) at frequency values that are consistent with expectations based on the measured mass of the platform (0.253 mg). The amplitude response (Fig. 45.17) displays the excellent linearity of the spring system for displacement amplitudes up to 80 μm (160 μm displacement range). Based on these near-direct-current (DC) (10 Hz) responses ($\approx 1.4\,\mu\text{m/mA}$) and a measured circuit resistance of 1.9 Ω, the power necessary for a 50 μm displacement amplitude is approximately equal to 2.5 mW.

Owing to limitations of the measurement technique, it was not possible to measure out-of-plane

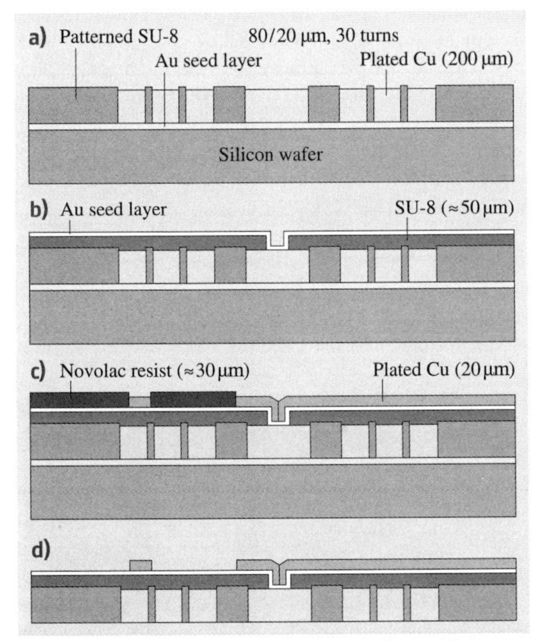

Fig. 45.15a–d Cross-section of the platform fabrication process. **(a)** Coils are electroplated through an SU-8 resist mask, which is retained as the body of the platform; **(b)** an insulator layer is deposited; **(c)** interconnects are electroplated; **(d)** the platform is released from the silicon substrate (after [45.34])

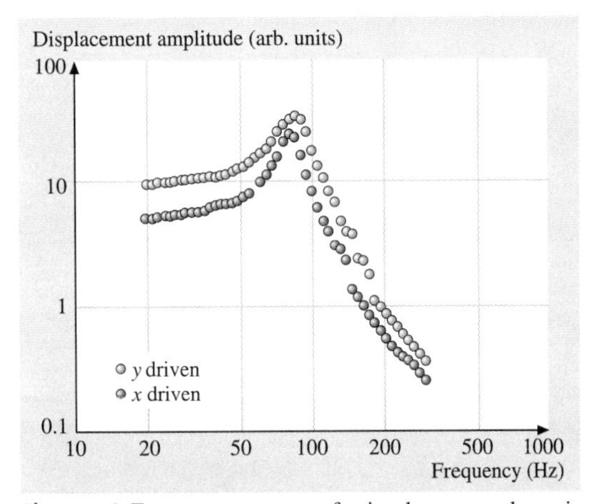

Fig. 45.16 Frequency response for in-plane x- and y-axis motion. The mechanical quality factors measured are between 3.3 and 4.6 (after [45.15])

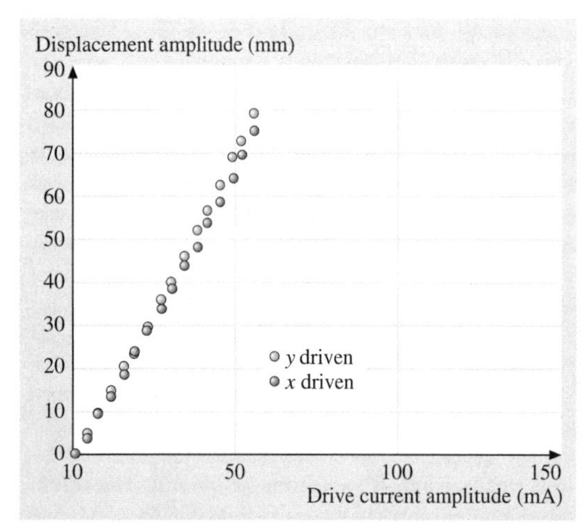

Fig. 45.17 In-plane displacement amplitude response for an alternating-current (AC) drive at 10 Hz (off resonance) (after [45.15])

displacements greater than 0.5 μm. However, the small-amplitude response for z-motion when all four corner coils are driven in-phase also displays good linearity over the range that can be measured (Fig. 45.18).

The electromagnetic scanner performs reliably and as predicted in terms of the scan range, device volume, and power requirements, achieving overall displacement ranges of 100 μm with ≈ 3 mW of power. The potential access time in the 100 μm storage field is on the order of a few ms. By being potentially cheap to manufacture, the integrated scanner presents a good alternative actuation system for many scanning probe applications.

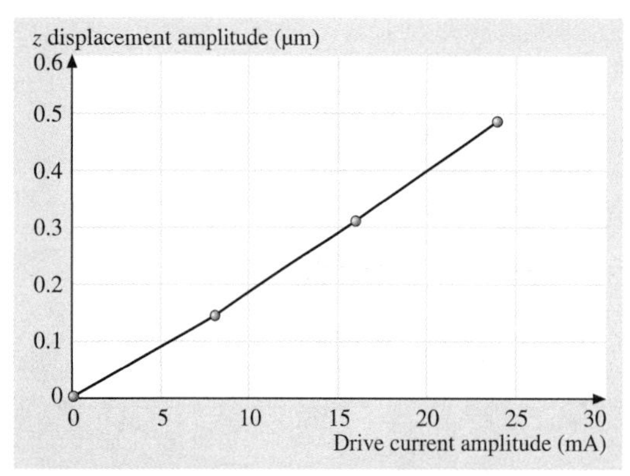

Fig. 45.18 Out-of-plane amplitude response for an AC drive at 3 Hz. The drive current is the total for all four corner coils, which are driven in phase (after [45.34])

An alternative design that has been realized for the displacement of the storage medium relative to the array of cantilevers is based on using a silicon microscanner with x, y-displacement capabilities of about 120 μm, i. e., ≈ 20% larger than the pitch between adjacent cantilevers in the array [45.42]. The scanner consists of a 6.8×6.8 mm^2 scan table and a pair of voice-coil-type actuators, all of which are supported by springs (Fig. 45.19a). The mechanical components of the scanner are fabricated from a 400 μm-thick silicon wafer using a deep-trench etching process. This scanner chip is then mounted on a silicon base plate that acts as the mechanical ground of the system. The base plate has been designed to provide a clearance of about 20 μm between its top surface and the bottom surface of the moving parts of the scanner. The scan table, which carries the polymer storage medium, can be displaced in two orthogonal directions (x and y) in the plane of the silicon wafer. Each voice-coil actuator consists of a pair of permanent magnets glued into a silicon frame, with a miniature 8.4 Ω coil mounted between them on the base plate. Actuation is achieved by applying a current to the coil, which generates a force on the magnets and induces a displacement of the actuator. This motion is coupled to the scan table by means of a mass-balancing scheme that provides a 1 : 1 translation of the motion while making the scanner robust against external shock and vibrations [45.32].

During motion in one of the two orthogonal directions in the plane of the silicon wafer, three forces act on the microscanner: the external electromagnetic force produced by the actuator, the restoring force produced by the springs, and the damping force that mainly re-

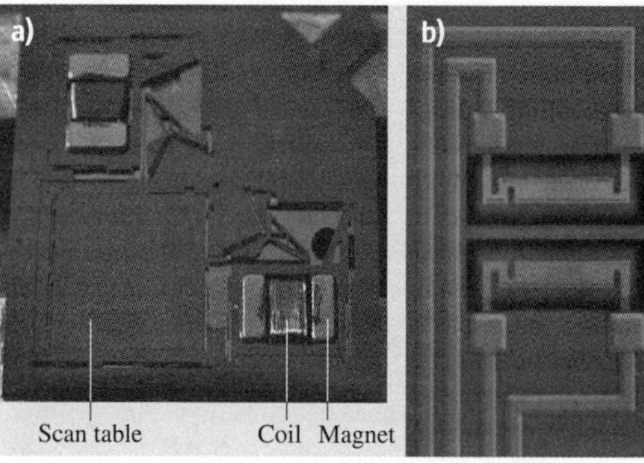

Fig. 45.19a,b Picture of (a) a microscanner, (b) thermal position sensors (after [45.41], © 2007 IEEE)

sults from the air. In the scanner, the force produced by each of the actuators is a linear function of the current applied to the actuator coil, and the resulting displacement of the scan table is a linear function of this force. Thus, assuming that there is no coupling between the horizontal (x-direction) and the vertical (y-direction) motion, the mechanical behavior of the scan table in the x and the y-direction can be approximated by a simple spring–damper–mass model, described by the second-order differential equation

$$m\ddot{x}(t)+b\dot{x}(t)+kx(t)=g(t)\,, \tag{45.1}$$

where m denotes the mass of scan table and actuator, b is the damping coefficient, and k is the spring constant. The external force is expressed by the product of the current $i(t)$ applied to the coil and the transducer gain g. The first and second derivatives of the position, denoted by $\dot{x}(t)$ and $\ddot{x}(t)$, respectively, indicate the velocity and acceleration of the scan table. The Laplace transform of (45.1) results in the second-order system

$$\frac{X(s)}{I(s)}=K_0\frac{\omega_0^2}{s^2+\left(\frac{\omega_0}{Q}\right)s+\omega_0^2}\,, \tag{45.2}$$

where $K_0=g/k$, $\omega_0=\sqrt{k/m}$, and $Q=\sqrt{km}/b$ denote the DC gain, the resonance frequency, and the quality factor, respectively. Estimates of the spring constant and the damping coefficient of the scanner are obtained by measuring the mass, the resonance frequency, and the quality factor. The transducer gain g is estimated by applying a constant current to the coil and measuring the resulting displacement at the equilibrium position.

The probe-based storage device described here uses two pairs of thermal position sensors to provide x, y-position information to the servo controller during closed-loop operation of the scanner [45.32, 41, 43]. These sensors are fabricated on the cantilever-array chip and positioned directly above the scan table (Fig. 45.19b). The sensors consist of thermally isolated, resistive strip heaters made from moderately doped silicon. Each sensor is positioned above an edge of the scan table and heated by application of a current. A fraction of this heat is conducted through the ambient air into the scan table, which acts as a heat sink. Displacement of the scan table translates into a change in the temperature of these heaters and thus a change in their electrical resistance. When driving the sensors with a constant voltage, the changes in resistance can be detected by measuring the resulting current. The time response of each sensor is well modeled by a first-order low-pass filter with a time constant of 100 ms. Therefore the cutoff frequency of the low-pass filter is typically set to 10 kHz. The thermal sensors operate over the entire travel range of the microscanner and hence provide global position information.

Although these devices are quite simple, the linearity, noise limit, and drift performance that can be achieved are surprisingly good. The measured responses for both axes are essentially linear over the entire 120 mm range. For a given scanner displacement amplitude, the signal-to-noise ratio of these devices depends on the heating power applied as well as on the bandwidth and the noise performance of the amplifier. The DC drift of these sensors in ambient laboratory conditions is also very low, with a typical standard deviation of less than 6 nm in 1 h. These performance characteristics are crucial for the overall performance of a servo controller [45.32, 41].

45.7 First Write/Read Results with the 32×32 Array Chip

We have built a prototype that includes all the basic building blocks of the millipede concept (Fig. 45.1) [45.44]. A $3\times3\,\text{mm}^2$ silicon substrate is spin-coated with the SU-8/PMMA polymer medium, as described in Sect. 45.3. This storage medium is attached to the x, y, z microscanner and approaching device. The magnetic z-approaching actuators bring the medium into contact with the tips of the array chip. The z-distance between the medium and the millipede chip is controlled by the approaching sensors in the corners of the array. The signals from these cantilevers are used to determine the forces on the z-actuators and, hence, the forces of the cantilever while it is in contact with the medium. This sensing/actuation feedback loop continues to operate during x, y-scanning of the medium. The personal computer (PC)-controlled write/read scheme addresses the 32 cantilevers of one row in parallel. Writing is performed by connecting the addressed row for 20 μs to a high negative voltage and simultaneously applying data inputs ("0" or "1") to the 32 column lines. The data input is a high positive voltage for a "1" and ground for a "0." This row-enabling and column-addressing scheme supplies a heater current to all cantilevers, but only those cantilevers with high

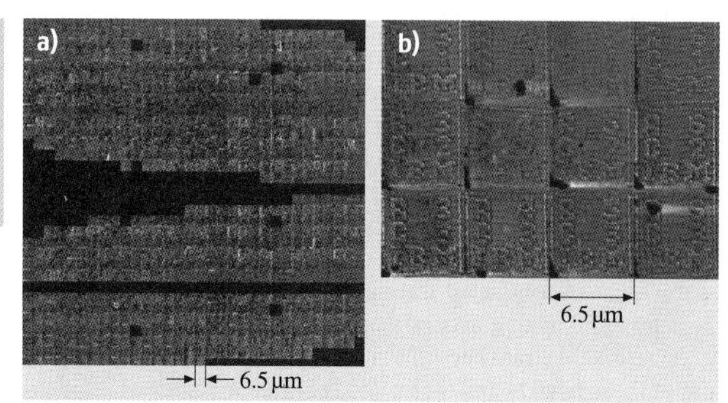

Fig. 45.20 (**a**) 1024 images, each one obtained from a cantilever of the 2-D array. (**b**) Enlarged view of typical images from (**a**). The *numbers* in the images indicate the row and column of each lever (after [45.44])

positive voltage generate an indentation ("1"). Those grounded are not hot enough to form an indentation and thus write a "0." When the scan stage has moved to the next logical mark position, the process is repeated, and this is continued until the line scan is finished. In the read process, the selected row line is connected to a moderate negative voltage, and the column lines are grounded via a protection resistor of about 10 kΩ, which keeps the cantilevers warm. During scanning, the voltages across the resistors are measured. Depending on the topography of the recording surface, the degree of cooling of each cantilever varies, thus changing the resistance and voltage across the series resistor and allowing written data to be read back.

The results of writing and reading in this fashion can be seen in Fig. 45.20, which shows 1024 images corresponding to the 1024 storage fields and associated cantilevers. Of the 1024 levers, 834 were able to write and read data back, i. e., a success rate of more than 80%. The sequence is as follows. First, a bit pattern is written simultaneously by each of the levers in row 1, then read back simultaneously, followed by row 2, etc. through row 32. The data sent to the levers is different, each lever writing its own row and column number in the array. The bit pattern consists of 64×64 bits, where odd-numbered bits are always set to 0. In this case, the area used is $6.5 \times 6.5\,\mu m^2$. The read-back image is a grey-scale bit map of 128×128 pixels. The distance between levers is 92 μm, so the images in Fig. 45.20 are also 92 μm apart. The areal density of the written information shown in Fig. 45.20 corresponds to $15-30\,Gb/inch^2$, depending on the coding scheme adopted. More recently, an areal density of $150-200\,Gb/inch^2$, at an array yield of about 60%, has been demonstrated.

Those levers that did not read back failed for one of four reasons:

(i) A defective chip connector rendered an entire column unusable;
(ii) A point defect occurred, meaning that a single lever or tip was broken;
(iii) There was a nonuniformity of the tip contact due to tip/lever variability or storage substrate bowing due to mounting;
(iv) There were thermal drifts.

The latter two reasons were the most likely and major failure sources. At present, there is clearly a tradeoff between the number of working levers and the density, which will most likely be resolved by a better substrate/chip mounting technique and lower thermal drifts.

The writing and read-back rates achieved with this system are 1 kb/s per lever, thus the total data rate is about 32 kb/s. This rate is limited by the rate at which data can be transferred over a typical PC bus, not by a fundamental time limitation of the read/write process.

45.8 Polymer Medium

The polymer storage medium plays a crucial role in millipede-like thermomechanical storage systems. The thin-film multilayer structure with PMMA as active layer (Fig. 45.2) is not the only possible choice, considering the almost unlimited range of polymer materials available. The ideal medium should be easily deformable for writing, yet indentations should be stable against tip wear and thermal degradation. Finally, one would also like to be able to erase and rewrite data repeatedly. In order to be able to address all important aspects properly, some understanding of the basic physical mechanism of thermomechanical writing and erasing is required.

45.8.1 Writing Mechanism

In a *gedanken* experiment we visualize writing of an indentation as the motion of a rigid body (the tip) in a viscous medium (the polymer melt). Let us initially assume that the polymer, i. e., PMMA, behaves like a simple liquid after it has been heated above the glass-transition temperature in a small volume around the tip. As viscous drag forces must not exceed the loading force applied to the tip during indentation, we can estimate an upper bound for the viscosity ζ of the polymer melt using Stokes's equation

$$F = 6\pi\zeta\varrho v\,. \tag{45.3}$$

In actual indentation formation, the tip loading force is on the order of $F = 50\,\mathrm{nN}$ and the radius of curvature at the apex of the tip is typically $\varrho = 20\,\mathrm{nm}$. Assuming a depth of the indentation of, say, $h = 50\,\mathrm{nm}$ and a heat pulse of duration $\tau_\mathrm{h} = 10\,\mu\mathrm{s}$, the mean velocity during indentation formation is on the order of $v = h/\tau_\mathrm{h} = 5\,\mathrm{mm/s}$. Note that thermal relaxation times are on the order of ms [45.20, 21] and, hence, the heating time can be equated to the time it takes to form an indentation. With these parameters we obtain $\zeta < 25\,\mathrm{Pa\,s}$, whereas typical values for the shear viscosity of PMMA are at least seven orders of magnitude larger, even at temperatures well above the glass-transition point [45.45].

This apparent contradiction can be resolved by considering that polymer properties are strongly dependent on the time scale of observation. At time scales on the order of 1 ms and below, entanglement motion is in effect frozen in and the PMMA molecules form a relatively static network. Deformation of the PMMA now proceeds by means of uncorrelated deformations of short molecular segments, rather than by a flow mechanism involving the coordinated motion of entire molecular chains. The price one has to pay is that elastic stress builds up in the molecular network as a result of the deformation (the polymer is in a so-called rubbery state). On the other hand, corresponding relaxation times are orders of magnitude smaller, giving rise to an effective viscosity at millipede time scales on the order of 10 Pa s [45.45], as required by our simple argument (45.3). Note that, unlike normal viscosity, this high-frequency viscosity is basically independent of the detailed molecular structure of the PMMA, i. e., chain length, tacticity, polydispersity, etc. In fact, we can even expect that similar high-frequency viscous properties can be found in a large class of other polymer materials, which makes thermomechanical writing a rather robust process in terms of material selection.

We have argued above that elastic stress builds up in the polymer film during the formation of an indentation, creating a corresponding reaction force on the tip on the order of $F_\mathrm{r} \sim 2\pi G\varrho^2$, where G denotes the elastic shear modulus of the polymer [45.46]. An important property for millipede operation is that the shear modulus drops by several orders of magnitude in the glass-transition regime, i. e., for PMMA from $\approx 1\,\mathrm{GPa}$ below Θ_g to ≈ 0.5–$1\,\mathrm{MPa}$ above Θ_g, where Θ_g denotes the glass-transition temperature [45.45]. The bulk modulus, on the other hand, retains its low-temperature value of several GPa. Hence, in this elastic regime, formation of an indentation above Θ_g constitutes a volume-preserving deformation. For proper indentation formation, the tip load must be balanced between the extremes of the elastic reaction force F_r for temperatures below and above Θ_g, i. e., $F \ll 2.5\,\mu\mathrm{N}$ for PMMA to prevent indentation of the polymer in the cold state and $F \gg 2.5\,\mathrm{nN}$ to overcome the elastic reaction force in the hot state. Unlike the deformation of a simple liquid, the indentation represents a metastable state of the entire deformed volume, which is under elastic tension. Recovery of the unstressed initial state is prevented by rapid quenching of the indentation below the glass-transition temperature with the tip in place. As a result, the deformation is frozen in, because below Θ_g motion of molecular-chain segments is, in effect, inhibited (Fig. 45.21).

This mechanism also allows indentations to be erased locally; it suffices to heat the deformed volume locally above Θ_g, whereupon the indented volume reverts to its unstressed flat state driven by internal elastic stress. In addition, erasing is promoted by surface tension forces, which give rise to a restoring surface pressure on the order of $\gamma(\pi/\varrho)^2 h \approx 25\,\mathrm{MPa}$, where $\gamma \approx 0.02\,\mathrm{N/m}$ denotes the polymer–air surface tension.

One question immediately arises from these speculations: If the polymer behavior can be determined from the macroscopic characteristics of the shear modulus as a function of time, temperature, and pressure, can the time–temperature superposition principle also be applied in this case? The time–temperature superposition principle is a powerful concept of polymer physics [45.47]. It basically states that the time scale and the temperature are interdependent variables that determine the polymer behavior such as the shear modulus. A simple transformation can be used to translate time into temperature-dependent data and vice versa. It is not clear, however, whether this principle can be applied in our case, i. e., under such extreme conditions (high pressures, short time scales, and nanometer-sized volumes, which are clearly below the radius of gyration of individual polymer molecules).

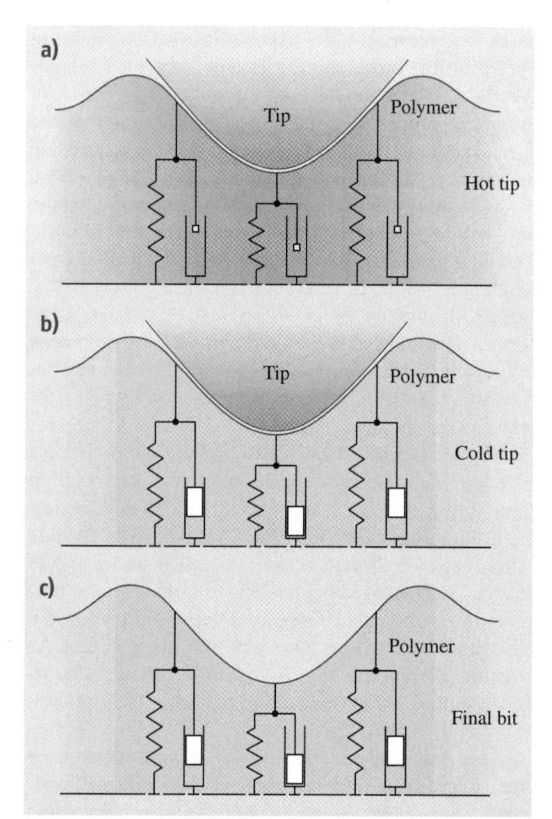

Fig. 45.21a–c Viscoelastic model of indentation writing. (**a**) The hot tip heats a small volume of polymer material to more than Θ_g. The shear modulus of the polymer drops drastically from GPa to MPa, which in turn allows the tip to indent the polymer. In response, elastic stress (represented as *compression springs*) builds up in the polymer. In addition, viscous forces (represented as *pistons*) associated with the relaxation time for the local deformation of molecular segments limit the indentation speed. (**b**) At the end of the writing process, the temperature is quenched on a microsecond time scale to room temperature: The stressed configuration of the polymer is frozen-in (represented by the *locked pistons*). (**c**) The final indentation corresponds to a metastable configuration. The original unstressed flat state of the polymer can be recovered by heating the indentation volume to more than Θ_g, which unlocks the compressed springs (after [45.15])

To test this, we varied the heating time, the heating temperature, and the loading force in indentation writing experiments on a standard PMMA sample. The results are summarized in Fig. 45.22. The minimum heater temperature at which the formation of an indentation starts for a given heating pulse length and loading force was determined. This so-called threshold temperature is plotted against the heating pulse length. Careful calibration of the heater temperature has to be done to allow comparison of the data. The heater temperature was determined by assuming proportionality between temperature and electrical power dissipated in the heater resistor at the end of the heating pulse when the tip has reached its maximum temperature. An absolute temperature scale is established using two well-defined reference points. One is room temperature, corresponding to zero electrical power. The other is provided by the point of turnover from positive to negative differential resistance (Fig. 45.8), which corresponds to a heater temperature of $\approx 550\,°C$. The general shape of the measured curves of threshold temperature versus heating time indeed shows the characteristics of time–

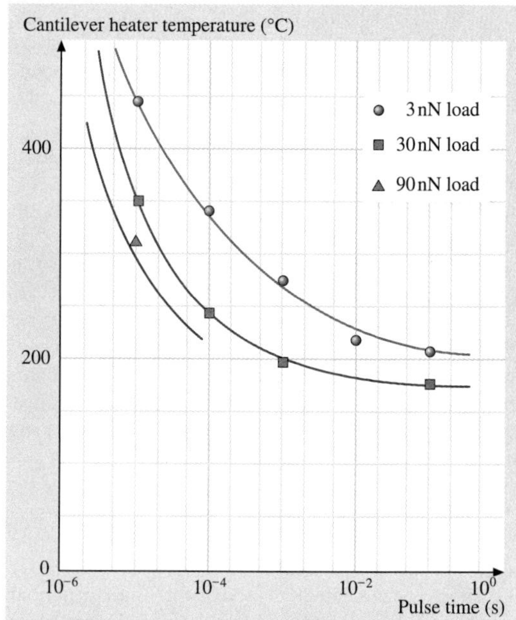

Fig. 45.22 Indentation writing threshold measurements. The load was controlled by pushing the cantilever tip into the sample with a controlled displacement and a known spring constant of the cantilever. When a certain threshold is reached, the indentations become visible in subsequent imaging scans (see also Fig. 45.21). The *solid lines* are guides to the eye. Curves of similar shape would be expected from the time–temperature superposition principle (after [45.15])

temperature superposition. In particular, the curves are identical up to a load-dependent shift with respect to the time axis. Moreover, we observe that the time it takes to form an indentation at constant heater temperature is inversely proportional to the tip load. This property is exactly what one would expect if internal friction (owing to the high-frequency viscosity) is the rate-limiting step in forming an indentation (45.3).

The time it takes to heat the indentation volume of polymer material higher than the glass-transition temperature is another potentially rate-limiting step. Here, the spreading resistance of the heat flow in the polymer and the thermal contact resistance are the most critical parameters. Simulations suggest [45.20, 21] that equilibration of temperature in the polymer occurs within less than 1 μs. Very little is known, however, about the thermal coupling efficiency across the tip–polymer interface. We have several indications that the heat transfer between tip and sample plays a crucial role, one of them being the asymptotic heater temperature for long writing times, which according to the graph in Fig. 45.22 is $\approx 200\,°\mathrm{C}$. The exact temperature of the polymer is unknown. However, the polymer temperature should approach the glass-transition temperature (around 120 °C for PMMA) asymptotically. Hence, the temperature drop between heater and polymer medium is substantial. Part of the temperature difference is due to a temperature drop along the tip, which according to heat-flow simulations [45.20, 21] is expected to be on the order of 30 °C at most. Therefore, a significant temperature gradient must exist in the tip–polymer contact zone. Further experiments on the heat transfer from tip to surface are needed to clarify this point.

We also find that the heat transfer for a nonspherical tip is anisotropic. As shown in Fig. 45.23, in the case of a pyramid-shaped tip, the indentation not only exhibits sharp edges, but also the region around the indentation, where polymer material is piled up, is anisotropic. The pile-up characteristics will be discussed in detail below. At this point we take it as an indication of the relevance of the heat transfer to the measurements.

One of the most striking conclusions of our model of the indentation writing process is that it should in principle work for most polymer materials. The general behavior of the mechanical properties as a function of temperature and frequency is similar for all polymers [45.47]. The glass-transition temperature Θ_g would then be one of the main parameters that determines the threshold writing temperature.

A verification of this was found experimentally by comparing various polymer films. The samples were prepared in the same way as the PMMA samples discussed above [45.18], i. e., by spin-casting thin films (10–30 nm) onto a silicon wafer with a photoresist buffer. Then, threshold measurements were done by applying heat pulses with increasing current (or temperature) to the tip while the load and the heating time were held constant (load about 10 nN and heating time 10 μs). Examples of such measurements are shown in Fig. 45.24, where the increasing size and depth of indentations can be seen for different heater temperatures. A threshold can be defined based on such data and compared with the glass-transition temperature of these materials. The results show a clear correlation between the threshold heater temperature and the glass-transition temperature (Fig. 45.25).

With our simple viscoelastic model of writing we are able to formulate a set of requirements that potential candidate materials for millipede data storage have to fulfill. First, the material should ideally exhibit a well-defined glass-transition point with a large drop of the shear modulus at Θ_g. Second, a rather high value of Θ_g (on the order of 150 °C) is preferred to facilitate thermal read-back of the data without destroying the information. We have investigated a number of materials to explore the Θ_g parameter space. The fact that all polymer types tested are suitable for forming small indentations leaves us free to choose which polymer type to optimize in terms of the technical requirements for a device such as lifetime of indentations, polymer endurance of the read and write process, power consumption, etc. These are subjects of ongoing research.

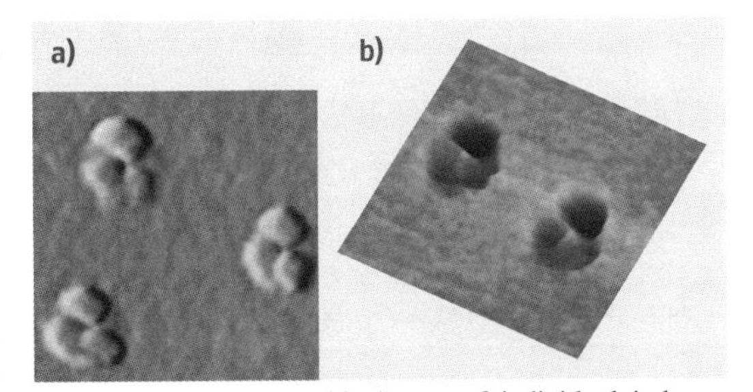

Fig. 45.23a,b Topographic image of individual indentations. **(a)** The region around the actual indentations clearly shows the threefold symmetry of the tip, here a three-sided pyramid. **(b)** The indentations themselves exhibit sharp edges, as can be seen from the inverted 3-D image. Image size is $2\times2\,\mu\mathrm{m}^2$ (after [45.15], © 2002 IEEE)

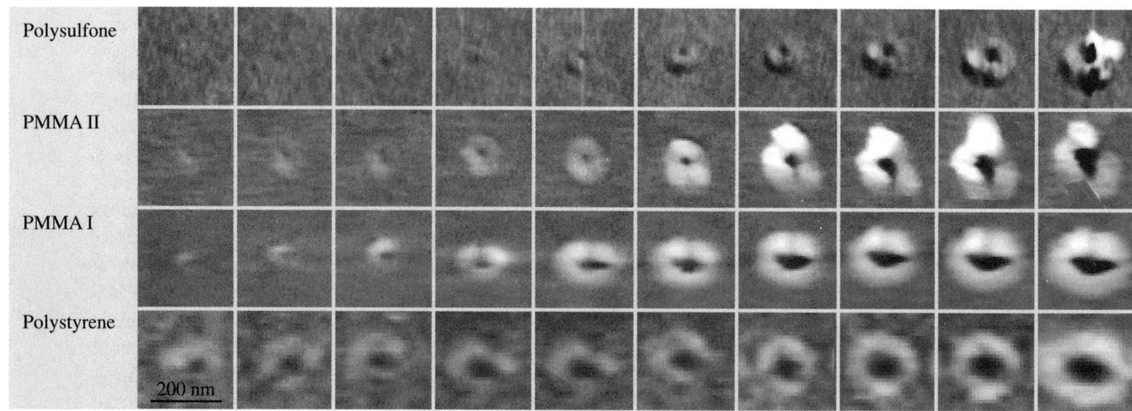

Fig. 45.24 Written indentations for different polymer materials. The heating pulse length was 10 μs, the load about 10 nN. The *grey scale* is the same for all images. The heater temperatures for the indentation on the *left-hand side* are 445 °C, 400 °C, 365 °C, and 275 °C for the polymers polysulfone, PMMA II (anionically polymerized PMMA, $M \approx 26$ k), PMMA I (Polymer Standard Service, Germany, $M \approx 500$ k), and polystyrene, respectively. The temperature increase between events on the horizontal axis is 14 °C, 22 °C, 20 °C, and 9 °C, respectively (after [45.15], © 2002 IEEE)

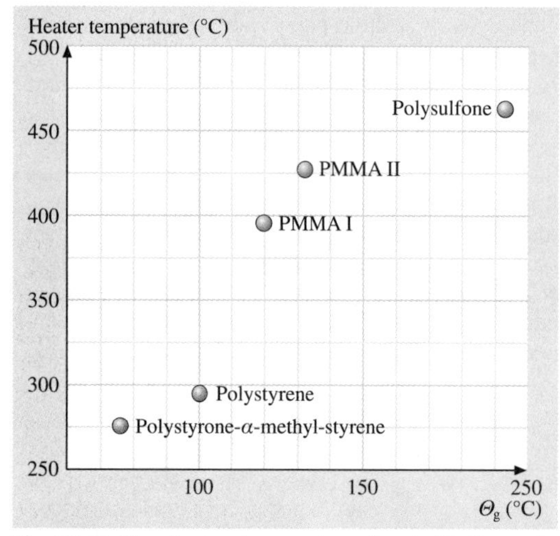

Fig. 45.25 The heater temperature threshold for writing indentations with the same parameters as in Fig. 45.21, plotted against the glass-transition temperature for these polymers, including poly-α-methyl-styrene (after [45.15])

45.8.2 Erasing Mechanism

It is worthwhile to look at the detailed shapes of the written indentations. The polymer material around an indentation appears piled up, as can be seen, for example, in Fig. 45.23. This is not only material that was pushed aside during indentation formation as a result of volume conservation. Rather, the flash heating by the tip and subsequent rapid cooling result in an increase of the specific volume of the polymer. This phenomenon, that the specific volume of a polymer can be increased by rapidly cooling a sample through the glass transition [45.47], is well known. Our system allows a cooling time on the order of microseconds, which is much faster than the highest rates that can be achieved with standard polymer analysis tools. However, a quantitative measurement of the specific volume change cannot be easily performed with our type of experiments. On the other hand, the pile-up effect serves as a convenient threshold thermometer. The outer perimeter of the pile-ups surrounding the indentations corresponds to the Θ_g isotherm, and the temperature in the enclosed area has certainly reached values greater than Θ_g during the indentation process. Based on our viscoelastic model, one would thus conclude that previously written indentations that overlap with the piled up region of a subsequently written indentation should be erased.

That this pile-up effect actually works against the formation of an indentation can clearly be seen in the line scans of a series of indentations written in polysulfone (Fig. 45.26). Here, the heating of the tip was accompanied by a rather high normal force. The force was high enough to create a small indentation, even if the tip was too cold to modify the polymer ("a" in Fig. 45.26). Then, with increasing tip heating, the indentations initially fill up in the piled-up region ("b" in Fig. 45.26) before they finally become deeper ("c" in Fig. 45.26).

The pile-up phenomenon turns out to be particularly beneficial for data-storage applications. The following example demonstrates the effect. If we look at the sequence of images in Fig. 45.27, taken on a standard PMMA sample, we find that the piled up regions can overlap each other without disturbing the indentations. However, if the piled-up region of an individual writing event extends over the indented area of a previously written "1", the depth of the corresponding indentation decreases markedly (Fig. 45.27d). This can be used for erasing written data. On the other hand, if the pitch between two successive indentations is decreased even further, this erasing process will no longer work. Instead, a broader indentation is formed, as shown in Fig. 45.27e. Hence, to exclude mutual interference, the minimum pitch between successive indentations, which we denote by the minimum indentation pitch (BP_{min}), must be larger than the radius of the piled-up area around an indentation.

In the example shown in Fig. 45.27, the temperature chosen was so high that the ring around the indentations was very large, whereas the depth of the indentation was limited by the stop layer underneath the PMMA material. Clearly, the temperature was too high here to form small indentations, the minimum pitch of which is around 250 nm. However, by carefully optimizing all parameters it is possible to achieve areal densities of up to 1 Tb/inch^2, as demonstrated in Fig. 45.3c.

The new erasing scheme based on this volume effect switches from writing to erasing merely by decreasing the pitch of writing indentations. This can be done in a very controlled fashion, as shown in Fig. 45.28, where individual lines or predefined subareas are erased. Hence, this new erasing scheme can be made to work in a way that is controlled on the scale of individual indentations. Compared with earlier global erasing schemes [45.23], this simplifies erasing significantly.

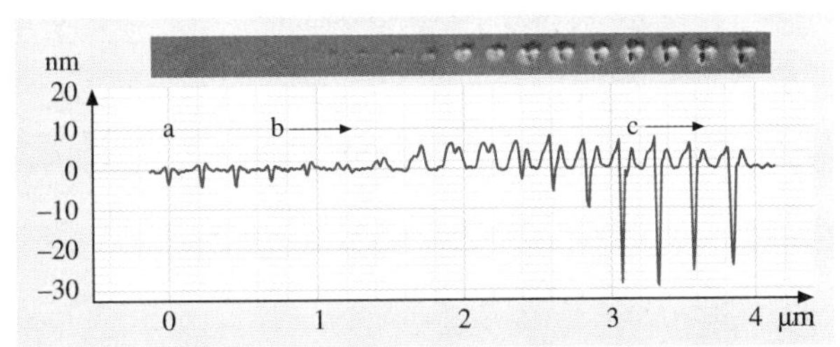

Fig. 45.26 Section through a series of indentations similar to those shown in Fig. 45.21. Here, a load of about 200 nN was applied before a heating pulse of 10 μs length was fired. The temperature of the heater at the end of the pulse has been increased from 430 °C to 610 °C in steps of about 10.6 °C. a – The load was sufficient to form a plastic indentation even if the polymer was not heated enough to come near the glass transition. b – With increasing heater temperature, the polymer swells. This eliminates the indentation, thus erasing previously written *cold* marks. c – As this process continues, the thermomechanical formation of indentations begins to dominate until, finally, normal thermomechanical indentation writing occurs (after [45.15])

45.8.3 Overwriting Mechanism

Overwriting data on some part of the storage medium can be achieved by first erasing the entire area and

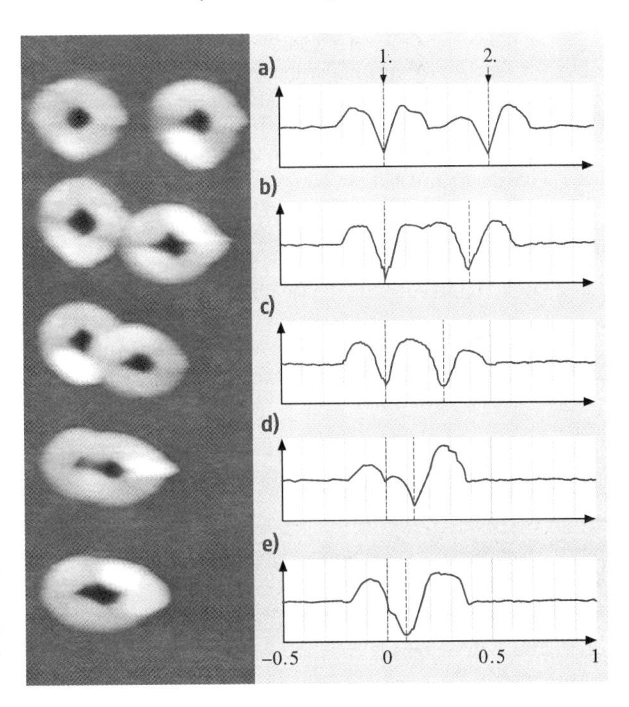

Fig. 45.27a–e Indentations in a PMMA film at several separations. The depth of the indentations is ≈ 15 nm, roughly the same as the thickness of the PMMA layer. The indentations on the *left-hand side* were written first, then a second series of indentations was made at decreasing separation from the first series, going from **(a)** to **(e)** (after [45.15]) ▶

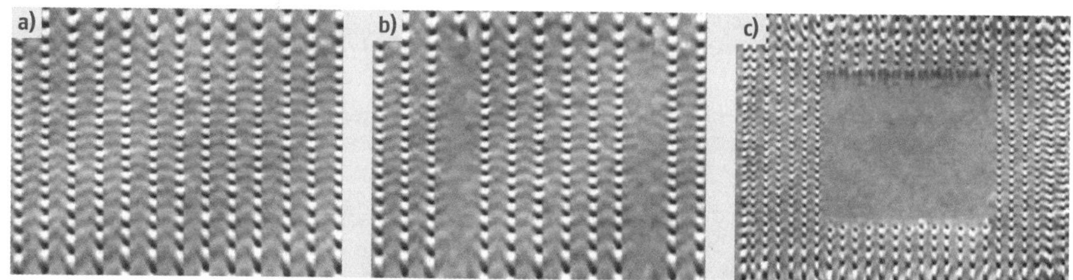

Fig. 45.28a–c Demonstration of the new erasing scheme: (**a**) A bit pattern recorded with variable pitch in the *vertical axis* (fast scan axis) and constant pitch in the *horizontal direction* (slow scan axis) was prepared. (**b**) Then two of the lines were erased by decreasing the pitch in the vertical direction by a factor of three, showing that the erasing scheme works for individual lines. One can also erase entire fields of indentations without destroying indentations at the edges of the fields. This is demonstrated in (**c**), where a field has been erased from an indentation field similar to the one shown in (**a**). The distance between the lines is 70 nm (after [45.15], © 2002 IEEE)

then writing the desired data on the erased surface. Although this process works well, it is time-consuming and dissipates a significant amount of power. In a millipede-based storage device, where data rate and power consumption are at a premium, such a two-step overwriting mechanism may be impractical. Instead, a one-step, direct overwriting process similar to those applied in magnetic hard-disk drives and rewritable optical drives is desired.

As discussed above, switching from writing to erasing may be achieved by decreasing the pitch of writing. It has been found experimentally that erasing can be performed effectively by halving the pitch of writing successive indentations, which is denoted as BP, provided that the condition $\mathrm{BP} \approx \mathrm{BP}_{\mathrm{min}}$ is satisfied. This suggests that the basic distance unit for combined write–erase operations should be BP/2. Written indentations are spaced n units apart, where $n \geq 2$. Let us recall that the presence of an indentation corresponds to a logical bit "1" and the absence of an indentation to a logical bit "0." Logical bits are then stored in the medium at the points of a regular lattice with minimum distance between points equal to BP/2 in the on-track direction, with successive "1"s separated by at least one "0." This condition is necessary in order to avoid mutual interference between successive "1"s. It is also the basis for an important category of codes known as (d, k) codes, which are described in Sect. 45.10.3. Coding can thus be used to enable direct data overwriting in an elegant way. Direct overwriting requires the simultaneous realization of two conditions: If previously written "1"s exist where "0"s are to be written, then these "1"s have to be *erased*. On the other hand, if "0"s exist where "1"s are to be written, then these "1"s have to be *written*. Writing an indentation is performed thermomechanically as described above. Erasing an existing indentation is done by writing another indentation next to it, at a distance of BP/2 units. However, as this operation creates a new indentation shifted by BP/2 with respect to the one erased, the erasing process must be performed repeatedly until the newly created indentation lies at a position corresponding to a "1" in the new data pattern. The basic principle of erasing is illustrated in Fig. 45.29a,b. The figure shows how the four bit strings 001, 010, 100, and 101 are modified

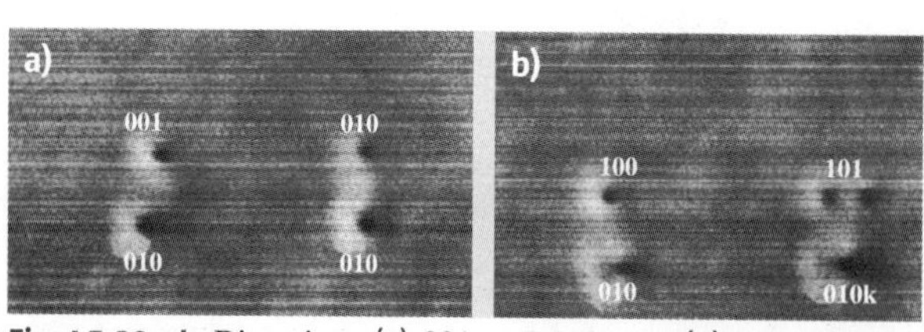

Fig. 45.29a,b Bit strings (**a**) 001 and 010, and (**b**) 100 and 101, overwritten on 010

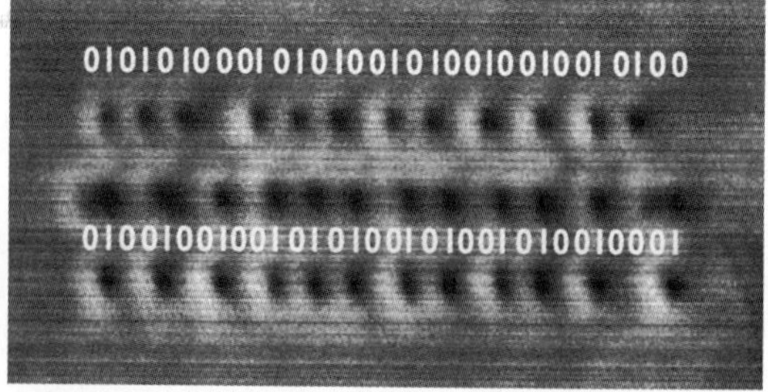

Fig. 45.30 Experimental result of overwriting a bit sequence

into the string 010. Figure 45.30 depicts the results of a rewriting experiment; the top track shows a prestored sequence, which is to be overwritten by another sequence, shown on the bottom track for comparison. The result of direct overwriting of the prestored sequence is shown on the middle track.

Comparison with the sequence on the bottom track, which is written on a clean surface, illustrates the effectiveness of the proposed procedure. Although the write/read quality of overwritten data is somewhat inferior to that of data written on a clean storage surface, detection of the newly written sequence is not affected. However, repeated overwriting may further degrade the quality of stored data. As the extent and rate of degradation are important characteristics of a storage system, this remains an area of ongoing investigation.

45.9 Read Channel Model

Let us now consider the read-back channel for a single cantilever that is scanning a storage field where bits are written as indentations or absence of indentations in the storage medium. As discussed above, a cantilever is modeled as a variable resistor that depends on the temperature at the cantilever tip. The model of the read channel, used for the design and analysis of the detection system, is illustrated in Fig. 45.31 [45.12, 26, 27, 48].

To evaluate the evolution of the temperature of a heated cantilever during the read process, we resort to a simple RC-equivalent thermal circuit, illustrated in Fig. 45.32, where $(1+\eta_x)R^{\Theta}$ and C^{Θ} denote the thermal resistance and capacitance, respectively. The parameter $\eta_x = \Delta R^{\Theta}(x)/R^{\Theta}$ indicates the relative variation of thermal resistance that results from the small change in air-gap width between the cantilever and the storage medium. The subscript "x" indicates the distance in the direction of scanning from the initial point. Therefore, the parameter η_x will assume the largest absolute value when the tip of the cantilever is located at the center of an indentation. The heating power that is dissipated in the cantilever heater region is expressed as

$$P^e[t, \Theta(t,x)] = \frac{V_C^2(t)}{R^e[\Theta(t,x)]} , \tag{45.4}$$

where $V_C(t)$ is the voltage across the cantilever, $\Theta(t,x)$ is the cantilever temperature, and $R^e[\Theta(t,x)]$ is the temperature-dependent cantilever resistance.

As the heat-transfer process depends on the value of thermal resistance and on the read pulse waveform, the cantilever temperature $\Theta(t,x)$ depends on time t and distance x. However, because the time it takes for the cantilever to move from the center of one logical mark to the next is greater than the duration of a read pulse, we assume that $\Theta(t,x)$ does not vary significantly as a function of x while a read pulse is being applied, and that it decays to the ambient temperature Θ_0 before the next pulse is applied. Therefore, the evolution of the cantilever temperature in response to a pulse applied at

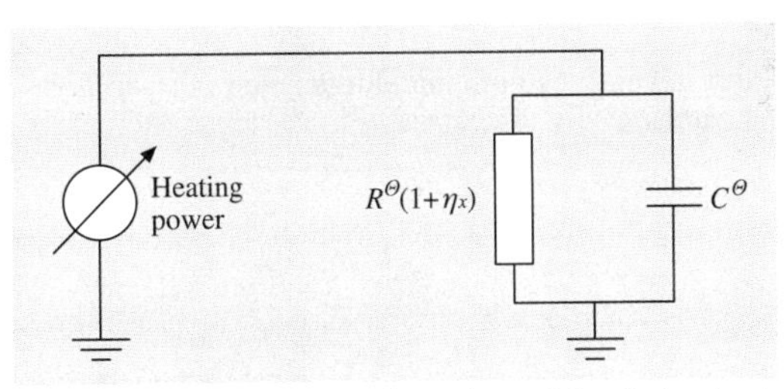

Fig. 45.32 RC-equivalent thermal model of the heat-transfer process (after [45.12])

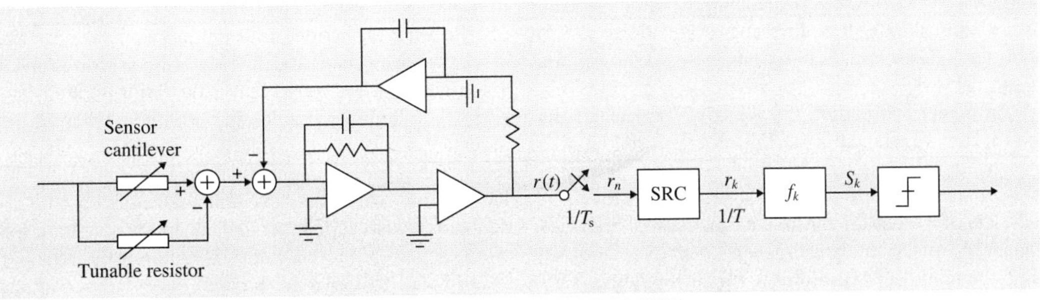

Fig. 45.31 Block diagram of the read signal path (after [45.48], © IEEE 2005)

time $t_0 = x_0/v$ at a certain distance x_0 from the initial point of scanning and for a certain constant velocity v of the scanner obeys a differential equation expressed as

$$\Theta'(t,x_0) + \frac{1}{(1+\eta_{x_0})R^{\Theta}C^{\Theta}}\left[\Theta(t,x_0) - \Theta_0\right] = \frac{1}{C^{\Theta}} \frac{V_C^2(t)}{R^e[\Theta(t,x_0)]}\,, \tag{45.5}$$

where $\Theta'(t,x_0)$ denotes the derivative of $\Theta(t,x_0)$ with respect to time.

With reference to the block diagram of the read channel illustrated in Fig. 45.31, the source generates the read pulse $V_P(t)$ applied to the cantilever. Clearly, because of the virtual ground at the operational amplifier input, the voltage $V_C(t)$ across the cantilever variable resistance is equal to $V_P(t)$. Furthermore, the active low-pass RC detector filter, where R_{lpf} and C_{lpf} denote the resistance and capacitance of the low-pass filter, respectively, is realized using an ideal operational amplifier that exhibits infinite input impedance, zero output impedance, and infinite frequency-independent gain. Therefore, the read-back signal $V_O(t,x_0)$ obtained at the low-pass filter output in response to the applied voltage $V_P(t) = A\,\mathrm{rect}\left[(t-T_0)/\tau\right]$, where

$$\mathrm{rect}\left(\frac{t}{\tau}\right) = \begin{cases} 1 & \text{if } 0 \le t \le \tau \\ 0 & \text{otherwise} \end{cases}\,, \tag{45.6}$$

and A denotes the pulse amplitude, obeys the differential equation

$$V_O'(t,x_0) = \frac{1}{R_{lpf}C_{lpf}} \times \left[-V_O(t,x_0) + \frac{R_{lpf}}{R^e\left[\Theta(t,x_0)\right]} V_P(t)\right]. \tag{45.7}$$

As the voltage at the output of the low-pass filter depends on the value of the variable resistance $R^e[\Theta(t,x_0)]$, the read-back signal is determined by solving jointly the differential equations (45.5) and (45.7), with initial conditions $\Theta(t_0,x_0) = \Theta_0$ and $V_O(t_0,x_0) = 0$.

As mentioned earlier, the relative variation of the read heater resistance is on the order of $10^{-5}\,\mathrm{nm}^{-1}$, and the information-carrying signal can be viewed as a small signal superimposed on a very large offset signal. Therefore one of the main functions of the read channel is to accurately estimate and remove the offset from the read-back waveform to eliminate the need for high-resolution analog-to-digital conversion. In the read channel model of Fig. 45.31, the read sensor of the cantilever is shown as a variable resistor. A second resistor with a tunable resistance is driven by the same voltage source as the read sensor. This resistor is tuned in such a way that it approximately matches the average value of the resistance of the read sensor when the latter is heated at the read temperature. As a result, the signal at the output of the first adder has a smaller dynamic range, as the bulk of the offset has been removed. The purpose of the second adder is to eliminate any residual offsets still present in the read-back signal by means of a feedback loop. Consequently, the signal $r(t)$ at the output of the second amplifier is offset free. An analog-to-digital converter (ADC) samples the signal $r(t)$ at the oversampling rate $1/T_s$. In the digital domain the sequence at the output of the ADC is synchronized to the symbol rate $1/T$ by a sampling rate converter (SRC), and filtered by a digital filter f_k to eliminate rapid DC fluctuations. Then the filter output signal is passed to a binary threshold detector to produce estimates of the stored binary symbols. A VLSI implementation of the analog front-end of the read channel is presented in [45.49, 50].

The continuous-time read-back signal, after low-pass filtering, sampling, and resynchronization to the symbol rate, is translated to a discrete-time sequence r_k that may be expressed as

$$r_k = \sum_{i=-\infty}^{\infty} a_i g_{k-i} + n_k\,, \tag{45.8}$$

where $a_k \in \{0, 1\}$ is the encoded information sequence, g_k is the compound impulse response of the write and read channels, and n_k is additive noise. The linear model of (45.8) appears to be quite accurate at low to moderate storage densities and/or recording power. However, at high storage densities and/or high recording power, the channel exhibits a specific nonlinear behavior, called partial erasing [45.28]. In particular, the depth of indentations decreases when other indentations are written in close proximity to them. In a series of consecutively written indentations at close spacing, all but the last written indentation exhibit this nonlinear amplitude loss. Moreover, the indentation written first suffers the largest reduction in amplitude, whereas the indentation written last exhibits an amplitude enhancement with regard to the nominal indentation amplitude. Indentations that do not have immediate neighboring indentations, also called *isolated* indentations, attain the nominal read-back amplitude. This nonlinear behavior is attributed to the write process. These observations suggest a simple two-step model of the read-back signal at high storage densities or high recording power: because of the nonlinear nature of the write process, the encoded binary information sequence a_k is mapped in a first step to a five-level sequence b_k through the following mapping

$$b_k = a_k + (\alpha - 1)a_k a_{k+1} + \beta a_k a_{k-1} , \quad (45.9)$$

where α and β are positive real parameters taking values in [0, 1]. This mapping essentially reflects the partial erasing effects that have been observed experimentally depending on whether a_k corresponds to an isolated indentation. In the second step, the read-back sequence is obtained as a linear transformation of b_k

$$r_k = \sum_{i=-\infty}^{\infty} b_i g_{k-i} + n_k . \quad (45.10)$$

The proposed nonlinear model belongs to the class of Volterra series models [45.51]. By substituting (45.9) into (45.10) and rearranging terms, one obtains a truncated Volterra series with a second-order next-neighbor nonlinearity expressed as

$$r_k = \sum_{i=-\infty}^{\infty} a_i g_{k-i} + \sum_{i=-\infty}^{\infty} a_i a_{i-1} h_{k-i} + n_k , \quad (45.11)$$

where $h_k = (\alpha - 1)g_{k+1} + \beta g_k$. As the nonlinear kernel is coupled to the linear kernel in the proposed model, it is actually a special case of the Volterra series with a smaller number of parameters.

Based on the above analysis, the response to a pulse applied to the cantilever at a distance x from some initial point can be calculated given the parameter $\eta_x = \Delta R^{\Theta}(x)/R^{\Theta}$. Recall that the value of η_x is proportional to the distance of the cantilever from the storage medium at the current location of the tip. Therefore, during tip displacement due to scanner motion, η_x is modulated from the topographical features of the storage surface such as written indentations, rings, and dust particles. This indicates that the modeling of the read-back signal is a two-step process. First, a model for the storage surface topography is developed, which directly determines η_x, and then the above procedure is used to calculate the read-back signal samples in response to pulses applied at selected points in the particular storage area.

In the absence of any imperfections during the manufacturing and the writing process, the storage surface would consist of completely flat regions interrupted by uniformly shaped indentations, possibly surrounded by polymer rings. A 1-D cross-section of an indentation along the scanning direction is modeled by a function with one main lobe and two side lobes, one on each side of the main lobe, the magnitude and the extent of which can be varied independently. The side lobes are of opposite sign to the main lobe and simulate the polymer rings around written indentations. By varying their magnitude and extent while keeping the total extent of the pulse fixed, one can simulate indentations/rings of varying width and asymmetric ring formation, phenomena that are caused by different recording conditions. In practice, however, no polymer surface is entirely flat and indentation shapes are far from uniform. The deviation of indentations from uniformity is simulated by scaling the amplitude of each pulse shape by a random number drawn from a Gaussian distribution with unit mean and adjustable variance. The surface roughness is in turn simulated by adding white Gaussian noise to the height of every point in the area of interest. Note that surface roughness is a medium-related effect and manifests itself in the read-back signal as a noise process, which is, however, of a very different nature than thermal noise. The advantage of the adopted two-step model is that it naturally decouples these unrelated noise sources, as well as the write and the read processes.

Figure 45.33 illustrates the experimental and synthetic read-back signals obtained along a data track. The

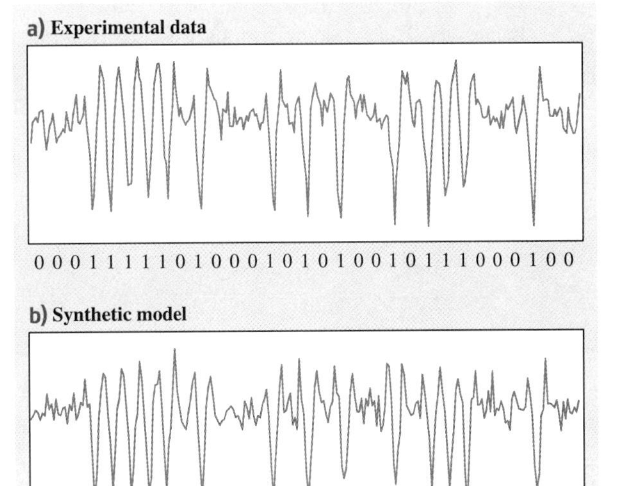

Fig. 45.33a,b Comparison between **(a)** the read-back signal obtained experimentally along a data track and **(b)** the read-back signal obtained by the synthetic model (after [45.12])

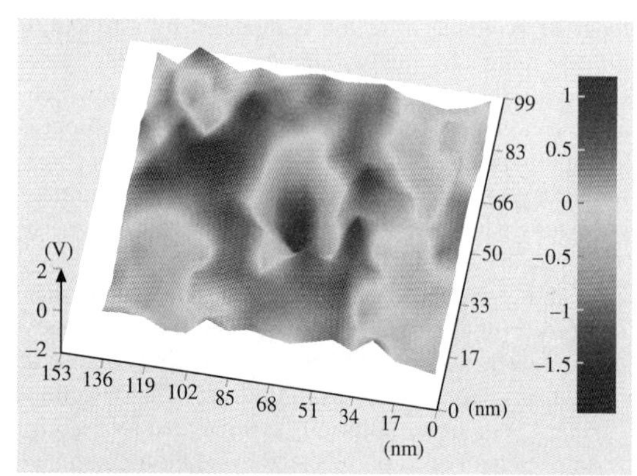

Fig. 45.34 Three-dimensional view of an isolated indentation obtained experimentally

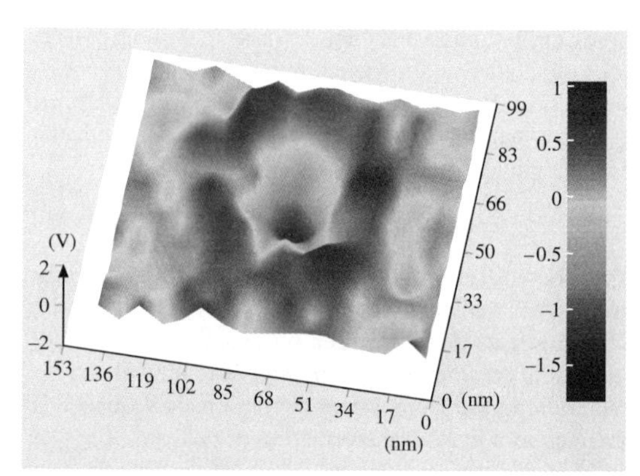

Fig. 45.35 Three-dimensional view of an isolated indentation obtained by the synthetic model

waveforms shown in Fig. 45.33 were obtained by applying pulses at the oversampling rate of q/T, where q denotes the oversampling factor. For a more detailed comparison between model and actual signals, Figs. 45.34 and 45.35 illustrate three-dimensional (3-D) views of isolated indentations from experimental and synthetic read-back signals, respectively. The dark regions in the center of both figures correspond to the indentation centers, whereas dark regions around them are due to rings. Note also the irregular height of the surrounding surface, which is attributed to the roughness of the storage medium.

45.10 System Aspects

In this section, we describe various aspects of a storage system based on our millipede concept. Each cantilever can write data to and read data from a dedicated area of the polymer surface, called a *storage field*. As mentioned above, in each storage field the presence (absence) of an indentation corresponds to a logical "1" ("0"). All indentations are nominally of equal depth and size. The logical marks are placed at a fixed horizontal distance from each other along a data track. We refer to this distance, measured from one logical mark center to the next, as the *bit pitch* (BP). The vertical (cross-track) distance between logical mark centers, the *track pitch* (TP), is also fixed. To read and write data the polymer medium is moved under the (stationary) cantilever array at a constant velocity.

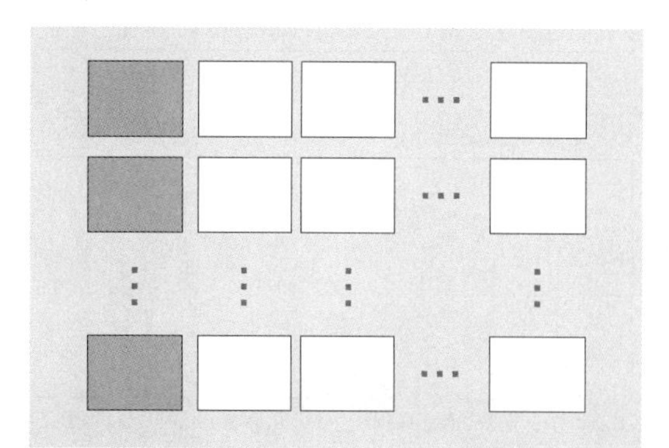

Fig. 45.36 Layout of data and servo/timing fields. *Dark boxes* represent dedicated servo/timing fields; *white boxes* represent data fields

A robust way to achieve synchronization and servo control in an x, y-actuated large 2-D array is by reserving a small number of storage fields exclusively for timing recovery and servo-control purposes, as illustrated in Fig. 45.36. Because of the large number of levers in the millipede, this solution is advantageous in terms of overhead compared with the alternative of timing and servo information being embedded in all data fields.

45.10.1 PES Generation for the Servo Loop

With logical marks as densely spaced as in the millipede, accurate track following becomes a critical issue.

Track following means controlling the position of each tip such that the tip is always positioned over the center of a desired track during reading. During writing, the tip position should be such that the written marks are aligned in a predefined way. In electromechanical systems, track following is performed in a servo loop, which is driven by an appropriate error signal called the position-error signal (PES). Ideally, its magnitude is a direct estimate of the vertical (cross-track) distance of the tip from the closest track centerline, and its polarity indicates the direction of this offset.

Several approaches exist to generate a PES for AFM-based storage devices [45.9]. However, based on the results reported, none of these methods can achieve the track-following accuracy required for the millipede system. The quality of the PES directly affects the stability and robustness of the associated tracking servo loop [45.52].

We describe a method for generating a uniquely decodable PES for the millipede system [45.12, 27]. The method is based on the concept of *bursts* that are vertically displaced with respect to each other, arranged in such a way as to produce two signals in quadrature, which can be combined to provide a robust PES. This concept is borrowed from magnetic recording [45.52]. However, servo marks, as opposed to magnetic transitions, are placed in bursts labeled A and B for the in-phase signal and C and D for the quadrature signal. The centers of servo marks in burst B are vertically offset from mark centers in burst A by d' units of length. This amount of vertical spacing is related to the diameter of the written marks. The same principle applies to marks in the quadrature bursts C and D, with the additional condition that mark centers in burst C are offset by $d'/2$ units from mark centers in burst A in the cross-track direction. The latter condition is required in order to generate a quadrature signal. The configuration of servo bursts is illustrated in Fig. 45.37 for the case where $\mathrm{TP} = 3d'/2$. Although each burst typically consists of many marks to enable averaging of the corresponding read-out signals, only two marks per burst are shown here to simplify the presentation. The solid horizontal lines depict track centerlines, and circles represent written marks, which are modeled here as perfect conical indentations on the polymer storage surface.

To illustrate the principle of PES generation, let us assume that marks in all bursts are spaced BP units apart in the longitudinal direction, and that sampling occurs exactly at mark centers, so that timing is perfect. The assumption of perfect timing is made only for the purpose of illustration. In actual operation, sampling is performed with the aid of a timing recovery loop, as described in Sect. 45.10.2. Referring to Fig. 45.37, let us further assume that the cantilever tip is located on the line labeled "0" and moves vertically toward line "3" in a line crossing the centers of the leftmost marks in burst A (shown as a brown dashed line). The tip moves from the edge of the top mark toward its center, then toward its bottom edge, then to a blank space, again to a mark, and so on. The read-out signal magnitude decreases linearly with the distance from the mark center and reaches a constant, background level value at a distance greater than the mark radius from the mark center, according to the adopted (conical mark) model. To synthesize the in-phase signal, the read-out signal is also captured as the tip (conceptually) moves in a vertical line crossing the mark centers of burst B (brown dashed line in Fig. 45.37). The in-phase signal is then formed

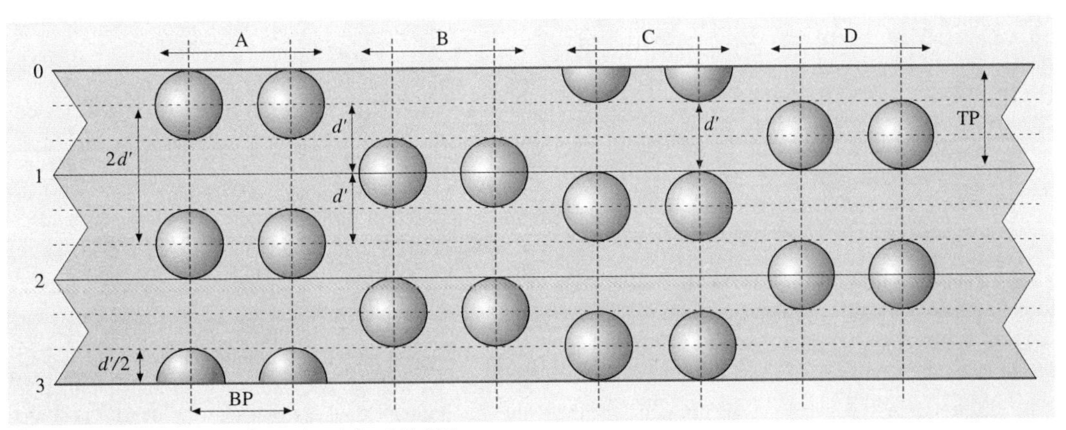

Fig. 45.37 Servo burst configuration (after [45.12])

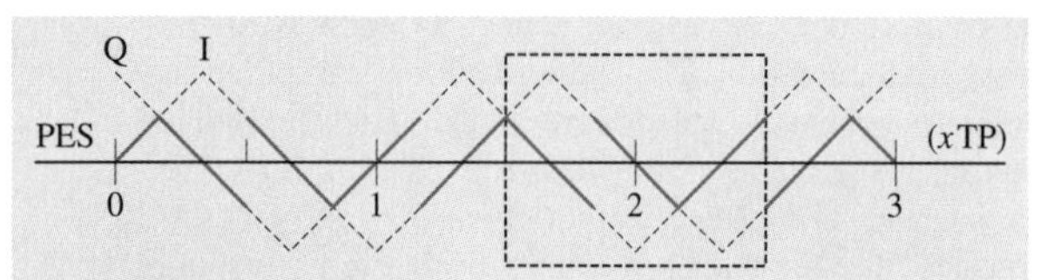

Fig. 45.38 Ideal position-error signal (after [45.12], © 2003 IEEE)

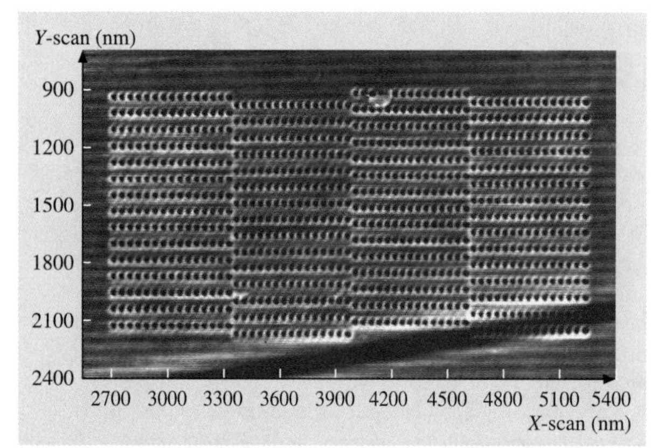

Fig. 45.39 Experimental A, B, C, and D servo bursts (BP = 42 nm) (after [45.12], © 2003 IEEE)

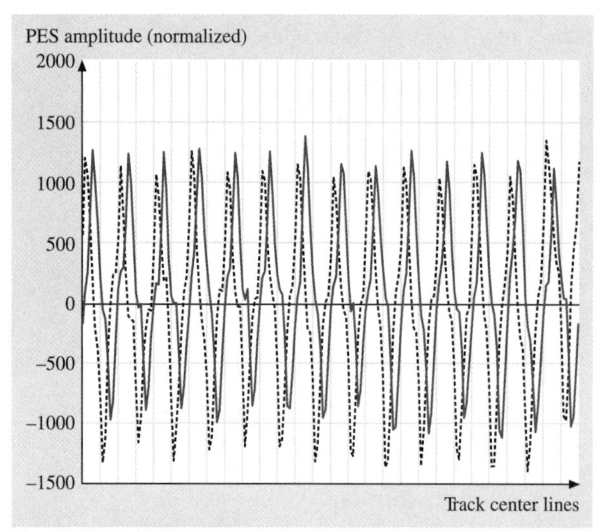

Fig. 45.40 Demodulated in-phase (*solid line*) and quadrature (*dashed line*) PES based on the servo burst of Fig. 45.38 (after [45.12], © 2003 IEEE)

as the difference $\bar{A}-\bar{B}$, where $\bar{A}$ and $\bar{B}$ stand for the measured signal amplitudes in bursts A and B, respectively. This signal is represented by the line labeled "I" in Fig. 45.38. It has zero-crossings at integer multiples of d', which do not generally correspond to track centers because we set $\text{TP} = 3d'/2$ in this example. Therefore, the signal I is not a valid PES in itself. This is why the quadrature (Q) signal becomes necessary in this case. The Q-signal is generated from the servo read-back signals of bursts C and D as $\bar{C}-\bar{D}$ and is also shown in Fig. 45.38 (curve labeled "Q"). Note that it exhibits zero-crossings at points where the signal I has local extrema.

A certain combination of the two signals (I and Q), shown as solid lines in Fig. 45.38, has zero-crossings at all track center locations and constant (absolute) slope, which qualifies it as a valid PES. However, this PES exhibits zero-crossings at all integer multiples of $d'/2$. For our example of $\text{TP} = 3d'/2$, three such zero-crossings exist in an area of width equal to TP around any track centerline. This fact, however, does not hamper unique position decoding. At even-numbered tracks, it is the zero of the *in-phase* signal that indicates the track center. The zeros of the quadrature signal, in turn, can be uniquely mapped into a position estimate by examining the polarity of the in-phase signal at the corresponding positions. This holds for any value of the combined PES within an area of width equal to TP around each current track centerline. The signals exchange roles for odd-numbered tracks. The current track number, which is known a priori from the seek operation, is used to determine the mode of operation for the position demodulation procedure.

The principle of PES generation based on servo marks has been verified experimentally. For this purpose, A, B, C, and D bursts were written by an AFM cantilever tip on an appropriate polymer medium consisting of a polymer coating on a silicon substrate. The bit pitch was set to 42 nm, and the track pitch was taken to be approximately equal to d', the cross-track distance between A (C) and B (D) bursts. An image created by reading the written pattern with the same cantilever is shown in Fig. 45.39. Shaded areas indicate indentations. The read-out signal from the cantilever was also used for servo demodulation, as described above. The resulting in-phase and quadrature signals are shown in Fig. 45.40. The track centerlines are indicated by the vertical lines in the graph.

It can be observed that the zero-crossings of the in-phase signal are closely aligned with the track centerlines, as well as with the minima and maxima of the quadrature signal, as required for unique position decoding across all possible cross-track positions, at least in cases where TP$\neq d'$. Moreover, the PES slope is nearly linear along a cross-track width of one track pitch around each track center, as TP $\approx d'$ in this case, although deviations from the ideal signal shape exist. These deviations occur mainly because written indentations do not have perfect conical shapes, and also because of medium noise due to the roughness of the recording medium. Nevertheless, the experimentally generated error signals indicate that the proposed concept is valid and promising. Specifically, the results indicate that servo self-writing is feasible, that servo demodulation is almost identical to data read-out and can be performed by any cantilever without special provisions, and that the PES generated closely approximates the desirable features described above.

45.10.2 Timing Recovery

Similar to obtaining servo information based on using dedicated servo fields, we employ separate dedicated clock fields for recovery of timing information [45.12, 26, 27]. The concept is to have continuous access to a pilot signal for synchronization after initial phase acquisition and gain estimation. The recovered clock is then distributed to all remaining storage fields to allow reliable detection of random data. Initial phase acquisition is obtained by a robust correlation algorithm, gain estimation is based on averaging of the read-back signal obtained from a predefined stored pattern, and tracking of the optimum sampling phase is achieved by a second-order digital loop.

At the beginning of the read process, several signal parameters have to be estimated prior to data detection. Besides the clock phase and frequency, it is necessary to estimate the gain of the overall read channel. To solve the problem of initial estimation of signal parameters prior to data detection, the sequence written in the clock field consists of a preamble, followed by a pattern of all "1"s for tracking the optimum sampling phase during the detection of random data. The transition between the preamble and the pattern of all "1"s must be detected reliably, as it indicates the start of data records to the remaining storage fields. Assuming that the initial frequency offset is within a small, predetermined range, usually 1000 ppm, we distinguish the tasks needed for timing recovery as follows:

(i) Acquisition of the optimum sampling phase
(ii) Estimation of the overall channel gain needed for threshold detection
(iii) Detection of the transition between the preamble and the pattern of all "1"s
(iv) Tracking of the optimum sampling phase.

At the beginning of the acquisition process, an estimate of the optimum sampling phase is obtained by resorting to a correlation method. We rely on the knowledge of the preamble and of an ideal reference-channel impulse response, which closely resembles the actual impulse response (Sect. 45.10). The channel output samples obtained at the oversampling rate q/T are first processed by removing the DC offset, then averaging is performed to reduce the noise level, and finally the resulting sequence is correlated with the reference impulse response to determine the phase estimate.

After determining the estimate of the optimum sampling phase, an estimate of the overall channel gain is obtained by averaging the amplitude of the channel output samples at the optimum sampling instants. The gain estimate is obtained from an initial segment of the preamble corresponding to an all-"1" binary pattern. As mentioned above, it is necessary that the end of the preamble is indicated by a *sync* pattern, which marks the transition between acquisition mode and tracking mode. Detection of the sync pattern is also based on a robust correlation method. After the sync pattern, an all-"1" pattern, as in the case of robust phase acquisition and gain estimation, is employed for tracking. The all-"1" pattern corresponds to regularly spaced indentations, which convey reliable timing information.

Tracking of the optimum sampling phase is achieved by the second-order loop configuration shown in Fig. 45.41. Assuming that data detection is performed at instants that correspond to integer multiples of the oversampling factor q, the deviation of the sampling phase from the optimum sampling phase is estimated

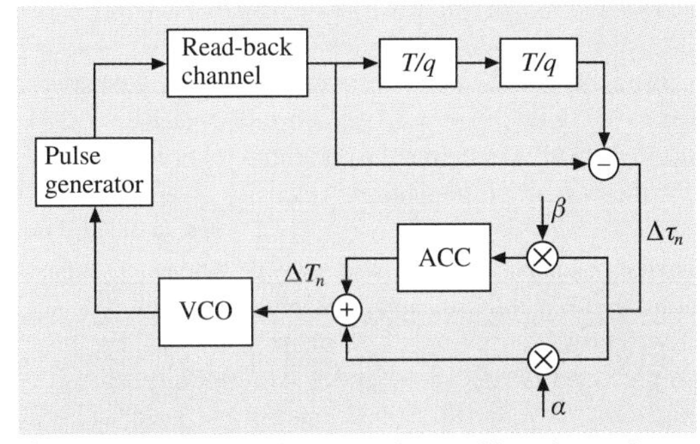

Fig. 45.41 Second-order loop for tracking the optimum sampling phase (after [45.12], © 2003 IEEE)

as

$$\Delta\tau_n = r(t_{s,nq+1}) - r(t_{s,nq-1}) . \quad (45.12)$$

This estimate of the phase deviation is input to a second-order loop filter, which provides an output given by

$$\Delta T_n = u_n + \alpha\,\Delta\tau_n , \quad (45.13)$$

where the discrete-time integrator is recursively updated as

$$u_{n+1} = u_n + \beta\,\Delta\tau_n . \quad (45.14)$$

The loop-filter output then determines the control signal for a voltage-controlled oscillator (VCO).

Note that a similar concept for timing recovery can also be applied if no separate clock field is available. In this case, the timing information is extracted from the random user data on each storage field.

45.10.3 Considerations on Capacity and Data Rate

The ultimate locality provided by nanometer-sharp tips represents the pathway to the high areal densities that will be needed in the foreseeable future. The intrinsic nonlinear interactions between closely spaced indentations, however, may limit the minimum distance between successive indentations and, hence, the areal density. The storage capacity of a millipede-based storage device can be further increased by applying modulation or constrained codes [45.12].

With modulation coding, a desired constraint is imposed on the data input sequence, so that the encoded data stream satisfies certain properties in the time or frequency domain. These codes are very important in digital recording devices and have become ubiquitous in all data-storage applications. The particular class of codes that imposes restrictions on the number of consecutive "1"s and "0"s in the encoded data sequence, generally known as run-length-limited (RLL) (d, k) codes [45.53], can be used to facilitate overwriting and also increase the effective areal density of a millipede-based storage device. The code parameters d and k are nonnegative integers with $k > d$, where d and k indicate the minimum and maximum number of "0"s between two successive "1"s, respectively. In the past, the precoded (RLL) (d, k) codes were mainly used for spreading the magnetic transitions further apart via the d-constraint, thereby minimizing intersymbol interference and nonlinear distortion, and for preventing loss of clock synchronization via the k-constraint. In optical recording, precoded RLL codes are primarily used for increasing the shortest pit length in order to improve the reliability of bit detection, as well as for limiting the number of identical symbols, so that useful timing information can be extracted from the read-back signal.

For the millipede application, where dedicated clock fields are used, the k-constraint does not really play an important role and, therefore can in principle be set to infinity, thereby facilitating the code design process. In a precoder-less RLL code design, where the presence or absence of an indentation represents a "1" or "0", respectively, the d-constraint is instrumental in limiting the interference between successive indentations, as well as in increasing the effective areal density of the storage device. In particular, the quantity $(d+1)R$, where R denotes the rate of the (d, k) code, is a direct measure of the increase in linear recording density. Clearly, the packing density can be made larger by increasing d. On the other hand, large values of d lead to codes with very low rate, which implies high recording symbol rates, thus rendering these codes impractical for storage systems that are limited by the clock speed. The choice of $d = 1$ and $k \geq 6$ guarantees the existence of a code with rate $R = 2/3$. Use of $(d = 1, k \geq 6)$ modulation coding reduces the bit distance by half while maintaining the pitch between "1"s constant, thereby increasing the linear density by a factor of 4/3. Similarly, the choice of $d = 2$ and $k > 6$ guarantees the existence of a code with rate $R = 1/2$. Use of $(d = 2, k > 6)$ modulation coding reduces the bit distance to a third while maintaining the pitch between "1"s constant, thereby increasing the linear density by a factor of 3/2. Figure 45.42 shows the areal density as a function of the indentation spacing for an uncoded system, as well as for systems coded with $(d = 1, k \geq 6)$ and $(d = 2, k > 6)$, where coding is applied only in the on-track direction.

Table 45.1 shows the achievable areal densities and storage capacities for a 32×32 cantilever array with 1024 storage fields, each having an area of $100 \times 100\,\mu\text{m}^2$, resulting in a total storage area of $3.2 \times 3.2\,\text{mm}^2$. The indentation pitch and the track pitch are set equal to 30 nm. Finally, for the computation of the storage capacity an overall efficiency of 85% has been assumed, taking into account the redundancy of the outer error-correction coding, as well as the presence of dedicated servo and clock fields.

Figure 45.43 shows the user data rate as a function of the total number of cantilevers accessed simultaneously, for various symbol rates and a $(d = 1, k \geq 6)$

Table 45.1 Areal density and storage capacity (after [45.12], © 2003 IEEE)

Coding	Linear density (kb/inch)	Track density (kt/inch)	Areal density (Gb/inch^2)	Capacity (Gb)
Uncoded	847	847	717	1.21
$(d=1, k \geq 6)$	1129	847	956	1.61
$(d=2, k>6)$	1269	847	1075	1.81

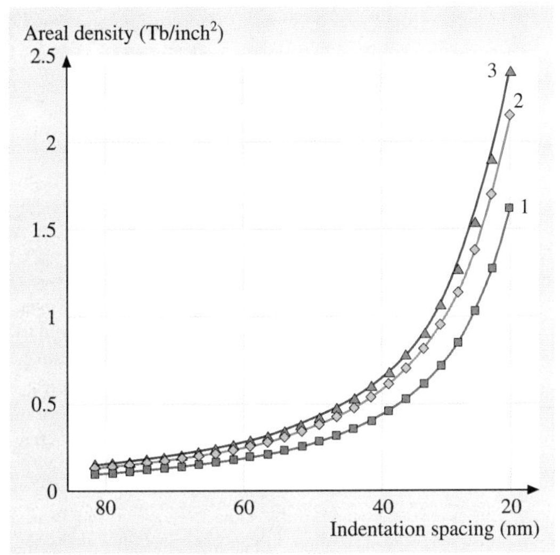

Fig. 45.42 Areal density versus indentation spacing. Curve 1: $d=0$; curve 2: $d=1$; and curve 3: $d=2$

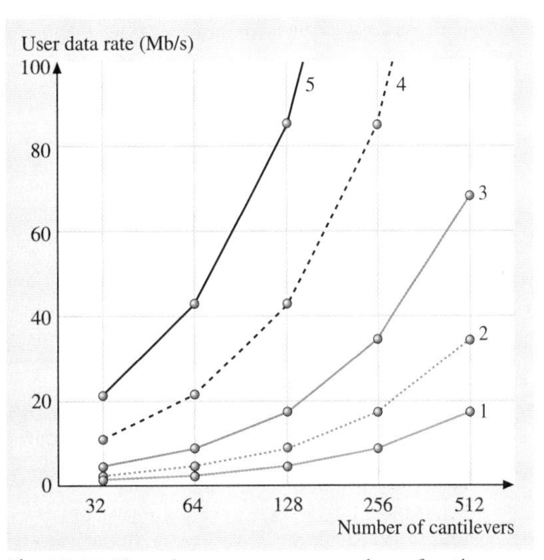

Fig. 45.43 User data rate versus number of active cantilevers for the $(d=1, k \geq 6)$ coding scheme. *Curve 1*: $T=20\,\mu s$; *curve 2*: $T=10\,\mu s$; *curve 3*: $T=5\,\mu s$; *curve 4*: $T=2\,\mu s$; and *curve 5*: $T=1\,\mu s$ (after [45.12])

modulation coding scheme. For example, for a 32×32 cantilever array, a system designed to access a maximum of 256 cantilevers every $T=5\,\mu s$ provides a user data rate of 34.1 Mb/s. Alternatively, by resorting to the row/column multiplexing scheme with $T=80\,\mu s$ a data rate of 8.5 Mb/s is achieved.

45.11 Conclusions

A very large 2-D array of AFM probes has been operated for the first time in a multiplexed/parallel fashion, and write/read/erase operations in a thin polymer medium have been successfully demonstrated at densities of or significantly higher than those achieved with current magnetic storage systems.

The millipede has the potential to achieve ultrahigh storage areal densities on the order of 1 Tb/inch^2 or higher. The high areal storage density, small form factor, and low power consumption render the millipede concept a very attractive candidate as a future storage technology for mobile applications, as it offers several Gb of capacity at data rates of several Mb/s. Dedicated servo and timing fields allow reliable system operation with a very small overhead. The read channel model provides the methodology for analyzing system performance and assessing various aspects of the detection and servo/timing algorithms that are key to achieving the system reliability required by the applications envisaged.

Although the first high-density storage operations with the largest 2-D AFM array chip ever built have been demonstrated, there are a number of issues that need further investigation, such as overall system reliability, including long-term stability of written indentations, tip and medium wear, limits of data rates, array and cantilever size, as well as tradeoffs between data rate and power consumption.

References

45.1 E. Grochowski, R.F. Hoyt: Future trends in hard disk drives, IEEE Trans. Magn. **32**, 1850–1854 (1996)

45.2 D.A. Thompson, J.S. Best: The future of magnetic data storage technology, IBM J. Res. Dev. **44**, 311–322 (2000)

45.3 G. Binnig, H. Rohrer, C. Gerber, E. Weibel: 7×7 reconstruction on Si(111) resolved in real space, Phys. Rev. Lett. **50**, 120–123 (1983)

45.4 G. Binnig, C.F. Quate, C. Gerber: Atomic force microscope, Phys. Rev. Lett. **56**, 930–933 (1986)

45.5 H.J. Mamin, D. Rugar: Thermomechanical writing with an atomic force microscope tip, Appl. Phys. Lett. **61**, 1003–1005 (1992)

45.6 R.P. Ried, H.J. Mamin, B.D. Terris, L.S. Fan, D. Rugar: 6-MHz 2-N/m piezoresistive atomic-force-microscope cantilevers with INCISIVE tips, J. Microelectromech. Syst. **6**, 294–302 (1997)

45.7 B.D. Terris, S.A. Rishton, H.J. Mamin, R.P. Ried, D. Rugar: Atomic force microscope-based data storage: Track servo and wear study, Appl. Phys. A **66**, S809–S813 (1998)

45.8 H.J. Mamin, B.D. Terris, L.S. Fan, S. Hoen, R.C. Barrett, D. Rugar: High-density data storage using proximal probe techniques, IBM J. Res. Dev. **39**, 681–699 (1995)

45.9 H.J. Mamin, R.P. Ried, B.D. Terris, D. Rugar: High-density data storage based on the atomic force microscope, Proc. IEEE **87**, 1014–1027 (1999)

45.10 L.R. Carley, J.A. Bain, G.K. Fedder, D.W. Greve, D.F. Guillou, M.S.C. Lu, T. Mukherjee, S. Santhanam, L. Abelmann, S. Min: Single-chip computers with microelectromechanical systems-based magnetic memory, J. Appl. Phys. **87**, 6680–6685 (2000)

45.11 G. Gibson, T.I. Kamins, M.S. Keshner, S.L. Neberhuis, C.M. Perlov, C.C. Yang: Ultra-high density storage device, (1996) US Patent 5557596

45.12 E. Eleftheriou, T. Antonakopoulos, G.K. Binnig, G. Cherubini, M. Despont, A. Dholakia, U. Dürig, M.A. Lantz, H. Pozidis, H.E. Rothuizen, P. Vettiger: Millipede – A MEMS-based scanning-probe data-storage system, IEEE Trans. Magn. **39**, 938–945 (2003)

45.13 G.K. Binnig, H. Rohrer, P. Vettiger: Mass-storage applications of local probe arrays, (1998) US Patent 5835477

45.14 P. Vettiger, J. Brugger, M. Despont, U. Drechsler, U. Dürig, W. Häberle, M. Lutwyche, H. Rothuizen, R. Stutz, R. Widmer, G. Binnig: Ultrahigh density, high-data-rate NEMS-based AFM data storage system, J. Microelectron. Eng. **46**, 11–17 (1999)

45.15 P. Vettiger, G. Cross, M. Despont, U. Drechsler, U. Dürig, B. Gotsmann, W. Häberle, M.A. Lantz, H.E. Rothuizen, R. Stutz, G.K. Binnig: The "millipede" – Nanotechnology entering data storage, IEEE Trans. Nanotechnol. **1**, 39–55 (2002)

45.16 M. Lutwyche, C. Andreoli, G. Binnig, J. Brugger, U. Drechsler, W. Häberle, H. Rohrer, H. Rothuizen, P. Vettiger: Microfabrication and parallel operation of 5×5 2D AFM cantilever array for data storage and imaging, Proc. IEEE 11th Int. Workshop MEMS, Heidelberg 1998 (IEEE, Piscataway 1998) pp. 8–11

45.17 M. Lutwyche, C. Andreoli, G. Binnig, J. Brugger, U. Drechsler, W. Häberle, H. Rohrer, H. Rothuizen, P. Vettiger, G. Yaralioglu, C. Quate: 5×5 2D AFM cantilever arrays: A first step towards a terabit storage device, Sens. Actuators A **73**, 89–94 (1999)

45.18 P. Vettiger, M. Despont, U. Drechsler, U. Dürig, W. Häberle, M.I. Lutwyche, H.E. Rothuizen, R. Stutz, R. Widmer, G.K. Binnig: The "millipede" – More than one thousand tips for future AFM data storage, IBM J. Res. Dev. **44**, 323–340 (2000)

45.19 B.W. Chui, H.J. Mamin, B.D. Terris, D. Rugar, K.E. Goodson, T.W. Kenny: Micromachined heaters with 1-μs thermal time constants for AFM thermomechanical data storage, Proc. IEEE Transducers, Chicago 1997 (IEEE, Piscataway 1997) 1085–1088

45.20 W.P. King, J.G. Santiago, T.W. Kenny, K.E. Goodson: Modelling and prediction of sub-micrometer heat transfer during thermomechanical data storage, 1999 Microelectromechanical Systems (MEMS). Proc. ASME Intl. Mech. Eng. Congr. Expo., ed. by A.P. Lee, L. Lin, F.K. Forster, Y.C. Young, K. Goodson, R.S. Keynton (ASME, New York 1999) pp. 583–588

45.21 W.P. King, T.W. Kenny, K.E. Goodson, G.L.W. Cross, M. Despont, U. Dürig, H. Rothuizen, G. Binnig, P. Vettiger: Design of atomic force microscope cantilevers for combined thermomechanical writing and thermal reading in array operation, J. Microelectromech. Syst. **11**, 765–774 (2002)

45.22 G.K. Binnig, M. Despont, W. Häberle, P. Vettiger: Method of forming ultrasmall structures and apparatus therefore, (March 1999) US Patent Office, Application No. 147865

45.23 G. Binnig, M. Despont, U. Drechsler, W. Häberle, M. Lutwyche, P. Vettiger, H.J. Mamin, B.W. Chui, T.W. Kenny: Ultra high-density AFM data storage with erase capability, Appl. Phys. Lett. **74**, 1329–1331 (1999)

45.24 G.K. Binnig, J. Brugger, W. Häberle, P. Vettiger: Investigation and/or manipulation device (March 1999) US Patent Office, Application No. 147867

45.25 S.M. Sze: *Physics of Semiconductors Devices* (Wiley, New York 1981)

45.26 G. Cherubini, T. Antonakopoulos, P. Bächtold, G.K. Binnig, M. Despont, U. Drechsler, A. Dholakia, U. Dürig, E. Eleftheriou, B. Gotsmann, W. Häberle, M.A. Lantz, T. Loeliger, H. Pozidis, H.E. Rothuizen, R. Stutz, P. Vettiger: The millipede, a very dense, highly parallel scanning-probe data-storage system, ESSCIRC – Proc. 28th Eur. Solid-State Circuits Conf., ed. by A. Baschirotto, P. Malcovati (Univ. Bologna, Bologna 2002) pp. 121–125

45.27 E. Eleftheriou, T. Antonakopoulos, G.K. Binnig, G. Cherubini, M. Despont, A. Dholakia, U. Dürig, M.A. Lantz, H. Pozidis, H.E. Rothuizen, P. Vettiger: "Millipede": A MEMS-based scanning-probe data-storage system, Digest of the Asia-Pacific Magnetic Recording Conference 2002, APMRC '02 (IEEE, Piscataway 2002) CE–2–1–CE2–2

45.28 H. Pozidis, W. Häberle, D. Wiesmann, U. Drechsler, M. Despont, T.R. Albrecht, E. Eleftheriou: Demonstration of thermomechanical recording at 641 Gbit/in^2, IEEE Trans. Magn. **40**, 2531–2536 (2004)

45.29 M. Despont, J. Brugger, U. Drechsler, U. Dürig, W. Häberle, M. Lutwyche, H. Rothuizen, R. Stutz, R. Widmer, G. Binnig, H. Rohrer, P. Vettiger: VLSI-NEMS chip for AFM data storage, Technical Digest 12th IEEE Int. Micro Electro Mech. Syst. Conf. "MEMS '99", Orlando 1999 (IEEE, Piscataway 1999) 564–569

45.30 T.S. Ravi, R.B. Marcus: Oxidation sharpening of silicon tips, J. Vac. Sci. Technol. B **9**, 2733–2737 (1991)

45.31 M. Despont, J. Brugger, U. Drechsler, U. Dürig, W. Häberle, M. Lutwyche, H. Rothuizen, R. Stutz, R. Widmer, G. Binnig, H. Rohrer, P. Vettiger: VLSI-NEMS chip for parallel AFM data storage, Sens. Actuators A **80**, 100–107 (2000)

45.32 A. Pantazi, M. Lantz, G. Cherubini, H. Pozidis, E. Eleftheriou: A servomechanism for a micro-electro-mechanical-system-based scanning-probe data storage device, Nanotechnology **15**, S612–S621 (2004)

45.33 M. Despont, U. Drechsler, R. Yu, H.B. Pogge, P. Vettiger: Wafer-scale microdevice transfer/interconnect: from a new integration method to its application in an AFM-based data-storage system, Technical Digest, Transducers '03 (IEEE, Piscataway 2003) pp. 1907–1910

45.34 H. Rothuizen, M. Despont, U. Drechsler, G. Genolet, W. Häberle, M. Lutwyche, R. Stutz, P. Vettiger: Compact copper/epoxy-based micromachined electromagnetic scanner for scanning probe applications, Technical Digest, 15th IEEE Int. Conf. on Micro Electro Mech. Syst. "MEMS 2002" (IEEE, Piscataway 2002) pp. 582–585

45.35 S.C. Minne, G. Yaralioglu, S.R. Manalis, J.D. Adams, A. Atalar, C.F. Quate: Automated parallel high-speed atomic force microscopy, Appl. Phys. Lett. **72**, 2340–2342 (1998)

45.36 M. Lutwyche, U. Drechsler, W. Häberle, R. Widmer, H. Rothuizen, P. Vettiger, J. Thaysen: Planar micromagnetic x/y/z scanner with five degrees of freedom. In: *Magnetic Materials, Processes, and Devices: Applications to Storage and Micromechanical Systems (MEMS)*, Vol. 98-20, ed. by L.T. Romankiw, S. Krongelb, C.H. Ahn (Electrochemical Society, Pennington 1999) pp. 423–433

45.37 H. Rothuizen, U. Drechsler, G. Genolet, W. Häberle, M. Lutwyche, R. Stutz, R. Widmer, P. Vettiger: Fabrication of a micromachined magnetic x/y/z scanner for parallel scanning probe applications, Microelectron. Eng. **53**, 509–512 (2000)

45.38 J.-J. Choi, H. Park, K.Y. Kim, J.U. Jeon: Electromagnetic micro x-y stage for probe-based data storage, J. Semicond. Technol. Sci. **1**, 84–93 (2001)

45.39 H. Lorenz, M. Despont, N. Fahrni, J. Brugger, P. Vettiger, P. Renaud: High-aspect-ratio, ultrathick, negative-tone near-UV photoresist and its applications for MEMS, Sens. Actuators A **64**, 33–39 (1998)

45.40 C.Q. Davis, D. Freeman: Using a light microscope to measure motions with nanometer accuracy, Opt. Eng. **37**, 1299–1304 (1998)

45.41 A. Pantazi, A. Sebastian, G. Cherubini, M.A. Lantz, H. Pozidis, H. Rothuizen, E. Eleftheriou: Control of MEMS-based scanning-probe data-storage devices, IEEE Trans. Control Syst. Technol. **15**, 824–841 (2007)

45.42 M. Despont, U. Drechsler, W. Häberle, M.A. Lantz, H. Rothuizen: A vibration resistant nanopositioner for mobile parallel-probe storage applications, J. Microelectromech. Syst. **16**, 130–139 (2007)

45.43 M.A. Lantz, G.K. Binnig, M. Despont, U. Drechsler: A micromechanical thermal displacement sensor with nanometer resolution, Nanotechnolology **16**, 1089–1094 (2005)

45.44 M.I. Lutwyche, M. Despont, U. Drechsler, U. Dürig, W. Häberle, H. Rothuizen, R. Stutz, R. Widmer, G.K. Binnig, P. Vettiger: Highly parallel data storage system based on scanning probe arrays, Appl. Phys. Lett. **77**, 3299–3301 (2000)

45.45 K. Fuchs, C. Friedrich, J. Weese: Viscoelastic properties of narrow-distribution poly(methyl metacrylates), Macromolecules **29**, 5893–5901 (1996)

45.46 U. Dürig, B. Gotsman: This estimate is based on a fluid dynamic deformation model of a thin film, private communication

45.47 J.D. Ferry: *Viscoelastic Properties of Polymers*, 3rd edn. (Wiley, New York 1980)

45.48 H. Pozidis, P. Bächtold, G. Cherubini, E. Eleftheriou, C. Hagleitner, A. Pantazi, A. Sebastian: Signal processing for probe storage, Proc. Int. Conf. Acoust. Speech Signal Process. "ICASSP 2005", Philadelphia 2005 (IEEE, Piscataway 2005) pp. 745–748

45.49 T. Loeliger, P. Bächtold, G.K. Binnig, G. Cherubini, U. Dürig, E. Eleftheriou, P. Vettiger, M. Uster, H. Jäckel: CMOS sensor array with cell-level analog-to-digital conversion for local probe data storage, ESSCIRC – Proc. 28th Eur. Solid-State Circuits Conf., ed. by A. Baschirotto, P. Malcovati (Univ. Bologna, Bologna 2002) pp. 623–626

45.50 C. Hagleitner, T. Bonaccio, H. Rothuizen, J. Lienemann, D. Wiesmann, G. Cherubini, J.G. Korvink, E. Eleftheriou: Modeling, design, and verification for the analog front-end of a MEMS-based parallel scanning-probe storage device, IEEE J. Solid State Circuits **42**, 1779–1789 (2007)

45.51 M. Schetzen: Nonlinear system modeling based on the Wiener theory, Proc. IEEE **69**, 1557–1573 (1981)

45.52 A.H. Sacks: Position signal generation in magnetic disk drives. Ph.D. Thesis (Carnegie Mellon University, Pittsburgh 1995)

45.53 K.A.S. Immink: *Coding Techniques for Digital Recorders* (Prentice Hall, Hemel 1991)

46. Nanorobotics

Bradley J. Nelson, Lixin Dong

Nanorobotics is the study of robotics at the nanometer scale, and includes robots that are nanoscale in size and large robots capable of manipulating objects that have dimensions in the nanoscale range with nanometer resolution. With the ability to position and orient nanometer-scale objects, nanorobotic manipulation is a promising way to enable the assembly of nanosystems including nanorobots.

This chapter overviews the state of the art of nanorobotics, outlines nanoactuation, and focuses on nanorobotic manipulation systems and their application in nanoassembly, biotechnology and the construction and characterization of nanoelectromechanical systems (NEMS) through a hybrid approach.

Because of their exceptional properties and unique structures, carbon nanotubes (CNTs) and SiGe/Si nanocoils are used to show basic processes of nanorobotic manipulation, structuring and assembly, and for the fabrication of NEMS including nano tools, sensors and actuators.

A series of processes of nanorobotic manipulation, structuring and assembly has been demonstrated experimentally. Manipulation of individual CNTs in 3-D free space has been shown by grasping using dielectrophoresis and placing with both position and orientation control for mechanical and electrical property characterization and assembly of nanostructures and devices. A variety of material property investigations can be performed, including bending, buckling, and pulling to investigate elasticity as well as strength and tribological characterization. Structuring of CNTs can be performed including shape modification, the exposure of nested cores, and connecting CNTs by van der Waals forces, electron-beam-induced deposition and mechanochemical bonding.

46.1 **Overview of Nanorobotics** 1634

46.2 **Actuation at Nanoscales** 1635
- 46.2.1 Electrostatics 1636
- 46.2.2 Electromagnetics 1636
- 46.2.3 Piezoelectrics 1636
- 46.2.4 Other Techniques 1637

46.3 **Nanorobotic Manipulation Systems** 1637
- 46.3.1 Overview 1637
- 46.3.2 Nanorobotic Manipulation Systems 1641

46.4 **Nanorobotic Assembly** 1642
- 46.4.1 Overview 1642
- 46.4.2 Carbon Nanotubes 1644
- 46.4.3 Nanocoils 1648

46.5 **Applications** 1651
- 46.5.1 Robotic Biomanipulation 1651
- 46.5.2 Nanorobotic Devices 1652

References 1654

Nanorobotics provides novel techniques for exploring the biodomain by manipulation and characterization of nanoscale objects such as cellular membranes, DNA and other biomolecules. Nano tools, sensors and actuators can provide measurements and/or movements that are calculated in nanometers, gigahertz, piconewtons, femtograms, etc., and are promising for molecular machines and bio- and nanorobotics applications. Efforts are focused on developing enabling technologies for nanotubes and other nanomaterials and structures for NEMS and nanorobotics. By combining bottom-up nanorobotic manipulation and top-down nanofabrication processes, a hybrid approach is demonstrated for creating complex 3-D nanodevices. Nanomaterial science, bionanotechnology, and nanoelectronics will benefit from advances in nanorobotics.

46.1 Overview of Nanorobotics

Progress in robotics over the past years has dramatically extended our ability to explore the world from perception, cognition and manipulation perspectives at a variety of scales extending from the edges of the solar system down to individual atoms (Fig. 46.1). At the bottom of this scale, technology has been moving toward greater control of the structure of matter, suggesting the feasibility of achieving thorough control of the molecular structure of matter atom by atom as *Richard Feynman* first proposed in 1959 in his prophetic article on miniaturization [46.1]:

> *What I want to talk about is the problem of manipulating and controlling things on a small scale... I am not afraid to consider the final question as to whether, ultimately – in the great future – we can arrange the atoms the way we want: the very atoms, all the way down!*

He asserted that:

> *At the atomic level, we have new kinds of forces and new kinds of possibilities, new kinds of effects. The problems of manufacture and reproduction of materials will be quite different. The principles of physics, as far as I can see, do not speak against the possibility of maneuvering things atom by atom.*

This technology is now labeled *nanotechnology*.

The *great future* of Feynman began to be realized in the 1980s. Some of the capabilities he dreamed of have been demonstrated, while others are being developed. Nanorobotics represents the next stage in miniaturization for maneuvering nanoscale objects. Nanorobotics is the study of robotics at the nanometer scale, and includes robots that are nanoscale in size, i. e., nanorobots, and large robots capable of manipulating objects that have dimensions in the nanoscale range with nanometer resolution, i. e., nanorobotic manipulators. The field of nanorobotics brings together several disciplines, including nanofabrication processes used for producing nanoscale robots, nanoactuators, nanosensors, and physical modeling at nanoscales. Nanorobotic manipulation technologies, including the assembly of nanometer-sized parts, the manipulation of biological cells or molecules, and the types of robots used to perform these types of tasks also form a component of nanorobotics.

As the 21st century unfolds, the impact of nanotechnology on the health, wealth, and security of humankind

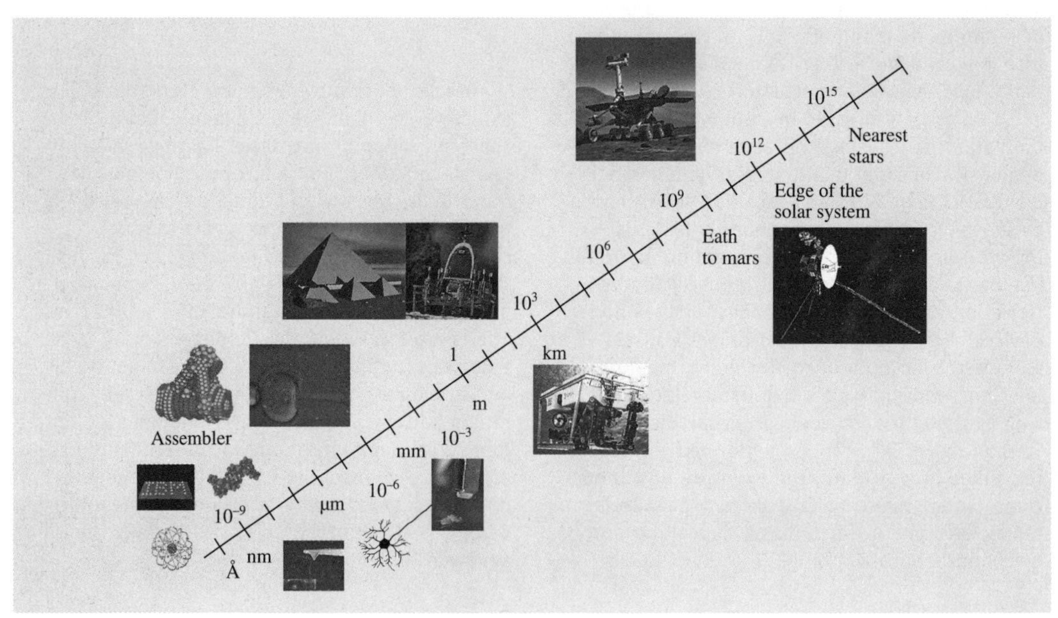

Fig. 46.1 Robotic exploration

is expected to be at least as significant as the combined influences in the 20th century of antibiotics, the integrated circuit, and human-made polymers. For example, *Lane* stated in 1998 [46.2]:

> *If I were asked for an area of science and engineering that will most likely produce the breakthroughs of tomorrow, I would point to nanoscale science and engineering.*

The great scientific and technological opportunities nanotechnology presents have stimulated extensive exploration of the nanoworld and initiated an exciting worldwide competition, which has been accelerated by the publication of the *National Nanotechnology Initiative* by the US government in 2000 [46.3]. Nanorobotics will play a significant role as an enabling nanotechnology and could ultimately be a core part of nanotechnology if *Drexler*'s machine-phase nanosystems based on self-replicative molecular assemblers via mechanosynthesis can be realized [46.4].

By the early 1980s, scanning tunneling microscopes (STMs) [46.5] radically changed the ways in which we interacted with and even regarded single atoms and molecules. The very nature of proximal probe methods encourages exploration of the nanoworld beyond conventional microscopic imaging. Scanned probes now allow us to perform engineering operations on single molecules, atoms, and bonds, thereby providing a tool that operates at the ultimate limits of fabrication. They have also enabled exploration of molecular properties on an individual nonstatistical basis.

STMs and other nanomanipulators are nonmolecular machines but use bottom-up strategies. Although performing only one molecular reaction at a time is obviously impractical for making large amounts of a product, it is a promising way to provide the next generation of nanomanipulators. Most importantly, it is possible to realize the directed assembly of molecules or supermolecules to build larger nanostructures through nanomanipulation. The products produced by nanomanipulation could be the first step of a bottom-up strategy in which these assembled products are used to self-assemble into nanomachines.

One of the most important applications of nanorobotic manipulation will be nanorobotic assembly. However, it appears that until assemblers capable of replication can be built, the parallelism of chemical synthesis and self-assembly are necessary when starting from atoms; groups of molecules can self-assemble quickly due to their thermal motion, enabling them to *explore* their environments and find (and bind to) complementary molecules. Given their key role in natural molecular machines, proteins are obvious candidates for early work in self-assembling artificial molecular systems. *Degrado* [46.6] demonstrated the feasibility of designing protein chains that predictably fold into solid molecular objects. Progress is also being made in artificial enzymes and other relatively small molecules that perform functions like those of natural proteins; the 1987 Nobel prize for chemistry went to *Cram* and *Lehn* for such work on supramolecular chemistry [46.7]. Several bottom-up strategies using self-assembly appear feasible [46.8]. *Fujita* et al.'s pioneering work has shown that self-assembly can be directed by adroitly exploiting the chemical and electrical bonds that hold natural molecules together, and hence get molecules to form desired nanometer-scale structures [46.9]. Chemical synthesis, self assembly, and supramolecular chemistry make it possible to provide building blocks at relatively large sizes beginning from the nanometer scale. Nanorobotic manipulation serves as the base for a hybrid approach to construct nanodevices by structuring these materials to obtain building blocks and assembling them into more complex systems.

This chapter focuses on nanorobotics including actuation, manipulation and assembly at the nanoscale. The main goal of nanorobotics is to provide an effective technology for the experimental exploration of the nanoworld, and to push the boundaries of this exploration from a robotics research perspective.

46.2 Actuation at Nanoscales

The positioning of nanorobots and nanorobotic manipulators depends largely on nanoactuators. While nanosized actuators for nanorobots are still under exploration and relatively far from implementation, microelectromechanical system (MEMS)-based efforts are focused on shrinking their sizes [46.10]. Nanometer-resolution motion has been extensively investigated and can be generated using various actuation principles. Electrostatics, electromagnetics, and piezoelectrics are the most common ways to realize actuation

Table 46.1 Actuation with MEMS

Actuation principle	Type of motion	Volume (mm^3)	Speed (s^{-1})	Force (N)	Stroke (m)	Resolution (m)	Power density (W/m^3)	Ref.
Electrostatic	Linear	400	5000	1×10^{-7}	6×10^{-6}	n/a	200	[46.12]
Magnetic	Linear	$0.4\times0.4\times0.5$	1000	2.6×10^{-6}	1×10^{-4}	n/a	3000	[46.13]
Piezoelectric	Linear	$25.4\times12.7\times1.6$	4000	350	1×10^{-3}	7×10^{-8}	n/a	[46.14]
Actuation principle	**Type of motion**	**Volume (mm^3)**	**Speed (rad/s)**	**Torque (N m)**	**Stroke (rad)**	**Resolution (rad)**	**Power density (W/m^3)**	**Ref.**
Electrostatic	Rotational	$\pi/4\times0.5^2\times3$	40	2×10^{-7}	2π	n/a	900	[46.15]
Magnetic	Rotational	$2\times3.7\times0.5$	150	1×10^{-6}	2π	$5/36\pi$	3000	[46.16]
Piezoelectric	Rotational	$\pi/4\times1.5^2\times0.5$	30	2×10^{-11}	0.7	n/a	n/a	[46.17]

at nanoscales. For nanorobotic manipulation, besides nanoresolution and compact sizes, actuators generating large strokes and high forces are best suited for such applications. The speed criteria are of less importance as long as the actuation speed is in the range of a couple of hertz and above. Table 46.1 provides a small selection of early works on actuators [46.11–16] suitable in actuation principle actuators suitable for nanorobotic applications (partially adapted from [46.10]).

Several extensive reviews on various actuation principles have been published [46.17–21]. During the design of an actuator, the tradeoffs among range of motion, force, speed (actuation frequency), power consumption, control accuracy, system reliability, robustness, load capacity, etc. must be taken into consideration. This section reviews basic actuation technologies and potential applications at nanometer scales.

46.2.1 Electrostatics

Electrostatic charge arises from a build up or deficit of free electrons in a material, which can exert an attractive force on oppositely charged objects, or a repulsive force on similarly charged objects. Since electrostatic fields arise and disappear rapidly, such devices will likewise demonstrate very fast operation speeds and be little affected by ambient temperatures.

Recent investigations have produced many examples of miniature devices using electrostatic force for actuation including silicon micro motors [46.22, 23], microvalves [46.24], and microtweezers [46.25]. This type of actuation is important for achieving nanosized actuation.

Electrostatic fields can exert great forces, but generally across very short distances. When the electric field must act over larger distances, a higher voltage will be required to maintain a given force. The extremely low-current consumption associated with electrostatic devices makes for highly efficient actuation.

46.2.2 Electromagnetics

Electromagnetism arises from electric current moving through a conducting material. Attractive or repulsive forces are generated adjacent to the conductor and proportional to the current flow. Structures can be built which gather and focus electromagnetic forces, and harness these forces to create motion.

Electromagnetic fields arise and disappear rapidly, thus permitting devices with very fast operation speeds. Since electromagnetic fields can exist over a wide range of temperatures, performance is primarily limited by the properties of the materials used in constructing the actuator.

One example of a microfabricated electromagnetic actuator is a microvalve which uses a small electromagnetic coil wrapped around a silicon micromachined valve structure [46.26]. However, the downward scalability of electromagnetic actuators into the micro- and nanorealm may be limited by the difficulty of fabricating small electromagnetic coils. Furthermore, most electromagnetic devices require perpendicularity between the current conductor and the moving element, presenting a difficulty for planar fabrication techniques commonly used to make silicon devices.

An important advantage of electromagnetic devices is their high efficiency in converting electrical energy into mechanical work. This translates into less current consumption from the power source.

46.2.3 Piezoelectrics

Piezoelectric motion arises from the dimensional changes generated in certain crystalline materials when

Table 46.2 Comparison of nanoactuators

Method	Efficiency	Speed	Power density
Electrostatic	Very high	Fast	Low
Electromagnetic	High	Fast	High
Piezoelectric	Very high	Fast	High
Thermomechanical	Very high	Medium	Medium
Phase change	Very high	Medium	High
Shape memory	Low	Medium	Very high
Magnetostrictive	Medium	Fast	Very high
Electrorheological	Medium	Medium	Medium
Electrohydrodynamic	Medium	Medium	Low
Diamagnetism	High	Fast	High

subjected to an electric field or to an electric charge. Structures can be built which gather and focus the force of the dimensional changes, and harness them to create motion. Typical piezoelectric materials include quartz (SiO_2), lead zirconate titanate (PZT), lithium niobate, and polymers such as polyvinyledene fluoride (PVDF).

Piezoelectric materials respond very quickly to changes in voltages and with great repeatability. They can be used to generate precise motions with repeatable oscillations, as in quartz timing crystals used in many electronic devices. Piezo materials can also act as sensors, converting tension or compression strains to voltages.

On the microscale, piezoelectric materials have been used in linear inchworm drive devices [46.27], and micropumps [46.28]. STMs and most nanomanipulators use piezoelectric actuators.

Piezo materials operate with high force and speed, and return to a neutral position when unpowered. They exhibit very small strokes (under 1 percent). Alternating electric currents produce oscillations in the piezo material, and operation at the sample's fundamental resonant frequency produces the largest elongation and highest power efficiency [46.29]. Piezo actuators working in the stick–slip mode can provide millimeter to centimeter strokes. Most commercially available nanomanipulators adopt this type of actuators, such as Picomotors from New Focus and Nanomotors from Klock.

46.2.4 Other Techniques

Other techniques include thermomechanical, phase change, shape memory, magnetostrictive, electrorheological, electrohydrodynamic, diamagnetism, magnetohydrodynamic, shape changing, polymers, biological methods (living tissues, muscle cells, etc.) and so on. Table 46.2 lists a comparison of these.

46.3 Nanorobotic Manipulation Systems

46.3.1 Overview

Nanomanipulation, or positional and/or force control at the nanometer scale, is a key enabling technology for nanotechnology by filling the gap between top-down and bottom-up strategies, and may lead to the appearance of replication-based molecular assemblers [46.4]. These types of assemblers have been proposed as general-purpose manufacturing devices for building a wide range of useful products as well as copies of themselves (self-replication).

Presently, nanomanipulation can be applied to the scientific exploration of mesoscopic physical phenomena, biology and the construction of prototype nanodevices. It is a fundamental technology for property characterization of nanomaterials, structures and mechanisms, for the preparation of nanobuilding blocks, and for the assembly of nanodevices such as nanoelectromechanical systems (NEMS).

Nanomanipulation was enabled by the inventions of the STM [46.5], atomic force microscope (AFM) [46.30], and other types of scanning probe microscope (SPM). Besides these, optical tweezers (laser trapping) [46.31] and magnetic tweezers [46.32] are also possible nanomanipulators. Nanorobotic manipulators (NRMs) [46.33, 34] are characterized by the capability of 3-D positioning, orientation control, independently actuated multiple end-effectors, and independent real-time observation systems, and can be integrated with scanning probe microscopes. NRMs largely extend the complexity of nanomanipulation.

A concise comparison of STM, AFM, and NRM technology is shown in Fig. 46.2. With its incomparable imaging resolution, an STM can be applied to particles as small as atoms with atomic resolution. However, limited by its 2-D positioning and available strategies for manipulations, standard STMs are ill-suited for complex manipulation and cannot be used in 3-D space. An AFM is another important type of nanomanipulator. There are three imaging modes for AFMs, i. e., contact mode, tapping mode (periodic contact mode), and non-contact mode. The latter two are also called dynamic modes and can attain higher imaging resolution than the contact mode. Atomic resolution is obtainable with non-contact mode. Manipulation with an AFM can be done in either contact or dynamic mode. Generally, manipulation with an AFM involves moving an object by touching it with a tip. A typical manipulation is like this: image a particle first in non-contact mode, then remove the tip oscillation voltage and sweep the tip across the particle in contact with the surface and with the feedback disabled. Mechanical pushing can exert larger forces on objects and, hence, can be applied for the manipulation of relatively larger objects. 1-D to 3-D objects can be manipulated on a 2-D substrate. However, the manipulation of individual atoms with an AFM remains a challenge. By separating the imaging and manipulation functions, nanorobotic manipulators can have many more degrees of freedom including rotation for orientation control, and, hence, can be used for the manipulation of 0-D (symmetric spheres) to 3-D objects in 3-D free space. Limited by the relative lower resolution of electron microscopes, NRMs are difficult to use for the manipulation of atoms. However, their general robotic capabilities including 3-D positioning, orientation control, independently actuated multiple end-effectors, separate real-time observation system, and integrations with SPMs inside makes NRMs quite promising for complex nanomanipulation.

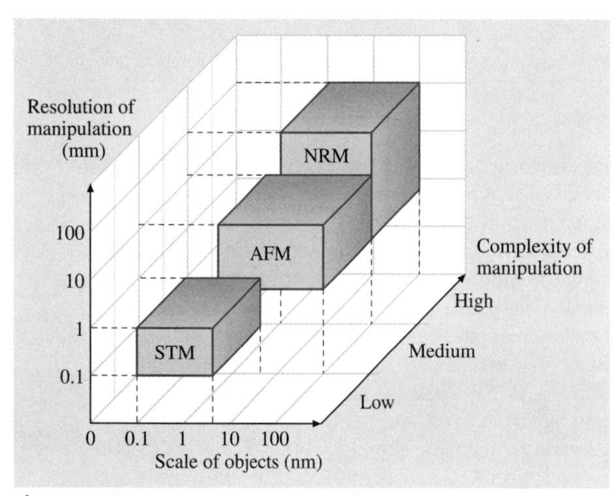

Fig. 46.2 Comparison of nanomanipulators

The first nanomanipulation experiment was performed by *Eigler* and *Schweizer* in 1990 [46.35]. They used an STM and materials at low temperatures (4 K) to position individual xenon atoms on a single-crystal nickel surface with atomic precision. The manipulation enabled them to fabricate rudimentary structures of their own design, atom by atom. The result is the famous set of images showing how 35 atoms were moved to form the three-letter logo *IBM*, demonstrating that matter could indeed be maneuvered atom by atom as *Feynman* suggested [46.1].

A nanomanipulation system generally includes nanomanipulators as the positioning device, microscopes as *eyes*, various end-effectors including probes and tweezers among others as its *fingers*, and types of sensors (force, displacement, tactile, strain, etc.) to facilitate the manipulation and/or to determine the properties of the objects. Key technologies for nanomanipulation include observation, actuation, measurement, system design and fabrication, calibration and control, communication, and human–machine interface.

Strategies for nanomanipulation are basically determined by the environment – air, liquid or vacuum – which is further decided by the properties and size of the objects and observation methods. Figure 46.3 depicts the microscopes, environments and strategies of nanomanipulation. In order to observe manipulated objects, STMs can provide subangstrom imaging resolution, whereas AFMs can provide atomic resolution. Both can obtain 3-D surface topology. Because AFMs can be used in an ambient environment, they provide a powerful tool for biomanipulation that may require a liquid environment. The resolution of scanning electron microscopes (SEM) is limited to about 1 nm, whereas field-emission SEMs (FESEM) can achieve higher resolutions. SEMs/FESEMs can be used for 2-D real-time observation for both the objects and end-effectors of manipulators, and large ultrahigh-

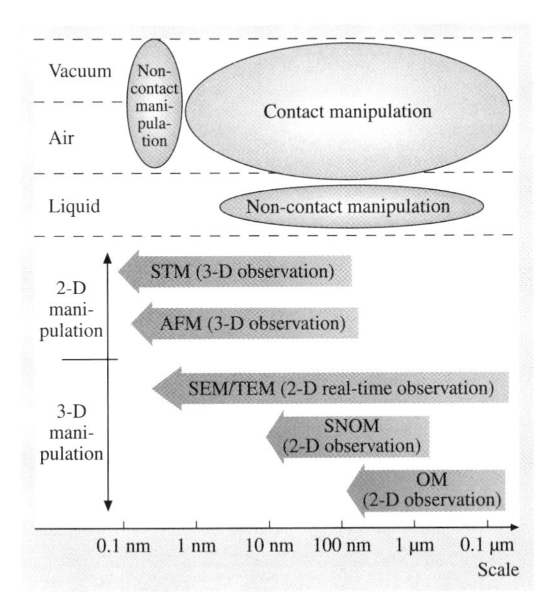

Fig. 46.3 Microscopes, environments and strategies of nanomanipulation

vacuum (UHV) sample chambers provide enough space to contain an NRM with many degrees of freedom (DOFs) for 3-D nanomanipulation. However, the 2-D nature of the observation makes positioning along the electron-beam direction difficult. High-resolution transmission electron microscopes (HRTEM) can provide atomic resolution. However, the narrow UHV specimen chamber makes it difficult to incorporate large manipulators. In principle, optical microscopes (OMs) cannot be used for nanometer-scale (smaller than the wavelength of visible lights) observation because of diffraction limits. Scanning near-field OMs (SNOMs) break this limitation and are promising as a real-time observation device for nanomanipulation, especially for ambient environments. SNOMs can be combined with AFMs, and potentially with NRMs for nanoscale biomanipulation.

Nanomanipulation processes can be broadly classified into three types: (1) lateral non-contact, (2) lateral contact, and (3) vertical manipulation. Generally, lateral non-contact nanomanipulation is mainly applied for atoms and molecules in UHV with an STM or bio-object in liquid using optical or magnetic tweezers. Contact nanomanipulation can be used in almost any environment, generally with an AFM, but is difficult for atomic manipulation. Vertical manipulation can be per-

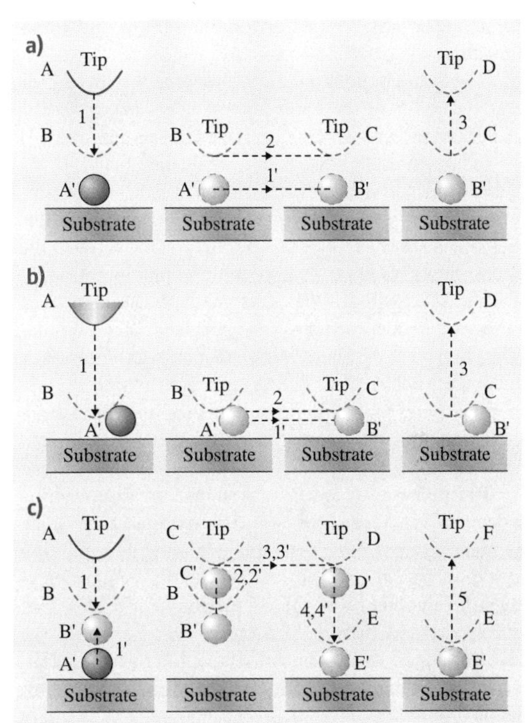

Fig. 46.4a–c Basic strategies of nanomanipulation. In the figure, A, B, C, ... represent the positions of end-effector (e.g., a tip), A′, B′, C′, ... the positions of objects, 1, 2, 3, ... the motions of end-effector, and 1′, 2′, 3′, ... the motions of objects. Tweezers can be used in pick-and-place to facilitate the picking-up, but are generally not necessarily helpful for placing. **(a)** Lateral non-contact nanomanipulation (sliding), **(b)** lateral contact nanomanipulation (pushing/pulling), **(c)** vertical nanomanipulation (picking and placing)

formed by NRMs. Figure 46.4 shows the processes of the three basic strategies.

Motion of the lateral noncontact manipulation processes are shown in Fig. 46.4a. Applicable effects [46.36] able to cause the motion include long-range van der Waals (vdW) forces (attractive) generated by the proximity of the tip to the sample [46.37], electric-field-induced fields caused by the voltage bias between the tip and the sample [46.38, 39], tunneling current local heating or inelastic tunneling vibration [46.40, 41]. With these methods, some nanodevices and molecules have been assembled [46.42, 43]. Laser trapping (optical tweezers) and magnetic tweezers are possible for non-contact manipulation of nanoorder biosamples, e.g. DNA [46.44, 45].

Non-contact manipulation combined with STMs has revealed many possible strategies for manipulating atoms and molecules. However, for the manipulation of CNTs no examples have been demonstrated.

Pushing or pulling nanometer objects on a surface with an AFM is a typical manipulation using this method as shown in Fig. 46.4b. Early work showed the effectiveness of this method for the manipulation of nanoparticles [46.46–49]. This method has also been shown in nanoconstruction [46.50] and biomanipulation [46.51]. A virtual-reality interface facilitates such manipulation [46.52, 53] and may create an opportunity for other types of manipulation. This technique has been used in the manipulation of nanotubes on a surface, and some examples will be introduced later in this chapter.

The pick-and-place task as shown in Fig. 46.4c is especially significant for 3-D nanomanipulation since its main purpose is to assemble prefabricated building blocks into devices. The main difficulty is in achieving sufficient control of the interaction between the tool and object and between the object and the substrate. Two strategies have been presented for micromanipulation [46.54] and have also proven to be effective for nanomanipulation [46.34, 55]. One strategy is to apply a dielectrophoretic force between a tool and an object

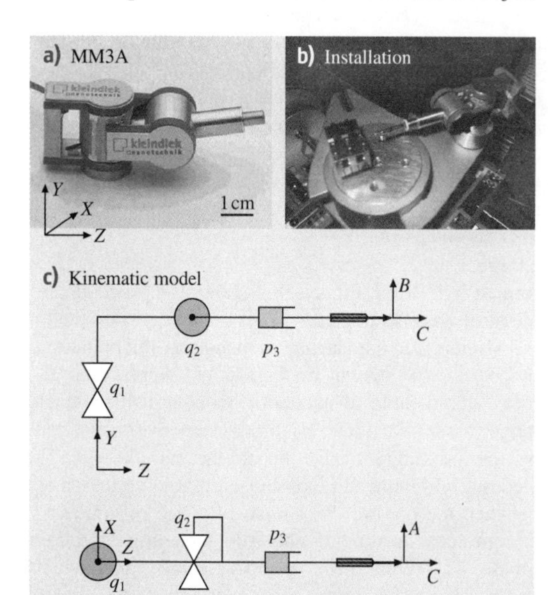

Fig. 46.5a–c Nanomanipulator (MM3A from Kleindiek) inside an SEM

Table 46.3 Specifications of MM3A

Item	Specification
Operating range q_1 and q_2	240°
Operating range Z	12 mm
Resolution A (horizontal)	10^{-7} rad (5 nm)
Resolution B (vertical)	10^{-7} rad (3.5 nm)
Resolution C (linear)	0.25 nm
Fine (scan) range A	20 μm
Fine (scan) range B	15 μm
Fine (scan) range C	1 μm
Speed A, B	10 mm/s
Speed C	2 mm/s

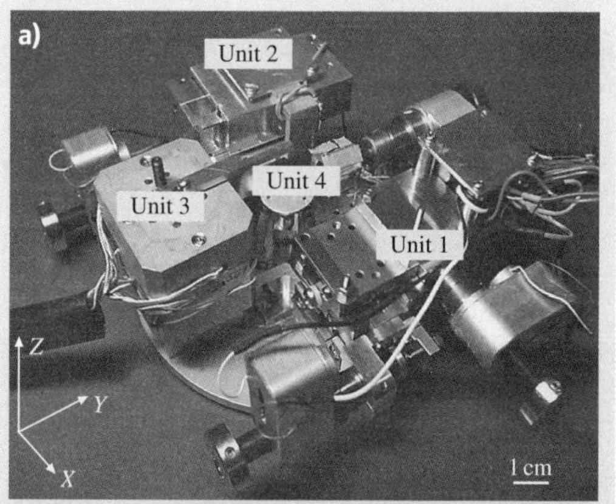

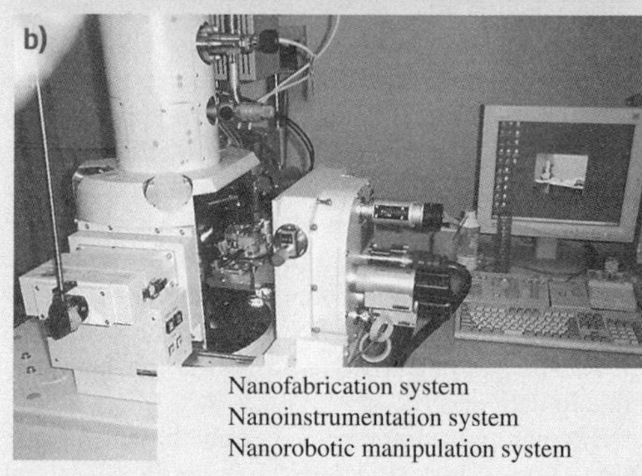

Fig. 46.6a,b Nanorobotic system. (**a**) Nanorobotic manipulators, (**b**) system setup

as a controllable additional external force by applying a bias between the tool and the substrate on which the object is placed. Another strategy is to modify the van der Waals and other intermolecular and surface forces between the object and the substrate. For the former,

Table 46.4 Specifications of a nanorobotic manipulation system

Item	Specification
Nanorobotic manipulation system	
DOFs	Total: 16 DOFs Unit 1: 3 DOFs (x, y and β; coarse) Unit 2: 1 DOF (z; coarse), 3-DOF (x, y and z; fine) Unit 3: 6 DOFs (x, y, z, α, β, γ; ultrafine) Unit 4: 3 DOFs (z, α, β; fine)
Actuators	4 Picomotors (Units 1 & 2) 9 PZTs (Units 2 & 3) 7 Nanomotors (Units 2 & 4)
End-effectors	3 AFM cantilevers + 1 substrate or 4 AFM cantilevers
Working space	18 mm × 18 mm × 12 mm × 360° (coarse, fine), 26 μm × 22 μm × 35 μm (ultrafine)
Positioning resolution	30 nm (coarse), 2 mrad (coarse), 2 nm (fine), sub-nm (ultrafine)
Sensing system	FESEM (imaging resolution: nm) and AFM cantilevers
Nanoinstrumentation system	
FESEM	Imaging resolution: 1.5 nm
AFM cantilever	Stiffness constant: 0.03 nN/nm
Nanofabrication system	
EBID	FESEM emitter: T-FE CNT emitter

an AFM cantilever is ideal as one electrode to generate a nonuniform electrical field between the cantilever and the substrate.

46.3.2 Nanorobotic Manipulation Systems

Nanorobotic manipulators are the core components of nanorobotic manipulation systems. The basic requirements for a nanorobotic manipulation system for 3-D manipulation include nanoscale positioning resolution, a relative large working space, enough DOFs including rotational ones for 3-D positioning and orientation control of the end-effectors, and usually multiple end-effectors for complex operations.

A commercially available nanomanipulator (MM3A from Kleindiek) installed inside a SEM (Carl Zeiss DSM962) is shown in Fig. 46.5. The manipulator has three degrees of freedom, and nanometer to subnanometer-scale resolution (Table 46.3). Calculations show that, when moving/scanning in A/B-direction by joint q_1/q_2, the additional linear motion in C is very small. For example, when the arm length is 50 mm, the additional motion in the C-direction is only 0.25–1 nm when moving in the A-direction for 5–10 μm; these errors can be ignored or compensated with an additional motion of the prismatic joint p_3, which has a 0.25 nm resolution.

Figure 46.6a shows a nanorobotic manipulation system that has 16 DOFs in total and can be equipped with three to four AFM cantilevers as end-effectors for both manipulation and measurement. Table 46.4 lists the specifications of the system. Table 46.5 shows the functions of the nanorobotic manipulation system for nanomanipulation, nanoinstrumentation, nanofabri-

Table 46.5 Functions of a nanorobotic manipulation system

Functions	Manipulations involved
Nanomanipulation	Picking up nanotubes by controlling intermolecular and surface forces, and positioning them together in 3-D space
Nanoinstrumentation	Mechanical properties: buckling or stretching Electrical properties: placing between two probes (electrodes)
Nanofabrication	EBID with a CNT emitter and parallel EBID Destructive fabrication: breaking Shape modification: deforming by bending and buckling, and fixing with EBID
Nanoassembly	Connecting with van der Waals Soldering with EBID Bonding through mechanochemical synthesis

cation and nanoassembly. The positioning resolution is subnanometer order and strokes are centimeter scale. The manipulation system is not only for nanomanipulation, but also for nanoassembly, nanoinstrumentation and nanofabrication. Four-probe semiconductor measurements are perhaps the most complex manipulation this system can perform, because it is necessary to actuate four probes independently by four manipulators. Theoretically, 24 DOFs are needed for four manipulators for general-purpose manipulations, i. e., 6 DOFs for each manipulator for complete control of three linear DOFs and three rotation DOFs. However, 16 DOFs are sufficient for this specific purpose. In general, two manipulators are sufficient for most tasks. More probes provide for more potential applications. For example, three manipulators can be used to assemble a nanotube transistor, a third probe can be applied to cut a tube supported on the other two probes, four probes can be used for four-terminal measurements to characterize the electric properties of a nanotube or a nanotube cross-junction. There are many potential applications for the manipulators if all four probes are used together. With the advancement of nanotechnology, one could shrink the size of nanomanipulators and insert more DOFs inside the limited vacuum chamber of a microscope, and, perhaps, the molecular version of manipulators such as that dreamed of by *Drexler* could be realized [46.4].

For the construction of multiwalled carbon nanotubes (MWNT)-based nanostructures, manipulators position and orient nanotubes for the fabrication of nanotube probes and emitters, for performing nanosoldering with electron-beam-induced deposition (EBID) [46.56], for the property characterization of single nanotubes for selection purposes and for characterizing junctions to test connection strength.

A nanolaboratory is shown in Fig. 46.6b, and its specifications are listed in Table 46.4. The nanolaboratory integrates a nanorobotic manipulation system with a nano analytical system and a nanofabrication system, and can be applied for manipulating nanomaterials, fabricating nanobuilding blocks, assembling nanodevices, and for in situ analysis of the properties of such materials, building blocks and devices. Nanorobotic manipulation within the nanolaboratory has opened a new path for constructing nanosystems in 3-D space, and will create opportunities for new nanoinstrumentation and nanofabrication processes.

46.4 Nanorobotic Assembly

46.4.1 Overview

Nanomanipulation is a promising strategy for nanoassembly. Key techniques for nanoassembly include the structuring and characterization of nanobuilding blocks, the positioning and orientation control of the building blocks with nanometer-scale resolution, and effective connection techniques. Nanorobotic manipu-

Table 46.6 Properties of carbon nanotubes

Property	Item	Data
Geometrical	Layers	Single/multiple
	Aspect ratio	10–1000
	Diameter	≈ 0.4 nm to >3 nm (SWNTs)
		≈ 1.1 to > 100 nm (MWNTs)
	Length	Several μm (rope up to cm)
Mechanical	Young's modulus	≈ 1 TPa (steel: 0.2 TPa)
	Tensile strength	45 GPa (steel: 2 GPa)
	Density	1.33–1.4 g/cm^3 (Al: 2.7 g/cm^3)
Electronic	Conductivity	Metallic/semiconductivity
	Current carrying	
	Capacity	≈ 1 TA/cm^3 (Cu: 1 GA/cm^3)
	Field emission	Activate phosphorus at 1–3 V
Thermal	Heat transmission	> 3 kW/(m K) (diamond: 2 kW/(m K))

lation, which is characterized by multiple DOFs with both position and orientation controls, independently actuated multi-probes, and a real-time observation system, have been shown to be effective for assembling nanotube-based devices in 3-D space.

The well-defined geometries, exceptional mechanical properties, and extraordinary electric characteristics, among other outstanding physical properties (as listed in Table 46.6), of CNTs [46.57] qualify them for many potential applications (as concisely listed in Table 46.7), especially in nanoelectronics [46.58–60], NEMS, and other nanodevices [46.61]. For NEMS, some of the most important characteristics of nanotubes include their nanometer diameter [46.62], large aspect ratio (10–1000) [46.63, 64], TPa-scale Young's modulus [46.65–71], excellent elasticity [46.33, 72], ultrasmall interlayer friction, excellent capability for field emission, various electric conductivities [46.73–75], high thermal conductivity [46.76], high current-carrying capability with essentially no heating [46.77, 78], sensitivity of conductance to various physical or chemical changes, and charge-induced bond-length change.

Helical 3-D nanostructures, or nanocoils, have been synthesized from various materials, including helical carbon nanotubes [46.79] and zinc oxide nanobelts [46.80]. A new method of creating structures with nanometer-scale dimensions has recently been

Table 46.7 Applications of carbon nanotubes

State	Device	Main properties applied
Bulk/array	Composite	High strength, conductivity, etc.
	Field emission devices: flat display, lamp, gas discharge tube, X-ray source, microwave generator, etc.	Field emission: stable emission, long lifetimes, and low emission threshold potentials, high current densities
	Electrochemical devices: supercapacitor, battery cathode, electromechanical actuator, etc.	Large surface area conductivity, high strength, high reversible component of storage capacity
	Fuel cell, hydrogen storage, etc.	Large surface area
Individual	Nanoelectronics: wire, diode, transistor, switch memory, etc.	Small sizes, semiconducting/metallic
	NEMS: probe, tweezers, scissors, sensor, actuator, bearing, gear, etc.	Well-defined geometrics, exceptional mechanical and electronic properties

presented [46.81] and can be fabricated in a controllable way [46.82]. The structures are created through a top-down fabrication process in which a strained nanometer-thick heteroepitaxial bilayer curls up to form 3-D structures with nanoscale features. Helical geometries and tubes with diameters between 10 nm and 10 μm have been achieved. Because of their interesting morphology, mechanical, electrical, and electromagnetic properties, potential applications of these nanostructures in NEMS include nanosprings [46.83], electromechanical sensors [46.84], magnetic-field detectors, chemical or biological sensors, generators of magnetic beams, inductors, actuators, and high-performance electromagnetic-wave absorbers.

NEMS based on individual single- or multi-walled carbon nanotubes (SWNTs [46.85, 86] or MWNTs [46.57]) and nanocoils are of increasing interest, indicating that capabilities for incorporating these individual building blocks at specific locations on a device must be developed. Random spreading [46.87], direct growth [46.88], self-assembly [46.89], dielectrophoretic assembly [46.90] and nanomanipulation [46.91] have been demonstrated for positioning as-grown nanotubes on electrodes for the construction of these devices. However, for nanotube-based structures, nanorobotic assembly is still the only technique capable of in situ structuring, characterization and assembly. Because the as-fabricated nanocoils are not free-standing from their substrate, nanorobotic assembly is virtually the only way to incorporate them into devices at present.

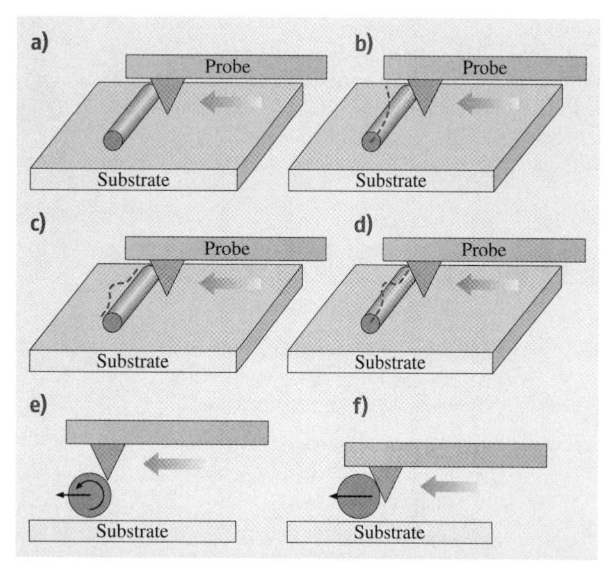

Fig. 46.7a–f 2-D manipulation of CNTs. Starting from the original state shown in (**a**), pushing the tube at a different site with a different force may cause the tube to deform as in (**b**) and (**c**), to break as in (**d**), or to move as in (**e**) and (**f**). (**a**) Original state, (**b**) bending, (**c**) kinking, (**d**) breaking, (**e**) rolling, (**f**) sliding

46.4.2 Carbon Nanotubes

Nanotube manipulation in two dimensions on a surface was first performed with an AFM by contact pushing on a substrate. Figure 46.7 shows the typical methods for 2-D pushing. Although similar to that shown in Fig. 46.4b, the same manipulation caused various results because nanotubes cannot be regarded as 0-D points. The first demonstration was given by *Lieber* and coworkers for measuring the mechanical properties of a nanotube [46.66]. They used the method shown in Fig. 46.7b, i. e., bending a nanotube by pushing on one end of it and fixing the other end. The same strategy was used for the investigation of the behaviors of nanotubes under large strain [46.92]. *Dekker* and coworkers applied the strategies shown in Fig. 46.7c,d to get a kinked junction and crossed nanotubes [46.93, 94]. *Avouris* and coworkers combined this technique with an inverse process, namely straightening by pushing along a bent tube, and realized the translation of the tube to another location [46.95] and between two electrodes to measure the conductivity [46.96], and to form a field-effect transistor (FET) [46.97]. This technique was also used to place a tube on another tube to form a single electron transistor (SET) with cross-junction of nanotubes [46.98]. Pushing-induced breaking (Fig. 46.7d) has also been demonstrated for an adsorbed nanotube [46.95] and a freely suspended SWNT rope [46.72]. The simple assembly of two bent tubes and a straight one formed the Greek letter θ. To investigate the dynamics of rolling at the atomic level, rolling and sliding of a nanotube (as shown in Fig. 46.7e,f) are performed on graphite surfaces using an AFM [46.99, 100].

Manipulation of CNTs in 3-D space is important for assembling CNTs into structures and devices. The basic techniques for the nanorobotic manipulation of carbon nanotubes are shown in Fig. 46.8 [46.101]. These serve as the basis for handling, structuring, characterizing and assembling NEMS.

The basic procedure is to pick up a single tube from nanotube soot, Fig. 46.8a. This has been shown first by using dielectrophoresis [46.34] through nanorobotic manipulation (Fig. 46.8b). By applying a bias between a sharp tip and a plane substrate, a nonuniform electric

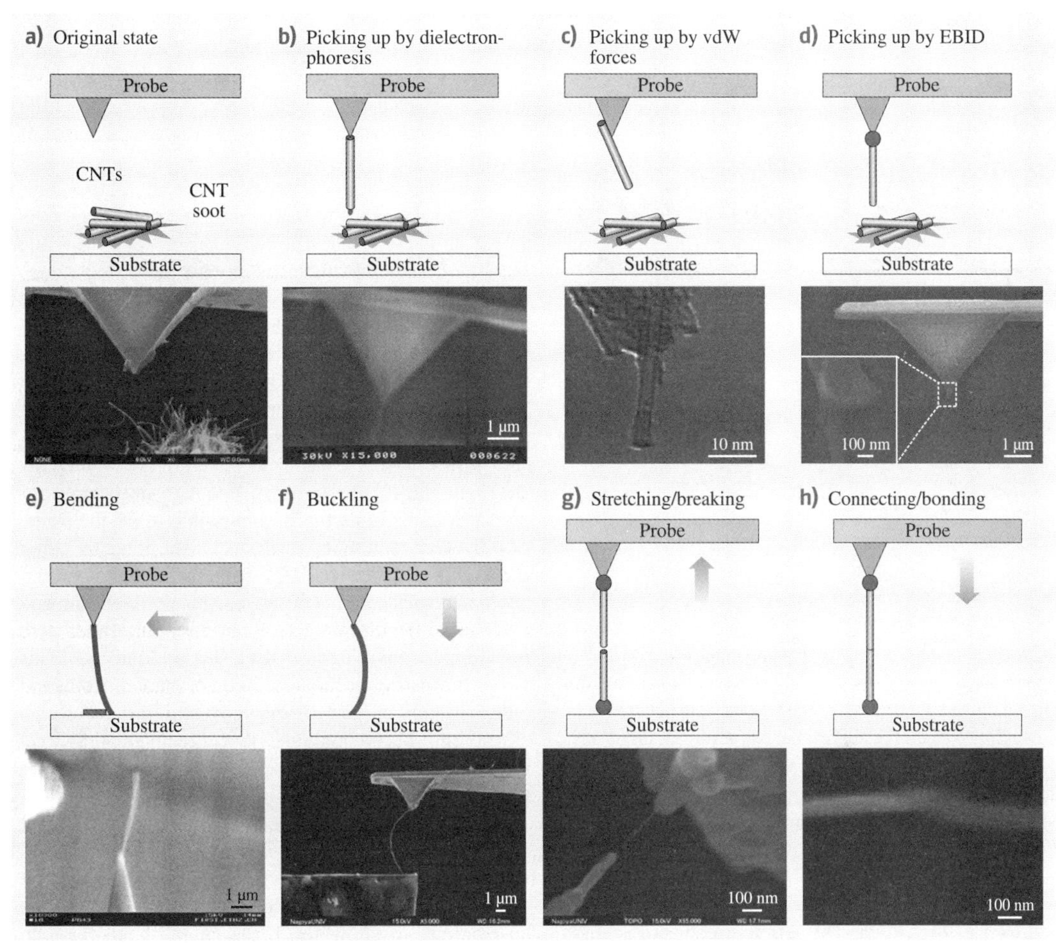

Fig. 46.8a–h Nanorobotic manipulation of CNTs. The basic technique is to pick up an individual tube from CNT soot (**a**) or from an oriented array; (**b**) shows a free-standing nanotube picked up by dielectrophoresis generated by a nonuniform electric field between the probe and substrate, (**c**) (after [46.102]) and (**d**) show the same manipulation by contacting a tube with the probe surface or fixing (e.g. with EBID) a tube to the tip (*inset* shows the EBID deposit). Vertical manipulation of nanotubes includes bending (**e**), buckling (**f**), stretching/breaking (**g**), and connecting/bonding (**h**). All examples with the exception of (**c**) are from the authors' work

field can be generated between the tip and the substrate with the strongest field near the tip. This field can cause a tube to orient along the field or further *jump* to the tip by electrophoresis or dielectrophoresis (determined by the conductivity of the objective tubes). Removing the bias, the tube can be placed at other locations at will. This method can be used for free-standing tubes on nanotube soot or on a rough surface on which surface van der Waals forces are generally weak. A tube strongly rooted in CNT soot or lying on a flat surface cannot be picked up in this way. The interaction between a tube and the atomically flat surface of the AFM cantilever tip has been shown to be strong enough to pick up a tube with the tip [46.102] (Fig. 46.8c). By using EBID, it is possible to pick up and fix a nanotube onto a probe [46.103] (Fig. 46.8d). For handling a tube, a weak connection between the tube and the probe is desired.

Bending and buckling a CNT as shown in Fig. 46.8e,f are important for in situ property char-

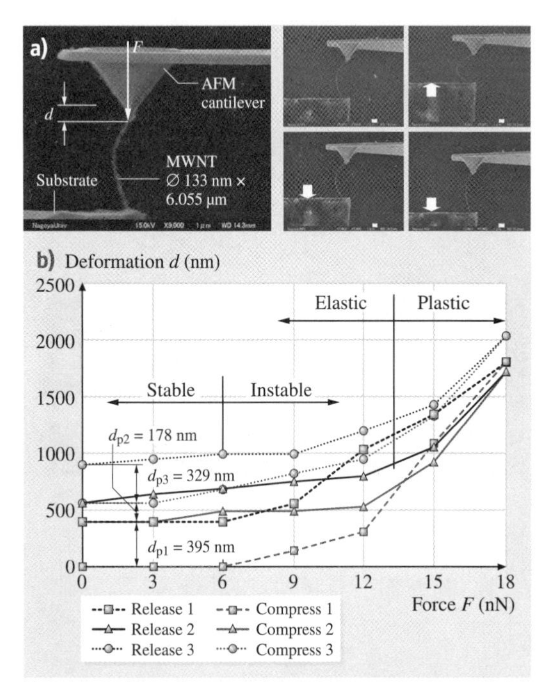

Fig. 46.9a,b In-situ mechanical property characterization of a nanotube by buckling it (scale bars: 1 µm). **(a)** Process of buckling, **(b)** elastic and plastic properties of an MWNT

acterization of a nanotube [46.104, 105], which is a simple way to get the Young's modulus of a nanotube without damaging the tube (if performed within its elastic range) and, hence, can be used for the selection of a tube with desired properties. The process is shown in Fig. 46.9a. The left figure shows an individual MWNT, whereas the right four show a bundle of MWNTs being buckled. Figure 46.9b depicts the property curve of the elastic and plastic deformations

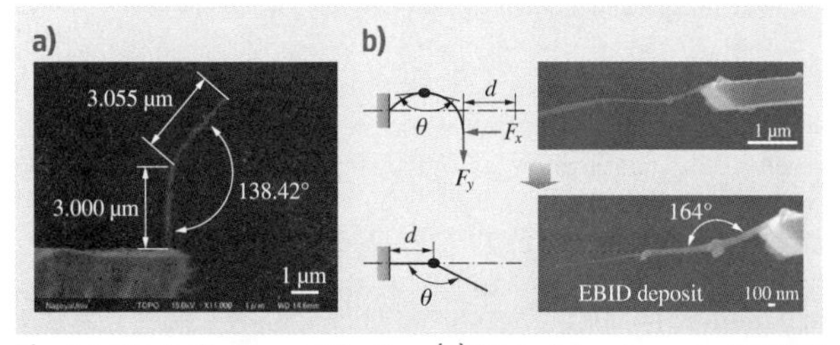

Fig. 46.10a,b Shape modification. **(a)** Kinked structure of a MWNT through plastic deformation, **(b)** shape modifications of a MWNT by elastic bending and buckling deformation and shape fixing through EBID

of the MWNT (⌀133 nm × 6.055 µm), where d and F are the axial deformation and buckling force, as shown in Fig. 46.9a (left). By using the model and analysis method presented in [46.55], the flexural rigidity of the MWNT bundle shown in Fig. 46.9a (right) is found to be $EI = 2.086 \times 10^{-19}$ [N m^2]. This result suggests that the diameter of the nanotube is 46.4 nm if the theoretical value of the Young's modulus of the nanotube $E = 1.26$ TPa is used. The SEM image shows that the bundle of nanotubes includes at least three single ones with diameters of 31, 34, and 41 nm. Hence, it can be determined that there must be damaged parts in the bundle because the stiffness is too low, and it is necessary to select another one without defects. By buckling an MWNT over its elastic limit, a kinked structure can be obtained. After three loading/releasing rounds, as shown in Fig. 46.9b, a kinked structure is obtained, as shown in Fig. 46.10a [46.106]. To obtain any desired angle for a kinked junction it is possible to fix the shape of a buckled nanotube within its elastic limit by using EBID (Fig. 46.10b) [46.107]. For a CNT, the maximum angular displacement will appear at the fixed left end under pure bending or at the middle point under pure buckling. A combination of these two kinds of loads will achieve a controllable position of the kink point and a desired kink angle θ. If the deformation is within the elastic limit of the nanotube, it will recover as the load is released. To avoid this, EBID can be applied at the kink point to fix the shape.

Stretching a nanotube between two probes or a probe and a substrate has generated several interesting results (Fig. 46.8g). The first demonstration of 3-D nanomanipulation of nanotubes took this as an example to show the breaking mechanism of an MWNT [46.33], and to measure the tensile strength of CNTs [46.71]. By breaking an MWNT in a controlled manner, interesting nanodevices have been fabricated. This technique – destructive fabrication – has been presented to get sharpened and layered structures of nanotubes and to improve control of the length of nanotubes [46.108]. Typically, a layered and a sharpened structure can be obtained from this process, similar to that achieved from electric pulses [46.109]. Bearing motion has also been observed in an incompletely broken MWNT (Fig. 46.11). As shown in Fig. 46.11a, an MWNT is supported between a substrate (left end) and an AFM cantilever (right end). Figure 46.11b shows a zoomed image of the centrally blocked part of Fig. 46.11a, and the inset shows its structure schematically. It can be found that the nanotube has a thinner neck (part B in Fig. 46.11b) that was formed by destructive fabrication, i.e., by moving the cantilever to the right. To move it further in the same direction, a motion like a linear bearing is observed, as shown in Fig. 46.11c and schematically by the inset. By comparing Fig. 46.11b,c, we find that part B remained unchanged in its length and diameter, while its two ends brought out two new parts I and II from parts A and B, respectively. Part II has uniform diameter (⌀22 nm), while part I is a tapered structure with a smallest diameter of ⌀25 nm. The interlayer friction has been predicted to be very small [46.110], but direct measurement of the friction remains a challenging problem.

The reverse process, namely the connection of broken tubes (Fig. 46.8h), has been demonstrated recently, and the mechanism is revealed as rebonding of unclosed dangling bonds at the ends of broken tubes [46.111]. Based on this interesting phenomenon, mechanochemical nanorobotic assembly has been performed [46.112].

Assembly of nanotubes is a fundamental technology for enabling nanodevices. The most important tasks include the connection of nanotubes and placing of nanotubes onto electrodes. Pure nanotube circuits [46.113–115] created by interconnecting nanotubes of different diameters and chirality could lead to further size reductions in devices. Nanotube intermolecular and intramolecular junctions are basic elements for such systems. An intramolecular kink junction behaving like a rectifying diode has been reported [46.116]. Room-temperature (RT) SETs [46.117] have been shown with a short ($\approx$ 20 nm) nanotube section that is created by inducing local barriers into the tube with an AFM, and Coulomb charging has been observed. With a cross-junction of two SWNTs (semiconducting/metallic), three- and four-terminal electronic devices have been made [46.118]. A suspended cross-junction can function as electromechanical nonvolatile memory [46.119].

Although some kinds of junctions have been synthesized with chemical methods, there is no evidence yet showing that a self-assembly-based approach can provide more complex structures. SPMs were also used to fabricate junctions, but they are limited to a 2-D plane. We have presented 3-D nanorobotic manipulation-based nanoassembly, which is a promising strategy, both for the fabrication of nanotube junctions and for the construction of more complex nanodevices with such junctions as well.

Nanotube junctions can be classified into different types by: the kind of components – SWNTs or MWNTs; geometric configuration – V (kink), I, X- (cross), T-, Y- (branch), and 3-D junctions; conductivity – metallic or semiconducting; and connection methods – intermolecular (connected with van der Waals force, EBID, etc.) or intramolecular (connected with chemical bonds) junctions. Here we show the fabrication of several kinds of MWNT junctions by emphasizing the connection methods. These methods will also be effective for SWNT junctions. Figure 46.12 shows CNT junctions constructed by connecting with van der Waals forces (a), joining by electron-beam-induced deposition (b), and bonding through mechanochemistry (c).

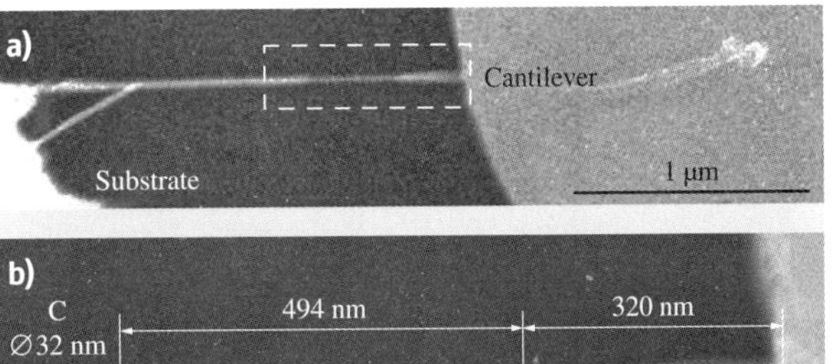

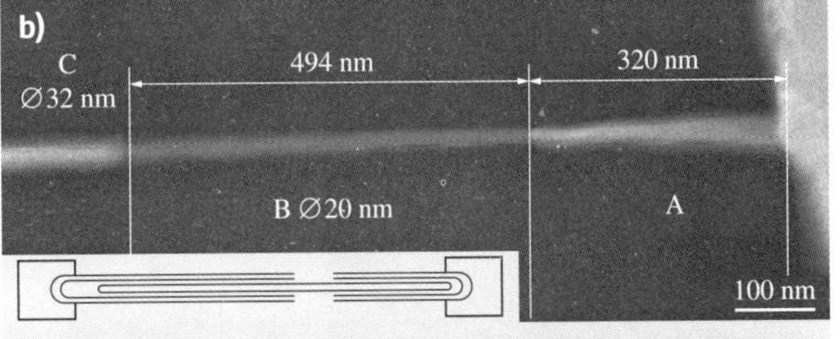

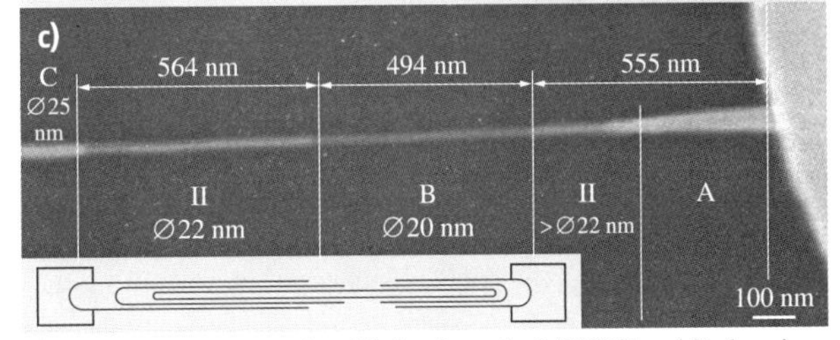

Fig. 46.11a–c Destructive fabrication of a MWNT and its bearing-like motion

MWNT junctions connected with van der Waals forces are the basic forms of junctions. To fabricate such junctions, the main process is to position two or more nanotubes together with nanometer resolution; they will then be connected naturally by intermolecular van der Waals forces. Such junctions are mainly for structures where contact rather than strength is emphasized. Placing them onto a surface can make them more stable. In some cases, when lateral movement along the surface of nanotubes is desired while keeping them in contact, van der Waals-type connections are the only ones that are suitable.

Figure 46.12a shows a T-junction connected with van der Waals forces, which is fabricated by position-

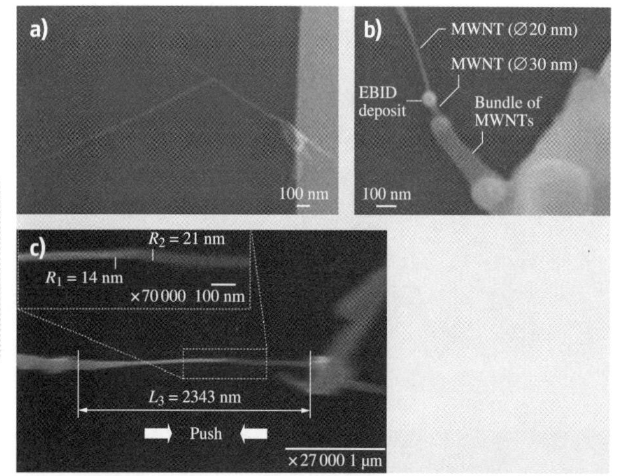

Fig. 46.12a–c MWNT junctions. **(a)** MWNTs connected with van der Waals force. **(b)** MWNTs joined with EBID. **(c)** MWNTs bonded with a mechanochemical reaction

ing the tip of an MWNT onto another MWNT until they form a bond. The contact is checked by measuring the shear connection force.

EBID provides a soldering method to obtain stronger nanotube junctions than those connected through van der Waals forces. Hence, if the strength of nanostructures is emphasized, EBID can be applied. Figure 46.12b shows an MWNT junction connected through EBID, in which the upper MWNT is a single one with a diameter of 20 nm and the lower one is a bundle of MWNTs with an extruded single CNT with ⌀30 nm. The development of conventional EBID has been limited by the expensive electron filament used and low productivity. We have presented a parallel EBID system by using CNTs as emitters because of their excellent field-emission properties [46.120, 121]. The feasibility of parallel EBID is presented. It is a promising strategy for large-scale fabrications of nanotube junctions. Similar to its macro counterpart, welding, EBID works by adding material to obtain stronger connections, but in some cases, added material might influence normal functions for nanosystems. So, EBID is mainly applied to nanostructures rather than nanomechanisms.

To construct stronger junctions without adding additional material, mechanochemical nanorobotic assembly is an important strategy. Mechanochemical nanorobotic assembly is based on solid-phase chemical reactions, or mechanosynthesis, which is defined as chemical synthesis controlled by mechanical systems operating with atomic-scale precision, enabling direct positional selection of reaction sites [46.4]. By picking up atoms with dangling bonds rather than natural atoms only, it is easier to form primary bonds, which provides a simple but strong connection. Destructive fabrication provides a way to form dangling bonds at the ends of broken tubes. Some of the dangling bonds may close with neighboring atoms, but generally a few bonds will remain dangling. A nanotube with dangling bonds at its end will bind more easily to another to form intramolecular junctions. Figure 46.12c shows such a junction. An MWNT (length $L_1 = 1329\,\text{nm}$, diameter $D_1 = 42\,\text{nm}$) is placed between a substrate and an AFM cantilever with a CNT tip, and the two ends are fixed. By pulling the two ends of the MWNT, it is broken into two parts. By pushing the two nanotubes head to head close enough, a new one is formed. To test the strength of this nanotube, it was broken again. The fact that the nanotube breaks at a different site suggests that the tensile strength of the connected nanotubes is not weaker than that of the original nanotube itself. We have determined that no type of connection based on van der Waals interactions can provide such a strong connection strength [46.112]. Also, we have shown that, from the measured tensile strength (1.3 TPa), chemical bonds must have been formed when the junction formed, and that these are most likely to be covalent bonds (sp^2-hybrid type, as in a nanotube).

3-D nanorobotic manipulation has opened a new route for structuring and assembly nanotubes into nanodevices. However, at present nanomanipulation is still performed in a serial manner with master–slave control, which is not a large-scale production-oriented technique. Nevertheless, with advances in the exploration of mesoscopic physics, better control of the synthesis of nanotubes, more accurate actuators, and effective tools for manipulation, high-speed and automatic nanoassembly will be possible. Another approach might be parallel assembly by positioning building blocks with an array of probes [46.122] and joining them together simultaneously, e.g., with the parallel EBID [46.103] approach we presented. Further steps might progress towards exponential assembly [46.123], and in the far future to self-replicating assembly [46.4].

46.4.3 Nanocoils

The construction of nanocoil-based NEMS involves the assembly of as-grown or as-fabricated nanocoils, which is a significant challenge from a fabrication stand-

Fig. 46.13a,b As-fabricated nanocoils with a thickness of $t = 20\,\text{nm}$ (without Cr layer) or 41 nm (with Cr layer). Diameter: $D = 3.4\,\mu\text{m}$ ▶

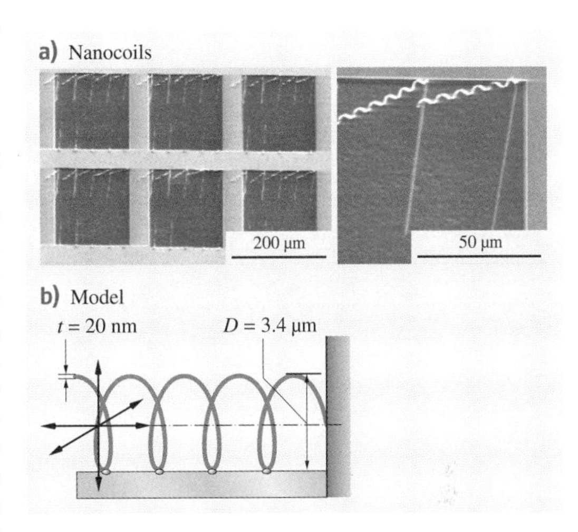

point. Focusing on the unique aspects of manipulating nanocoils due to their helical geometry, high elasticity, single end fixation, and strong adhesion to the substrate from wet etching, a series of new processes is presented using a manipulator (MM3A, Kleindiek) installed in an SEM (Zeiss DSM962). As-fabricated SiGe/Si bilayer nanocoils are shown in Fig. 46.13. Special tools have been fabricated including a nanohook prepared by controlled *tip-crashing* of a commercially available tungsten sharp probe (Picoprobe T-4-10-1 mm and T-4-10) onto a substrate, and a *sticky* probe prepared by tip dipping into a double-sided SEM silver conductive tape (Ted Pella, Inc.). As shown in Fig. 46.14, experiments demonstrate that nanocoils can be released from a chip by lateral pushing, picked up with a nanohook or a *sticky* probe, and placed between the probe/hook and another probe or an AFM cantilever (Nanoprobe, NP-S). Axial pulling/pushing, radial compressing/releasing, and bending/buckling have also been demonstrated. These processes have shown the effectiveness of manipulation

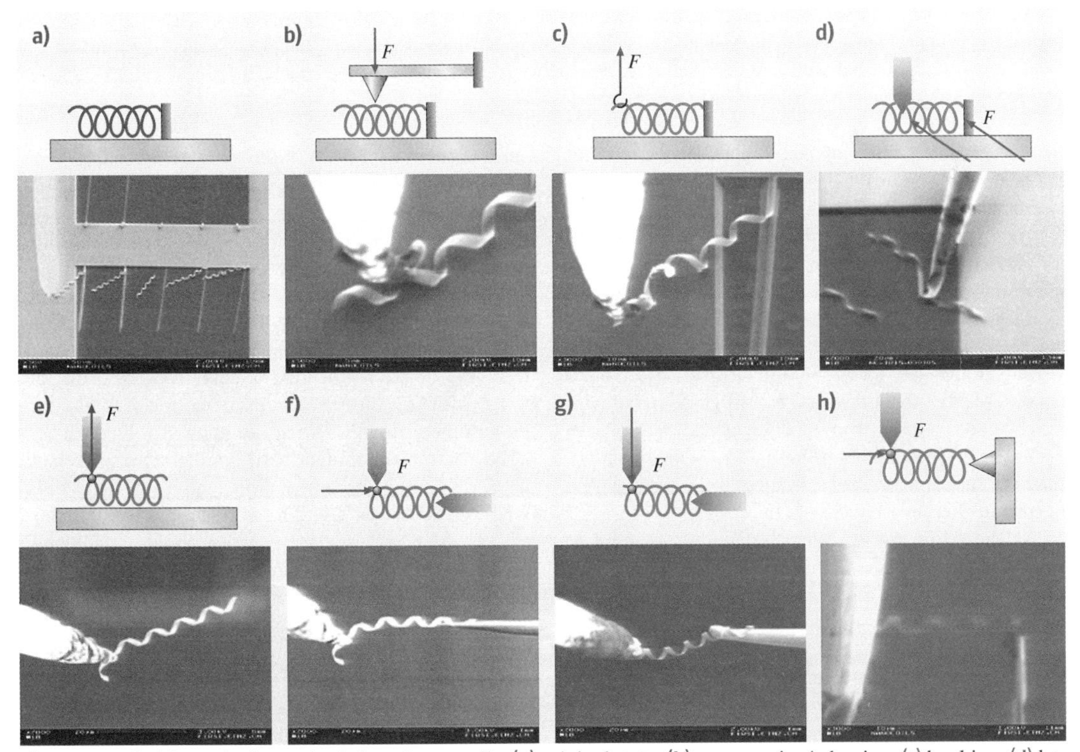

Fig. 46.14a–h Nanorobotic manipulation of nanocoils: (**a**) original state, (**b**) compressing/releasing, (**c**) hooking, (**d**) lateral pushing/breaking, (**e**) picking, (**f**) placing/inserting, (**g**) bending, and (**h**) pushing and pulling

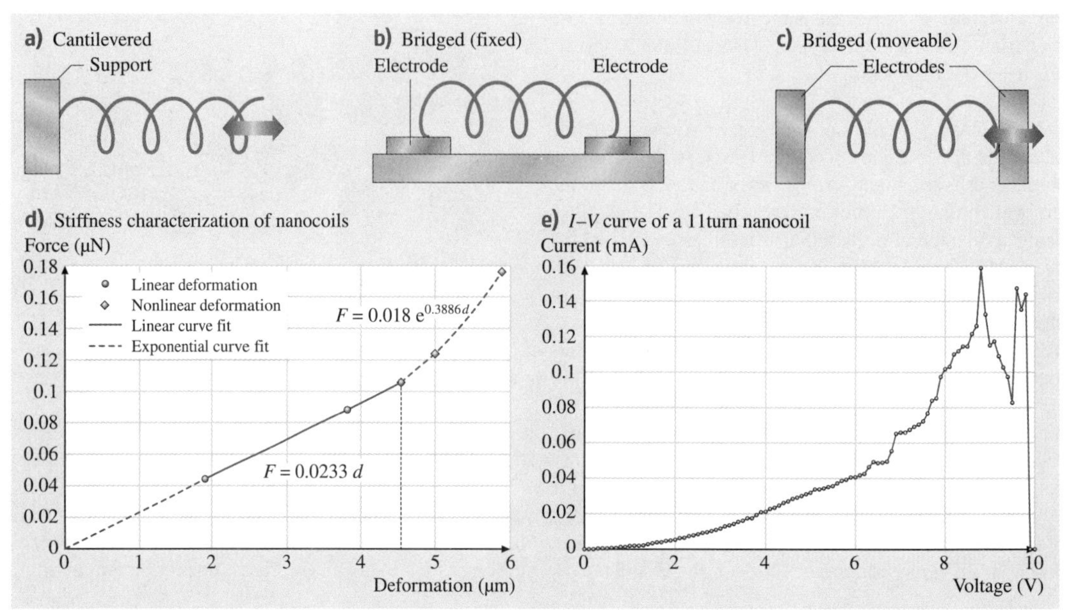

Fig. 46.15a–e Nanocoil-based devices. Cantilevered nanocoils (**a**) can serve as nanosprings. Nanoelectromagnets, chemical sensors, and nanoinductors involve nanocoils bridged between two electrodes (**b**). Electromechanical sensors can use a similar configuration but with one end connected to a moveable electrode (**c**). Mechanical stiffness (**d**) and electric conductivity (**e**) are basic properties of interest for these devices

for the characterization of coil-shaped nanostructures and their assembly for NEMS, which have been otherwise unavailable.

Configurations of nanodevices based on individual nanocoils are shown in Fig. 46.15. Cantilevered nanocoils as shown in Fig. 46.15a can serve as nanosprings. Nanoelectromagnets, chemical sensors and nanoinductors involve nanocoils bridged between two electrodes, as shown in Fig. 46.15b. Electromechanical sensors can use a similar configuration but with one end connected to a moveable electrode, as shown in Fig. 46.15c. Mechanical stiffness and electric conductivity are fundamental properties for these devices that must be further investigated.

As shown in Fig. 46.14h, axial pulling is used to measure the stiffness of a nanocoil. A series of SEM images are analyzed to extract the AFM tip displacement and the nanospring deformation, i. e. the relative displacement of the probe from the AFM tip. From this displacement data and the known stiffness of the AFM cantilever, the tensile force acting on the nanospring versus the nanospring deformation was plotted. The deformation of the nanospring was measured relative to the first measurement point. This was necessary because proper attachment of the nanospring to the AFM cantilever must be verified. Afterwards, it was not possible to return to the point of zero deformation. Instead, the experimental data, as presented in Fig. 46.15d, has been shifted such that, with the calculated linear elastic spring stiffness, the line begins at zero force and zero deformation. From Fig. 46.15d, the stiffness of the spring was estimated to be 0.0233 N/m. The linear elastic region of the nanospring extends to a deformation of 4.5 μm. An exponential approximation was fitted to the nonlinear region. When the applied force reached 0.176 μN, the attachment between the nanospring and the AFM cantilever broke. Finite element simulation (ANSYS 9.0) was used to validate the experimental data [46.84]. Since the exact region of attachment cannot be identified according to the SEM images, simulations were conducted for 4, 4.5, and 5 turns to get an estimate of the possible range, given that the apparent number of turns of the nanospring is between 4 and 5. The nanosprings in the simulations were fixed on one end and had an axial load of 0.106 μN applied to the other end. The simulation results for the spring with 4 turns yield a stiffness of 0.0302 N/m; for the nanospring with 5 turns it is 0.0191 N/m. The measured stiffness falls into this range at 22.0% above the minimum value and 22.8% below the maximum value, and very close to the stiffness of a 4.5 turn nanospring, which has a stiffness of 0.0230 N/m according to the simulation.

Figure 46.15e shows the results from electrical characterization experiments on a nanospring with 11 turns using the configuration as shown in Fig. 46.14g. The I–V curve is nonlinear, which may be caused by the resistance change of the semiconductive bilayer due to ohmic heating. Another possible reason is the decrease in contact resistance caused by thermal stress. The maximum current was found to be 0.159 mA under an 8.8 V bias. Higher voltage causes the nanospring to *blow off*. From the fast scanning screen of the SEM, the extension of the nanospring on the probes was observed around the peak current so that the current does not drop abruptly. At 9.4 V, the extended nanospring is broken down, causing an abrupt drop in the I–V curve.

From fabrication and characterization results, the helical nanostructures appear to be suitable to function as inductors. They would allow further miniaturization compared to state-of-the-art microinductors. For this purpose, higher doping of the bilayer and an additional metal layer would result in the required conductance. Conductance, inductance, and quality factor can be further improved if, after curling up, additional metal is electroplated onto the helical structures. Moreover, a semiconductive helical structure, when functionalized with binding molecules, can be used for chemical sensing using the same principle as demonstrated with other types of nanostructures [46.124]. With bilayers in the range of a few monolayers, the resulting structures would exhibit a very high surface-to-volume ratio with the whole surface exposed to an incoming analyst.

46.5 Applications

Material science, biotechnology, electronics, and mechanical sensing and actuation will benefit from advances in nanorobotics. Research topics in bionanorobotics include the autonomous manipulation of single cells or molecules, the characterization of biomembrane mechanical properties using nanorobotic systems with integrated vision and force-sensing modules, and more. The objective is to obtain a fundamental understanding of single-cell biological systems and provide characterized mechanical models of biomembranes for deformable cell tracking during biomanipulation and cell injury studies. Robotic manipulation at nanometer scales is a promising technology for structuring, characterizing and assembling nanobuilding blocks into NEMS. Combined with recently developed nanofabrication processes, a hybrid approach is realized to build NEMS and other nanorobotic devices from individual carbon nanotubes and SiGe/Si nanocoils.

46.5.1 Robotic Biomanipulation

Biomanipulation – Autonomous Robotic Pronuclei DNA Injection

To improve the low success rate of manual operation, and to eliminate contamination, an autonomous robotic system (shown in Fig. 46.16) has been developed to deposit DNA into one of the two nuclei of a mouse embryo without inducing cell lysis [46.125, 126]. The laboratory's experimental results show that the success rate for the autonomous embryo pronuclei DNA injection is dramatically improved over manual conventional injection methods. The autonomous robotic system features a hybrid controller that combines visual servoing and precision position control, pattern recognition for detecting nuclei, and a precise autofocusing scheme. Figure 46.17 illustrates the injection process.

To realize large-scale injection operations, a MEMS cell holder was fabricated using anodic wafer-bonding

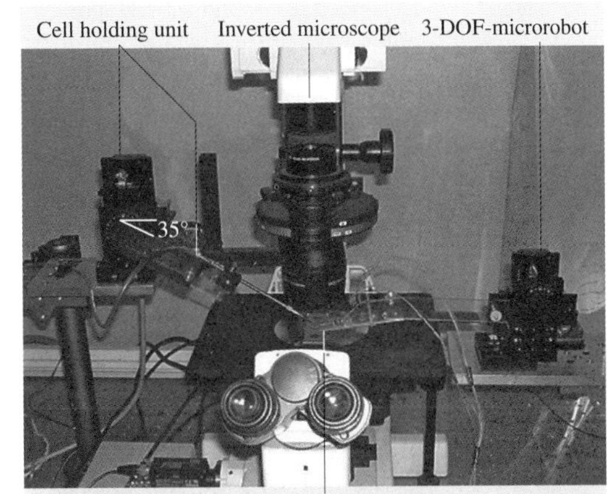

Fig. 46.16 Robotic biomanipulation system with vision and force feedback

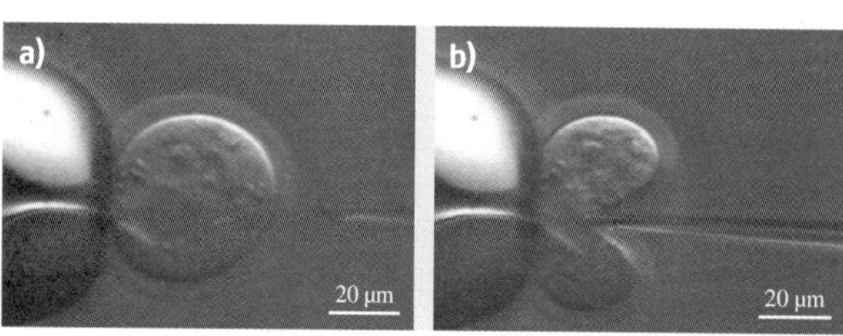

Fig. 46.17a,b Cell injection process. **(a,b)** Cell injection of a mouse oocyte

techniques. Arrays of holes are aligned on the cell holder, which are used to contain and fix individual cells for injection. When well-calibrated, the system with the cell holder makes it possible to inject large numbers of cells using position control. The cell-injection operation can be conducted in a move–inject–move manner.

Successful injection is determined greatly by injection speed and trajectory, and the forces applied to cells (Figure 46.17). To further improve the robotic system's performance, a multi-axial MEMS-based capacitive cellular force sensor is being designed and fabricated to provide real-time force feedback to the robotic system. The MEMS cellular force sensor also aids our research in biomembrane mechanical property characterization.

MEMS-Based Multi-Axis Capacitive Cellular Force Sensor

The MEMS-based two-axis cellular force sensor [46.127] shown in Fig. 46.18 is capable of resolving normal forces applied to a cell as well as tangential forces generated by improperly aligned cell probes. A high-yield microfabrication process was developed to form the 3-D high-aspect-ratio structure using deep reactive ion etching (DRIE) on silicon-on-insulator (SOI) wafers. The constrained outer frame and the inner movable structure are connected by four curved springs. A load applied to the probe causes the inner structure to move, changing the gap between each pair of interdigitated comb capacitors. Consequently, the total capacitance change resolves the applied force. The interdigitated capacitors are orthogonally configured to make the force sensor capable of resolving forces in both the x- and y-directions. The cellular force sensors used in the experiments are capable of resolving forces up to 25 µN with a resolution of 0.01 µN.

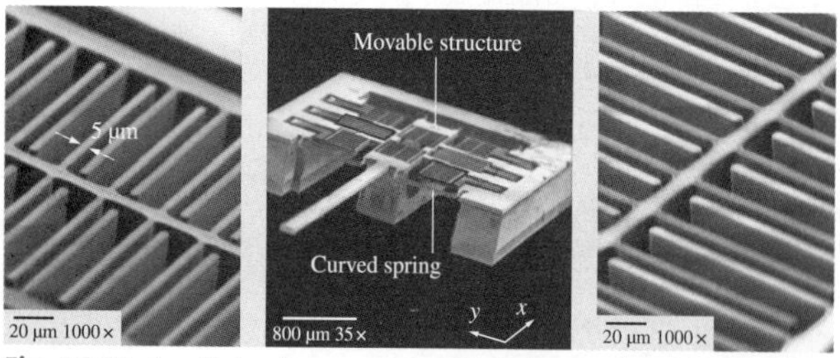

Fig. 46.18 A cellular force sensor with orthogonal comb drives detailed

Tip geometry affects the quantitative force measurement results. A standard injection pipette (Cook K-MPIP-1000-5) tip section with a tip diameter of 5 µm is attached to the probe of the cellular force sensors.

The robotic system and high-sensitivity cellular force sensor are also applied to biomembrane mechanical property studies [46.128]. The goal is to obtain a general parameterized model describing cell membrane deformation behavior when an external load is applied. This parameterized model serves two chief purposes. First, in robotic biomanipulation, it allows online parameter recognition so that cell membrane deformation behavior can be predicted. Second, for a thermodynamic model of membrane damage in cell injury and recovery studies, it is important to appreciate the mechanical behavior of the membranes. This allows the interpretation of such reported phenomena as mechanical resistance to cellular volume reduction during dehydration, and its relationship to injury. The establishment of such a biomembrane model will greatly facilitate cell injury studies.

Experiments demonstrate that robotics and MEMS technology can play important roles in biological studies such as automating biomanipulation tasks. Aided by robotics, the integration of vision and force-sensing modules, and MEMS design and fabrication techniques, investigations are being conducted in biomembrane mechanical property modeling, deformable cell tracking, and single-cell and biomolecule manipulation.

46.5.2 Nanorobotic Devices

Nanorobotic devices involve tools, sensors, and actuators at the nanometer scale. Shrinking device size makes it possible to manipulate nanosized objects with nanosized tools, measure mass in femtogram ranges, sense force at piconewton scales, and induce GHz motion, among other amazing advancements.

Top-down and bottom-up strategies for manufacturing such nanodevices have been independently investigated by a variety of researchers. Top-down strategies are based on nanofabrication and include technologies such as nanolithography, nanoimprinting, and chemical etching. Presently, these are 2-D fabrication processes with relatively low resolution. Bottom-up strategies are assembly-based techniques. At present, these strategies include such techniques as self-assembly, dip-pen lithography, and directed self-assembly. These techniques can generate regular nanopatterns at large scales. With the ability to position and orient nanometer-scale objects, nanorobotic manipulation is an enabling technology for structuring, characterizing and assembling many types of nanosystems [46.101]. By combining bottom-up and top-down processes, a hybrid nanorobotic approach (as shown in Fig. 46.19) based on nanorobotic manipulation provides a third way to fabricate NEMS by structuring as-grown nanomaterials or nanostructures. This new nanomanufacturing technique can be used to create complex 3-D nanodevices with such building blocks. Nanomaterial science, bionanotechnology, and nanoelectronics will also benefit from advances in nanorobotic assembly.

The configurations of nanotools, sensors, and actuators based on individual nanotubes that have been experimentally demonstrated are summarized as shown in Fig. 46.20.

For detecting deep and narrow features on a surface, cantilevered nanotubes (Fig. 46.20a, [46.103]) have been demonstrated as probe tips for an AFM [46.129], an STM and other types of SPM. Nanotubes provide ultrasmall diameters, ultralarge aspect ratios, and excellent mechanical properties. Manual assembly, direct growth [46.130] and nanoassembly [46.131] have proven effective for their construction. Cantilevered nanotubes have also been demonstrated as probes for the measurement of ultrasmall physical quantities, such as femtogram masses [46.67], mass flow sensors [46.132], and piconewton-order force sensors [46.133] on the basis of their static deflections or change of resonant frequencies detected within an electron microscope. Deflections cannot be measured from micrographs in real time, which limits the application of this kind of sensor. Interelectrode distance changes cause emission current variation of a nanotube emitter and may serve as a candidate to replace microscope images.

Bridged individual nanotubes (Fig. 46.20b, [46.134]) have been the basis for electric characterization. A nanotube-based gas sensor adopted this configuration [46.135].

Opened nanotubes (Fig. 46.20c, [46.136]) can serve as an atomic or molecular container. A thermometer based on this structure has been shown by monitor-

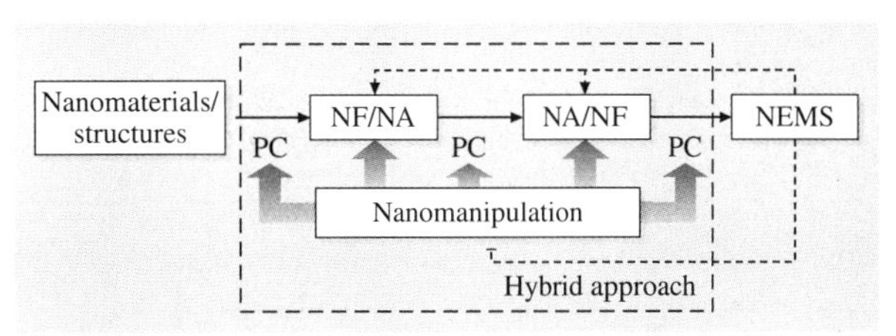

Fig. 46.19 Hybrid approach to NEMS (PC: property characterization, NF: nanofabrication, NA: nanoassembly)

ing the height of the gallium inside the nanotube using TEM [46.137].

Bulk nanotubes can be used to fabricate actuators based on charge-injection-induced bond-length change [46.138], and, theoretically, individual nanotubes also work on the same principle. Electrostatic deflection of a nanotube has been used to construct a relay [46.139]. A new family of nanotube actuators can be constructed by taking advantage of the ultralow inter-layer friction of a multiwalled nanotube. Linear bearings based on telescoping nanotubes have been demonstrated [46.110]. Recently, a micro actuator with a nanotube as a rotation bearing has been demonstrated [46.140]. A preliminary experiment on a promising nanotube linear motor with field emission current serving as position feedback has been shown with nanorobotic manipulation (Fig. 46.20d, [46.136]).

Cantilevered dual nanotubes have been demonstrated as nanotweezers [46.141] and nanoscissors (Fig. 46.20e, [46.91]) by manual and nanorobotic assembly, respectively.

Based on electric resistance change under different temperatures, nanotube thermal probes (Fig. 46.20f) have been demonstrated for measuring the temperature at precise locations. These thermal probes are more advantageous than nanotube-based thermometers because the thermometers require TEM imaging. The probes also have better reproducibility than devices based on dielectrophoretically assembled bulk nanotubes [46.142]. Gas sensors and hot-wire-based mass-flow sensors can also be constructed in this configuration rather than a bridged one.

The integration of these devices can be realized using the configurations shown in Figs. 46.20g [46.143] and 46.20h [46.90]. Arrays of individual nanotubes can also be used to fabricate nanosensors, such as position encoders [46.144].

Nanotube-based NEMS remains a rich research field with a large number of open problems. New materials and effects at the nanoscale will enable a new

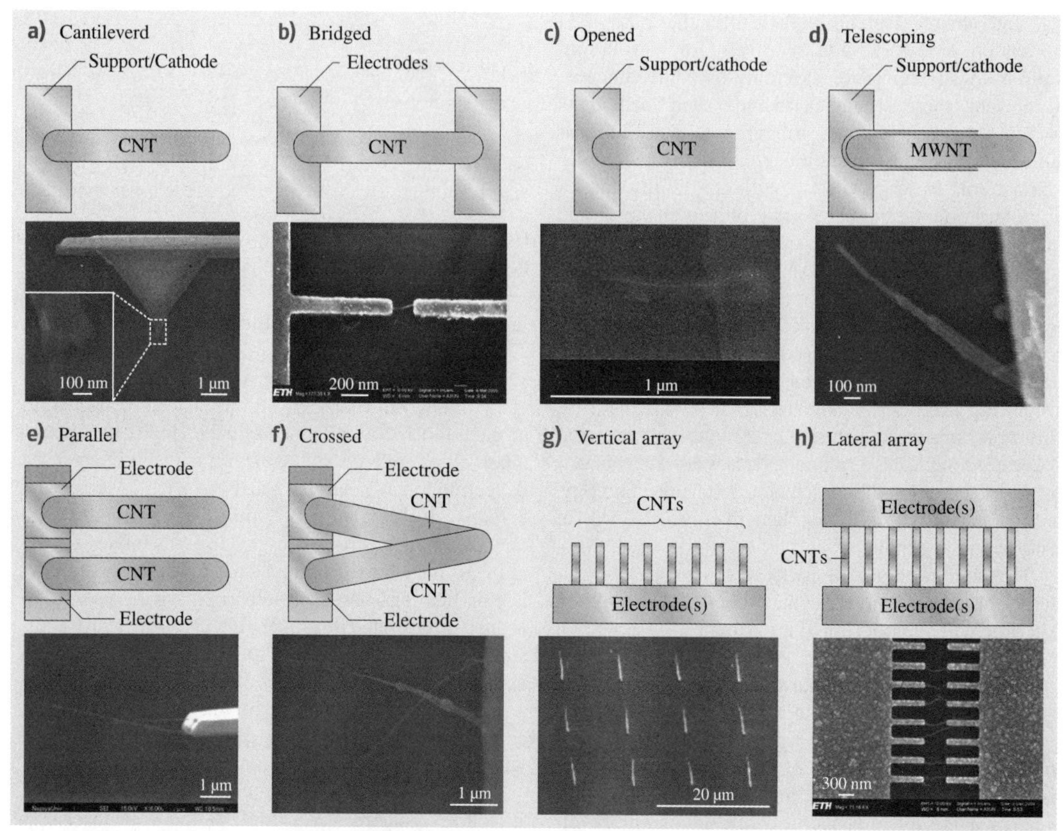

Fig. 46.20a–h Configurations of individual nanotube-based NEMS. Scale bars: (**a**) 1 μm (*inset*: 100 nm), (**b**) 200 nm, (**c**) 1 μm, (**d**) 100 nm, (**e**) and (**f**) 1 μm, (**g**) 20 μm, and (**h**) 300 nm. All examples are from the authors' work

family of sensors and actuators for the detection and actuation of ultrasmall quantities or objects with ultrahigh precision and frequencies. Through random spreading, direct growth, and nanorobotic manipulation, prototypes have been demonstrated. However, for integration into NEMS, self-assembly processes will become increasingly important. Among them, we believe that dielectrophoretic nanoassembly will play a significant role for large-scale production of 2-D regular structures.

References

46.1 R.P. Feynman: There's plenty of room at the bottom, Caltech Eng. Sci. **23**, 22–36 (1960)

46.2 M.C. Roco, R.S. Williams, P. Alivisatos: Nanotechnology research directions, Interag. Work. Group Nanosci. Eng. Technol. (IWGN) Workshop Rep. (Kluwer, Dordrecht 2000)

46.3 Committee on Technology, M.L. Downey, D.T. Moore, G.R. Bachula, D.M. Etter, E.F. Carey, L.A. Perine: *National Nanotechnology Initiative: Leading to the Next Industrial Revolution, A Report by the Interagency Working Group on Nanoscience, Engineering and Technology* (National Science and Technology Council, Washington 2000)

46.4 K. Drexler: *Nanosystems: Molecular Machinery, Manufacturing and Computation* (Wiley Interscience, New York 1992)

46.5 G. Binnig, H. Rohrer, C. Gerber, E. Weibel: Surface studies by scanning tunneling microscopy, Phys. Rev. Lett. **49**, 57–61 (1982)

46.6 W.F. Degrado: Design of peptides and proteins, Adv. Protein Chem. **39**, 51–124 (1998)

46.7 J.-M. Lehn: *Supramolecular Chemistry: Concepts and Perspectives* (VCH, Weinheim 1995)

46.8 G.M. Whitesides, B. Grzybowski: Self-assembly at all scales, Science **295**, 2418–2421 (2002)

46.9 M. Fujita, N. Fujita, K. Ogura, K. Yamaguchi: Spontaneous assembling of ten small components into a three-dimensionally interlocked compound consisting of the same two cage frameworks, Nature **400**, 52–55 (1999)

46.10 T. Ebefors, G. Stemme: Microrobotics. In: *The MEMS Handbook*, ed. by M. Gad-el-Hak (CRC, Boca Raton 2002)

46.11 C.-J. Kim, A.P. Pisano, R.S. Muller: Silicon-processed overhanging microgripper, IEEE/ASME J. MEMS **1**, 31–36 (1992)

46.12 C. Liu, T. Tsao, Y.-C. Tai, C.-M. Ho: Surface micromachined magnetic actuators, Proc. 7th IEEE Int. Conf. Micro Electro Mech. Syst., Oiso (IEEE, Piscataway 1994) pp. 57–62

46.13 J. Judy, D.L. Polla, W.P. Robbins: A linear piezoelectric stepper motor with submicron displacement and centimeter travel, IEEE Trans. Ultrason. Ferroelectr. Freq. Control **37**, 428–437 (1990)

46.14 K. Nakamura, H. Ogura, S. Maeda, U. Sangawa, S. Aoki, T. Sato: Evaluation of the micro wobbler motor fabricated by concentric buildup process, Proc. 8th IEEE Int. Conf. Micro Electro Mech. Syst., Amsterdam (IEEE, Piscataway 1995) pp. 374–379

46.15 A. Teshigahara, M. Watanabe, N. Kawahara, I. Ohtsuka, T. Hattori: Performance of a 7-mm microfabricated car, IEEE/ASME J. MEMS **4**, 76–80 (1995)

46.16 K.R. Udayakumar, S.F. Bart, A.M. Flynn, J. Chen, L.S. Tavrow, L.E. Cross, R.A. Brooks, D.J. Ehrlich: Ferroelectric thin-film ultrasonic micromotors, Proc. 4th IEEE Int. Conf. Micro Electro Mech. Syst., Nara (IEEE, Piscataway 1991) pp. 109–113

46.17 W. Trimmer, R. Jebens: Actuators for micro robots, Proc. 1989 IEEE Int. Conf. Robot. Autom., Scottsdale (IEEE, Piscataway 1989) pp. 1547–1552

46.18 P. Dario, R. Valleggi, M.C. Carrozza, M.C. Montesi, M. Cocco: Review – Microactuators for microrobots: A critical survey, J. Micromech. Microeng. **2**, 141–157 (1992)

46.19 I. Shimoyama: Scaling in microrobots, Proc. IEEE/RSJ Intell. Robot. Syst., Pittsburgh (IEEE, Piscataway 1995) pp. 208–211

46.20 R.S. Fearing: Powering 3-dimensional microrobots: power density limitations. In: *Tutorial on Micro Mechatronics and Micro Robotics, Proc. IEEE Int. Conf. Robot. Autom, Leuven*, ed. by G. Girait, P. Dario (IEEE, Piscataway 1998)

46.21 R.G. Gilbertson, J.D. Busch: A survey of microactuator technologies for future spacecraft missions, J. Br. Interplanet. Soc. **49**, 129–138 (1996)

46.22 M. Mehregany, P. Nagarkar, S.D. Senturia, J.H. Lang: Operation of microfabricated harmonic and ordinary side-drive motors, Proc. 3th IEEE Int. Conf. Micro Electro Mech. Syst., Napa Valley (IEEE, Piscataway 1990) pp. 1–8

46.23 Y.C. Tai, L.S. Fan, R.S. Muller: IC-processed micromotors: design, technology, and testing, Proc. 2nd IEEE Int. Conf. Micro Electro Mech. Syst., Salt Lake City (IEEE, Piscataway 1989) pp. 1–6

46.24 T. Ohnstein, T. Fukiura, J. Ridley, U. Bonne: Micromachined silicon microvalve, Proc. 3rd IEEE Int. Conf. Micro Electro Mech. Syst., Napa Valley (IEEE, Piscataway 1990) pp. 95–99

46.25 L.Y. Chen, S.L. Zhang, J.J. Yao, D.C. Thomas, N.C. MacDonald: Selective chemical vapor deposition of tungsten for microdynamic structures, Proc. 2nd IEEE Int. Conf. Micro Electro Mech. Syst., Salt Lake City (IEEE, Piscataway 1989) pp. 82–87

46.26 K. Yanagisawa, H. Kuwano, A. Tago: An electromagnetically driven microvalve, Proc. 7th Int. Conf. Solid-State Sens. Actuators, Yokohama (IEEE, Piscataway 1993) pp. 102–105

46.27 S. Brand: New applications of piezo-electric actuators, Proc. 3rd Int. Conf. New Actuators (Messe Bremen GmbH, Bremen 1992) p. 59

46.28 M. Esashi, S. Shoji, A. Nakano: Normally close microvalve and micropump fabricated on a silicon wafer, Proc. 4th IEEE Int. Conf. Micro Electro Mech. Syst., Salt Lake City (IEEE, Piscataway 1989) pp. 29–34

46.29 R. Petrucci, K. Simmons: An introduction to piezoelectric crystals, Sens. Mag. **May**, 26–31 (1994)

46.30 G. Binnig, C.F. Quate, C. Gerber: Atomic force microscope, Phys. Rev. Lett. **56**, 93–96 (1986)

46.31 A. Ashkin, J.M. Dziedzic: Optical trapping and manipulation of viruses and bacteria, Science **235**, 1517–1520 (1987)

46.32 F.H.C. Crick, A.F.W. Hughes: The physical properties of cytoplasm: A study by means of the magnetic particle method, Part I: Experimental, Exp. Cell Res. **1**, 37–80 (1950)

46.33 M.F. Yu, M.J. Dyer, G.D. Skidmore, H.W. Rohrs, X.K. Lu, K.D. Ausman, J.R. Von Ehr, R.S. Ruoff: Three-dimensional manipulation of carbon nanotubes under a scanning electron microscope, Nanotechnology **10**, 244–252 (1999)

46.34 L.X. Dong, F. Arai, T. Fukuda: 3-D nanorobotic manipulation of nano-order objects inside SEM, Proc. 2000 Int. Symp. Micromechatron. Hum. Sci., Nagoya (IEEE, Piscataway 2000) pp. 151–156

46.35 D.M. Eigler, E.K. Schweizer: Positioning single atoms with a scanning tunneling microscope, Nature **344**, 524–526 (1990)

46.36 P. Avouris: Manipulation of matter at the atomic and molecular levels, Acc. Chem. Res. **28**, 95–102 (1995)

46.37 M.F. Crommie, C.P. Lutz, D.M. Eigler: Confinement of electrons to quantum corrals on a metal surface, Science **262**, 218–220 (1993)

46.38 L.J. Whitman, J.A. Stroscio, R.A. Dragoset, R.J. Celota: Manipulation of adsorbed atoms and creation

of new structures on room-temperature surfaces with a scanning tunneling microscope, Science **251**, 1206–1210 (1991)
46.39 I.-W. Lyo, P. Avouris: Field-induced nanometer-scale to atomic-scale manipulation of silicon surfaces with the STM, Science **253**, 173–176 (1991)
46.40 G. Dujardin, R.E. Walkup, P. Avouris: Dissociation of individual molecules with electrons from the tip of a scanning tunneling microscope, Science **255**, 1232–1235 (1992)
46.41 T.-C. Shen, C. Wang, G.C. Abeln, J.R. Tucker, J.W. Lyding, P. Avouris, R.E. Walkup: Atomic-scale desorption through electronic and vibrational-excitation mechanisms, Science **268**, 1590–1592 (1995)
46.42 M.T. Cuberes, R.R. Schittler, J.K. Gimzewski: Room-temperature repositioning of individual C_{60} molecules at Cu steps: Operation of a molecular counting device, Appl. Phys. Lett. **69**, 3016–3018 (1996)
46.43 H.J. Lee, W. Ho: Single-bond formation and characterization with a scanning tunneling microscope, Science **286**, 1719–1722 (1999)
46.44 T. Yamamoto, O. Kurosawa, H. Kabata, N. Shimamoto, M. Washizu: Molecular surgery of DNA based on electrostatic micromanipulation, IEEE Trans. IA **36**, 1010–1017 (2000)
46.45 C. Haber, D. Wirtz: Magnetic tweezers for DNA micromanipulation, Rev. Sci. Instrum. **71**, 4561–4570 (2000)
46.46 D.M. Schäfer, R. Reifenberger, A. Patil, R.P. Andres: Fabrication of two-dimensional arrays of nanometer-size clusters with the atomic force microscope, Appl. Phys. Lett. **66**, 1012–1014 (1995)
46.47 T. Junno, K. Deppert, L. Montelius, L. Samuelson: Controlled manipulation of nanoparticles with an atomic force microscope, Appl. Phys. Lett. **66**, 3627–3629 (1995)
46.48 P.E. Sheehan, C.M. Lieber: Nanomachining, manipulation and fabrication by force microscopy, Nanotechnology **7**, 236–240 (1996)
46.49 C. Baur, B.C. Gazen, B. Koel, T.R. Ramachandran, A.A.G. Requicha, L. Zini: Robotic nanomanipulation with a scanning probe microscope in a networked computing environment, J. Vac. Sci. Technol. B **15**, 1577–1580 (1997)
46.50 R. Resch, C. Baur, A. Bugacov, B.E. Koel, A. Madhukar, A.A.G. Requicha, P. Will: Building and manipulating 3-D and linked 2-D structures of nanoparticles using scanning force microscopy, Langmuir **14**, 6613–6616 (1998)
46.51 J. Hu, Z.-H. Zhang, Z.-Q. Ouyang, S.-F. Chen, M.-Q. Li, F.-J. Yang: Stretch and align virus in nanometer scale on an atomically flat surface, J. Vac. Sci. Technol. B **16**, 2841–2843 (1998)
46.52 M. Sitti, S. Horiguchi, H. Hashimoto: Controlled pushing of nanoparticles: Modeling and experiments, IEEE/ASME Trans. Mechatron. **5**, 199–211 (2000)
46.53 M. Guthold, M.R. Falvo, W.G. Matthews, S. Paulson, S. Washburn, D.A. Erie, R. Superfine, F.P. Brooks Jr., R.M. Taylor II: Controlled manipulation of molecular samples with the nanoManipulator, IEEE/ASME Trans. Mechatron. **5**, 189–198 (2000)
46.54 F. Arai, D. Andou, T. Fukuda: Micro manipulation based on micro physics – strategy based on attractive force reduction and stress measurement, Proc. IEEE/RSJ Int. Conf. Intell. Robot. Syst., Pittsburgh (IEEE, Piscataway 1995) pp. 236–241
46.55 L.X. Dong, F. Arai, T. Fukuda: 3-D nanorobotic manipulations of nanometer scale objects, J. Robot. Mechatron. **13**, 146–153 (2001)
46.56 H.W.P. Koops, J. Kretz, M. Rudolph, M. Weber, G. Dahm, K.L. Lee: Characterization and application of materials grown by electron-beam-induced deposition, Jpn. J. Appl. Phys. **33**, 7099–7107 (1994)
46.57 S. Iijima: Helical microtubules of graphitic carbon, Nature **354**, 56–58 (1991)
46.58 S.J. Tans, A.R.M. Verchueren, C. Dekker: Room-temperature transistor based on a single carbon nanotube, Nature **393**, 49–52 (1998)
46.59 Y. Huang, X.F. Duan, Y. Cui, L.J. Lauhon, K.-H. Kim, C.M. Lieber: Logic gates and computation from assembled nanowire building blocks, Science **294**, 1313–1317 (2001)
46.60 A. Bachtold, P. Hadley, T. Nakanishi, C. Dekker: Logic circuits with carbon nanotube transistors, Science **294**, 1317–1320 (2001)
46.61 R.H. Baughman, A.A. Zakhidov, W.A. de Heer: Carbon nanotubes – the route toward applications, Science **297**, 787–792 (2002)
46.62 N. Wang, Z.K. Tang, G.D. Li, J.S. Chen: Single-walled 4 Å carbon nanotube arrays, Nature **408**, 50–51 (2000)
46.63 Z.W. Pan, S.S. Xie, B.H. Chang, C.Y. Wang, L. Lu, W. Liu, W.Y. Zhou, W.Z. Li, L.X. Qian: Very long carbon nanotubes, Nature **394**, 631–632 (1998)
46.64 H.W. Zhu, C.L. Xu, D.H. Wu, B.Q. Wei, R. Vajtai, P.M. Ajayan: Direct synthesis of long single-walled carbon nanotube strands, Science **296**, 884–886 (2002)
46.65 M.J. Treacy, T.W. Ebbesen, J.M. Gibson: Exceptionally high Young's modulus observed for individual carbon nanotubes, Nature **381**, 678–680 (1996)
46.66 E.W. Wong, P.E. Sheehan, C.M. Lieber: Nanobeam mechanics: Elasticity, strength, and toughness of nanorods and nanotubes, Science **277**, 1971–1975 (1997)
46.67 P. Poncharal, Z.L. Wang, D. Ugarte, W.A. de Heer: Electrostatic deflections and electromechanical resonances of carbon nanotubes, Science **283**, 1513–1516 (1999)
46.68 M.F. Yu, O. Lourie, M.J. Dyer, K. Moloni, T.F. Kelley, R.S. Ruoff: Strength and breaking mechanism of

multiwalled carbon nanotubes under tensile load, Science **287**, 637–640 (2000)
46.69 A. Krishnan, E. Dujardin, T.W. Ebbesen, P.N. Yianilos, M.M.J. Treacy: Young's modulus of single-walled nanotubes, Phys. Rev. B **58**, 14-013–14-019 (1998)
46.70 J.P. Salvetat, G.A.D. Briggs, J.-M. Bonard, R.R. Bacsa, A.J. Kulik, T. Stockli, N.A. Burnham, L. Forro: Elastic and shear moduli of single-walled carbon nanotube ropes, Phys. Rev. Lett. **82**, 944–947 (1999)
46.71 M.F. Yu, B.S. Files, S. Arepalli, R.S. Ruoff: Tensile loading of ropes of single wall carbon nanotubes and their mechanical properties, Phys. Rev. Lett. **84**, 5552–5555 (2000)
46.72 D.A. Walters, L.M. Ericson, M.J. Casavant, J. Liu, D.T. Colbert, K.A. Smith, R.E. Smalley: Elastic strain of freely suspended single-wall carbon nanotube ropes, Appl. Phys. Lett. **74**, 3803–3805 (1999)
46.73 R. Saito, M. Fujita, G. Dresselhaus, M.S. Dresselhaus: Electronic structure of graphene tubules based on C_{60}, Phys. Rev. B **46**, 1804–1811 (1992)
46.74 T.W. Ebbesen, H.J. Lezec, H. Hiura, J.W. Bennett, H.F. Ghaemi, T. Thio: Electrical conductivity of individual carbon nanotubes, Nature **382**, 54–56 (1996)
46.75 H.J. Dai, E.W. Wong, C.M. Lieber: Probing electrical transport in nanomaterials: conductivity of individual carbon nanotubes, Science **272**, 523–526 (1996)
46.76 P. Kim, L. Shi, A. Majumdar, P.L. McEuen: Thermal transport measurements of individual multiwalled nanotubes, Phys. Rev. Lett. **87**, 215502 (2001)
46.77 W.J. Liang, M. Bockrath, D. Bozovic, J.H. Hafner, M. Tinkham, H. Park: Fabry–Perot interference in a nanotube electron waveguide, Nature **411**, 665–669 (2001)
46.78 S. Frank, P. Poncharal, Z.L. Wang, W.A. de Heer: Carbon nanotube quantum resistors, Science **280**, 1744–1746 (1998)
46.79 X.B. Zhang, D. Bernaerts, G. Van Tendeloo, S. Amelincks, J. Van Landuyt, V. Ivanov, J.B. Nagy, P. Lambin, A.A. Lucas: The texture of catalytically grown coil-shaped carbon nanotubules, Europhys. Lett. **27**, 141–146 (1994)
46.80 X.Y. Kong, Z.L. Wang: Spontaneous polarization-induced nanohelixes, nanosprings, and nanorings of piezoelectric nanobelts, Nano Lett. **3**, 1625–1631 (2003)
46.81 S.V. Golod, V.Y. Prinz, V.I. Mashanov, A.K. Gutakovsky: Fabrication of conducting GeSi/Si micro- and nanotubes and helical microcoils, Semicond. Sci. Technol. **16**, 181–185 (2001)
46.82 L. Zhang, E. Deckhardt, A. Weber, C. Schönenberger, D. Grützmacher: Controllable fabrication of SiGe/Si and SiGe/Si/Cr helical nanobelts, Nanotechnology **16**, 655–663 (2005)
46.83 D.J. Bell, L.X. Dong, Y. Sun, L. Zhang, B.J. Nelson, D. Grützmacher: Manipulation of nanocoils for nanoelectromagnets, Proc. 5th IEEE Conf. Nanotechnol., Nagoya (IEEE, Piscataway 2005) pp. 149–152
46.84 D.J. Bell, Y. Sun, L. Zhang, L.X. Dong, B.J. Nelson, D. Grützmacher: Three-dimensional nanosprings for electromechanical sensors, Proc. 13th Int. Conf. Solid-State Sens. Actuators Microsyst., Seoul (IEEE, Piscataway 2005) pp. 15–18
46.85 S. Iijima, T. Ichihashi: Single-shell carbon nanotubes of 1-nm diameter, Nature **363**, 603–605 (1993)
46.86 D.S. Bethune, C.H. Kiang, M.S. de Vries, G. Gorman, R. Savoy, J. Vazquez, R. Beyers: Cobalt-catalysed growth of carbon nanotubes with single-atomic-layer walls, Nature **363**, 605–607 (1993)
46.87 R. Martel, T. Schmidt, H.R. Shea, T. Hertel, P. Avouris: Single- and multi-wall carbon nanotube field-effect transistors, Appl. Phys. Lett. **73**, 2447–2449 (1998)
46.88 N.R. Franklin, Y.M. Li, R.J. Chen, A. Javey, H.J. Dai: Patterned growth of single-walled carbon nanotubes on full 4-inch wafers, Appl. Phys. Lett. **79**, 4571–4573 (2001)
46.89 T. Rueckes, K. Kim, E. Joselevich, G.Y. Tseng, C.-L. Cheung, C.M. Lieber: Carbon nanotube-based non-volatile random access memory for molecular computing, Science **289**, 94–97 (2000)
46.90 A. Subramanian, B. Vikramaditya, L.X. Dong, D. Bell, B.J. Nelson: Micro and nanorobotic assembly using dielectrophoresis. In: *Robotics: Science and Systems I*, ed. by S. Thrun, G.S. Sukhatme, S. Schaal, O. Brock (MIT Press, Cambridge 2005) pp. 327–334
46.91 T. Fukuda, F. Arai, L.X. Dong: Assembly of nanodevices with carbon nanotubes through nanorobotic manipulations, Proc. IEEE **91**, 1803–1818 (2003)
46.92 M.R. Falvo, G.J. Clary, R.M. Taylor, V. Chi, F.P. Brooks, S. Washburn, R. Superfine: Bending and buckling of carbon nanotubes under large strain, Nature **389**, 582–584 (1997)
46.93 H.W.C. Postma, A. Sellmeijer, C. Dekker: Manipulation and imaging of individual single-walled carbon nanotubes with an atomic force microscope, Adv. Mater. **12**, 1299–1302 (2000)
46.94 H.W.C. Postma, M. de Jonge, Z. Yao, C. Dekker: Electrical transport through carbon nanotube junctions created by mechanical manipulation, Phys. Rev. B **62**, R10653–R10656 (2000)
46.95 T. Hertel, R. Martel, P. Avouris: Manipulation of individual carbon nanotubes and their interaction with surfaces, J. Phys. Chem. B **102**, 910–915 (1998)
46.96 P. Avouris, T. Hertel, R. Martel, T. Schmidt, H.R. Shea, R.E. Walkup: Carbon nanotubes: nanomechanics, manipulation, and electronic devices, Appl. Surf. Sci. **141**, 201–209 (1999)
46.97 L. Roschier, J. Penttila, M. Martin, P. Hakonen, M. Paalanen, U. Tapper, E.I. Kauppinen, C. Journet, P. Bernier: Single-electron transistor made

of multiwalled carbon nanotube using scanning probe manipulation, Appl. Phys. Lett. **75**, 728–730 (1999)
46.98 M. Ahlskog, R. Tarkiainen, L. Roschier, P. Hakonen: Single-electron transistor made of two crossing multiwalled carbon nanotubes and its noise properties, Appl. Phys. Lett. **77**, 4037–4039 (2000)
46.99 M.R. Falvo, R.M. Taylor II, A. Helser, V. Chi, F.P. Brooks Jr, S. Washburn, R. Superfine: Nanometre-scale rolling and sliding of carbon nanotubes, Nature **397**, 236–238 (1999)
46.100 M.R. Falvo, J. Steele, R.M. Taylor II, R. Superfine: Gearlike rolling motion mediated by commensurate contact: Carbon nanotubes on HOPG, Phys. Rev. B **62**, R10665–R10667 (2000)
46.101 L.X. Dong: Nanorobotic manipulations of carbon nanotubes. Ph.D. Thesis (Nagoya University, Nagoya 2003)
46.102 J.H. Hafner, C.-L. Cheung, T.H. Oosterkamp, C.M. Lieber: High-yield assembly of individual single-walled carbon nanotube tips for scanning probe microscopies, J. Phys. Chem. B **105**, 743–746 (2001)
46.103 L.X. Dong, F. Arai, T. Fukuda: Electron-beam-induced deposition with carbon nanotube emitters, Appl. Phys. Lett. **81**, 1919–1921 (2002)
46.104 L.X. Dong, F. Arai, T. Fukuda: 3-D nanorobotic manipulations of multi-walled carbon nanotubes, Proc. 2001 IEEE Int. Conf. Robot. Autom. (ICRA2001), Seoul (IEEE, Piscataway 2001) pp. 632–637
46.105 L.X. Dong, F. Arai, T. Fukuda: Three-dimensional nanorobotic manipulations of carbon nanotubes, J. Robot. Mechatron. JSME **14**, 245–252 (2002)
46.106 L.X. Dong, F. Arai, T. Fukuda: Inter-process measurement of MWNT rigidity and fabrication of MWNT junctions through nanorobotic manipulations, Conf. Proc. Nanonetw. Mater. Fuller. Nanotub. Relat. Mater., Vol. 590 (AIP, Melville 2001) pp. 71–74
46.107 L.X. Dong, F. Arai, T. Fukuda: Shape modification of carbon nanotubes and its applications in nanotube scissors, Proc. IEEE Int. Conf. Nanotechnol. (IEEE-NANO2002), Washington (IEEE, Piscataway 2002) pp. 443–446
46.108 L.X. Dong, F. Arai, T. Fukuda: Destructive constructions of nanostructures with carbon nanotubes through nanorobotic manipulations, IEEE/ASME Trans. Mechatron. **9**, 350–357 (2004)
46.109 J. Cumings, P.G. Collins, A. Zettl: Peeling and sharpening multiwall nanotubes, Nature **406**, 58 (2000)
46.110 J. Cumings, A. Zettl: Low-friction nanoscale linear bearing realized from multiwall carbon nanotubes, Science **289**, 602–604 (2000)
46.111 L.X. Dong, F. Arai, T. Fukuda: 3-D nanoassembly of carbon nanotubes through nanorobotic manipulations, Proc. 2002 IEEE Int. Conf. Robot. Autom. (ICRA2002), Washington (IEEE, Piscataway 2002) pp. 1477–1482
46.112 L.X. Dong, F. Arai, T. Fukuda: Mechanochemical nanorobotic manipulations of carbon nanotubes, Jpn. J. Appl. Phys. **42**, 295–298 (2003)
46.113 R. Saito, G. Dresselhaus, M.S. Dresselhaus: Tunneling conductance of connected carbon nanotubes, Phys. Rev. B **53**, 2044–2050 (1996)
46.114 L. Chico, V.H. Crespi, L.X. Benedict, S.G. Louie, M.L. Cohen: Pure carbon nanoscale devices: nanotube heterojunctions, Phys. Rev. Lett. **76**, 971–974 (1996)
46.115 M. Menon, D. Srivastava: Carbon nanotube 'T junctions': nanoscale metal-semiconductor-metal contact devices, Phys. Rev. Lett. **79**, 4453–4456 (1997)
46.116 Z. Yao, H.W.C. Postma, L. Balents, C. Dekker: Carbon nanotube intramolecular junctions, Nature **402**, 273–276 (1999)
46.117 H.W.C. Postma, T. Teepen, Z. Yao, M. Grifoni, C. Dekker: Carbon nanotube single-electron transistors at room temperature, Science **293**, 76–79 (2001)
46.118 M.S. Fuhrer, L. Shih, M. Forero, Y.-G. Yoon, M.S.C. Mazzoni, H.J. Choi, J. Ihm, S.G. Louie, A. Zettl, P.L. McEuen: Crossed nanotube junctions, Science **288**, 494–497 (2000)
46.119 T. Rueckes, K. Kim, E. Joselevich, G.Y. Treng, C.L. Cheung, C.M. Lieber: Carbon nanotube-based nonvolatile random access memory for molecular computing science, Science **289**, 94–97 (2000)
46.120 A.G. Rinzler, J.H. Hafner, P. Nikolaev, L. Lou, S.G. Kim, D. Tománek, P. Nordlander, D.T. Colbert, R.E. Smalley: Unraveling nanotubes: field emission from an atomic wire, Science **269**, 1550–1553 (1995)
46.121 Y. Saito, K. Hamaguchi, K. Hata, K. Uchida, Y. Tasaka, F. Ikazaki, M. Yumura, A. Kasaya, Y. Nishina: Conical beams from open nanotubes, Nature **389**, 554 (1997)
46.122 S.C. Minne, G. Yaralioglu, S.R. Manalis, J.D. Adams, J. Zesch, A. Atalar, C.F. Quate: Automated parallel high-speed atomic force microscopy, Appl. Phys. Lett. **72**, 2340–2342 (1998)
46.123 G.D. Skidmore, E. Parker, M. Ellis, N. Sarkar, R. Merkle: Exponential assembly, Nanotechnology **11**, 316–321 (2001)
46.124 Y. Cui, Q.Q. Wei, H.K. Park, C.M. Lieber: Nanowire nanosensors for highly sensitive and selective detection of biological and chemical species, Science **293**, 1289–1292 (2001)
46.125 Y. Sun, B.J. Nelson: Microrobotic cell injection, Proc. 2001 IEEE Int. Conf. Robot. Autom. (ICRA2001), Seoul (IEEE, Piscataway 2001) pp. 620–625
46.126 Y. Sun, B.J. Nelson: Autonomous injection of biological cells using visual servoing, Int. Symp. Exp. Robot. (ISER) (Lecture Notes Control Inform. Sci., Hawaii 2000) pp. 175–184

46.127 Y. Sun, B.J. Nelson, D.P. Potasek, E. Enikov: A bulk microfabricated multi-axis capacitive cellular force sensor using transverse comb drives, J. Micromech. Microeng. **12**, 832–840 (2002)
46.128 Y. Sun, K. Wan, K.P. Roberts, J.C. Bischof, B.J. Nelson: Mechanical property characterization of mouse zona pellucida, IEEE Trans. Nanobiosci. **2**, 279–286 (2003)
46.129 H.J. Dai, J.H. Hafner, A.G. Rinzler, D.T. Colbert, R.E. Smalley: Nanotubes as nanoprobes in scanning probe microscopy, Nature **384**, 147–150 (1996)
46.130 J.H. Hafner, C.L. Cheung, C.M. Lieber: Growth of nanotubes for probe microscopy tips, Nature **398**, 761–762 (1999)
46.131 H. Nishijima, S. Kamo, S. Akita, Y. Nakayama, K.I. Hohmura, S.H. Yoshimura, K. Takeyasu: Carbon-nanotube tips for scanning probe microscopy: preparation by a controlled process and observation of deoxyribonucleic acid, Appl. Phys. Lett. **74**, 4061–4063 (1999)
46.132 L.X. Dong, F. Arai, M. Nakajima, P. Liu, T. Fukuda: Nanotube devices fabricated in a nano laboratory, Proc. 2003 IEEE Int. Conf. Robot. Autom. (ICRA2003), Taipei (IEEE, Piscataway 2003) pp. 3624–3629
46.133 M. Nakajima, F. Arai, L.X. Dong, T. Fukuda: Pico-Newton order force measurement using a calibrated carbon nanotube probe inside a scanning electron microscope, J. Robot. Mechatron. JSME **16**(2), 155–162 (2004)
46.134 A. Subramanian, B.J. Nelson, L.X. Dong, D. Bell: Dielectrophoretic nanoassembly of nanotube-based NEMS with nanoelectrodes, 6th IEEE Int. Symp. Assem. Task Plan., Montréal (IEEE, Piscataway 2005) pp. 200–205
46.135 J. Kong, N.R. Franklin, C.W. Zhou, M.G. Chapline, S. Peng, K.J. Cho, H.J. Dai: Nanotube molecular wires as chemical sensors, Science **287**, 622–625 (2000)
46.136 L.X. Dong, B.J. Nelson, T. Fukuda, F. Arai: Towards linear nano servomotors with integrated position sensing, Proc. 2005 IEEE Int. Conf. Robot. Autom., Barcelona (IEEE, Piscataway 2005) pp. 867–872
46.137 Y.H. Gao, Y. Bando: Carbon nanothermometer containing gallium, Nature **415**, 599 (2002)
46.138 R.H. Baughman, C.X. Cui, A.A. Zakhidov, Z. Iqbal, J.N. Barisci, G.M. Spinks, G.G. Wallace, A. Mazzoldi, D. De Rossi, A.G. Rinzler, O. Jaschinski, S. Roth, M. Kertesz: Carbon nanotube actuators, Science **284**, 1340–1344 (1999)
46.139 S.W. Lee, D.S. Lee, R.E. Morjan, S.H. Jhang, M. Sveningsson, O.A. Nerushev, Y.W. Park, E.E.B. Campbell: A three-terminal carbon nanorelay, Nano Lett. **4**, 2027–2030 (2004)
46.140 A.M. Fennimore, T.D. Yuzvinsky, W.-Q. Han, M.S. Fuhrer, J. Cumings, A. Zettl: Rotational actuators based on carbon nanotubes, Nature **424**, 408–410 (2003)
46.141 P. Kim, C.M. Lieber: Nanotube nanotweezers, Science **286**, 2148–2150 (1999)
46.142 C.K.M. Fung, V.T.S. Wong, R.H.M. Chan, W.J. Li: Dielectrophoretic batch fabrication of bundled carbon nanotube thermal sensors, IEEE Trans. Nanotech. **3**, 395–403 (2004)
46.143 A. Subramanian, L.X. Dong, B.J. Nelson: Selective eradication of individual nanotubes from vertically aligned arrays, Proc. 2005 IEEE/ASME Int. Conf. Adv. Intell. Mechatron., Monterey (IEEE, Piscataway 2005) pp. 105–110
46.144 L.X. Dong, A. Subramanian, B.J. Nelson: Nano encoders based on arrays of single nanotube emitter, Proc. 5th IEEE Conf. Nanotechnol., Nagoya (IEEE, Piscataway 2005) pp. 211–214

Micro-/N

Part H

Part H Micro-/Nanodevice Reliability

47 MEMS/NEMS and BioMEMS/BioNEMS: Materials, Devices, and Biomimetics
Bharat Bhushan, Columbus, USA

48 Friction and Wear in Micro- and Nanomachines
Maarten P. de Boer, Pittsburgh, USA
Alex D. Corwin, Niskayuna, USA
Frank W. DelRio, Gaithersburg, USA
W. Robert Ashurst, Auburn, USA

49 Failure Mechanisms in MEMS/NEMS Devices
W. Merlijn van Spengen, Leiden, The Netherlands
Robert Modliñski, Terrassa, Spain
Robert Puers, Leuven, Belgium
Anne Jourdain, Leuven, Belgium

50 Mechanical Properties of Micromachined Structures
Harold Kahn, Cleveland, USA

51 High-Volume Manufacturing and Field Stability of MEMS Products
Jack Martin, Foxborough, USA

52 Packaging and Reliability Issues in Micro-/Nanosystems
Yu-Chuan Su, Hsinchu, Taiwan
Jongbaeg Kim, Seoul, South Korea
Yu-Ting Cheng, Hsinchu, Taiwan, R.O.C.
Mu Chiao, Vancouver, Canada
Liwei Lin, Berkeley, USA

47. MEMS/NEMS and BioMEMS/BioNEMS: Materials, Devices, and Biomimetics

Bharat Bhushan

Micro-/nanoelectromechanical systems (MEMS/NEMS) need to be designed to perform expected functions in short durations, typically in the millisecond to picosecond range. The expected life of devices for high-speed contacts can vary from a few hundred thousand to many billions of cycles, e.g., over a hundred billion cycles for digital micromirror devices (DMDs), which puts serious requirements on materials. The surface-area-to-volume ratio in MEMS/NEMS is large, and in systems involving relative motion, surface forces such as adhesion, friction, and meniscus and viscous forces become very large compared with inertial and electromagnetic forces. There is a need for fundamental understanding of adhesion, friction/stiction, wear, lubrication, and the role of surface contamination and environment, all on the nanoscale. Most mechanical properties are known to be scale dependent, therefore the properties of nanoscale structures need to be measured. For bioMEMS/bioNEMS, adhesion between biological molecular layers and the substrate, and friction and wear of biological layers, can be important. Component-level studies are required to provide a better understanding of the tribological phenomena occurring in MEMS/NEMS. The emergence of the fields of nanotribology and nanomechanics, and atomic force microscopy (AFM)-based techniques, has provided researchers with a viable approach to address these problems. The emerging field of biomimetics holds promise for the development of biologically inspired nanomaterials and nanotechnology products. One example is the design of surfaces with roughness-induced superhydrophobicity, self-cleaning, and low adhesion based on the so-called lotus effect. This chapter presents an overview of nanoscale adhesion, friction, and wear studies of materials and lubrication for MEMS/NEMS and bioMEMS/bioNEMS, and component-level studies of stiction phenomena in MEMS/NEMS devices, as well as hierarchical nanostructured surfaces for superhydrophobicity, self-cleaning, and low adhesion.

47.1 MEMS/NEMS Basics ... 1664
47.1.1 Introduction to MEMS ... 1665
47.1.2 Introduction to NEMS ... 1667
47.1.3 Introduction to BioMEMS/BioNEMS ... 1668
47.1.4 Nanotribology and Nanomechanics Issues in MEMS/NEMS and BioMEMS/BioNEMS ... 1668

47.2 Nanotribology and Nanomechanics Studies of Silicon and Related Materials ... 1683
47.2.1 Virgin and Treated/Coated Silicon Samples ... 1684
47.2.2 Nanotribological and Nanomechanical Properties of Polysilicon Films and SiC Film ... 1689

47.3 Lubrication Studies for MEMS/NEMS ... 1691
47.3.1 Perfluoropolyether Lubricants ... 1691
47.3.2 Self-Assembled Monolayers (SAMs) 1694
47.3.3 Hard Diamond-Like Carbon (DLC) Coatings ... 1697

47.4 Nanotribological Studies of Biological Molecules on Silicon-Based and Polymer Surfaces and Submicron Particles for Therapeutics and Diagnostics ... 1698
47.4.1 Adhesion, Friction, and Wear of Biomolecules on Si-Based Surfaces 1698
47.4.2 Adhesion of Coated Polymer Surfaces ... 1705
47.4.3 Submicron Particles for Therapeutics and Diagnostics ... 1706

47.5 Surfaces with Roughness-Induced Superhydrophobicity, Self-Cleaning, and Low Adhesion ... 1708
47.5.1 Modeling of Contact Angle for a Liquid Droplet in Contact with a Rough Surface ... 1710

47.5.2 Fabrication and Characterization of Microstructures, Nanostructures, and Hierarchical Structures........... 1711
47.5.3 Summary 1717

47.6 **Component-Level Studies**...................... 1717
47.6.1 Surface Roughness Studies of Micromotor Components 1717
47.6.2 Adhesion Measurements of Microstructures 1719
47.6.3 Microtriboapparatus for Adhesion, Friction, and Wear of Microcomponents 1719
47.6.4 Static Friction Force (Stiction) Measurements in MEMS................ 1723
47.6.5 Mechanisms Associated with Observed Stiction Phenomena in Digital Micromirror Devices (DMD) and Nanomechanical Characterization.......................... 1725

47.7 **Conclusions**... 1728

47.A **Micro-Nanofabrication Techniques**......... 1729
47.A.1 Top-Down Techniques 1729
47.A.2 Bottom-Up Fabrication (Nanochemistry).......................... 1732

References 1733

47.1 MEMS/NEMS Basics

Microelectromechanical systems (MEMS) refer to microscopic devices that have a characteristic length of < 1 mm but > 100 nm and that combine electrical and mechanical components. Nanoelectromechanical systems (NEMS) refer to nanoscopic devices that have a characteristic length of < 100 nm and that combine electrical and mechanical components. In mesoscale devices, if the functional components are on the micro- or nanoscale, they may be referred to as MEMS or NEMS, respectively. These are referred to as intelligent miniaturized systems comprising sensing, processing, and/or actuating functions and combining electrical and mechanical components. The acronym MEMS originated in the USA. The term commonly used in Europe is microsystem technology (MST), and in Japan it is micromachines. Another term generally used is micro-nanodevices. The terms MEMS/NEMS are also now used in a broad sense to include electrical, mechanical, fluidic, optical, and/or biological functions. MEMS/NEMS for optical applications are referred to as micro-/nanooptoelectromechanical systems (MOEMS/NOEMS). MEMS/NEMS for electronic applications are referred to as radiofrequency MEMS/NEMS (RF-MEMS/RF-NEMS). MEMS/NEMS for biological applications are referred to as bioMEMS/bioNEMS.

To put the characteristic dimensions and weights of MEMS/NEMS and bioNEMS into perspective, see Fig. 47.1 and Table 47.1. NEMS and bioNEMS shown in the figure range in size from 2 to 300 nm, and the size of MEMS is 12 000 nm. For comparison, individual atoms are typically a fraction of a nanometer in diameter, deoxyribonucleic acid (DNA) molecules are ≈ 2.5 nm wide, biological cells are in the range of thousands of nm in diameter, and human hair is $\approx 75\,\mu$m in diameter. NEMS can be built with weight as low as 10^{-20} N with cross sections of about 10 nm, and a micromachined silicon structure can have a weight as low as 1 nN. For comparison, the weight of a drop of water is $\approx 10\,\mu$N and the weight of an eyelash is ≈ 100 nN.

Micro-nanofabrication techniques include top-down methods, in which one builds down from the large to the small, and bottom-up methods, in which one builds up from the small to the large. Top-down methods include micro-nanomachining methods and methods based on lithography as well as nonlithographic miniaturization, mostly for MEMS and fabrication of a few NEMS devices. In bottom-up methods, also referred to as nanochemistry, devices and systems are assembled from their elemental constituents for NEMS fabrication, much as nature uses proteins and other macromolecules to construct complex biological systems. The bottom-up approach has the potential to go far beyond the limits of top-down technology by producing nanoscale features through synthesis and subsequent assembly. Furthermore, the bottom-up approach offers the potential to produce structures with enhanced and/or completely new functions. It allows a combination of materials with distinct chemical composition, structure, and morphology. For a brief overview of fabrication techniques, see Appendix A.

MEMS/NEMS and bioMEMS/bioNEMS are expected to have a major impact on our lives, comparable to that of semiconductor technology, information technology, or cellular and molecular biology [47.1–4]. They are used in electromechanical, electronics,

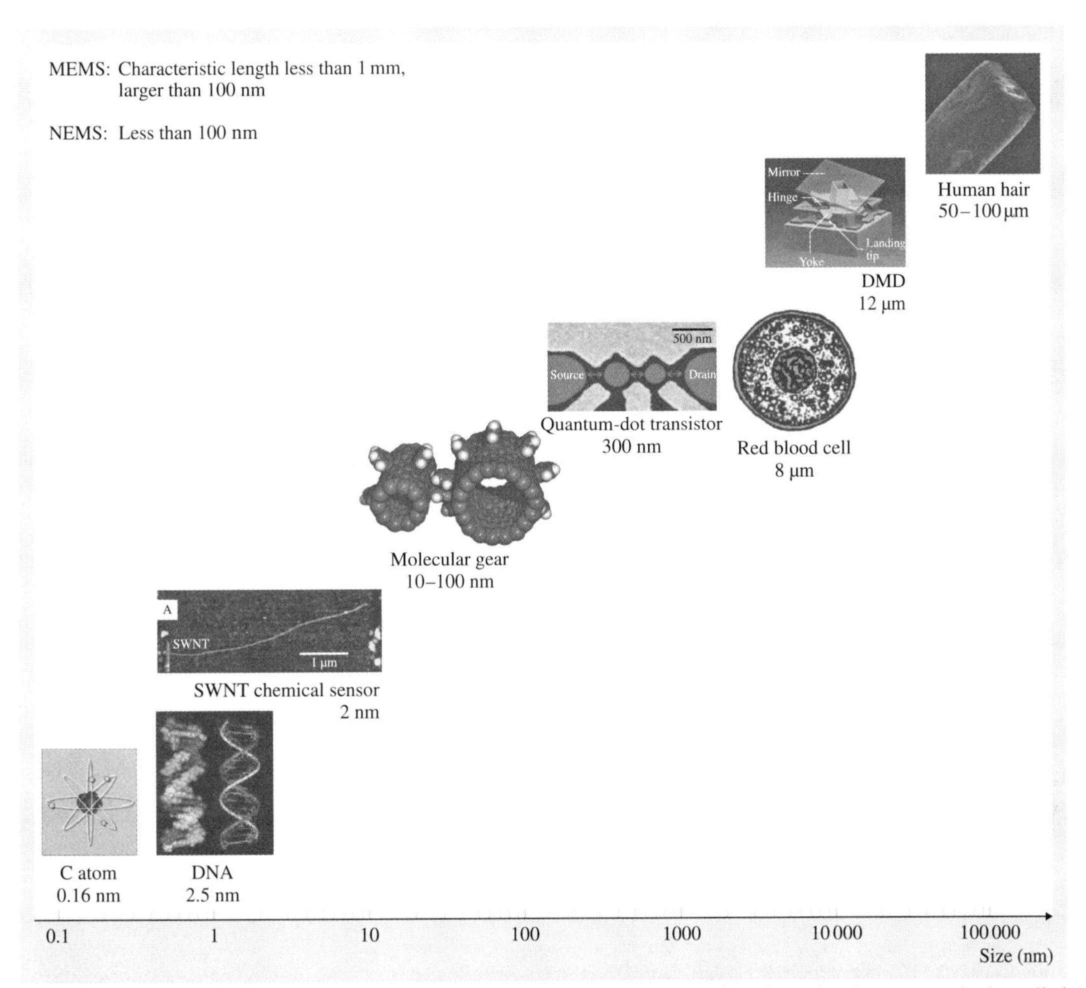

Fig. 47.1 Characteristic dimensions of MEMS/NEMS and bioNEMS in perspective. Examples shown are a single-walled carbon nanotube (SWNT) chemical sensor [47.7], molecular dynamic simulations of carbon-nanotube-based gears [47.8], quantum-dot transistor obtained from *van der Wiel* et al. [47.9], and DMD (DLP Texas Instruments). For comparison, dimensions and weights of various biological objects found in nature are also presented

information/communication, chemical, and biological applications. The MEMS industry in 2004 was worth $\approx$ US$ 4.5 billion, with a projected annual growth rate of 17% [47.5]. Growth of Si-based MEMS is slowing down, while nonsilicon MEMS are picking up. The NEMS industry was worth $\approx$ US$ 10 billion dollars in 2004, mostly in nanomaterials [47.6]. It is expected to expand in nanomaterials and biomedical applications as well as in nanoelectronics or molecular electronics. Due to the enabling nature of these systems and because of the significant impact they can have on both commercial and defense applications, industry as well as federal governments have taken special interest in seeing growth nurtured in this field. MEMS/NEMS and bioMEMS/bioNEMS are the next logical step in the *silicon revolution*.

47.1.1 Introduction to MEMS

The advances in silicon photolithographic process technology since the 1960s led to the development of MEMS in the early 1980s. More recently, lithographic

Table 47.1 Characteristic dimensions and weights in perspective

Characteristic dimensions in perspective	
NEMS characteristic length	< 100 nm
MEMS characteristic length	< 1 mm and > 100 nm
Single-walled carbon nanotube (SWNT) chemical sensor	≈ 2 nm
Molecular gear	≈ 10 nm
Quantum-dot transistor	300 nm
Digital micromirror	12 000 nm
Individual atoms	typically a fraction of a nm in diameter
DNA molecules	≈ 2.5 nm wide
Biological cells	in the range of thousands of nm in diameter
Human hair	≈ 75 000 nm in diameter
Weights in perspective	
NEMS built with cross-sections of ≈ 10 nm	as low as 10^{-20} N
Micromachines, silicon structure	as low as 1 nN
Eyelash	≈ 100 nN
Water droplet	≈ 10 μN

processes have also been developed to process nonsilicon materials. Lithographic processes are being complemented with nonlithographic processes for fabrication of components or devices made from plastics or ceramics. Using these fabrication processes, researchers have fabricated a wide variety of devices with dimensions in the submicron range to a few thousand microns (see e.g., [47.10–20]). MEMS for mechanical applications include acceleration, pressure, flow, and gas sensors, linear and rotary actuators, and other microstructures or microcomponents such as electric motors, gear chains, gas turbine engines, fluid pumps, fluid valves, switches, grippers, and tweezers. MEMS for chemical applications include chemical sensors and various analytical instruments. MOEMS devices include optical components, such as micromirror arrays for displays, infrared image sensors, spectrometers, barcode readers, and optical switches. RF-MEMS include inductors, capacitors, antennas, and RF switches. High-aspect-ratio MEMS (HARMEMS) have also been introduced.

A variety of MEMS devices have been produced and some are in commercial use [47.11, 13–16, 18–20]. A variety of sensors are used in industrial, consumer, defense, and biomedical applications. The largest "killer" industrial applications include accelerometers, pressure sensors, thermal and piezoelectric inkjet printheads, and digital micromirror devices. Integrated capacitive-type silicon accelerometers have been used in airbag deployment in automobiles since 1991 [47.21, 22]; some 90 million units were installed in vehicles in 2004. Accelerometer technology was over a billion-dollar-a-year industry in 2004, dominated by Analog Devices followed by Freescale Semiconductor (formerly Motorola) and Bosch. It is expected to grow with an annual growth exceeding 30%. Tri-axis accelerometers are needed to describe three-dimensional motion. Multi-axis accelerometers are being used for many other applications such as vehicle stability, rollover control, and gyro sensors for automotive applications, and various consumer applications including handheld devices, e.g., laptops for free-fall detection (2003), cellular phones (2004), and personal digital assistants (PDAs) for menu navigation, gaming, image rotation, and free-fall detection. Silicon-based piezoresistive pressure sensors were launched in 1990 by GE NovaSensor for manifold absolute pressure (MAP) sensing for engines and for disposable blood-pressure sensors; their annual sales were more than 30 million units and more than 25 million units, respectively, in 2004. MAP sensors measure the pressure in the intake manifold, which is fed to a computer that determines the optimum air–fuel mixture to maximize fuel economy. Most vehicles have these as part of the electronic engine control system. Capacitive pressure sensors for tire pressure measurements were launched by Freescale Semiconductor (formerly Motorola) in early 2000 and are also manufactured by Infineon/SensoNor and GE Novasensor (2003). Piezoresistive-type sensors are also used, manufactured by various companies such as EnTire Solutions (2003). The sensing module is located inside the rim of the wheel and relays the information via radiofrequency to a central processing unit (CPU) in order to display it to the driver. In 2005, ≈ 9.2 million vehicles were equipped with sensors, which translated to ≈ 37 million units. Their sales have grown rapidly, as they are now required in automobiles in the USA (starting in 2008), which affects 17 million vehicles (with one device in each tire) sold every year. Pressure sensors can be used to detect altitude by measuring air pressure. For example, global positioning systems (GPS) used for navigation have good lateral resolution but poor vertical resolution, which creates problems in identifying the level in the case of multiple roads stacked up. A pressure sensor is needed to determine the level (altitude) by measuring air pressure.

Thermal inkjet printers were developed independently by HP and Canon and commercialized in 1984 [47.23–26] and today are made by Canon, Epson, HP, Lexmark, Xerox, and others. They typically cost less initially than dry-toner laser printers and are the solution of choice for low-volume print runs. Annual sales of thermal inkjet printheads with microscale functional components were > 500 million units in 2004.

Micromirror arrays are used for displays. Commercial digital light processing (DLP) equipment, using digital micromirror devices (DMD), were launched in 1996 by Texas Instruments (TI) for digital projection displays in computer projectors, high-definition television (HDTV) sets, and movie projectors (DLP cinema) [47.27–29]. Several million projectors had been sold by 2004 ($\approx$ US$ 700 million revenue by TI in 2004). Electrostatically actuated, membrane-type or cantilever-type microswitches have been developed for direct-current (DC), RF, and optical applications [47.30]. There exists two basic forms of RF microswitches: the metal-to-metal contact microswitch (ohmic) and the capacitive microswitch. RF microswitches can be used in a variety of RF applications, including cellular phones, phase shifters, smart antennas, multiplexers for data acquisition, etc. [47.31]. Optical microswitches are finding applications in optical networking, telecommunications, and wireless technologies [47.30, 32].

Other applications of MEMS devices include chemical/biological and gas sensors [47.20, 33], microresonators, infrared detectors and focal-plane arrays for Earth observation, space science, and missile defense applications, picosatellites for space applications, fuel cells, and many hydraulic, pneumatic, and other consumer products. MEMS devices are also being pursued in magnetic storage systems [47.34], where they are being developed for supercompact and ultrahigh-recording-density magnetic disk drives. Several integrated head–suspension microdevices have been fabricated for contact recording applications [47.35]. High-bandwidth servo-controlled microactuators have been fabricated for ultrahigh-track-density applications, where they serve as the fine-position control element of a two-stage coarse–fine servo system, coupled with a conventional actuator [47.36, 37].

Micro-nanoinstruments and micro-nanomanipulators are used to move, position, probe, pattern, and characterize nanoscale objects and nanoscale features [47.38]. Miniaturized analytical equipments include gas chromatography and mass spectrometry. Other instruments include micro-scanning tunneling microscope (micro-STM).

In some cases, MEMS devices are used primarily for their miniature size, while in others, as in the case of airbags, because of their low-cost manufacturing techniques. This latter fact has been possible since semiconductor processing costs have reduced drastically over the last decade, allowing the use of MEMS in many fields.

47.1.2 Introduction to NEMS

NEMS are produced by nanomachining in a typical top-down approach (from large to small) and bottom-up approach (from small to large), largely relying on nanochemistry (see, e.g., [47.39–45]). The NEMS field, in addition to the fabrication of nanosystems, has provided impetus for the development of experimental and computation tools. Examples of NEMS include microcantilevers with integrated sharp nanotips for STM and atomic force microscopy (AFM) [47.46, 47], quantum corrals formed using STM by placing atoms one by one [47.48], AFM cantilever arrays (millipede) for data storage [47.49], STM and AFM tips for nanolithography, dip-pen nanolithography for printing molecules, nanowires, carbon nanotubes, quantum wires (QWRs), quantum boxes (QBs), quantum transistors [47.9], nanotube-based sensors [47.50, 51], biological (DNA) motors, molecular gears formed by attaching benzene molecules to the outer walls of carbon nanotubes [47.8], devices incorporating nm-thick films [e.g., in giant-magnetoresistive (GMR) read/write magnetic heads and magnetic media for magnetic rigid disk and magnetic tape drives], nanopatterned magnetic rigid disks, and nanoparticles (e.g., nanoparticles in magnetic tape substrates and nanomagnetic particles in magnetic tape coatings) [47.34,52]. More than 2 billion read/write magnetic heads were shipped for magnetic disk and tape drives in 2004.

Nanoelectronics can be used to build computer memory using individual molecules or nanotubes to store bits of information [47.53], molecular switches, molecular or nanotube transistors, nanotube flat-panel displays, nanotube integrated circuits, fast logic gates, switches, nanoscopic lasers, and nanotubes as electrodes in fuel cells.

47.1.3 Introduction to BioMEMS/BioNEMS

BioMEMS/bioNEMS are increasingly used in commercial and defense applications (see, e.g., [47.54–61]). They are used for chemical and biochemical analyses (biosensors) in medical diagnostics (e.g., DNA, RNA, proteins, cells, blood pressure and assays, and toxin identification) [47.61, 62], tissue engineering [47.63–65], and implantable pharmaceutical drug delivery [47.66–68]. Biosensors, also referred to as biochips, deal with liquids and gases. There are two types of biosensors. A large variety of biosensors are based on micro-nanofluidics [47.61, 69–71]. Micro-nanofluidic devices offer the ability to work with smaller reagent volumes and shorter reaction times, and perform analyses of multiple types at once. The second type of biosensors includes micro-nanoarrays which perform one type of analysis thousands of times [47.72–75].

A chip, called lab-on-a-CD, with micro-nanofluidic technology embedded on the disk can test thousands of biological samples rapidly and automatically [47.69]. An entire laboratory can be integrated onto a single chip, called a lab-on-a-chip [47.61, 70, 71]. Silicon-based disposable blood-pressure sensor chips were introduced in early 1990s by GE NovaSensor for blood-pressure monitoring ($\approx$ 25 million units in 2004). A blood-sugar monitor, referred to as GlucoWatch, was introduced in 2002. It automatically checks blood sugar every 10 min by detecting glucose through the skin, without having to draw blood. If glucose is out of the acceptable range, it sounds an alarm so the diabetic patient can address the problem quickly. A variety of biosensors, many using plastic substrates, are manufactured by various companies including ACLARA, Agilent Technologies, Calipertech, and I-STAT.

The second type of biochips – micro-nanoarrays – is a tool used in biotechnology research to analyze DNA or proteins to diagnose diseases or discover new drugs. Also called DNA arrays, they can identify thousand of genes simultaneously [47.57, 72]. They include a microarray of silicon nanowires, roughly a few nm in size, to selectively bind and detect even a single biological molecule such as DNA or protein by using nanoelectronics to detect the slight electrical charge caused by such binding, or a microarray of carbon nanotubes to detect glucose electrically.

After the tragedy of September 11, 2001, concern about biological and chemical warfare has led to the development of handheld units with biological and chemical sensors for detection of biological germs, chemical or nerve agents, and mustard agents, and their chemical precursors, to protect subways, airports, water supplies, and the population at large [47.76].

BioMEMS/bioNEMS are also being developed for minimally invasive surgery, including endoscopic surgery, laser angioplasty, and microscopic surgery. Implantable artificial organs can also be produced. Other applications include: implantable drug-delivery devices, e.g., micro-nanoparticles with drug molecules encapsulated in functionalized shells for site-specific targeting applications, and a silicon capsule with a nanoporous membrane filled with drugs for long-term delivery [47.66, 77–79]; nanodevices for sequencing single molecules of DNA in the Human Genome Project [47.61]; cellular growth using carbon nanotubes for spinal-cord repair; nanotubes for nanostructured materials for various applications such as spinal fusion devices; organ growth; and growth of artificial tissues using nanofibers.

47.1.4 Nanotribology and Nanomechanics Issues in MEMS/NEMS and BioMEMS/BioNEMS

Tribological issues are important in MEMS/NEMS and bioMEMS/bioNEMS requiring intended and/or unintended relative motion. In these devices, various forces associated with the device scale down with the size. When the length of the machine decreases from 1 mm to 1 μm, the surface area decreases by a factor of a million, and the volume decreases by a factor of a billion. As a result, surface forces such as adhesion, friction, meniscus forces, viscous forces, and surface tension that are proportional to surface area become a thousand times larger than the forces proportional to the volume, such as inertial and electromagnetic forces. In addition to the consequence of large surface-to-volume ratios, the small tolerances for which these devices are designed make physical contacts more likely, thereby making them particularly vulnerable to adhesion between adjacent components. Slight particulate or chemical contamination present at the interface can be detrimental. Furthermore, the small start-up forces and the torques available to overcome retarding forces are small, and the increase in resistive forces such as adhesion and friction become a serious tribological concern that limits the durability and reliability of MEMS/NEMS [47.13]. A large lateral force required to initiate relative motion between two surfaces, i.e., large static friction, is referred to as *stiction* and has been studied extensively in the tribology of magnetic storage systems [47.34, 46, 80–84]. The source of stiction is generally liquid-mediated adhesion, with the source of liquid being process fluid or capillary condensation of water vapor from the environment. Adhesion, friction/stiction (static friction), wear, and surface contamination affect MEMS/NEMS and bioMEMS/bioNEMS performance and, in some cases, can even prevent devices from working. Some examples of devices that experience nanotribological problems follow.

Nanomechanical properties are scale dependent, therefore these should be measured at relevant scales.

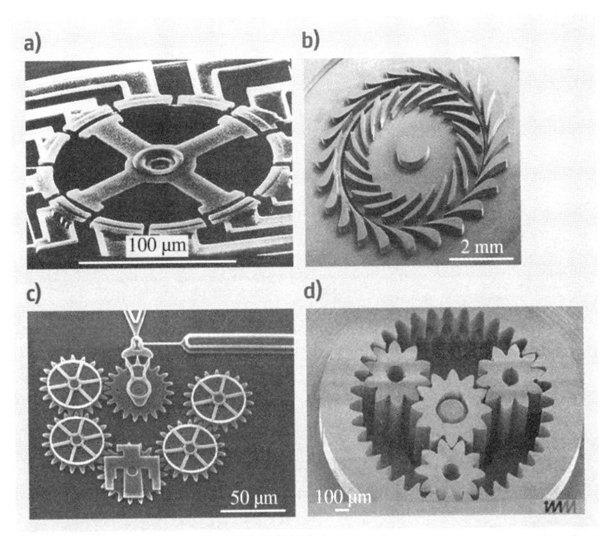

Fig. 47.2a–d Examples of MEMS devices and components that experience tribological problems. (**a**) Electrostatic micromotor (after [47.85]). (**b**) Microturbine bladed rotor and nozzle vanes on the stator (after [47.86]). (**c**) Six-gear chain (Sandia). (**d**) Ni/Fe Wolfrom-type gear system produced by LIGA (after [47.87])

MEMS

Figure 47.2 shows examples of several microcomponents that can encounter the above-mentioned tribological problems. The polysilicon electrostatic micromotor has 12 stators and a four-pole rotor and is produced by surface micromachining. The rotor diameter is 120 μm, and the air gap between the rotor and stator is 2 μm [47.85]. It is capable of continuous rotation at up to 100 000 rpm. The intermittent contact at the rotor–stator interface and physical contact at the rotor–hub flange interface result in wear issues, and high stiction between the contacting surfaces limits repeatability of operation or may even prevent operation altogether. Next, a bulk micromachined silicon stator–rotor pair is shown with a bladed rotor and nozzle guide vanes on the stator with dimensions < 1 mm [47.86, 88]. These are being developed for a high-temperature micro gas-turbine engine with rotor dimension of 4–6 mm in diameter and operating speed of up to 1 million rpm (with a sliding velocity in excess of 500 m/s, comparable to velocities of large turbines operating at high velocities) to achieve high specific power, up to a total of ≈ 10 W. Erosion of blades and vanes and design of the microbearings required to operate at the extremely high speeds used in the turbines are some of the concerns. Ultrashort, high-speed micro hydrostatic gas journal bearings with length-to-diameter ratio (L/D) of < 0.1 are being developed for operation at surface speeds of the order of 500 m/s, which results in unique design challenges [47.89]. Microfabrica Inc. in the USA is developing microturbines with outer diameter as low as 0.9 mm to be used as power sources for medical devices. They plan to use precision ball bearings.

Next in Fig. 47.2 is a scanning electron microscopy (SEM) micrograph of a surface-micromachined polysilicon six-gear chain from Sandia National Lab. (For more examples of an early version, see [47.90].) As an example of nonsilicon components, a milligear system produced using the LIGA process for a DC brushless permanent magnet millimotor (diameter = 1.9 mm, length = 5.5 mm) with an integrated milligear box [47.87, 91, 92] is also shown. The gears are made of metal (electroplated Ni-Fe) but can also be made from injected polymer materials (e.g., polyoxy-methylene or POM) using the LIGA process. Even though the torque transmitted at the gear teeth is small, of the order of a fraction of nN m, because of the small dimensions of gear teeth, the bending stresses are large where the teeth mesh. Tooth breakage and wear at the contact of gear teeth is a concern.

Figure 47.3 shows an optical micrograph of a microengine driven by an electrostatically activated comb drive connected to the output gear by linkages, for operation in the kHz frequency range, which can be used as

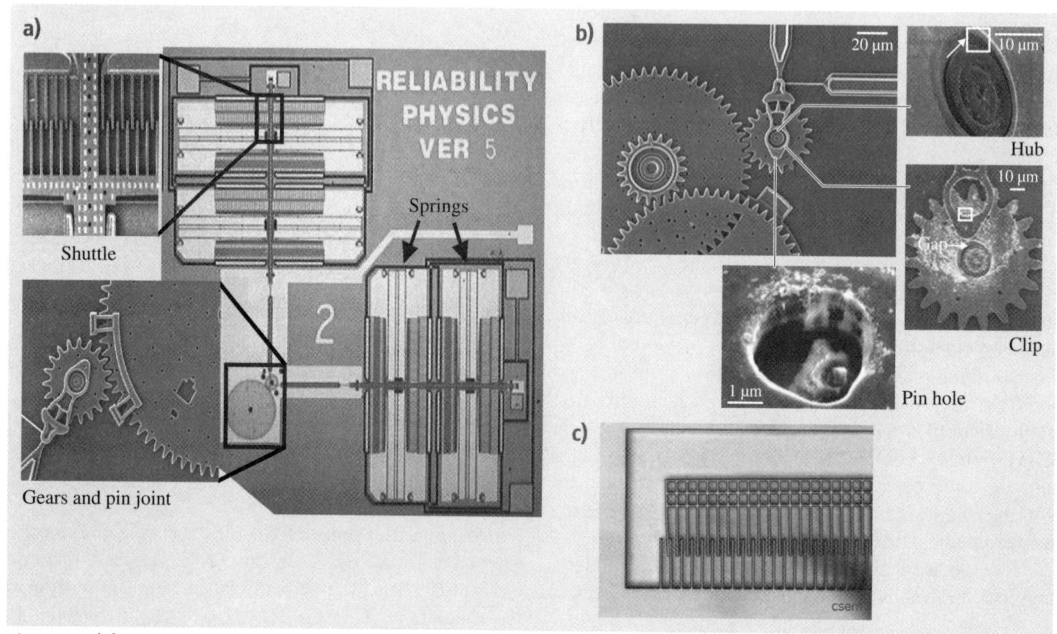

Fig. 47.3 (a) Optical micrograph of a microengine driven by an electrostatically actuated comb drive (microengine), fabricated by Sandia Summit Technologies (after [47.93]). (b) The polysilicon microgear unit can be driven at speeds of up to 250 000 rpm. Various sliding components are shown after laboratory wear test for 6000 cycles at 1.8% relative humidity (after [47.94]). (c) Stuck comb drive (CSEM)

a general drive and power source to drive micromechanisms [47.93]. Parts are fabricated from polysilicon. A microgear unit is used to convert reciprocating motion from a linear actuator into circular motion. Another drive linkage oriented at 90° to the original linkage, driven by another linear actuator, allows continuous motion to be maintained. The linkages are connected to the output gear through pin joints that allow relative motion.

One inset shows a polysilicon, multiple microgear speed reduction unit and its components after laboratory wear tests conducted for 6,000 cycles at 1.8% relative humidity (RH) [47.94]. Wear of various components is clearly observed in the figure. Humidity was shown to be a strong factor in the wear of rubbing surfaces. In order to improve the wear characteristics of rubbing surfaces, 20 nm-thick tungsten (W) coating deposited at 450 °C using chemical vapor deposition (CVD) technique was used [47.95]. Tungsten-coated microengines tested for reliability showed improved wear characteristics with longer lifetimes than polysilicon microengines. However, these coatings have poor yield. Instead, vapor-deposited self-assembled monolayers of fluorinated (dimethylamino)silane are used [47.96]. They can be deposited with high yield; however, durability is not as good. The second inset shows a comb drive with a deformed frame, which results in some fingers coming into contact. The contacting fingers can result in stiction.

Figure 47.4a shows a schematic of a micromachined flow modulator. Several micromachined flow channels are integrated in series with electrostatically actuated microvalves [47.97]. The flow channels lead to a central gas outlet hole drilled into the glass substrate. Gas enters the device through a bulk micromachined gas inlet hole in the silicon cap. The gas, after passing through an open microvalve, flows parallel to the glass substrate through flow channels and exits the device through an outlet. The normally open valve structure consists of a freestanding double-end-clamped beam, which is positioned beneath the gas inlet orifice. When electrostatically deflected upwards, the beam seals against the inlet orifice and the valve is closed. In these microvalves used for flow control, the mating valve surfaces should be smooth enough to seal while maintaining a minimum roughness to ensure low adhesion [47.80–82, 98].

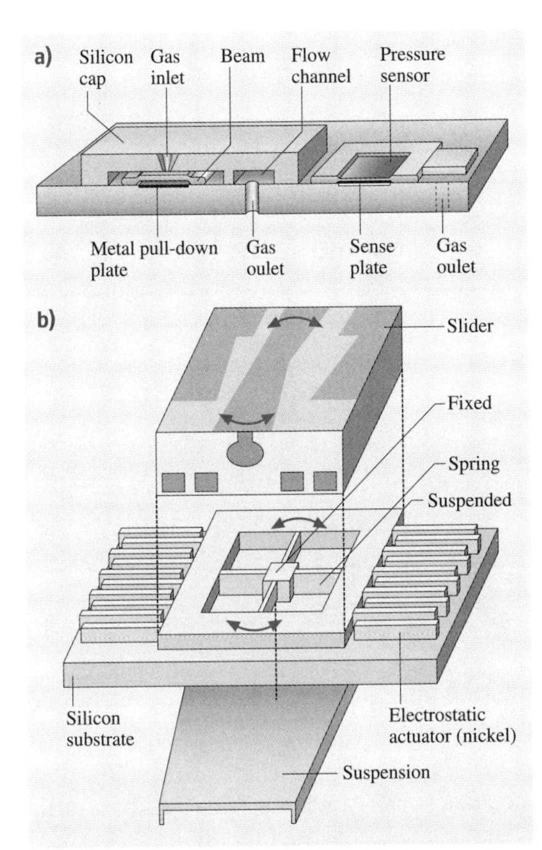

Fig. 47.4a,b Examples of MEMS devices that experience tribological problems. (**a**) Low-pressure flow modulator with electrostatically actuated microvalves (after [47.97]). (**b**) Electroplated-nickel rotary microactuator for magnetic disc drives (after [47.37])

The second MEMS device shown (Fig. 47.4b) is an electrostatically driven rotary microactuator for a magnetic disk drive, surface-micromachined by a multilayer electroplating method [47.37]. This high-bandwidth servo-controlled microactuator, located between a slider and a suspension, is being developed for ultrahigh-track-density applications, which serves as the fine-position and high-bandwidth control element of a two-stage coarse–fine servo system when coupled with a conventional actuator [47.36, 37]. A slider is placed on top of the central block of a microactuator, which provides rotational motion to the slider. The bottom of the silicon substrate is attached to the suspension. The radial flexure beams in the central block give rotational freedom of motion to the suspended mass (slider), and the electrostatic actuator drives the suspended mass. Actuation is accomplished via interdigitated, cantilevered electrode fingers, which are alternatingly attached to the central body of the moving part and to the stationary substrate to form pairs. A voltage applied across these electrodes results in an electrostatic force which rotates the central block. The interelectrode gap width is $\approx 2\,\mu m$. Any unintended contacts between the moving and stationary electroplated-nickel electrodes may result in wear and stiction.

Commercially available MEMS devices also exhibit tribological problems. Figure 47.5a shows an integrated capacitive-type silicon accelerometer fabricated using surface micromachining by Analog Devices, a couple of mm in dimension, which is used for airbag deployment in automobiles, and more recently for various other consumer electronic markets [47.21, 99]. The central suspended beam mass ($\approx 0.7\,\mu g$) is supported on the four corners by spring structures. The central beam has interdigitated cantilevered electrode fingers ($\approx 125\,\mu m$ long and $3\,\mu m$ thick) on all four sides that alternate with those of the stationary electrode fingers as shown, with about a $1.3\,\mu m$ gap. Lateral motion of the central beam causes a change in the capacitance between these electrodes, which is used to measure the acceleration. Stiction between the adjacent electrodes as well as stiction of the beam structure with the underlying substrate, under isolated conditions, is detrimental to the operation of the sensor [47.21, 99]. Wear during unintended contact of these polysilicon fingers is also a problem. A molecularly thick diphenyl siloxane lubricant film, resistant to high temperatures and oxidation, is applied by a vapor-deposition process on the electrodes to reduce stiction and wear [47.100]. For deposition, a small amount of liquid is dispensed into each package before it is sealed. As the package is heated in the furnace, the liquid evaporates and coats the sensor surface. As sensors are required to sense low-g accelerations, they need to be more compliant and stiction becomes an even bigger concern.

Figure 47.5b shows a cross-sectional view of a typical piezoresistive-type pressure sensor, which is used for various applications including manifold absolute pressure (MAP) and tire pressure measurements in automotive applications, and disposable blood-pressure measurements. The sensing material is a diaphragm formed on a silicon substrate, which bends with applied pressure [47.101, 102]. The deformation causes a change in the band structure of the piezoresistors that are placed on the diaphragm, leading to a change in the

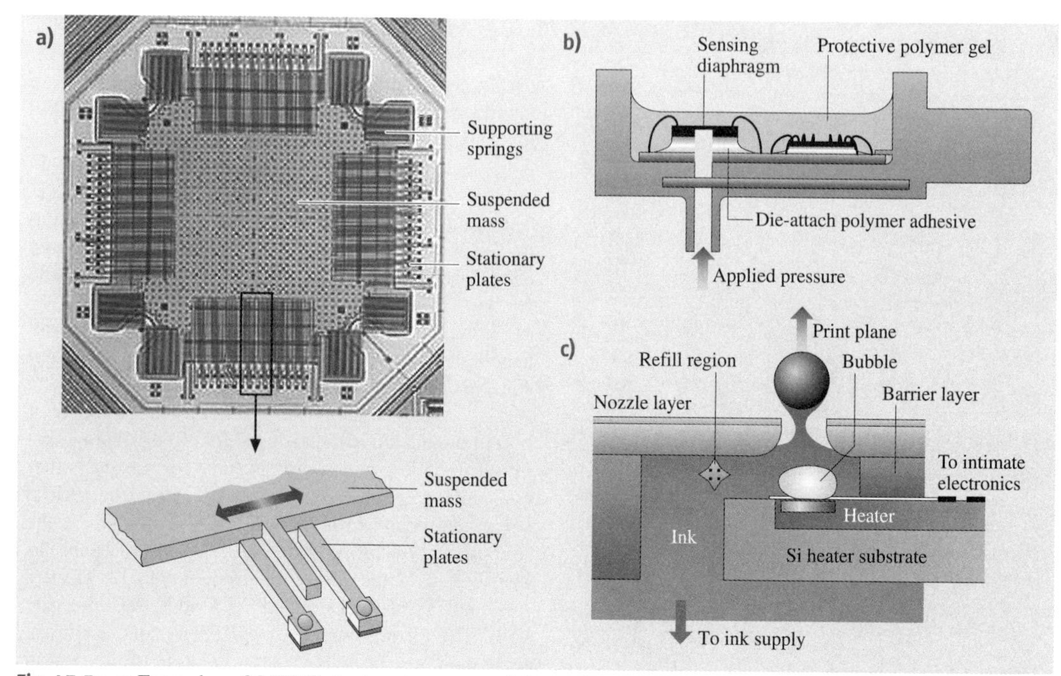

Fig. 47.5a–c Examples of MEMS devices in commercial use that experience tribological problems. **(a)** Capacitive-type silicon accelerometer for automotive sensory applications (after [47.99]). **(b)** Piezoresistive-type pressure sensor (after [47.102]). **(c)** Thermal inkjet printhead (after [47.25])

resistivity of the material. MAP sensors are subjected to drastic conditions – extreme temperatures, vibrations, sensing fluid, and thermal shock. Fluid under extreme conditions could cause corrosive wear. Fluid cavitation could cause erosive wear. The protective gel encapsulant generally used can react with the sensing fluid and result in swelling or dissolution of the gel. Silicon cannot deform plastically, therefore any pressure spikes leading to deformation past its elastic limit will result in fracture and crack propagation. Pressure spikes could also cause the diaphragm to delaminate from the support substrate. Finally, cyclic loading of the diaphragm during use can lead to fatigue and wear of the silicon diaphragm or delamination.

The schematic in Fig. 47.5c shows a cross-sectional view of a thermal printhead chip (of the order of $10–50\,cm^3$ in volume) used in inkjet printers [47.25]. They comprise an ink supply and an array of elements with microscopic heating resistors on a substrate mated to a matching array of injection orifices or nozzles ($\approx 70\,\mu m$ in diameter) [47.23, 24, 26]. In each element, a small chamber is heated by the resistor, where a brief electrical impulse vaporizes part of the ink and creates a tiny bubble. The heaters operate at several kHz and are therefore capable of high-speed printing. As the bubble expands, some of the ink is pushed out of the nozzle onto the paper. When the bubble pops, a vacuum is created and this causes more ink from the cartridge to move into the printhead. Clogged ink ports are the major failure mode. There are various tribological concerns [47.23]. The surface of the printhead from where the ink is ejected towards the paper can become scratched or damaged as a result of countless trips back and forth across the pages, which are somewhat rough. As a result of repeated heating and cooling, the heated resistors expand and contract. Over time, these elements will experience fatigue and may eventually fail. Bubble formation in the ink reservoir can lead to cavitation erosion of the chamber, which occurs when bubbles formed in the fluid become unstable and implode against the surface of the solid and impose impact energy on that surface. Fluid flow through nozzles may cause erosion and ink particles may also cause abrasive wear. Corrosion of the ink reservoir surfaces can also occur as a result of exposure of ink to high temperatures as well as due to

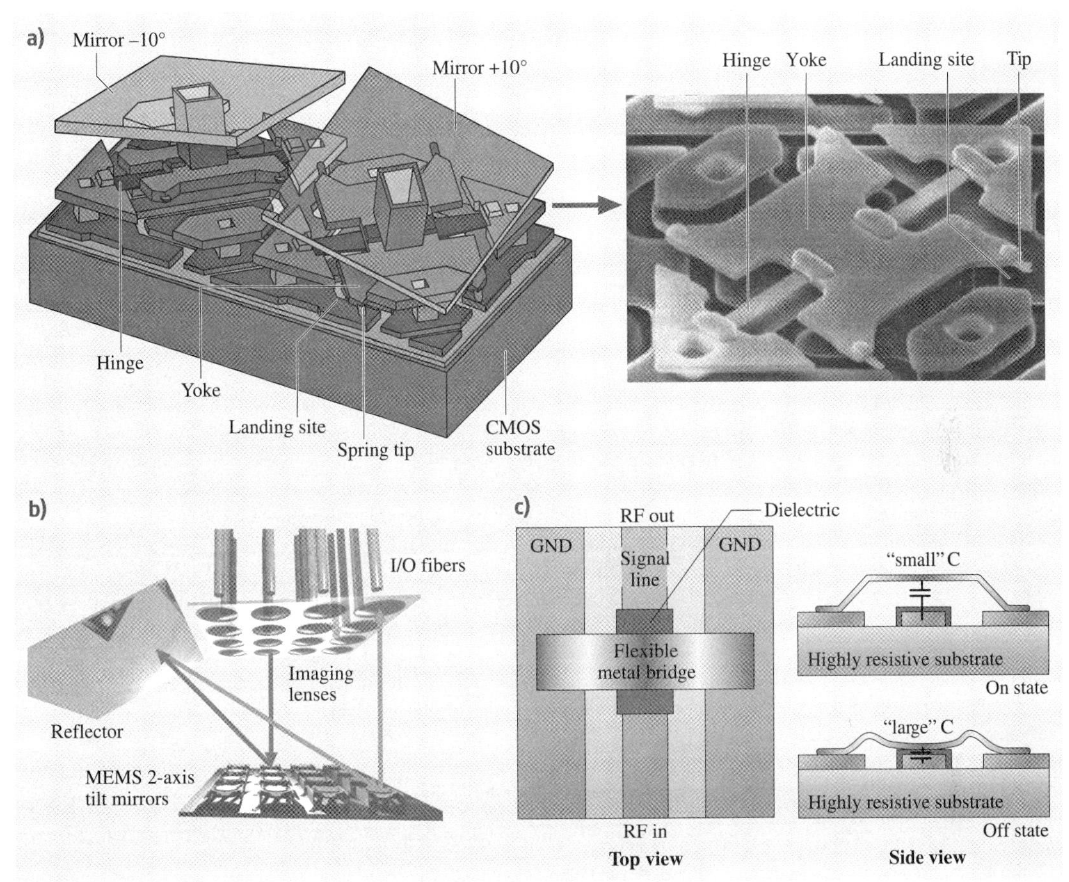

Fig. 47.6a–c Examples of two commercial MOEMS (**a**,**b**) and one RF-MEMS (**c**) device that experience tribological problems. (**a**) Digital micromirror devise for displays (after [47.28]). (**b**) Tilt mirror arrays for switching optical signal input and output fiber arrays in optical cross-connect for telecommunications (after [47.104]). (**c**) RF microswitch (© IMEC, Belgium)

ink pH. The substrate of the chip consists of silicon with a thermal barrier layer followed by a thin film of resistive material and then conducting material. The conductor and resister layers are generally protected by an overcoat layer of a plasma-enhanced chemical vapor deposition (PECVD) α-SiC:H layer, 200–500 nm thick [47.103].

Figure 47.6a shows two digital micromirror device (DMD) pixels used in digital light processing (DLP) technology for digital projection displays in computer projectors, high-definition television (HDTV) sets, and movie projectors [47.27–29]. The entire array (chip set) consists of a large number of oscillating aluminum alloy micromirrors as digital light switches which are fabricated on top of a complementary metal–oxide–semiconductor (CMOS) static random-access memory integrated circuit. The surface-micromachined array consists of half a million to more than two million of these independently controlled reflective micromirrors, each $\approx 12\,\mu$m square and with $13\,\mu$m pitch, which flip backward and forward at a frequency of of the order of 5000–7000 times a second as a result of electrostatic attraction between the micromirror structure and the underlying electrodes. For binary operation, the micromirror–yoke structure mounted on torsional hinges is oscillated $\pm 10°$ (with respect to the plane of the chip set), limited by a mechanical stop. Contact between cantilevered spring tips at the end of the yoke (four present on each yoke) and the underlying stationary landing sites is required for true digital (binary) operation. Stiction and wear during contact between aluminum-alloy spring tips and landing sites, hinge memory (metal creep at high operating temperatures), hinge fatigue, shock and vibration failure, and sensitivity to particles in the chip package and operating environment are some of the important issues affecting the reliable operation of a micromirror device [47.105–109]. A vapor-phase-deposited self-assembled monolayer of the fatty acid perfluorodecanoic acid (PFDA) on surfaces of the tip and landing sites is used to reduce stiction and wear [47.110, 111]. However, these films are susceptible to moisture, and to keep moisture out and create a background pressure of PFDA, a hermetic chip package is used. The spring tip is used in order to use the stored spring energy to pop up the tip during pull-off. A lifetime estimate of over 100 000 h operation with no degradation in image quality is the norm. At a mirror modulation frequency of 7 kHz, each micromirror element needs to switch ≈ 2.5 trillion cycles.

Figure 47.6b shows a schematic of a 256×256-port large optical cross-connect, introduced in 2000 by Glimmerglass (Hayward, CA) for optical telecommunication networks in order to be able to manipulate a larger number of optical signals rapidly [47.104]. This optical microswitch uses 256 or more movable mirrors on a chip for switching a light beam from an input fiber to a few output fibers. The mirrors are made of gold-coated polysilicon and are $\approx 500\,\mu m$ in diameter. Reliability concerns are the same as those described above for DMDs. To minimize stiction, the chipset is hermetically sealed in dry nitrogen (90% N_2, 10% He).

Figure 47.6c shows a schematic of an electrostatically-actuated capacitive-type RF microswitch for switching of RF signals at microwave and low frequencies [47.112]. It is of membrane type and consists of a flexible metal (Al) bridge that spans the RF transmission line in the center of a coplanar waveguide. When the bridge is up, the capacitance between the bridge and RF transmission line is small, and the RF signal passes without much loss. When a DC voltage is applied between the RF transmission line and the bridge, the latter is pulled down until it touches a dielectric isolation layer. The large capacitance thus created shorts the RF signal to ground. The failure modes include creep in the metal bridge, fatigue of the bridge, charging and degradation of the dielectric insulator, and stiction of the bridge to the insulator [47.30, 112]. Stiction occurs due to capillary condensation of water vapor from the environment, van der Waals forces, and/or charging effects. If the restoring force in the bridge of the switch is not large enough to pull the bridge up again after the actuation voltage has been removed, the device fails due to stiction. Humidity-induced stiction can be avoided by hermetically sealing the microswitch. Some roughness of the surfaces reduces the probability of stiction. Selected actuation waveforms can be used to minimize charging effects.

NEMS

Probe-based data recording technologies are being developed for ultrahigh-areal-density recording, where the probe tip is expected to be scanned at velocities up to 100 mm/s. There are three major techniques being developed: thermomechanical [47.49], phase change [47.113], and ferroelectric recording [47.114, 115]. We discuss the tribological issues with two of the widely pursued techniques [47.116].

Figure 47.7a shows the thermomechanical recording system which uses arrays of 1,024 silicon microcantilevers and playback on an ≈ 40 nm-thick polymer medium with a harder Si substrate [47.49]. The cantilevers consist of integrated tip heaters with tips of nanoscale dimensions. (The sharp tips themselves are also example of NEMS.) Thermomechanical recording is a combination of applying a local force to the polymer layer and softening it by local heating. The tip, heated to $\approx 400\,°C$, is brought into contact with the polymer for recording. Readings are done using the cantilever heater, originally used for recording, as a thermal readback sensor by exploiting its temperature-dependent resistance. The principle of thermal sensing is based on the fact that the thermal conductivity between the heater and the storage substrate changes according to the spacing between them. When the spacing between the heater and sample is reduced as the tip moves into a bit, the heater's temperature and hence its resistance will decrease. Thus, changes in temperature of the continuously heated resistor are monitored while the cantilever is scanned over data bits, providing a means of detecting the bits. Erasing for subsequent rewriting is carried out by thermal reflow of the storage field by heating the medium to 150 °C for a few seconds. The smoothness of the reflown medium allows multiple rewriting of the same storage field. Bit sizes ranging between 10 and 50 nm have been achieved by using a 32×32 (1,024) array write/read chip (3 mm×3 mm). It has been reported that tip wear occurs due to contact between tip and Si substrate during writing. Tip wear is considered a major concern for the device reliability.

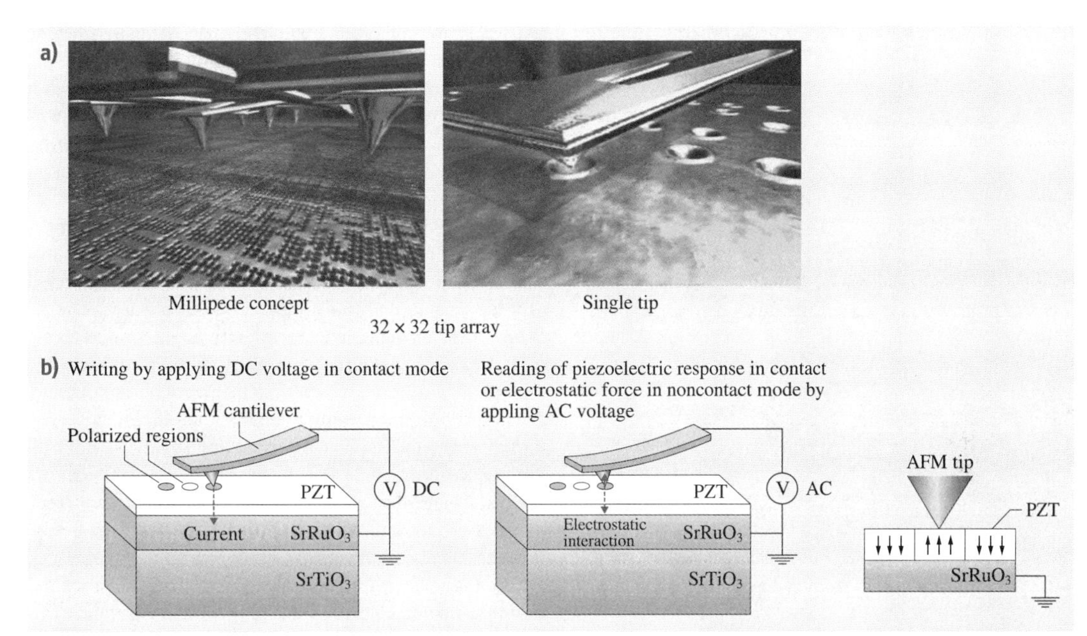

Fig. 47.7a,b Two example of NEMS devices: (**a**) thermomechanical recording, and (**b**) ferroelectric recording, which experience tribological problems

Figure 47.7b shows a schematic of domain writing and reading in a ferroelectric film. The electrically conductive AFM tips are placed in contact with a piezoelectric (lead zirconate titanate, PZT) film-coated medium [47.114, 115]. Ferroelectric domains on the PZT film are polarized by applying short voltage pulses ($\approx 10\,V$, $\approx 100\,\mu s$) that exceed the coercive field of the PZT layer, resulting in local, nonvolatile changes in the electronic properties of the underlying film. The temperature rise during recording is expected to be of the order of 80 °C. Reading out of the polarization states in the ferroelectric film can be carried out using two different methods. In one method, the static surface charge, proportional to the normal component of polarization, can be detected by electrostatic force microscopy in the noncontact mode. In the second method, an AFM is operated in contact mode and the piezoresponse force is measured by applying an alternating-current (AC) voltage. Wear of the conducting tip and the PZT layer at high scanning velocities is a major concern for device reliability. Various lubricant films are being developed to minimize wear [47.113–119].

In magnetic data storage, magnetic recording is accomplished by relative motion between the magnetic head slider and a magnetic rigid disk [47.34]. Magnetic rigid disks and heads used today for magnetic data storage consist of nanostructured films a few nm thick. Figure 47.8a shows a sectional view of a conventional multigrain magnetic rigid disk. The superparamagnetic effect poses a serious challenge for the ever-increasing areal density of disk drives. One of the promising methods to circumvent the density limitations imposed by this effect is the use of a nanopatterned disk (Fig. 47.8b). In a conventional disk, the thin magnetic layer forms a random mosaic of nanometer-scale grains, and each recorded bit consists of many tens of these random grains. In a patterned disk, the magnetic layer is created as an ordered array of highly uniform islands, each island capable of storing an individual bit. These islands may be one or a few grains, rather than a collection of random decoupled grains. This increases the density by a couple of orders of magnitude. Figure 47.8c shows a schematic of an inductive write/giant-magnetoresistive (GMR) read head structure. These are constructed from a variety of materials: magnetic alloys, metal conductors, ceramic, and polymer insulators in a complex three-dimensional structure. The multilayered thin-film structure used to construct the sensor and individual films are only a few nm thick. The head slider surface, which flies over the

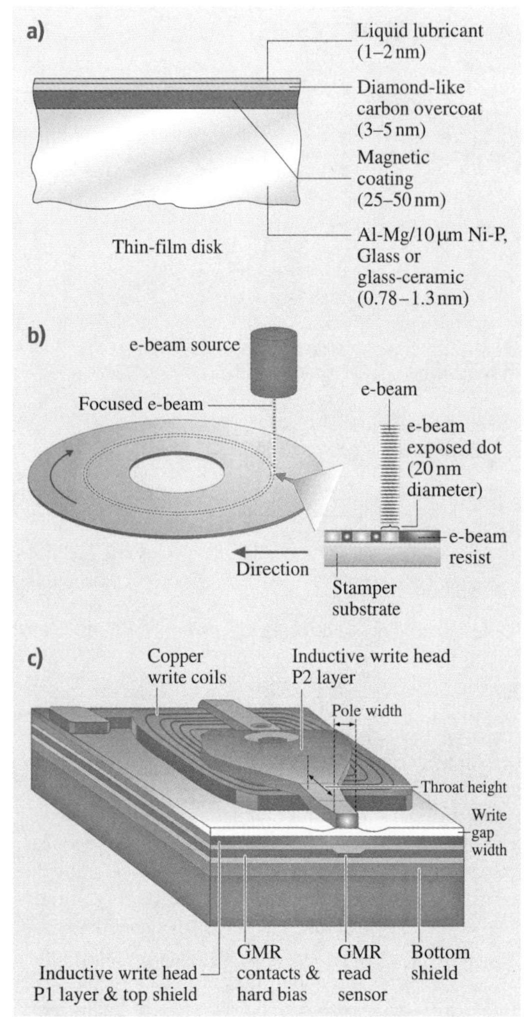

Fig. 47.8a–c Schematic of (**a**) sectional view of a conventional multigrain magnetic rigid disk, (**b**) nanopatterned magnetic rigid disk, and (**c**) an inductive write/GMR read magnetic head structure for magnetic data storage (Hitachi)

disk surface, is coated with ≈ 3 nm-thick diamond-like carbon coatings to protect the thin-film structure from electrostatic discharge. Any isolated contacts between the disk and sensor and lubricant pickup pose tribological concerns [47.34].

BioMEMS

An example of a wristwatch-type biosensor based on microfluidics, referred to as a lab-on-a-chip system, is shown in Fig. 47.9a [47.61, 70]. These systems are designed either to detect a single or a class of (bio)chemicals or for system-level analytical capabilities for a broad range of (bio)chemical species (known as a micro total analysis system, mTAS), and have the advantage of incorporating sample handling, separation, detection, and data analysis onto one platform. The chip relies on microfluidics and involves manipulation of tiny amounts of fluids in microchannels using microvalves. The test fluid is injected into the chip, generally using an external pump or syringe, for analysis. Some chips have been designed with an integrated electrostatically actuated diaphragm-type micropump. The sample, which can have volume measured in nanoliters, flows through microfluidic channels via an electric potential and capillary action using microvalves (having various designs, including membrane type) for various analyses. The fluid is preprocessed and then analyzed using a biosensor. Another example of a biosensor is the cassette-type biosensor used for human genomic DNA analysis; integrated biological sample preparation is shown in Fig. 47.9b [47.71]. The implementation of micropumps and microvalves allows for fluid manipulation and multiple sample processing steps in a single cassette. Blood or other aqueous solutions can be pumped into the system, where various processes are performed.

Microvalves, which are found in most microfluidic components of bioMEMS, can be classified in two categories: active microvalves (with an actuator) for flow regulation in microchannels and passive microvalves integrated with micropumps. Active microvalves consist of a valve seat and a diaphragm actuated by an external actuator [47.62, 120, 121]. Different types of actuators are based on piezoelectric, electrostatic, thermopneumatic, electromagnetic, and bimetallic materials, shaped-memory alloys, and solenoid plungers. An example of an electrostatic cantilever-type active microvalve is shown in Fig. 47.9c [47.120]. Passive microvalves used in micropumps include mechanical check valves and a diffuser/nozzle [47.62, 121–124]. Check valves consist of a flap or membrane that is capable of opening and closing with changes in pressure; see Fig. 47.9c for schematics. A diffuser/nozzle uses an entirely different principle and only works with the presence of a reciprocating diaphragm. When one convergent channel works simultaneously with another convergent channel oriented in a specific direction, a change in pressure is possible.

There are four main types of mechanical micropumps, which include a diaphragm micropump that

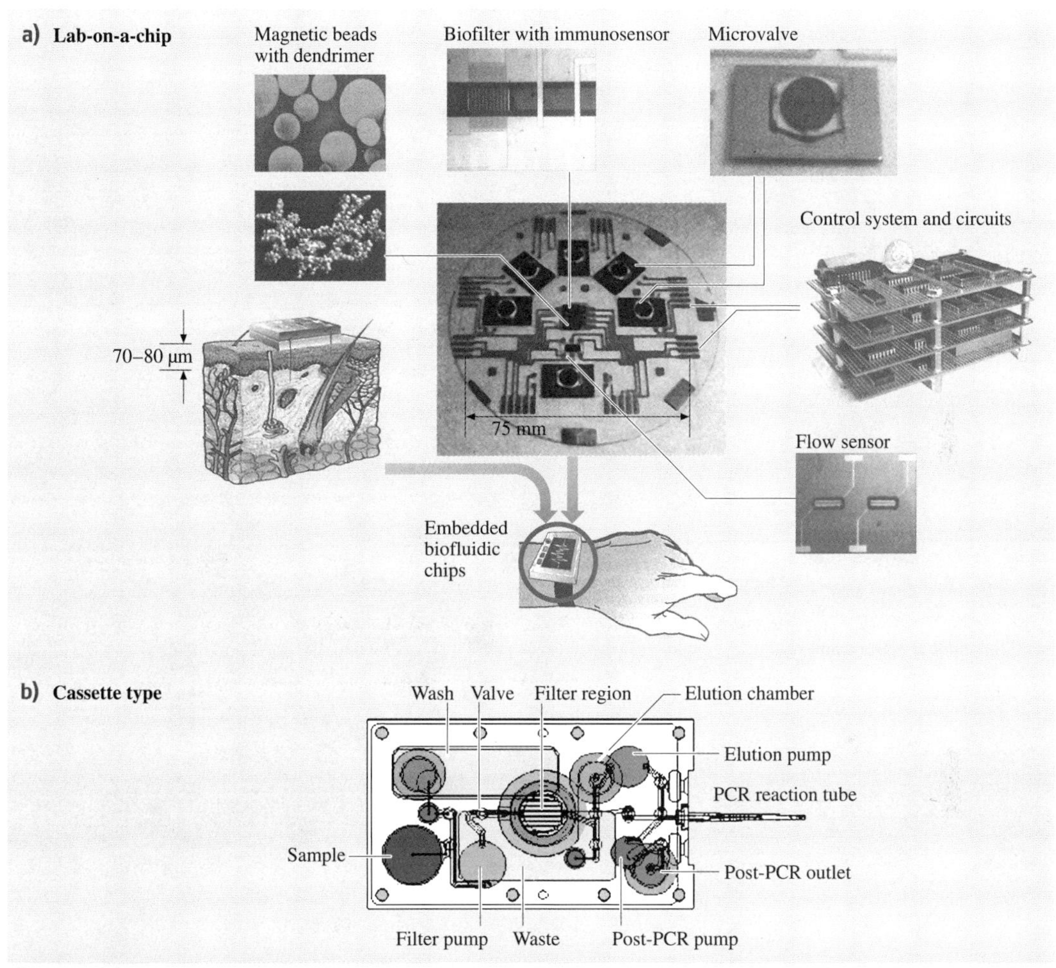

Fig. 47.9 **(a)** MEMS-based biofluidic chip, commonly known as a lab-on-a-chip, that can be worn like a wristwatch (after [47.70]). **(b)** Cassette-type biosensor used for human genomic DNA analysis (after [47.71]). PCR, polymerase chain reaction

involves mechanical check valves, valveless rectification pumps that use diffuser/nozzle type valves, valveless pumps without a diffuser/nozzle, electrostatic micropumps, and rotary micropumps [47.62, 121–124]. Diaphragm micropumps consist of a reciprocating diaphragm which can be piezoelectrically driven, working in synchronization with two check valves (Fig. 47.9c) [47.123]. Electrostatic micropumps have a diaphragm as well, but it is driven using two electrodes (Fig. 47.9c) [47.125]. Valveless micropumps also consist of a diaphragm, which is piezoelectrically driven, but do not incorporate passive mechanical valves. Instead, these pumps use an elastic buffer or variable-gap mechanism. Finally, a rotary micropump has a rotating rotor that simply adds momentum to the fluid by the fast moving action of the blades (Fig. 47.9c) [47.126, 127]. Rotary micropumps can be driven using an integrated electromagnetic motor or by the presence of an external electric field. All of these micropumps can be made of silicon or a polymer material.

During the operation of the microvalves and micropumps discussed above, adhesion and friction properties become important when contacts occur due to relative motion. During operation, active mechanical

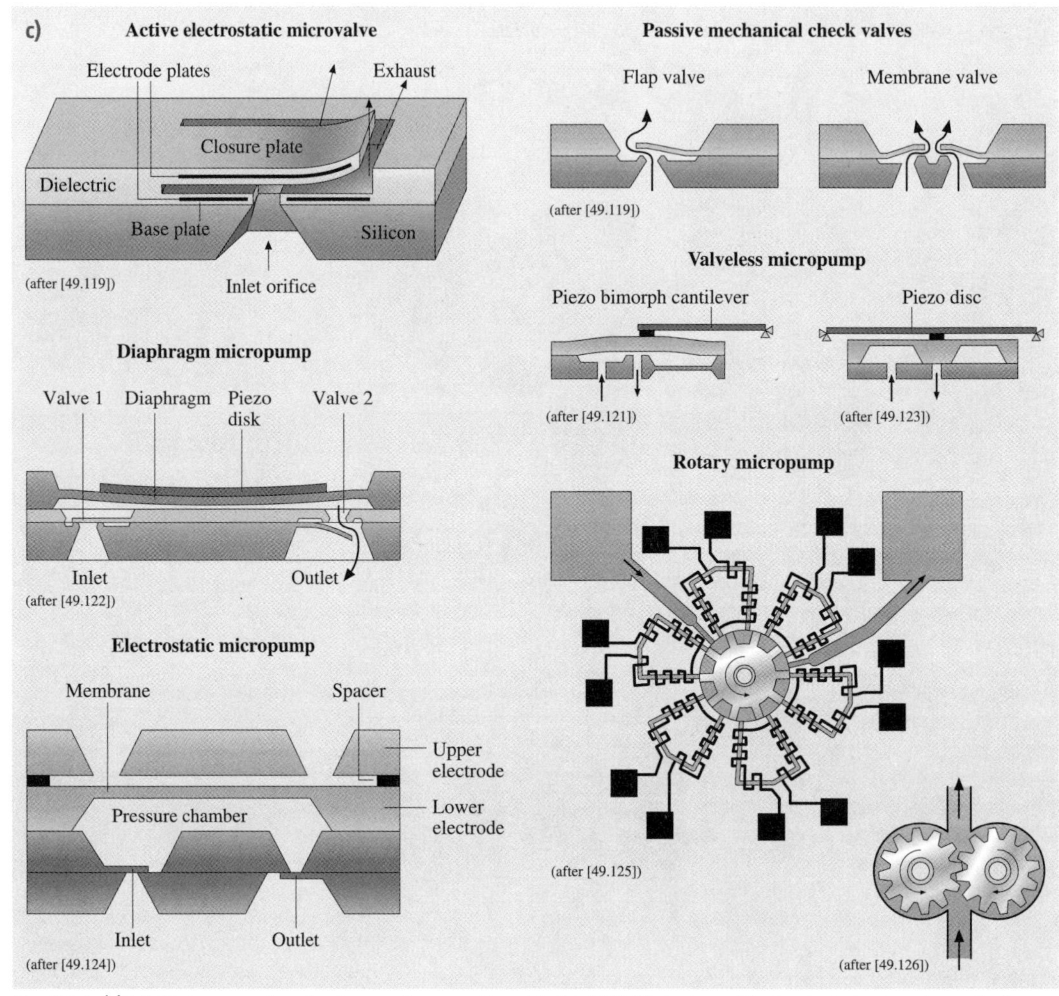

Fig. 47.9 (c) Multiple examples of valves and pumps found in bioMEMS devices. Mechanical check valves, diaphragm micropump, valveless micropump, and rotary micropump

microvalves have an externally actuated diaphragm which comes into contact with a valve seat to restrict fluid flow. Adhesion between the diaphragm and valve seat will affect the operation of the microvalve. In the diaphragm micropumps, two passive mechanical check valves are incorporated into the design. Passive mechanical check valves also exhibit adhesion when the flap or membrane comes into contact with the valve seat when fluid flow is prevented. Adhesion also occurs during the operation of valveless micropumps when the diaphragm, which is piezoelectrically driven, comes into contact with the rigid outlet. Finally, adhesion and friction can also be seen during the operation of rotary micropumps when the gears rotate and come into contact and rub against one another.

If the adhesion between the microchannel surface and the biofluid is high, biomolecules will stick to the microchannel surface and restrict flow. In order to facilitate flow, microchannel surfaces with low bioadhesion are required. Fluid flow in polymer channels can produce triboelectric surface potential, which may affect the flow. Polymers are known to generate surface potential, and the magnitude of the potential varies from one polymer to another [47.128–130]. Conductive surface layers on the polymer channels can be deposited to reduce triboelectric effects.

As just mentioned, the microfluidic biosensor shown in Fig. 47.9a required the use of micropumps and microvalves. For example, a microdevice with 1,000 channels requires 1,000 micropumps and 2,000 microvalves, which makes it bulky and poses reliability concerns. Two methods can be used for driving the flow of fluids in microchannels: pressure and electrokinetic drive. Electrokinetic flow is based on the movement of molecules in an electric field due to their charges. There are two components to electrokinetic flow: electrophoresis, which results from the accelerating force due to the charge of a molecule in an electric field, and electroosmosis, which uses electrically controlled surface tension to drive the uniform liquid flow. Biosensors based on electrokinetic flow have also been developed. In so-called *digital-based microfluidics*, based on the electroosmosis process, electrically controlled surface tension is used to drive liquid droplets, thus eliminating the need for valves and pumps [47.133, 134]. These microdevices consist of a rectangular grid of gold nanoelectrodes instead of micro-nanochannels. An externally applied electric field enables manipulation of a few nanoliter samples through the capillary circuitry.

An example of a microarray-type biosensor (under development in our laboratory) is based on a field-effect transistor (FET) and is shown in Fig. 47.10a [47.73, 135]. FETs are sensitive to the electrical field produced due to the charge at the surface of the gate insulator. In this sensor, the gate metal electrode of a metal–oxide-semiconductor field-effect transistor (MOSFET) is removed and replaced with a protein (receptor layer) whose cognate is the analyte (e.g., virus or bacteria) that is meant to be sensed. Various proteins may have 1–25 (positive or negative) charges per molecule. The binding of the receptor layer with the analyte produces a change in the effective charge, which creates a change in the electrical field. This electrical field change may produce a measurable change in the current flow through the device. Adhesion between the protein layer and silica substrate affect the reliability of the biosensor. In the case of implanted biosensors, they come into contact with exterior environment, such as tissues and fluids, and any relative motion of the sensor surface with respect to the exterior environment, such as tissues or fluids, may result in surface damage. A schematic of friction and wear points of generation, when an implanted biosensor surface comes into contact with a living tissue, is shown in Fig. 47.10b [47.131]. The friction, wear, and adhesion of the biosensor surface may be critical in these applications [47.73, 131, 136, 137].

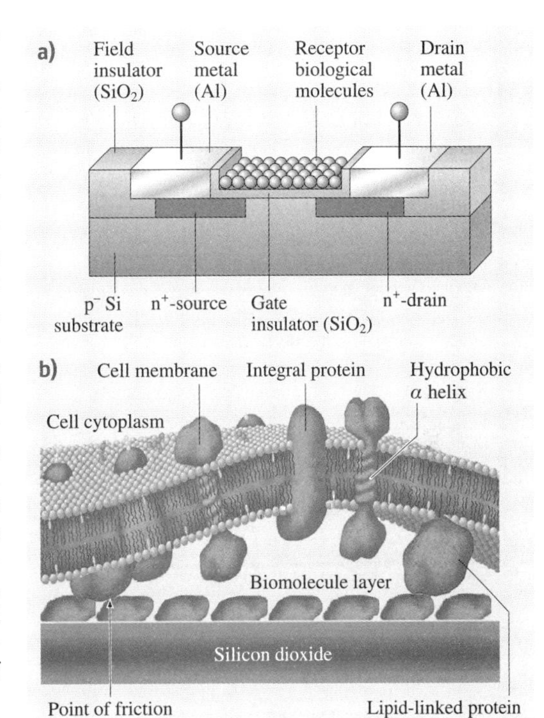

Fig. 47.10 (a) Schematic of a bioFET sensor (after [47.73]), and (b) schematic showing the generation of friction and wear points due to interaction of implanted biomolecule layer on a biosensor with living tissue (after [47.131])

Polymer bioMEMS are designed to measure cellular surfaces. Two examples are shown in Fig. 47.11 [47.132].

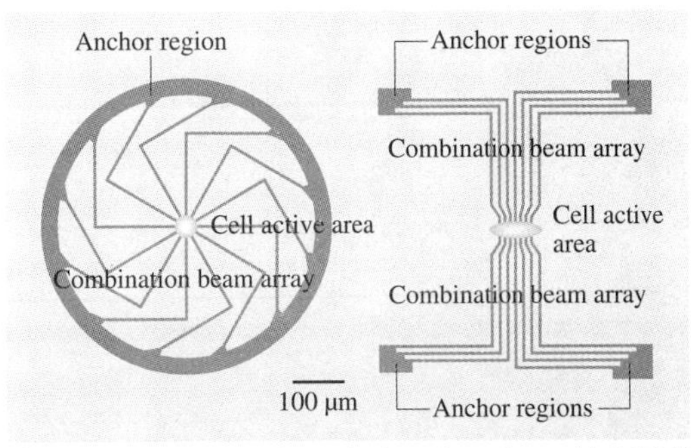

Fig. 47.11 Schematic of two designs for polymer bioMEMS structures to measure cellular forces (after [47.132])

The device on the left shows cantilevers anchored at the periphery of the circular structure, while the device on the right has cantilevers anchored at the two corners on the top and the bottom. The cell adheres to the center of the structure, and the contractile forces generated in the cell's cytoskeleton cause the cantilever to deflect. The deflection of the compliant polymer cantilevers is measured optically and related to the magnitude of the forces generated by the cell. Adhesion between cells and the polymer beam is desirable. In order to design the sensors, micro- and nanoscale mechanical properties of polymer structures are needed.

BioNEMS

Micro-nanofluidic devices provide a powerful platform for electrophoretic separations for a variety of biochemical and chemical analysis. Electrophoresis is a versatile analytical method which is used for separation of small ions, neutral molecules, and large biomolecules. Figure 47.12 shows an interdigitated micro-nanofluidic silicon array with nanochannels for a separation process. Figure 47.13a shows a schematic of an implantable, immunoisolation submicroscopic biocapsule, aimed at drug delivery in order to treat significant medical conditions such as type I diabetes [47.77, 78]. The purpose of the immunoisolation biocapsule is to create an implantable device capable of supporting foreign living cells that can be transplanted into humans. It is a silicon capsule consisting of two nanofabricated membranes bonded together with the drug (e.g., encapsulated insulin-producing islet cells) contained within the cavities for long-term delivery. The pores or nanochannels in a semipermeable membrane as small as 6 nm are used as flux regulators for long-term release of drugs. The nanomembrane also protects therapeutic substances from attack by the body's immune system. The pores are large enough to provide the flow of nutrients (e.g., glucose molecules) and drug (e.g., insulin), but small enough to block natural antibodies. Antibodies have the capability to penetrate any orifice > 18 nm. The 50 nm pores in silicon were etched by using sacrificial-layer lithography, described in Appendix A [47.78].

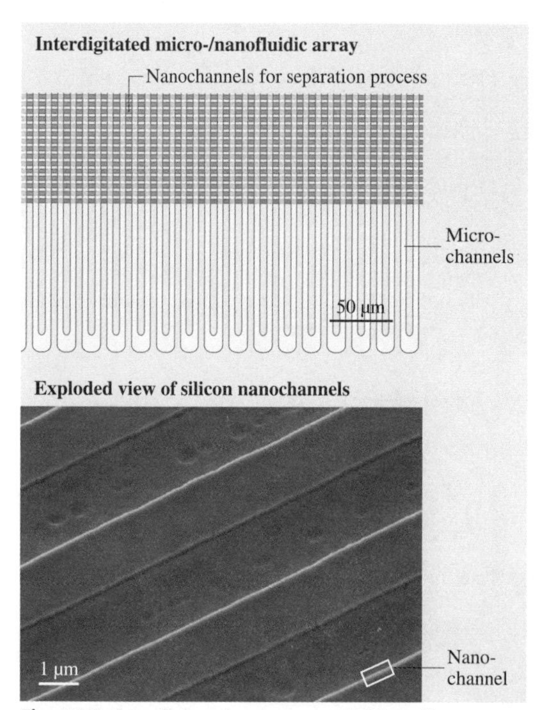

Fig. 47.12 Interdigitated micro-nanofluidic silicon array for a separation process (after [47.135, 138])

The main reliability concerns in the micro-nanofluidic silicon array and implantable biocapsules are biocompatibility and potential biofouling (undesirable accumulation of microorganisms) of the channels/membrane due to protein and cell adsorption from biological fluids. Biofouling can also result in clogging of the nanochannels/nanopores, which could potentially render the device ineffective. The adhesion of proteins and cells to an implanted device can also cause detrimental results such as inflammation and excessive fibrosis. Deposition of self-assembled monolayers of selected organic molecules onto the channels of the implants, which makes them hydrophobic, presents an innovative solution to combat the adverse effects of biological fluids [47.138–141].

Figure 47.13b shows a conceptual model of an intravascular drug-delivery device: nanoparticles used to search and destroy disease (tumor) cells [47.79]. (The tumor cells have one or two orders of magnitude higher density of receptors than normal cells and lower pH. Some receptors are only expressed on tumor cells.) With lateral dimensions of 1 µm or less, the particles are smaller than any blood cells. These particles can be injected into the blood stream and travel freely through the circulatory system. In order to direct these drug-delivery nanoparticles to cancer sites, their external surfaces are chemically modified to carry molecules that have lock-and-key binding specificity with molecules that support a growing cancer mass. As the particles come into close proximity with diseased cells, the ligands on the particle surfaces attach

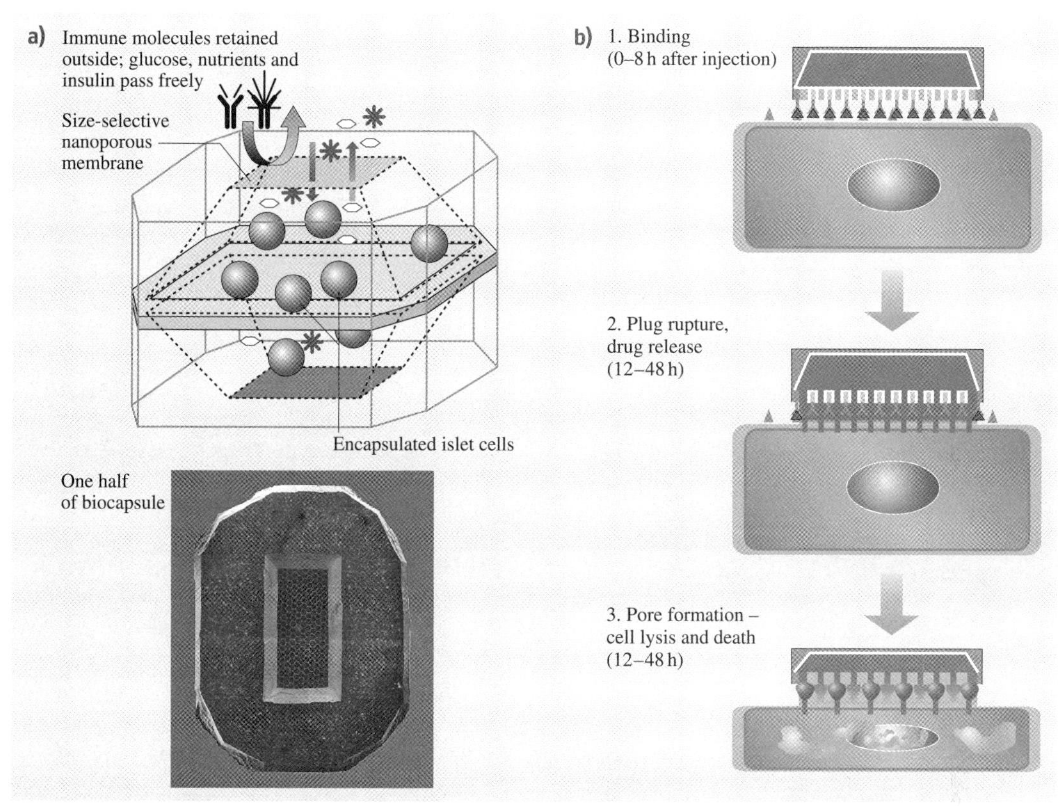

Fig. 47.13a,b Schematics of (a) implantable, immunoisolation submicroscopic biocapsules (drug-delivery device) (after [47.78]), and (b) intravascular nanoparticles for search and destroy of diseased blood cells (after [47.79])

to the receptors on the cells. As soon as the particles dock onto the cells, a compound is released that forms a pore on the membrane of the cells, which leads to cell death and ultimately to that of the cancer mass that was being nourished by the blood vessel. The adhesive interactions are regulated by specific (ligand–receptor binding) and nonspecific (short-range van der Waals, electrostatic, and steric) interactions [47.142]. Adhesion between nanoparticles and disease cells is required. Furthermore, the particles should travel close to the endothelium lining of vascular arteries to facilitate the interaction between the particles and diseased cells. *Decuzzi* et al. [47.142] analyzed the margination of a particle circulating in the blood stream and calculated the speed and time for margination (drifting of particles towards the blood vessel walls) as a function of the density and diameter of the particle, based on various forces present between the circulating particle and the endothelium lining. Human capillaries can have radii as small as 4–5 μm. They reported that particles used for drug delivery should have a radius smaller than a critical value in the range of 100 nm.

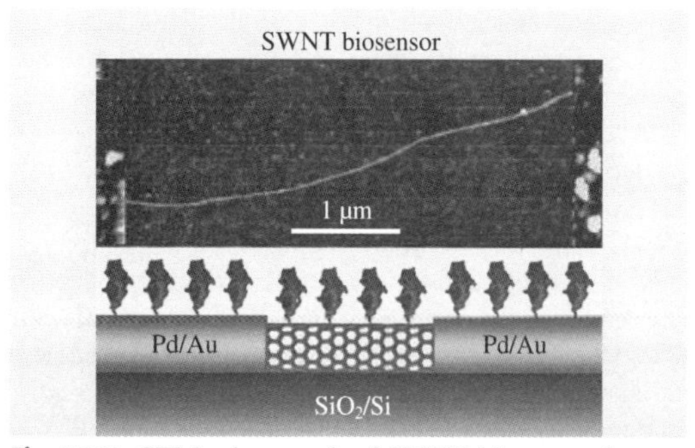

Fig. 47.14 SEM micrograph of SWNT biosensor; bottom schematic shows adsorption of protein molecules to the SWNT (after [47.7])

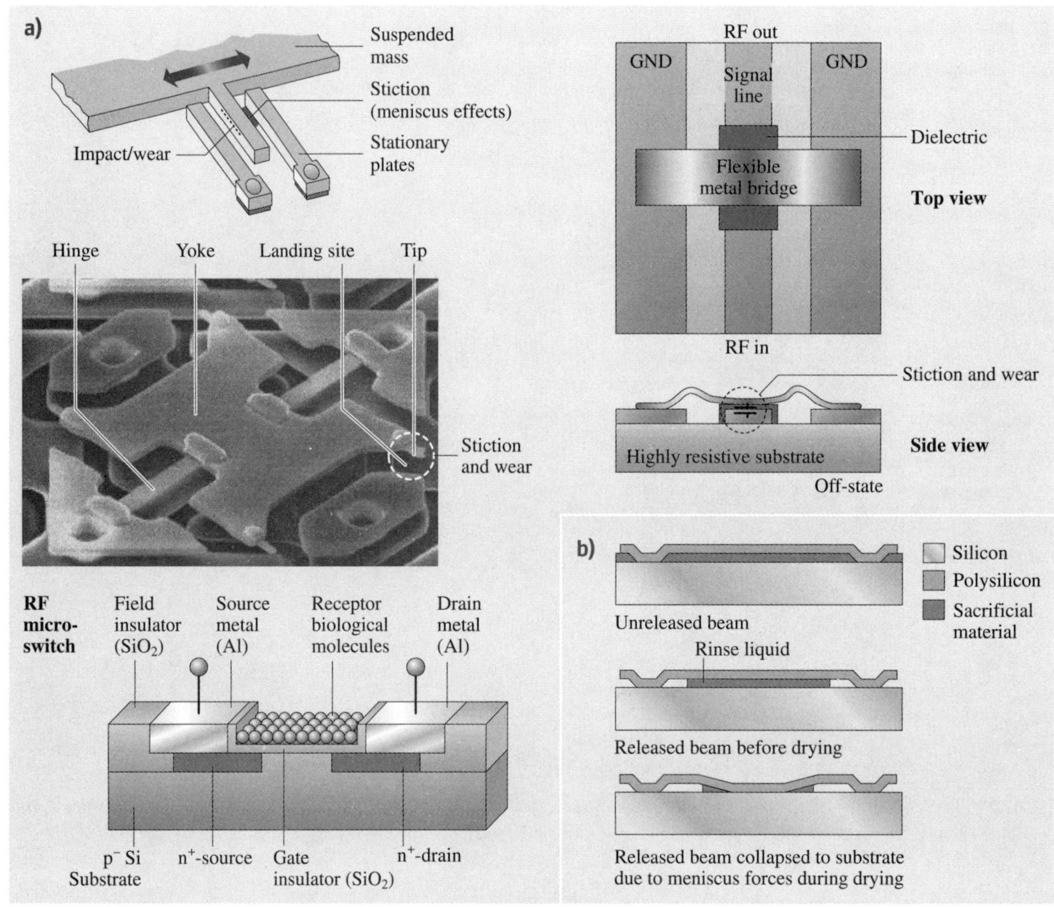

Fig. 47.15 (**a**) Summary of tribological issues in MEMS, MOEMS, RF-MEMS, and bioMEMS device operation (after [47.73, 139]), and (**b**) in microfabrication by surface micromachining

Recent studies show that a lateral force on the particles assists them in faster margination towards the endothelium walls. Thus, nonspherical particles are more desirable.

Because of their unique mechanical and electrical properties, single- and multiwalled carbon nanotubes (SWNT and MWNT) are being used for thermal management of high-power devices, reinforced composites and superstrong fiber and sheets, chemical and biological sensors, electromechanical devices, field-emission devices, and molecular electronics and computing [47.51, 143]. Figure 47.14 shows a SWNT biosensor [47.7]. The conductance of carbon nanotube (CNT) devices changes when proteins adsorb on the surface. The change in electrical resistance is a measure of protein adsorption. For high performance, adhesion should be strong between the adsorbent and SWNT.

In summary, adhesion, stiction/friction, and wear clearly limit the lifetimes and compromise the performance and reliability of MEMS/NEMS and bioMEMS/bioNEMS. Figure 47.15a summarizes the tribological problems encountered in some of the MEMS, MOEMS, RF-MEMS, and bioMEMS devices just discussed.

Microfabrication Processes

In addition to in-use stiction, stiction issues are also present in some processes used for the fabrication of MEMS/NEMS. For example, the last step in surface micromachining involves the removal of sacrificial layer(s), called release since the microstructures are released from the surrounding sacrificial layer(s). Release is accomplished by an aqueous chemical etch, rinsing, and drying processes. Due to meniscus effects as a result of wet processes, suspended structures can sometimes collapse and permanently adhere to the underlying substrate, as shown in Fig. 47.15b [47.144]. Adhesion is caused by water molecules adsorbed on the adhering surfaces and/or because of formation of adhesive bonds by silica residues that remain on the surfaces after the water has evaporated. This so-called release stiction is overcome by using dry release methods, such as CO_2 critical-point drying or sublimation drying [47.145]. CO_2 at high pressure is in a supercritical state and becomes liquid. Liquid CO_2 is used to remove wet etchant, and then it is converted back to gas phase.

Tribological Needs

Various MEMS/NEMS are designed to perform expected functions in millisecond to picosecond range. The expected life of the devices for high-speed contacts can vary from a few hundred thousand to many billions of cycles, e.g., over a hundred billion cycles for DMDs, which places stringent requirements on materials [47.13, 94, 146–149]. Adhesion between a biological molecular layer and the substrate (referred to as *bioadhesion*), reduction of friction and wear of biological layers, biocompatibility, and biofouling for bioMEMS/bioNEMS are important. Most mechanical properties are known to be scale dependent [47.150]. The properties of nanoscale structures need to be measured [47.151]. There is a need for the development of fundamental understanding of adhesion, friction/stiction, wear, and the role of surface contamination and environment [47.13]. MEMS/NEMS materials need to possess good mechanical and tribological properties on the micro-nanoscale. There is a need to develop lubricants and identify lubrication methods that are suitable for MEMS/NEMS. Methods need to be developed to enhance adhesion between biomolecules and the device substrate, referred to as bioadhesion. Component-level studies are required to provide better understanding of the tribological phenomena occurring in MEMS/NEMS.

The emergence of the field of nanotribology and nanomechanics, and atomic-force microscopy-based techniques, has provided researchers with a viable approach to address these problems [47.46, 47, 84, 152–155]. This chapter presents an overview of nanoscale adhesion, friction, and wear studies of materials and lubrication for MEMS/NEMS and bioMEMS/bioNEMS, and component-level studies of stiction phenomena in MEMS/NEMS devices. The emerging field of biomimetics holds promise for the development of biologically inspired nanomaterials and nanotechnology products [47.156]. One example includes the design of roughness induced surfaces with superhydrophobicity, self-cleaning, and low adhesion based on the so-called lotus effect. An overview of hierarchical nanostructured surfaces with superhydrophobicity, self-cleaning, and low adhesion is also presented.

47.2 Nanotribology and Nanomechanics Studies of Silicon and Related Materials

Materials of most interest for planar fabrication processes using silicon as the structural material are undoped and boron-doped (p^+-type) single-crystal silicon for bulk micromachining, and phosphorus (n^+-type) doped and undoped low-pressure chemical vapor deposition (LPCVD) polysilicon films for surface micromachining. Since silicon-based devices lack high-temperature capabilities with respect to both mechanical and electrical properties, SiC is being developed as a structural material for high-temperature microsensor and microactuator applications [47.157, 158]. SiC can also be desirable for high-frequency micromechanical resonators, in the GHz range, because of its high modulus of elasticity to density ratio and consequently high resonance frequency. Table 47.2 compares selected bulk properties of SiC and Si(100). Researchers have found low-cost techniques of producing single-crystalline 3C-SiC (cubic or β-SiC) films via epitaxial growth on large-area silicon substrates for bulk micromachining [47.159] and polycrystalline 3C-SiC films on polysilicon and silicon dioxide layers for surface micromachining of SiC [47.160]. Single-crystalline 3C-SiC piezoresistive pressure sensors have been fabricated using bulk micromachining for high-temperature gas-turbine applications [47.161]. Surface-micromachined polycrystalline SiC micromotors have been fabricated and have been reported to provide satisfactory operation at high temperatures [47.162].

Table 47.2 Selected bulk properties[a] of 3C (β- or cubic) SiC and Si(100)

Sample	Density (kg/m^3)	Hardness (GPa)	Elastic modulus (GPa)	Fracture toughness ($MPa\,m^{1/2}$)	Thermal conductivity[b] (W/(m K))	Coeff. of thermal expansion[b] ($\times 10^{-6}\,K^{-1}$)	Melting point (°C)	Bandgap (eV)
β-SiC	3210	23.5–26.5	440	4.6	85–260	4.5–6	2830	2.3
Si(100)	2330	9–10	130	0.95	155	2–4.5	1410	1.1

[a] Unless otherwise stated, data shown were obtained from [47.163]
[b] Obtained from [47.164]

As will be shown, bare silicon exhibits inadequate tribological performance and needs to be coated with a solid and/or liquid overcoat or be surface treated (e.g., oxidation and ion implantation, commonly used in semiconductor manufacturing), which reduces friction and wear. SiC films exhibit good tribological performance. Both macroscale and microscale tribological properties of virgin and treated/coated silicon, polysilicon films, and SiC are presented next.

47.2.1 Virgin and Treated/Coated Silicon Samples

Nanotribological and Nanomechanical Properties of Silicon and Effect of Ion Implantation

Friction and wear of single-crystalline and polycrystalline silicon samples have been studied, and the effect of ion implantation with various doses of C^+, B^+, N_2^+, and Ar^+ ion species at 200 keV energy to improve their friction and wear properties has been studied [47.165–167]. The coefficient of macroscale friction and the wear factor of virgin single-crystal silicon and C^+-implanted silicon samples as a function of ion dose are presented in Fig. 47.16 [47.165]. The macroscale friction and wear tests were conducted using a ball-on-flat tribometer. Each data bar represents the average value of four to six measurements. The coefficient of friction and wear factor for bare silicon are very high and decrease drastically with ion dose. Silicon samples bombarded with an ion dose above $10^{17}\,C^+\,cm^{-2}$ exhibit extremely low values of coefficient of friction (typically 0.03–0.06 in air) and wear factor (reduced by as much as four orders of magnitude). *Gupta* et al. [47.165] reported that a decrease in the coefficient of friction and wear factor of silicon as a result of C^+ ion bombardment occurred because of the formation of silicon carbide rather than amorphization of silicon. *Gupta* et al. [47.166] also reported an improvement in friction and wear with B^+ ion implantation.

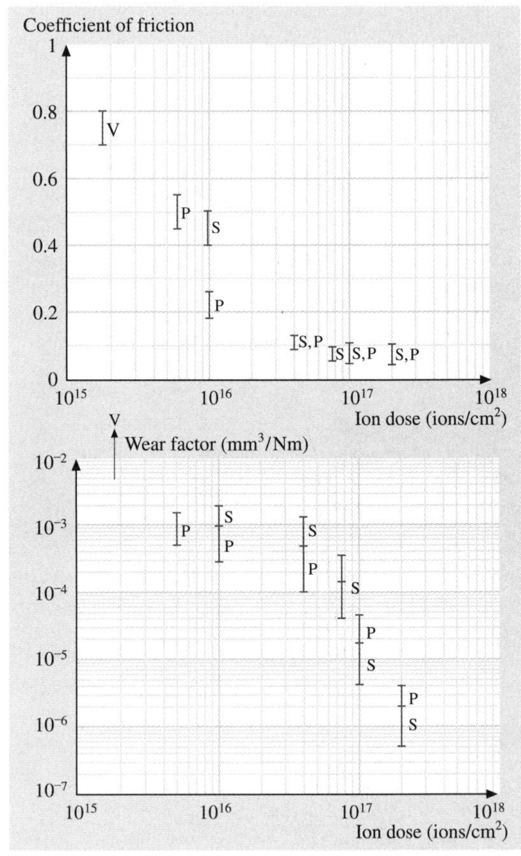

Fig. 47.16 Influence of ion doses on the coefficient of friction and wear factor on C^+-ion bombarded single-crystal and polycrystalline silicon slid against an alumina ball. V corresponds to virgin single-crystal silicon, while S and P denote tests for doped single- and polycrystalline silicon, respectively (after [47.165])

Table 47.3 Surface roughness and micro- and macroscale coefficients of friction of selected samples

Material	RMS roughness (nm)	Coefficient of microscale friction[a]	Coefficient of macroscale friction[b]
Si(111)	0.11	0.03	0.33
C^+-implanted Si(111)	0.33	0.02	0.18

[a] Versus Si_3N_4 tip, tip radius of 50 nm in the load range of 10–150 nN (2.5–6.1 GPa) at a scanning speed of 5 μm/s over a scan area of 1 μm × 1 μm in an AFM

[b] Versus Si_3N_4 ball, ball radius of 3 mm at a normal load of 0.1 N (0.3 GPa) at an average sliding speed of 0.8 mm/s using a tribometer

Microscale friction measurements were performed using an atomic force/friction force microscope (AFM/FFM) [47.46, 47, 84, 153]. Table 47.3 presents values of surface roughness and coefficients of macroscale and microscale friction for virgin and doped silicon. There is a decrease in the coefficients of microscale and macroscale friction values as a result of ion implantation. When measured for the small contact areas and very low loads used in microscale studies, indentation hardness and elastic modulus are higher than at the macroscale. This, added to the effect of the small apparent area of contact reducing the number of trapped particles on the interface, results in less plowing contribution and lower friction in the case of microscale friction measurements. Results of microscale wear resistance studies of ion-implanted silicon samples studied using a diamond tip in an AFM [47.168] are shown in Fig. 47.17a,b. For tests conducted at various loads on Si(111) and C^+-implanted Si(111), it is noted that the wear resistance of the implanted sample is slightly poorer than that of virgin silicon up to ≈ 80 μN. Above 80 μN, the wear resistance of implanted Si improves. As one continues to run tests at 40 μN for a larger number of cycles, the implanted sample, which forms hard and tough silicon carbide, exhibits higher wear resistance than the unimplanted sample. Damage from the implantation in the top layer results in poorer wear resistance; however, the implanted zone at the subsurface is more wear resistant than the virgin silicon.

Hardness values of virgin and C^+-implanted Si(111) at various indentation depths (normal loads) are presented in Fig. 47.17c [47.168]. The hardness at a small indentation depth of 2.5 nm is 16.6 GPa, and it drops to a value of 11.7 GPa at a depth of 7 nm and a normal load of 100 μN. Higher hardness values obtained in low-load indentation may arise from the observed pressure-induced phase transformation during nanoindentation [47.169, 170]. Additional increase in the hardness at an even lower indentation depth of 2.5 nm reported here may arise from the contribution of complex chemical films (not from native oxide films) present on the silicon surface. At small volumes there is a lower probability of encountering material defects (dislocations, etc.). Furthermore, according to the strain gradient plasticity theory advanced by *Fleck* et al. [47.171], large strain gradients inherent to small indentations lead to the accumulation of geometrically necessary dislocations that cause enhanced hardening. These are some of the plausible explanations for an increase in hardness at smaller volumes. If the silicon material were to be used at very light loads, such as in microsystems, the high hardness of surface films would protect the surface until it is worn.

From Fig. 47.17c, the hardness value of C^+-implanted Si(111) at normal load of 50 μN is 20.0 GPa for an indentation depth of ≈ 2 nm, which is comparable to the hardness value of 19.5 GPa at 70 μN, whereas the measured hardness value for virgin silicon at an indentation depth of ≈ 7 nm (normal load of 100 μN) is only ≈ 11.7 GPa. Thus, ion implantation with C^+ results in an increase in hardness in silicon. Note that the surface layer of the implanted zone is much harder compared with the subsurface and may be brittle, leading to higher wear on the surface. The subsurface of the implanted zone (SiC) is harder than the virgin silicon, resulting in higher wear resistance, which is also observed in the results of macroscale tests conducted at high loads.

Effect of Oxide Films on Nanotribological Properties of Silicon

Macroscale friction and wear experiments have been performed using a magnetic disk drive with bare, oxidized, and implanted pins sliding against amorphous-carbon-coated magnetic disks lubricated with a thin layer of perfluoropolyether lubricant [47.172–175]. Representative profiles for the variation of the coefficient of friction with number of sliding cycles for an Al_2O_3-TiC slider and bare and dry-oxidized silicon

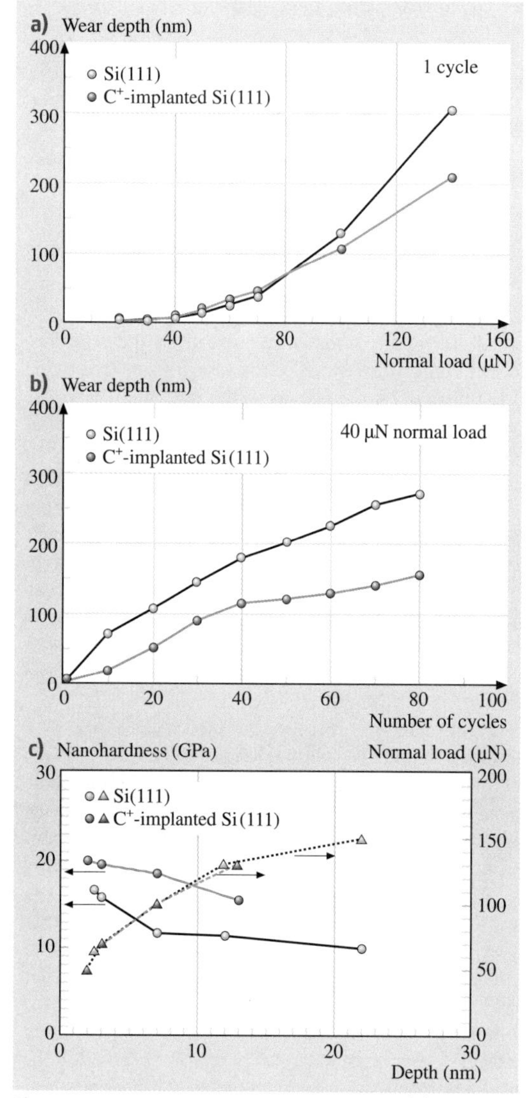

Fig. 47.17a–c Wear depth as a function of (a) load (after one cycle), and (b) cycles (normal load = 40 mN) for Si(111) and C^+-implanted Si(111). (c) Nanohardness and normal load as functions of indentation depth for virgin and C^+-implanted Si(111) (after [47.168])

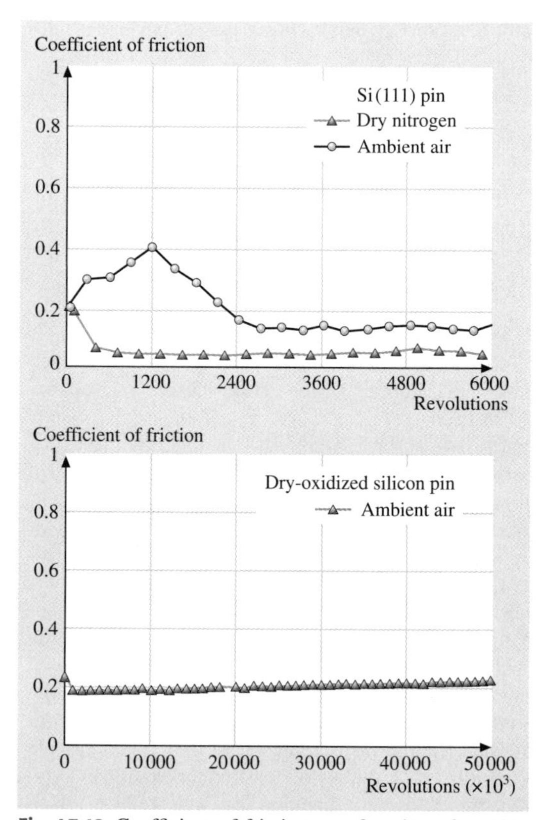

Fig. 47.18 Coefficient of friction as a function of number of sliding revolutions in ambient air for a Si(111) pin in ambient air and dry nitrogen, and a dry-oxidized silicon pin in ambient air (after [47.172])

pins are shown in Fig. 47.18. For bare Si(111), after an initial increase in the coefficient of friction, it drops to a steady state value of 0.1, as seen in Fig. 47.18. The rise in the coefficient of friction for the Si(111) pin is associated with the transfer of amorphous carbon from the disk to the pin and oxidation-enhanced fracture of pin material, followed by tribochemical oxidation of the transfer film, while the drop is associated with the formation of a transfer coating on the pin. Dry-oxidized Si(111) exhibits excellent characteristics, and no significant increase was observed over 500 cycles (Fig. 47.18). This behavior has been attributed to the chemical passivity of the oxide and the lack of transfer of diamond-like carbon (DLC) from the disk to the pin. The behavior of PECVD oxide (data are not presented here) was comparable to that of dry oxide, but for the wet oxide there was some variation in the coefficient of friction (0.26 to 0.4). The difference between the dry and wet oxide was attributed to increased porosity of the wet oxide [47.172]. Since tribochemical oxidation was determined to be a significant factor, experiments were conducted in dry nitrogen [47.173, 174].

Table 47.4 RMS, microfriction, microscratching/microwear, and nanoindentation hardness data for various virgin, coated, and treated silicon samples

Material	RMS roughness[a] (nm)	Coefficient of microscale friction[b]	Scratch depth[c] at 40 μN (nm)	Wear depth[c] at 40 μN (nm)	Nanohardness[c] at 100 μN (GPa)
Si(111)	0.11	0.03	20	27	11.7
Si(110)	0.09	0.04	20	–	–
Si(100)	0.12	0.03	25	–	–
Polysilicon	1.07	0.04	18	–	–
Polysilicon (lapped)	0.16	0.05	18	25	12.5
PECVD-oxide coated Si(111)	1.50	0.01	8	5	18.0
Dry-oxidized Si(111)	0.11	0.04	16	14	17.0
Wet-oxidized Si(111)	0.25	0.04	17	18	14.4
C^+-implanted Si(111)	0.33	0.02	20	23	18.6

[a] Scan size of 500 nm × 500 nm using AFM

[b] Versus Si_3N_4 tip in AFM/FFM, radius 50 nm; at 1 μm × 1 μm scan size

[c] Measured using an AFM with a diamond tip of radius of 100 nm

The variation of the coefficient of friction for a silicon pin sliding against a thin-film disk in dry nitrogen is shown in Fig. 47.18. It is seen that, in a dry nitrogen environment, the coefficient of friction of Si(111) sliding against a disk decreased from an initial value of about 0.2 to 0.05 with continued sliding. Based on SEM and chemical analysis, this behavior has been attributed to the formation of a smooth amorphous-carbon/lubricant transfer patch and suppression of oxidation in a dry nitrogen environment. Based on macroscale tests using disk drives, it is found that the friction and wear performance of bare silicon is not adequate. With dry-oxidized or PECVD SiO_2-coated silicon, no significant friction increase or interfacial degradation was observed in ambient air.

Table 47.4 and Fig. 47.19 show surface roughness, microscale friction, and scratch data, and nanoindentation hardness for the various silicon samples [47.168]. Scratch experiments were performed using a diamond tip in an AFM. Results on polysilicon samples are also shown for comparison. Coefficients of microscale friction values for all the samples are about the same. These samples could be scratched at a 10 μN load. Scratch depth increased with normal load. Crystalline orientation of silicon has little influence on scratch resistance because natural oxidation of silicon in ambient masks the expected effect of crystallographic orientation. PECVD-oxide samples showed the best scratch resistance, followed by dry-oxidized, wet-oxidized, and ion-implanted samples. Ion implantation with C^+ does not appear to improve scratch resistance.

Wear data on the silicon samples are also presented in Table 47.4 [47.168]. PECVD-oxide samples showed superior wear resistance followed by dry-oxidized, wet-oxidized, and ion-implanted samples. This agrees with the trends seen in scratch resistance. In PECVD, ion bombardment during the deposition improves the coating properties such as suppression of columnar growth, freedom from pinhole, decrease in crystalline size, and increase in density, hardness, and substrate-coating adhesion. These effects may help in improving the mechanical integrity of the sample surface. Coatings and treatments improved the nanohardness of

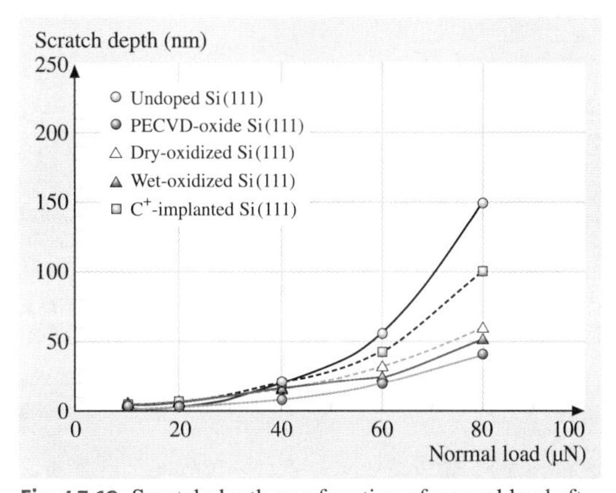

Fig. 47.19 Scratch depth as a function of normal load after ten cycles for various silicon samples: virgin, treated, and coated (after [47.168])

Table 47.5 Summary of micro-/nanotribological properties of the sample materials

Sample	RMS roughness[a] (nm)	P–V distance[a] (nm)	Coefficient of friction Micro[b]	Coefficient of friction Macro[c]	Scratch depth[d] (nm)	Wear depth[e] (nm)	Nano-hardness[f] (GPa)	Young's modulus[f] (GPa)	Fracture toughness[g], K_{IC} MPa m$^{1/2}$
Undoped Si(100)	0.09	0.9	0.06	0.33	89	84	12	168	0.75
Undoped polysilicon film (as deposited)	46	340	0.05	–	–	–	–	–	–
Undoped polysilicon film (polished)	0.86	6	0.04	0.46	99	140	12	175	1.11
n^+-Type polysilicon film (as deposited)	12	91	0.07	–	–	–	–	–	–
n^+-Type polysilicon film (polished)	1.0	7	0.02	0.23	61	51	9	95	0.89
SiC film (as deposited)	25	150	0.03	–	–	–	–	–	–
SiC film (polished)	0.89	6	0.02	0.20	6	16	25	395	0.78

[a] Measured using AFM over a scan size of 10 μm × 10 μm
[b] Measured using AFM/FFM over a scan size of 10 μm × 10 μm
[c] Obtained using a 3 mm-diameter sapphire ball in a reciprocating mode at a normal load of 10 mN and average sliding speed of 1 mm/s after 4 m sliding distance
[d] Measured using AFM at a normal load of 40 μN for ten cycles, scan length of 5 μm
[e] Measured using AFM at normal load of 40 μN for one cycle, wear area of 2 μm × 2 μm
[f] Measured using nanoindenter at a peak indentation depth of 20 nm
[g] Measured using microindenter with Vickers indenter at a normal load of 0.5 N

silicon. Note that dry-oxidized and PECVD films are harder than wet-oxidized films as these films may be porous. High hardness of oxidized films may be responsible for the measured high scratch/wear resistance.

47.2.2 Nanotribological and Nanomechanical Properties of Polysilicon Films and SiC Film

Studies have also been conducted on undoped polysilicon film, heavily doped (n^+-type) polysilicon film, heavily doped (p^+-type) single-crystal Si(100), and 3C-SiC (cubic or β-SiC) film [47.176–178]. The polysilicon films studied here are different from the ones discussed previously.

Table 47.5 presents a summary of the tribological studies conducted on polysilicon and SiC films. Values for single-crystal silicon are also shown for comparison. Polishing of the as-deposited polysilicon and SiC films drastically affect the roughness as the values reduce by two orders of magnitude. Si(100) appears to be the smoothest, followed by polished undoped polysilicon and SiC films, which have comparable roughness. The doped polysilicon film shows higher roughness than the undoped sample, which is attributed to the doping process. Polished SiC film shows the lowest friction, followed by polished and undoped polysilicon film, which strongly supports the candidacy of SiC films for use in MEMS/NEMS devices. Macroscale friction measurements indicate that SiC film exhibits one of the lowest friction values as compared with the other samples. Doped polysilicon sample shows low friction on the macroscale as compared with the undoped polysilicon sample, possibly due to the doping effect.

Figure 47.20a shows a plot of scratch depth versus normal load for various samples [47.176, 177]. Scratch depth increases with increasing normal load. Figure 47.21 shows three-dimensional (3-D) AFM maps and averaged two-dimensional (2-D) profiles of the scratch marks on the various samples. It is observed that scratch depth increases almost linearly with normal load. Si(100) and the doped and undoped polysilicon film show similar scratch resistance. From the data, it is clear that the SiC film is much more scratch resistant than the other samples. Figure 47.20b shows results from microscale wear tests on the various films. For all the materials, the wear depth increases almost linearly with increasing number of cycles. This suggests that the material is removed layer by layer in all the materials. Here also, SiC film exhibits lower wear depths than the other samples. Doped polysilicon film wears less than the undoped film. Higher fracture toughness and higher hardness of SiC as compared with Si(100) is responsible for its lower wear. Also the higher thermal conductivity of SiC (Table 47.2) as compared with the other materials leads to lower interface temperatures, which generally results in less degradation of the surface [47.34, 80, 82]. Doping of

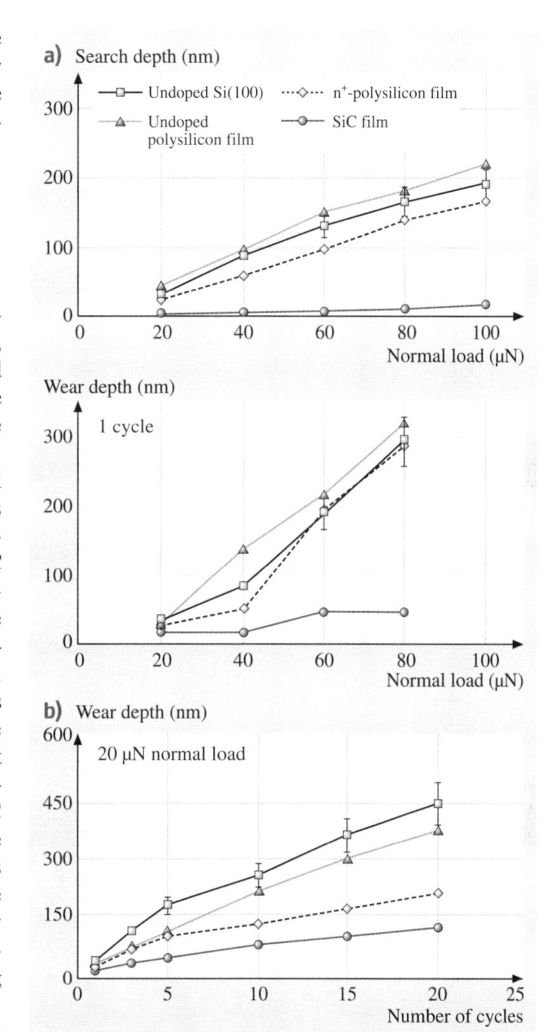

Fig. 47.20 (a) Scratch depths for ten cycles as a function of normal load, and (b) wear depths as a function of normal load and as a function of number of cycles for various samples (after [47.176])

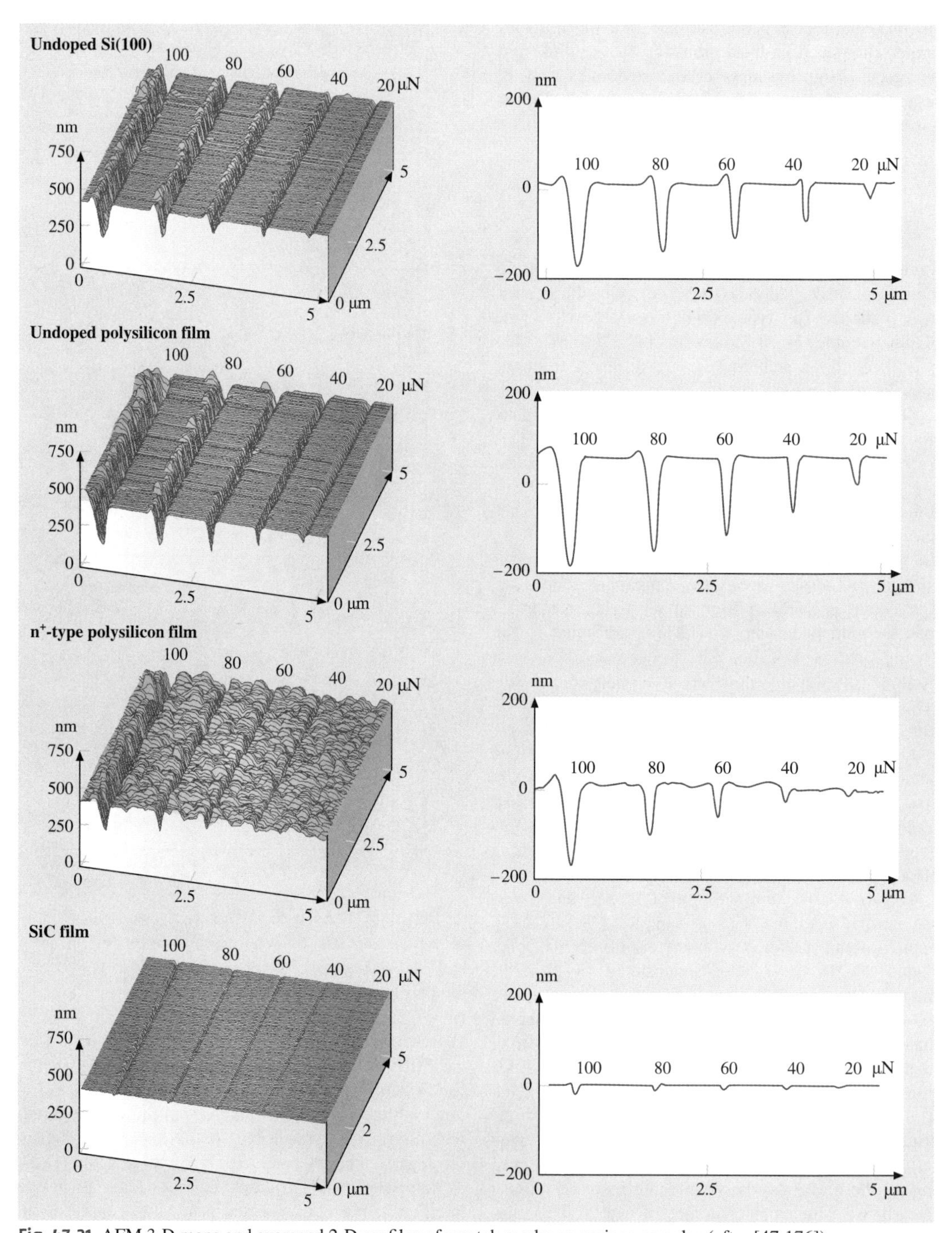

Fig. 47.21 AFM 3-D maps and averaged 2-D profiles of scratch marks on various samples (after [47.176])

the polysilicon does not affect the scratch/wear resistance and hardness much. The measurements made on the doped sample are affected by the presence of grain boundaries. These studies indicate that SiC film exhibits desirable tribological properties for use in MEMS devices.

47.3 Lubrication Studies for MEMS/NEMS

Several studies of liquid perfluoropolyether (PFPE) lubricant films, self-assembled monolayers (SAMs), and hard diamond-like carbon (DLC) coatings have been carried out for the purpose of minimizing adhesion, friction, and wear [47.46, 47, 80–84, 152–155, 175]. Many variations of these films are hydrophobic (low surface tension and high contact angle) and have low shear strength, which provides low adhesion, friction, and wear. Relevant details are presented below.

47.3.1 Perfluoropolyether Lubricants

The classical approach to lubrication uses freely supported multimolecular layers of liquid lubricants [47.46, 80, 82, 84]. The liquid lubricants are sometimes chemically bonded to improve their wear resistance. Partially chemically bonded, molecularly thick perfluoropolyether (PFPE) lubricants are widely used for lubrication of magnetic storage media, because of their thermal stability and extremely low vapor pressure [47.34], and are found to be suitable for MEMS/NEMS devices.

Adhesion, friction, and durability experiments have been performed on virgin Si(100) surfaces and silicon surfaces lubricated with various PFPE lubricants [47.46, 47, 179–185]. Results of the following two PFPE lubricants will be presented here: Z-15 (with $-CF_3$ nonpolar end groups), $CF_3-O-(CF_2-CF_2-O)_m-(CF_2-O)_n-CF_3$ ($m/n \approx 2/3$); and Z-DOL (with $-OH$ polar end groups), $HO-CH_2-CF_2-O-(CF_2-CF_2-O)_m-(CF_2-O)_n-CF_2-CH_2-OH$ ($m/n \approx 2/3$). Z-DOL film was thermally bonded at 150 °C for 30 min, and the unbonded fraction was removed by a solvent (fully bonded) [47.34]. The thicknesses of Z-15 and Z-DOL (fully bonded) films were 2.8 and 2.3 nm, respectively. Nanoscale measurements were made using an AFM. The adhesive forces of Si(100), Z-15, and Z-DOL (fully bonded) measured by force calibration plot, and friction force versus normal load plot, are summarized in Fig. 47.22. The results measured by these two methods are in good agreement. Figure 47.22 shows that the presence of mobile Z-15 lubricant film increases the adhesive force as compared with that of Si(100) by meniscus formation [47.80, 82, 186], whereas the presence of solid-phase Z-DOL (fully bonded) film reduces the adhesive force as compared with that of Si(100) because of the absence of mobile liquid. The schematic (bottom) in Fig. 47.22 shows the relative size and sources of meniscus. It is well known that the native oxide layer (SiO_2) on the top of Si(100) wafer exhibits hydrophilic properties, and some water molecules can be adsorbed on this surface. The condensed water will form a meniscus as the tip approaches the sample surface. The larger adhesive force in Z-15 is not just caused by the Z-15 meniscus; the nonpolarized Z-

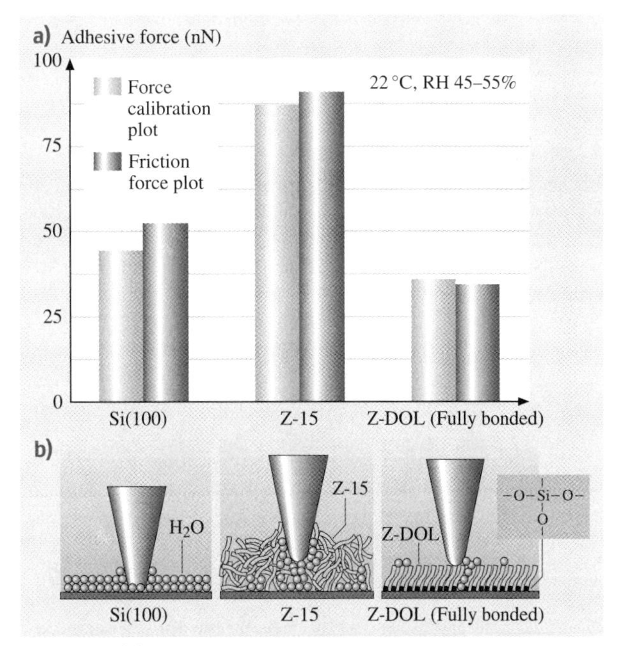

Fig. 47.22 **(a)** Summary of the adhesive forces of Si(100), Z-15, and Z-DOL (fully bonded) films measured by force calibration plots and friction force versus normal load plots in ambient air. **(b)** Schematic showing the effect of meniscus, formed between the AFM tip and the surface sample, on the adhesive and friction forces (after [47.179])

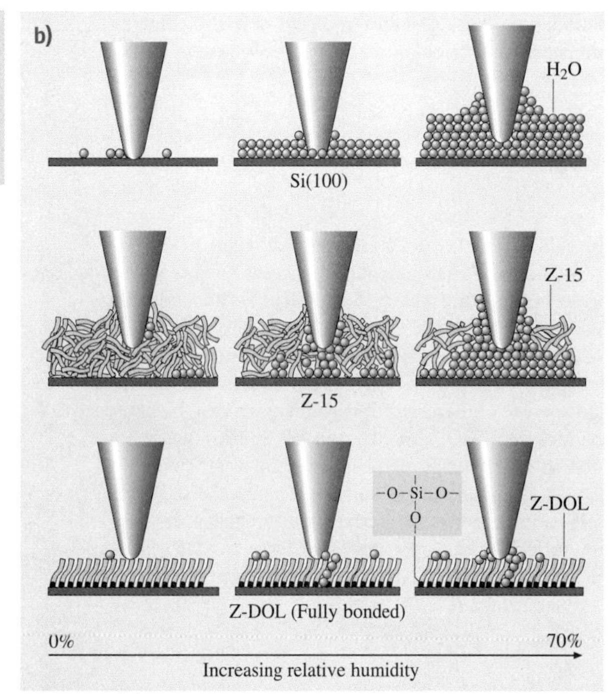

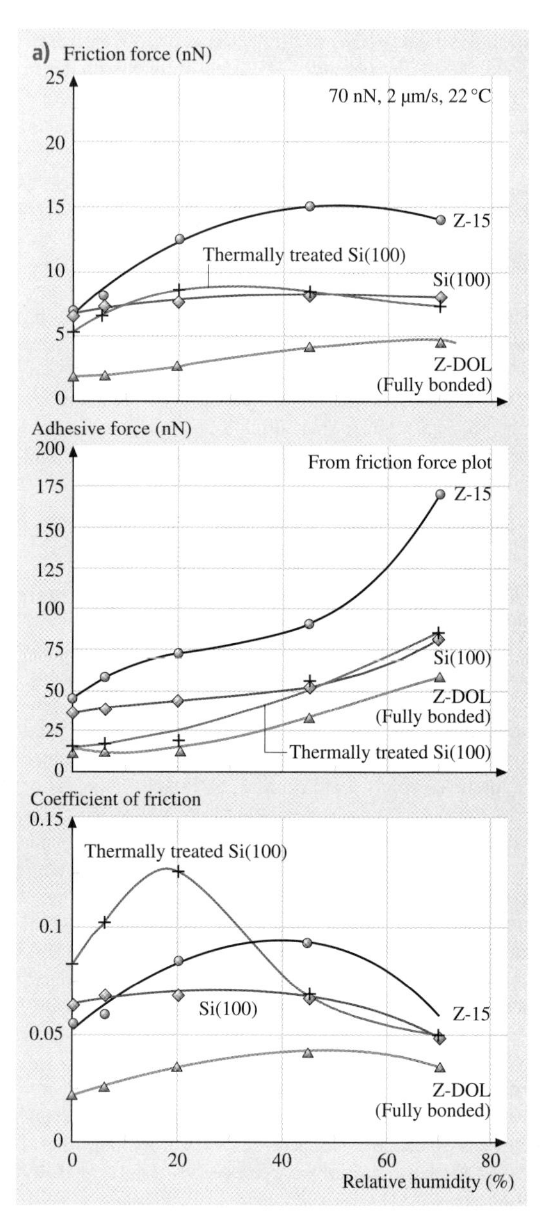

Fig. 47.23 (**a**) Influence of relative humidity on the friction force, adhesive force, and coefficient of friction of Si(100), Z-15, and Z-DOL (fully bonded) films at 70 nN, 2 µm/s, and in 22 °C air. (**b**) Schematic showing the change of meniscus on increasing the relative humidity. In this figure, thermally treated Si(100) is Si(100) wafer that was baked at 150 °C for 1 h in an oven (in order to remove the adsorbed water) just before it was placed in the 0% RH chamber (after [47.179])

15 liquid does not have good wettability and strong bonding with Si(100). Consequently, in the ambient environment, condensed water molecules from the environment permeate through the liquid Z-15 lubricant film and compete with the lubricant molecules present on the substrate. The interaction of the liquid lubricant with the substrate is weakened, and a boundary layer of the liquid lubricant forms puddles [47.180,181]. This dewetting allows water molecules to be adsorbed onto the Si(100) surface as aggregates along with Z-15 molecules, and both of them can form a meniscus while the tip approaches the surface. Thus, the dewetting of liquid Z-15 film results in higher adhesive force and poorer lubrication performance. In addition, as the Z-15 film is fairly soft compared with the solid Si(100) surface, penetration of the tip into the film occurs while pushing the tip down. This leads to a large area of the tip being involved in forming the meniscus at the tip–liquid (mixture of Z-15 and water) interface. It should also be noted that Z-15 has a higher viscosity compared with water, therefore Z-15 film provides higher resistance to lateral motion and coefficient of friction. In the case of Z-DOL (fully bonded) film, the active groups of Z-DOL molecules are mostly bonded onto Si(100) sub-

strate, thus the Z-DOL (fully bonded) film has low free surface energy and cannot be displaced readily by water molecules or readily adsorb water molecules. Thus, the use of Z-DOL (fully bonded) can reduce the adhesive force.

To study the effect of relative humidity on friction and adhesion, the variation of friction force, adhesive force, and coefficient of friction of Si(100), Z-15, and Z-DOL (fully bonded) as a function of relative humidity are shown in Fig. 47.23. It shows that, for Si(100) and Z-15 film, the friction force increases with a relative humidity increase up to 45% and then shows a slight decrease with further increase in relative humidity. Z-DOL (fully bonded) has smaller friction force than Si(100) and Z-15 in the whole testing range, and its friction force shows an apparent relative increase for relative humidity > 45%. For Si(100), Z-15, and Z-DOL (fully bonded), their adhesive forces increase with relative humidity, and their coefficients of friction increase with a relative humidity up to 45%, after which they decrease with further increasing of the relative humidity. It is also observed that the effect of humidity on Si(100) really depends on the history of the Si(100) sample. As the surface of the Si(100) wafer readily adsorbs water in air, without any pretreatment the Si(100) used in our study almost reaches its saturated stage of adsorbed water, and is responsible for less effect during increasing relative humidity. However, once the Si(100) wafer was thermally treated by baking at 150 °C for 1 h, a greater effect was observed.

The schematic (right) in Fig. 47.23 shows that Si(100), because of its high free surface energy, can adsorb more water molecules with increasing relative humidity. As discussed earlier, for the Z-15 film in the humid environment, the condensed water competes with the lubricant film present on the sample surface, and interaction of the liquid lubricant film with the silicon substrate is weakened and a boundary layer of the liquid lubricant forms puddles. This dewetting allows water molecules to be adsorbed onto the Si(100) substrate mixed with Z-15 molecules [47.180, 181]. Obviously, more water molecules can be adsorbed on the Z-15 surface with increasing relative humidity. The more adsorbed water molecules in the case of Si(100), along with lubricant molecules in the Z-15 film case, form a bigger water meniscus, which leads to an increase of friction force, adhesive force, and coefficient of friction of Si(100) and Z-15 with humidity. However, at very high humidity of 70%, large quantities of adsorbed water can form a continuous water layer that separates the tip and sample surface, acting as a kind of lubricant, which causes a decrease in the friction force and coefficient of friction. For Z-DOL (fully bonded) film, because of its hydrophobic surface properties, water molecules can be adsorbed at humidity > 45% and cause an increase in the adhesive force and friction force.

To study the durability of lubricant films at the nanoscale, the friction force of Si(100), Z-15, and Z-DOL (fully bonded) as a function of the number of scanning cycles is shown in Fig. 47.24. As observed earlier, the friction force of Z-15 is higher than that of Si(100), with the lowest values for Z-DOL(fully bonded). During cycling, the friction force of Si(100) shows a slight decrease during the initial few cycles, then remains constant. This is related to the removal of the top adsorbed layer. In the case of the Z-15 film, the friction force shows an increase during the initial few

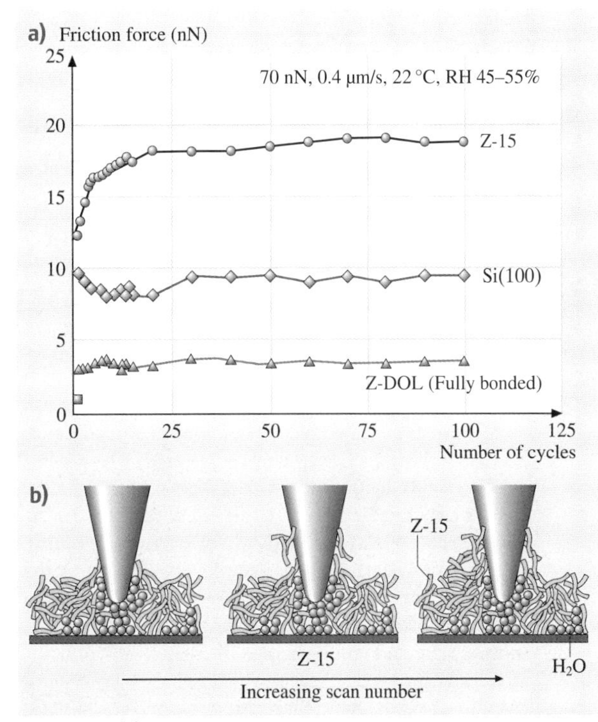

Fig. 47.24 (a) Friction force versus number of sliding cycles for Si(100), Z-15, and Z-DOL (fully bonded) films at 70 nN, 0.4 µm/s, and in ambient air. (b) Schematic (*bottom*) showing that some liquid Z-15 molecules can be attached to the tip. The molecular interaction between the molecules attached to the tip and the Z-15 molecules in the film results in an increase of the friction force on multiple scanning (after [47.179])

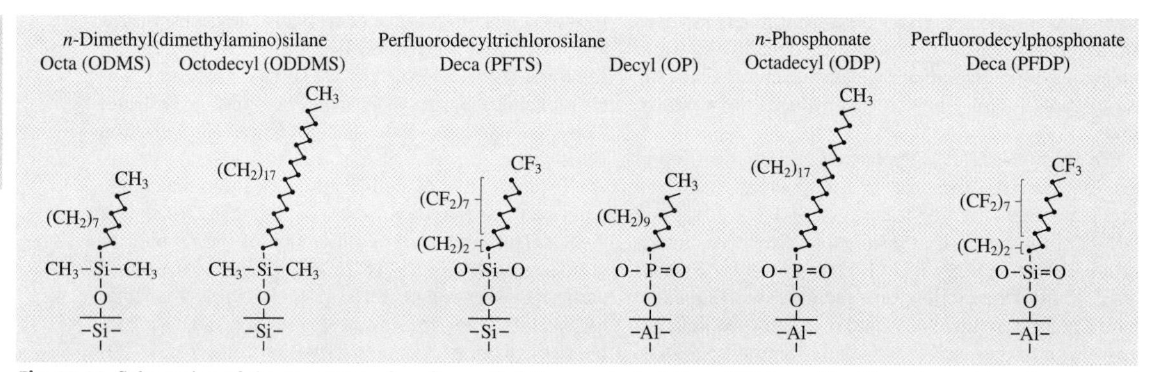

Fig. 47.25 Schematics of the structures of perfluoroalkylsilane and alkylsilane SAMs on Si with native oxide substrates, and perfluoroalkylphosphonate and alkylphosphonate SAMs on Al with native oxide substrates

cycles and then approaches higher and stable values. This is believed to be caused by attachment of the Z-15 molecules to the tip. The molecular interaction between these molecules attached to the tip and molecules on the film surface is responsible for the increase in friction. However, after several scans, this molecular interaction reaches equilibrium, after which the friction force and coefficient of friction remain constant. In the case of the Z-DOL (fully bonded) film, the friction force and coefficient of friction start out low and remain low during the entire test for 100 cycles. This suggests that Z-DOL (fully bonded) molecules do not become attached or displaced as readily as those of Z-15.

47.3.2 Self-Assembled Monolayers (SAMs)

For lubrication of MEMS/NEMS, another effective approach involves the deposition of organized and dense molecular layers of long-chain molecules. Two common methods to produce monolayers and thin films are the Langmuir–Blodgett (L–B) deposition and self-assembled monolayers (SAMs) by chemical grafting of molecules. L–B films are physically bonded to the substrate by a weak van der Waals attraction, while SAMs are chemically bonded via covalent bonds to the substrate. Because of the choice of chain length and terminal linking group that SAMs offer, they hold great promise for boundary lubrication of MEMS/NEMS. A number of studies have been conducted on the tribological properties of various SAMs [47.138–141, 187–193].

Bhushan and *Liu* [47.188] studied the effect of film compliance on adhesion and friction. Based on friction and stiffness measurements, SAMs with high-compliance long carbon chains exhibit low friction; chain compliance is desirable for low friction. The friction mechanism of SAMs is explained by a so-called *molecular-spring* model. According to this model, chemically adsorbed self-assembled molecules on a substrate are just like assembled molecular springs anchored to the substrate. An asperity sliding on the surface of SAMs is like a tip sliding on the top of *molecular springs* or a *molecular brush*. The molecular-spring assembly has compliant features and can experience orientation and compression under load. The orientation of the molecular springs or brush under normal load reduces the shearing force at the interface, which in turn reduces the friction force. The orientation is determined by the spring constant of a single molecule as well as the interaction between the neighboring molecules, which can be reflected by packing density or packing energy. It should be noted that orientation can lead to conformational defects along the molecular chains, which leads to energy dissipation. SAMs with high-compliance long carbon chains also exhibit the best wear resistance [47.188, 190]. In wear experiments, curves of wear depth as a function of normal load show a critical normal load. Below the critical normal load, SAMs undergo orientation. At the critical load, SAMs wear away from the substrate due to weak interface bond strengths, while above the critical normal load severe wear takes place on the substrate.

Bhushan et al. [47.139, 191], *Kasai* et al. [47.141], *Tambe* and *Bhushan* [47.192], and *Tao* and *Bhushan* [47.193] studied various SAMs which were vapor-phase-deposited on Si and Al substrates with native oxide films (Fig. 47.25). Perfluorodecyltricholorosilane (PFTS), *n*-octyldimethyl (dimethylamino) silane (ODMS) ($n = 7$), and *n*-octadecylmethyl (dimethy-

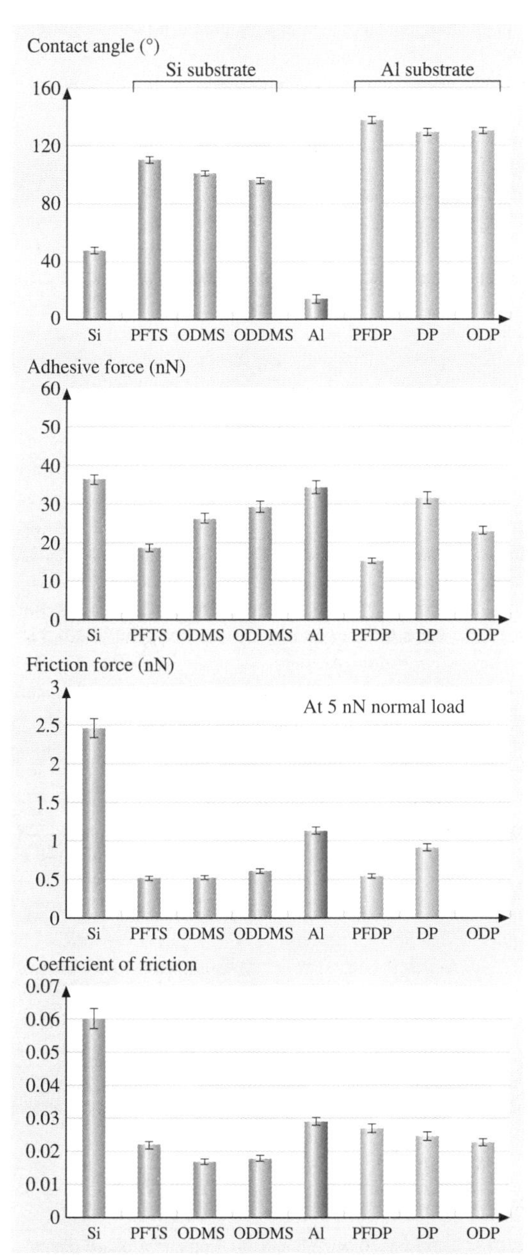

Fig. 47.26 The static contact angle, adhesive force, friction force, and coefficient of friction measured using an AFM for various SAMs on Si and Al substrates (after [47.131, 139, 191])

lamino)silane (ODDMS) ($n = 17$) were deposited on Si substrate. Perfluorodecylphosphonate (PFDP), n-decylphosphonate (DP) ($n = 7$), and n-octadecylphosphonate (ODP) ($n = 17$) were deposited on Al substrate. Figure 47.26 shows the static contact angle, adhesive force, friction force, and coefficient of friction of the two substrates with various SAMs under ambient conditions [47.139, 191]. Based on the data, all films exhibit higher contact angle and lower adhesive force and coefficient of friction as compared with corresponding substrates. Among the various films, PFTS/Si exhibits higher contact angle and lower adhesive force as compared with ODMS/Si and ODDMS/Si. Longer-chain film ODDMS/Si has superior performance than shorter-chain film ODMS. Trends for films on Al substrate are similar to that on Si substrate. Thus, substrate had little effect. The coefficients of friction of various SAMs are comparable.

The effect of relative humidity for various SAMs on Si substrate on adhesion and friction was studied. Adhesive force, friction force at 5 nN of normal load, coefficient of friction, and microwear data are presented in Fig. 47.27 [47.141, 191]. The result of adhesive force for silicon showed an increase with relative humidity. This is expected since the surface of silicon is hydrophilic, as shown in Fig. 47.26. More condensation of water at the tip–sample interface at higher humidity increases the adhesive force due to the capillary effect. On the other hand, the adhesive force for the SAMs showed a very weak dependency on humidity. This occurs since the surface of the SAMs is hydrophobic. The adhesive force of ODMS/Si and ODDMS/Si showed a slight increase from 75% to 90% RH. Such increase was absent for PFTS/Si, possibly because of the hydrophobicity of PFTS/Si. The friction force of silicon showed an increase with relative humidity up to $\approx 75\%$ RH and a slight decrease beyond this point. The initial increase possibly results from the increase in adhesive force. The decrease in friction force at higher humidity could be attributed to the lubricating effect of the water layer. This effect is more pronounced in the coefficient of friction. Since the adhesive force increased and the coefficient of friction decreased in this range, those effects cancel out each other, and the resulting friction force showed slight changes. On the other hand, the friction force and coefficient of friction of SAMs showed very small changes with relative humidity, like that found for adhesive force. This suggests that the adsorbed water layer on the surface maintained

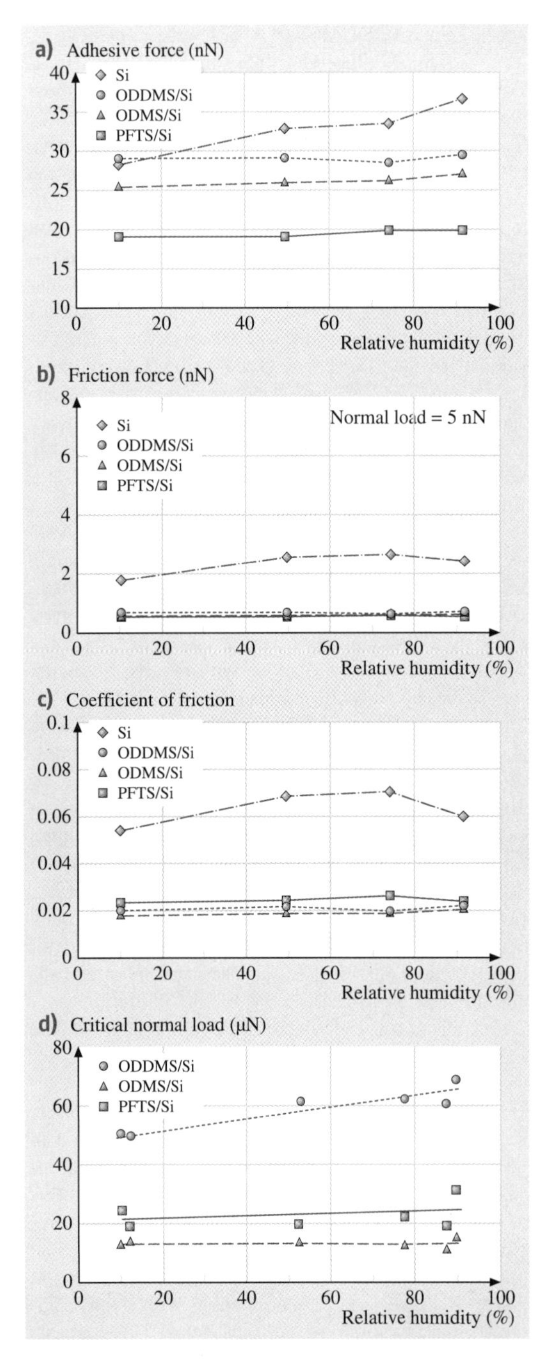

Fig. 47.27a–d Effect of relative humidity on (**a**) adhesive force, (**b**) friction force, (**c**) coefficient of friction, and (**d**) microwear for various SAMs on Si substrate (after [47.141, 191]) ◄

a similar thickness throughout the tested relative humidity range. The differences among the SAM types were small, within the measurement error; however, a closer look at the coefficient of friction for ODMS/Si showed a slight increase from 75% to 90% RH as compared with PFTS/Si, possibly due to the same reason as for the adhesive force increment. The inherent hydrophobicity of SAMs means that they did not show much relative humidity dependence.

Figure 47.28a shows the relationship between the decrease in surface height as a function of the normal load during wear tests for various SAMs on Si and Al substrates [47.141, 191]. As shown in the figure, the SAMs exhibit a critical normal load, beyond which the surface height drastically decreases. Figure 47.28a also shows the wear behavior of the Al and Si substrates. Unlike the SAMs, the substrates show a monotonic decrease in surface height with increasing normal load, with wear initiating from the very beginning, i. e., even for low normal loads. Si (Young's modulus of elasticity, $E = 130\,\text{GPa}$ [47.194, 195], hardness, $H = 11\,\text{GPa}$ [47.175]) is relatively hard in comparison with Al ($E = 77\,\text{GPa}$, $H = 0.41\,\text{GPa}$), and hence the decrease in surface height for Al is much larger than that for Si for similar normal loads.

The critical loads corresponding to the sudden failure of SAMs are shown in Fig. 47.28b. Amongst all the SAMs, ODDMS shows the best performance in the wear tests, which is believed to be because of the effect of the longer chain length. Fluorinated SAMs – PFTS and PFDP – show higher critical load as compared with ODMS and DP with similar chain lengths. ODP shows higher critical load as compared with DP because of its longer chain length. The mechanism of failure of compliant SAMs during wear tests was presented earlier. It is believed that the SAMs fail mostly due to shearing of the molecule at the head group, that is, by means of shearing of the molecules off the substrate.

To study the effect of relative humidity, wear tests were performed at various humidities. Figure 47.27d shows critical normal load as a function of relative humidity. The critical normal load showed weak dependency on relative humidity for ODMS/Si and PFTS/Si, and was larger for ODMS/Si than PFTS/Si throughout the humidity range. This suggests that water molecules could penetrate into the ODDMS, which might work as a lubricant [47.190]. This effect was absent for PFTS/Si and ODMS/Si.

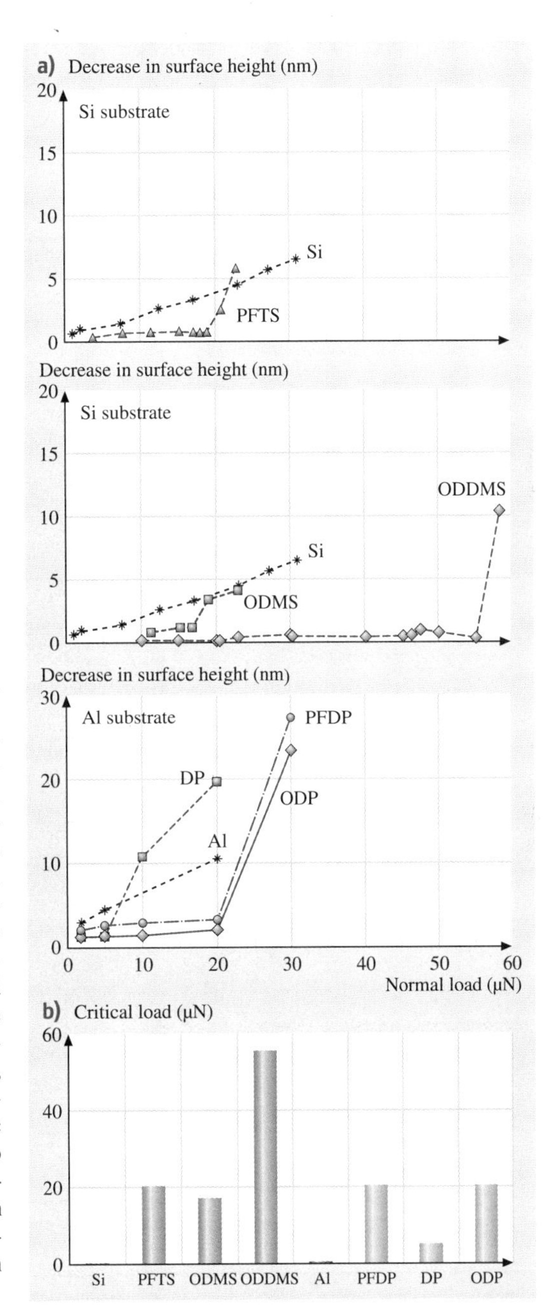

Fig. 47.28 (a) Decrease in surface height as a function of normal load after one scan cycle for various SAMs on Si and Al substrates. (b) Comparison of critical loads for failure during wear tests for various SAMs (after [47.141, 191]) ▶

Bhushan et al. [47.140] and *Lee* et al. [47.138] studied various fluoropolymer multilayers and fluorosilane monolayers on Si and a selected fluorosilane on PDMS surfaces. For nanoscale devices, such as in nanochannels, monolayers are preferred. They reported that all fluorosilane films increased the contact angle. The fluorosilane monolayer – 1H,1H,2H,2H-perfluorodecyltriethoxysilane (PFDTES) – resulted in a contact angle of $\approx 100°$.

Based on these studies, a perfluoro SAM with a compliant layer should have optimized tribological performance for MEMS/NEMS and bioMEMS/bioNEMS applications.

47.3.3 Hard Diamond-Like Carbon (DLC) Coatings

Hard amorphous carbon (a-C), commonly known as DLC (implying high hardness), coatings are deposited by a variety of deposition techniques, including filtered cathodic arc (FCA), ion beam, electron cyclotron resonance chemical vapor deposition (ECR-CVD), plasma-enhanced chemical vapor deposition (PECVD), and sputtering [47.163, 175]. These coatings are used in a wide range of applications, including tribological, optical, electronic, and biomedical applications. Ultrathin coatings (3–10 nm thick) are employed to protect against wear and corrosion in magnetic storage applications – thin-film rigid disks, metal evaporated tapes, and thin-film read/write head –, Gillette Mach 3 razor blades, glass windows, and sunglasses. The coatings exhibit low friction, high hardness and wear resistance, chemical inertness to both acids and alkalis, lack of magnetic response, and optical bandgap ranging from zero to a few electron-volts, depending upon the deposition technique and its conditions. Selected data on DLC coatings relevant for MEMS/NEMS applications are presented in a following section on adhesion measurements.

47.4 Nanotribological Studies of Biological Molecules on Silicon-Based and Polymer Surfaces and Submicron Particles for Therapeutics and Diagnostics

47.4.1 Adhesion, Friction, and Wear of Biomolecules on Si-Based Surfaces

Proteins on silicon-based surfaces are of extreme importance in various applications, including silicon microimplants, various bioMEMS such as biosensors, and therapeutics. Silicon is a commonly used substrate in microimplants, but it can have undesired interactions with the human immune system. Therefore, to mimic a biological surface, protein coatings are used on silicon-based surfaces as passivation layers, so that these implants are compatible with the body, avoiding rejection. Whether this surface treatment is applied to a large implant or a bioMEMS, the function of the protein passivation is obtained from the nanoscale 3-D structural conformation of the protein. Proteins are also used in bioMEMS because of their function specificity. For biosensor applications, the extensive array of protein activities provides a rich supply of operations that may be performed at the nanoscale. Many antibodies (proteins) have an affinity to specific protein antigens. For example, pathogens (disease-causing agents, e.g., viruses or bacteria) trigger the production of antigens which can be detected when bound to a specific antibody on the biosensor. The specific binding behavior of proteins that has been applied to laboratory assays may also be redesigned for in vivo use as sensing elements of a bioMEMS. The epitope-specific binding properties of proteins to various antigens are useful in therapeutics. Adhesion between the protein and substrate affects the reliability of an application. Among other things, the morphology of the substrate affects the adhesion. Furthermore, for in vivo environments, the proteins on the biosensor surface should exhibit high wear resistance during direct contact with tissue and circulatory blood flow without washing off.

Bhushan et al. [47.73] studied the step-by-step morphological changes and the adhesion of a model protein – streptavidin (STA) – on silicon-based surfaces. (Also see [47.135, 137, 196].) Figure 47.29a presents a flowchart showing the sequential modification of a silicon surface. In addition to physical adsorption, they also used nanopatterning and chemical linker methods to improve adhesion. A nanopatterned surface contains large edge surface area, leading to high surface energy which results in high adhesion. In the chemical linker method, sulfo-N-hydroxysuccinimido-biotin

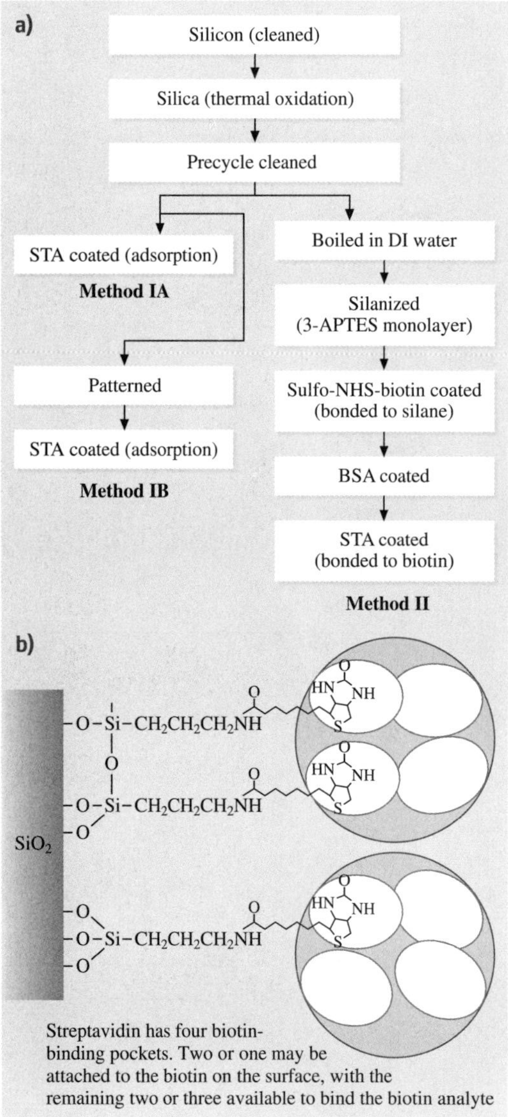

Fig. 47.29 (a) Flowchart showing the samples used and their preparation technique. (b) Chemical structure showing streptavidin protein binding to the silica substrate by the chemical linker method ◀

(sulfo-NHS-biotin) was used as a cross-linker because the bonds between the STA and the biotin molecule are one of the strongest noncovalent bonds known (Fig. 47.29b). It was connected to the silica surface through a silane linker, 3-aminopropyltriethoxysilane (3-APTES). In order to make a bond between the silane linker and the silica surface, the silica surface was hydroxylated. Bovine serum albumin (BSA) was used before STA in order to block nonspecific binding sites of the STA protein with silica surface. Figure 47.30 shows the step-by-step morphological changes in the silica surface during the deposition process using the chemical linker method. There is an increase in roughness [σ and peak-to-valley (P–V) distance] of the silica surface boiled in deionized (DI) water compared with the bare silica surface. After the silanization process, there are many free silane links on the surface, which caused higher roughness. Once biotin was coated on the silanized surface, the surface became smoother. Finally, after the deposition of STA, the surface shows large and small clumps. Presumably, the large clumps represent BSA, and the smaller ones represent STA. To measure adhesion between STA and the corresponding substrates, an STA-coated tip (or functionalized tip) was used and all measurements were made in phosphate-buffered saline (PBS) solution, a medium commonly used in protein analysis to simulate body fluid. Figure 47.31 shows the adhesion values of various surfaces. The adhesion value between biotin and STA was higher than that for other samples, as expected. Edges of patterned silica also exhibited high adhesion values. It appears that both nanopatterned surfaces and the chemical linker method increase adhesion with STA.

Bhushan et al. [47.131] studied friction and wear of STA deposited by physical adsorption and the chemical linker method. (Also see [47.137].) Figure 47.32 shows the coefficient of friction between the Si_3N_4 tip and various samples. The coefficient of friction is less for STA-coated silica samples compared with an uncoated sample. The streptavidin coating acts as a lubricant film. The coefficient of friction is found to be dependent upon the concentration of STA, and it decreases with increasing concentration. *Bhushan* et al. [47.73] have reported that the density and distribution of the biomolecules

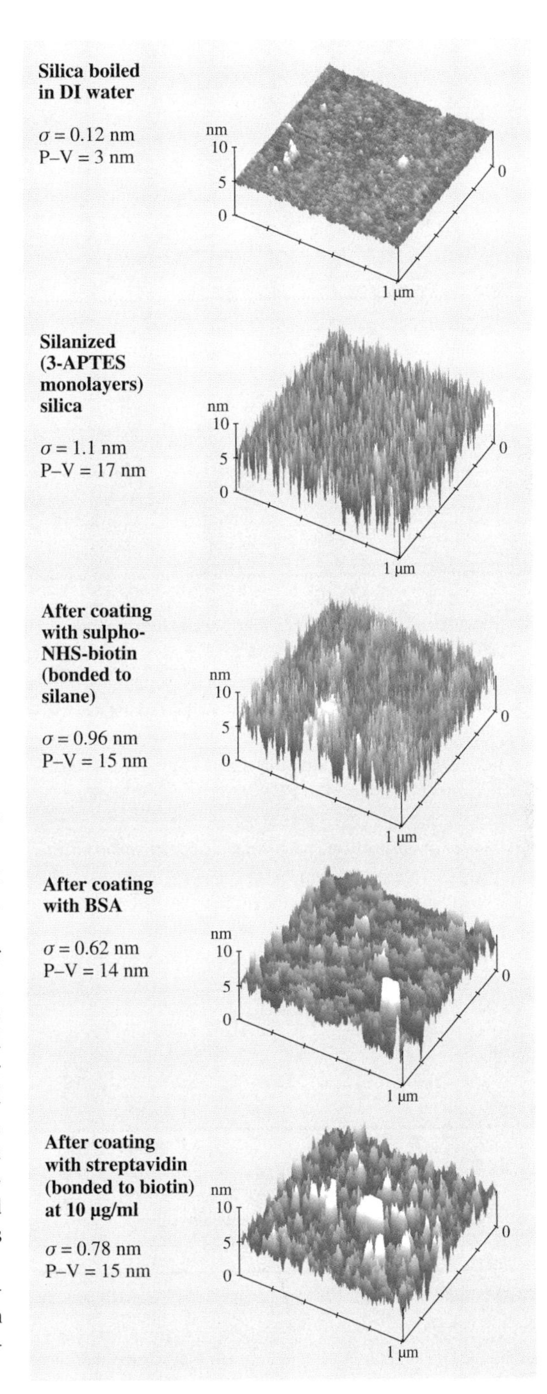

Fig. 47.30 Morphological changes in a silica surface during functionalization by the chemical linker, imaged in PBS. Streptavidin is covalently bonded at 10 μg/ml concentration (after [47.73]) ▶

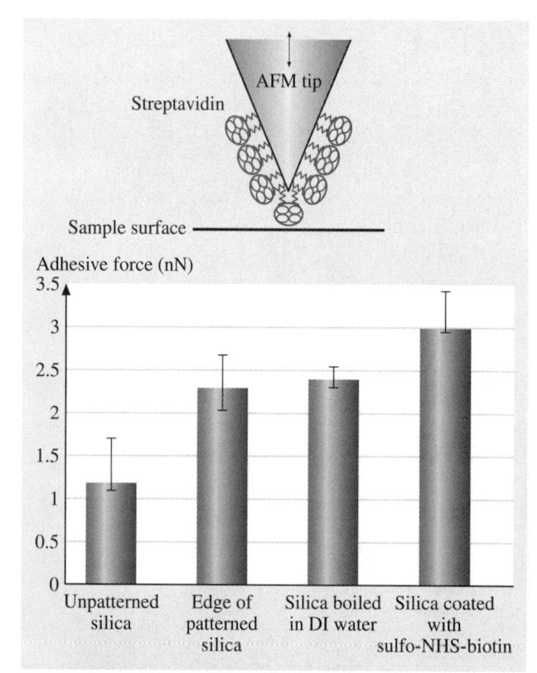

Fig. 47.31 Adhesion measurements of silica, patterned silicon, silica boiled in DI water, and sulfo-NHS-biotin using functionalized (with streptavidin) tips obtained from force–distance curves, captured in PBS

vary with concentration. At higher concentration of the solution, the coated layer is more uniform, and the silica substrate surface is more highly covered with biomolecules than at lower concentration. This means that the surface forms a continuous lubricant film at higher concentration.

In the case of samples prepared by the chemical linker method, the coefficient of friction increases with increasing biomolecular chain length due to increased compliance. When a normal load is applied to the surface, the surface gets compressed, resulting in a larger contact area between the AFM tip and the biomolecules. Besides that, the size of STA is much larger than that of APTES and biotin. This results in a tightly packed surface with the biomolecules, which results in very little lateral deflection of the linker in the case of STA-coated biotin. Due to this high contact area and low lateral deflection, the friction force increases for the same applied normal load compared with the directly adsorbed surface. These tests reveal that surfaces coated with biomolecules reduce the friction, but if the biomolecular coating of the surface is too thick or the surface has some cushioning effect as seen in the chemical linker method, this will increase the coefficient of friction.

Figure 47.33 shows the surface height maps and phase images of wear marks on STA deposited by physical adsorption after wear tests at three normal loads. The wear depth increases with increasing normal load. An increase in normal load causes partial damage to the folding structure of the streptavidin molecules. It is unlikely that the chemical (covalent) bonds within the streptavidin molecule are broken; instead, the folding structure is damaged, leading to wear marks. When the load is high, $\approx 30\%$ of free amplitude (≈ 8 nN), the molecules may have been removed by the AFM tip due to indentation effect. Because of this, there is significant increase in the wear depth from 50% of free amplitude (≈ 6 nN) to 30% of free amplitude (≈ 8 nN). The data show that biomolecules will be damaged during sliding.

In summary, for samples prepared using nanopatterning and the chemical linker methods, adhesion is higher compared with those prepared by the direct adsorption method. The coefficient of friction is lowre for STA-coated silica prepared using the direct adsorption method as compared with an uncoated silica sample. Coefficient of friction decreases with increasing concentration of STA in the solution because protein acts as a lubricating film. Friction increases for the STA sample prepared using the chemical linker method due to the cushioning effect and low lateral deformation. Wear of STA increases with the increasing load.

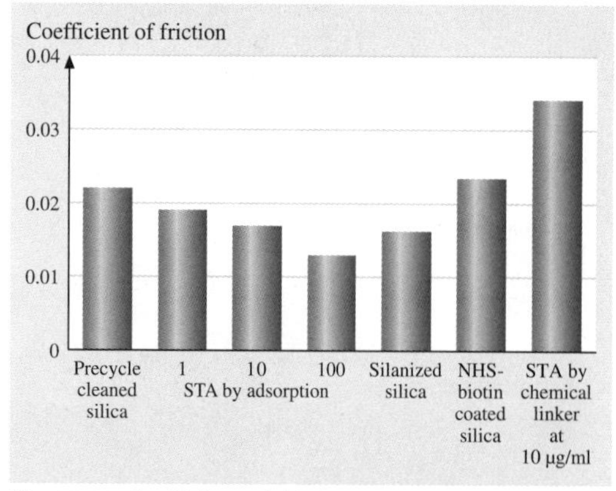

Fig. 47.32 Coefficient of friction in PBS for various surfaces with and without biomolecules (after [47.131])

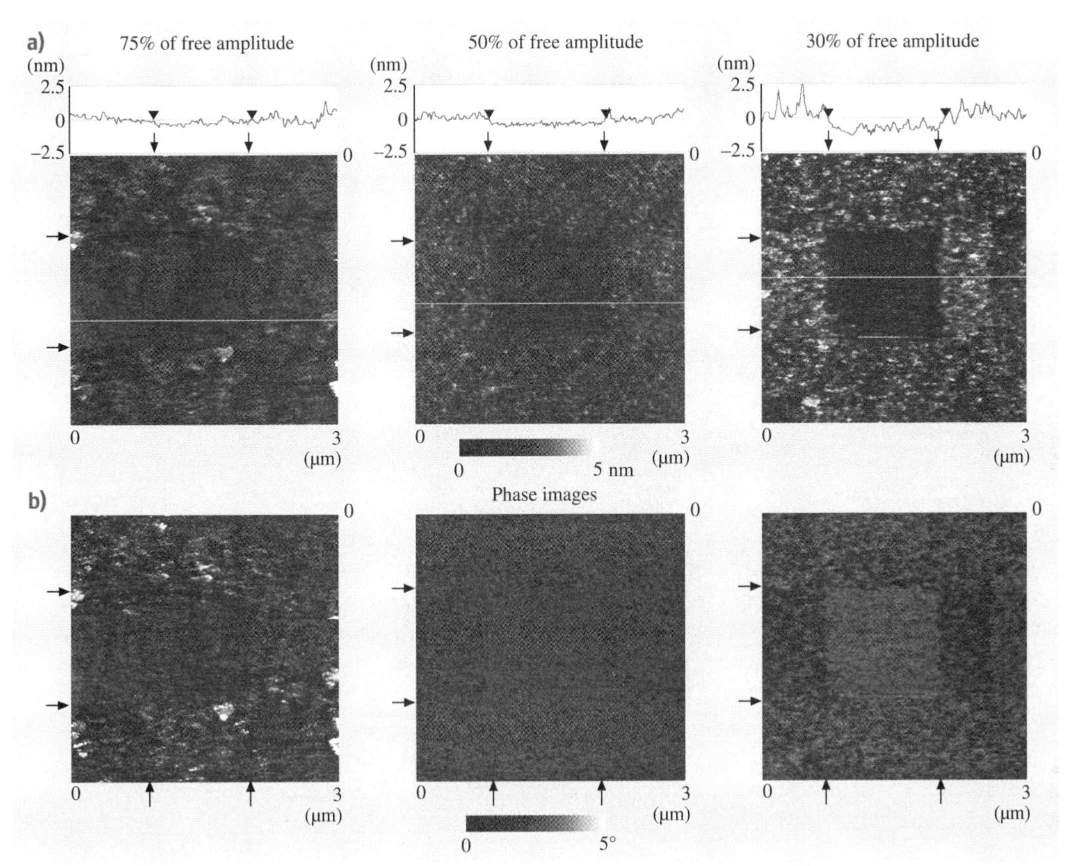

Fig. 47.33a,b Surface height maps and cross-sectional profiles (**a**) and phase images (**b**) of wear marks on precycle cleaned silica coated with streptavidin by physical adsorption after wear tests at three normal loads (increasing from *left* to *right*). The 75, 50 and 30% of free amplitudes correspond to equivalent normal loads of 3, 6 and 8 nN, respectively (after [47.131])

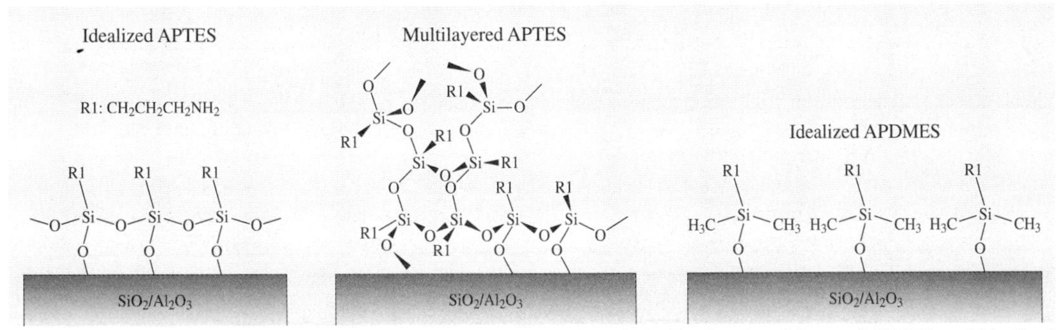

Fig. 47.34 Chemical structure of silane polymer linker. Schematics of idealized and multilayered APTES and idealized APDMES silane polymer films on silicon/SiO_2 and aluminum/Al_2O_3 substrates. Idealized film corresponds to a self-assembled monolayer (after [47.137])

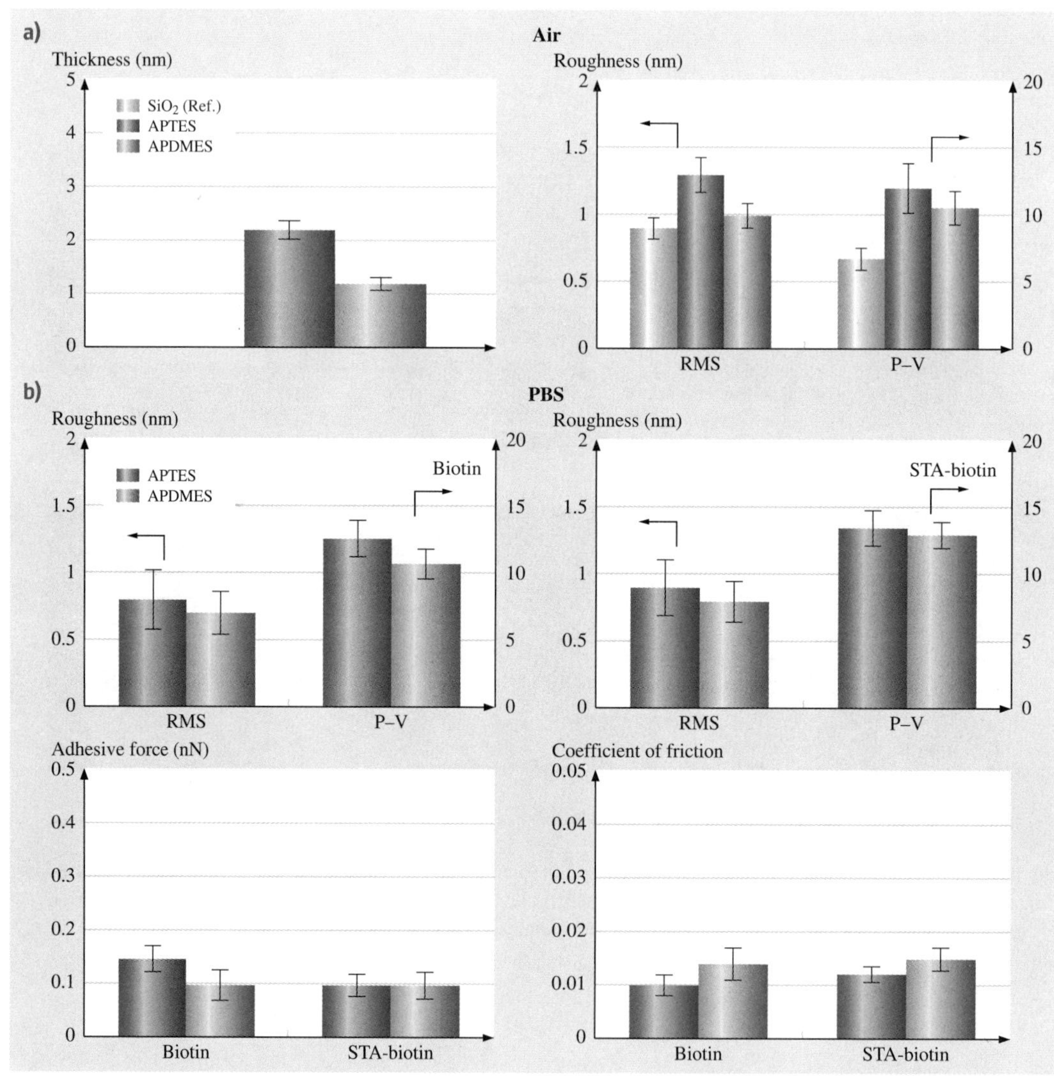

Fig. 47.35 (a) Summary of film thickness and surface roughness (RMS and P–V distance) for APTES and APDMES in air. **(b)** Summary of surface roughness (RMS and P–V distance), adhesive force, and coefficient of biotin and STA–biotin on APTES and APDMES films, all in PBS buffer solution. *Error bars* represent $\pm 1\sigma$ (after [47.137])

APDMES to Improve Adhesion, Friction, and Wear of Biomolecular Films

APTES films used in the just-reported study were not very smooth, and P–V distances were substantially greater than the summed bond strengths of APTES. The biotin/STA deposited on APTES films was not very robust. APTES films are commonly described as SAMs, though this is often inaccurate [47.197–199]. APTES monomers can form infinite, cross-linked siloxane polymer lattices because APTES is a multivalent (trivalent) silane and can form a multilayered structure due to intermolecular polymerization with significant cross-linking between monomers. It should consist of sparse cross-links between the polymer and the substrate (Fig. 47.34) [47.137]. The propensity to form multilayers and its low mechanical strength makes

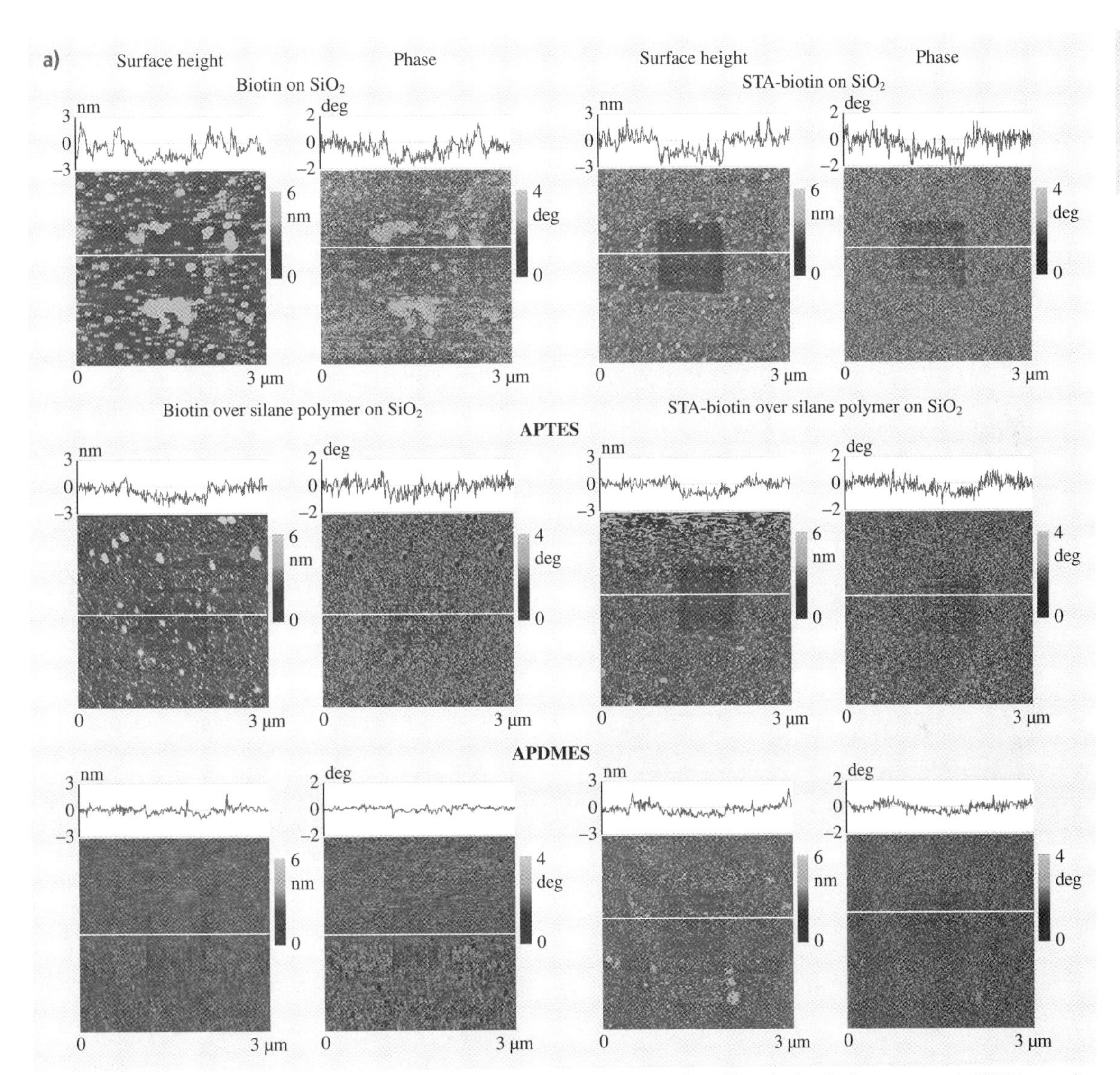

Fig. 47.36 (a) AFM surface height and phase-angle images and cross-sectional profiles obtained after wear test in PBS in tapping mode at 50% of free amplitude (≈ 2 nN) on biotin and STA–biotin on SiO_2 and APTES and APDMES films on SiO_2. The *white lines* indicate the locations of the cross sections

APTES an undesirable interface material. On the other hand, 3-aminopropyldimethlethoxysilane (APDMES) cannot polymerize into extensive networks because it is monovalent, forming only siloxane dimers or linkages to substrate oxides (Fig. 47.34). APDMES should therefore produce thinner films of greater mechanical robustness than APTES, and provide robustness to biomolecular layers deposited on top of it. Thinner interfaces would also theoretically increase sensor sensitivity to analyte.

Bhushan et al. [47.137] examined the thickness and durability of APDMES deposited on SiO_2 and Al_2O_3 substrates, with biotin and biotin/STA bound to them. Figure 47.35a shows the thickness and surface rough-

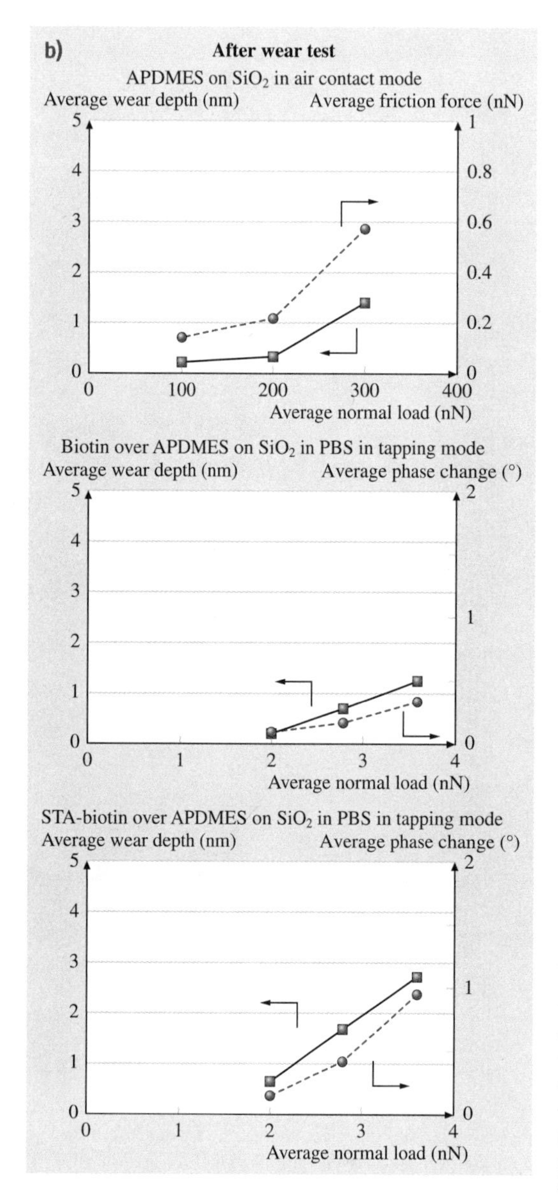

Fig. 47.36 (**b**) Plot of average wear depth and average friction force/phase angle as a function of average normal load for APDMES on SiO_2 in air in contact mode, biotin over APDMES on SiO_2 in PBS in tapping mode, and on STA–biotin over APDMES on SiO_2 in PBS in tapping mode (after [47.137]) ◀

ness (RMS and P–V) of APTES and APDMES and SiO_2 substrate for reference. Figure 47.35b shows the surface roughness, adhesive force. and coefficient of friction for biotin and STA–biotin deposited on APTES and APDMES films. The data show that APTES film is not very smooth. It was shown in Fig. 47.30 that biotin and STA deposited on APTES were also not very smooth, with high P–V values. The thickness of the APTES layer is larger than the expected monolayer thickness. The unexpected thickness of the film represents multilayering by APTES. APDMES films produced the thinner film, with a thickness comparable to the summed bond lengths of the APDMES polymer (Fig. 47.35a). *Bhushan* et al. [47.137] reported that the APDMES film was more uniform, smoother, and nearly continuous, and that it exhibited higher contact angle and lower adhesive force as compared with the APTES film. The surface roughness and adhesive force of biotin and STA–biotin on APDMES are also slightly lower than that on APTES film. The coefficient of friction on APDMES appears to be slightly higher than that on APTES (Fig. 47.35b).

Bhushan et al. [47.137] also studied wear properties of various films. Contact mode always immediately stripped the surface at low loads; consequently wear experiments were performed in tapping mode at various loads. AFM surface height and phase-angle images and cross-sectional profiles obtained after wear tests for biotin and STA–biotin on APTES and APDMES in PBS are shown in Fig. 47.36a. As controls, wear experiments of biotin and STA–biotin on SiO_2 without SAM were also carried out. The biomolecular films on APDMES were more robust than on APTES. Given that each molecule of APDMES must be bonded to a surface oxide group and that intrapolymer cross-links are not possible in APDMES, it has a higher density of siloxane linkages to the substrate oxide, making it more robust. *Bhushan* et al. [47.137] also studied the effect of load on the wear of APDMES and biotin and STA–biotin deposited on APDMES (Fig. 47.36b). As expected, the wear increased with increasing load. The relationship between the average wear depth (and the coefficient of friction) and the average normal load is generally linear. The slope of the wear depth (and coefficient of friction) against load is steepest for the interface to which STA–biotin was bound because of the cushioning effect of the thick film, as suggested earlier.

In summary, APDMES film is more uniform, smoother, and nearly continuous as compared with APTES film. These properties of APDMES provide a good interfacial material for biomolecular films, providing a smooth and robust structure.

47.4.2 Adhesion of Coated Polymer Surfaces

As mentioned in Appendix A, PMMA, PDMS, and other polymers are used in the construction of micro-nanofluidic-based biodevices. Adhesion between the moving parts needs to be minimized. Furthermore, if the adhesion between the microchannel surface and the biofluid is high, the biomolecules will stick to the microchannel surface and restrict flow. In order to facilitate flow, a surface with low bioadhesion is required.

Tambe and *Bhushan* [47.201, 202] and *Bhushan* and *Burton* [47.203] have reported adhesive force data for PMMA and PDMS against an AFM Si_3N_4 tip and a silicon ball. *Tokachichu* and *Bhushan* [47.200] measured contact angle and adhesion of bare PMMA and PDMS and coated with a perfluoro SAM of perfluorodecyltriethoxysilane (PFDTES). Oxygen plasma treatment was used for hydroxylation of the surface to enhance chemical bonding of the SAM to the polymer surface. They made measurements in ambient, in PBS, and fetal bovine serum (FBS); the latter is a blood component. Figures 47.37 and 47.38 show the contact angle and adhesion data. SAM-coated surfaces have high contact angles (Fig. 47.37), as expected. The adhesion value of PDMS in ambient is high because of the electrostatic charge present on the surface. The adhesion values of PDMS are higher than PMMA because PDMS is softer than PMMA (elastic modulus = 5 GPa and hardness = 410 MPa [47.132]), resulting in higher contact area between the PDMS surface and the AFM tip, and PMMA does not develop electrostatic charge. When SAM is coated on PMMA and PDMS surfaces, the adhesion values are similar, which shows that electrostatic charge on virgin PDMS plays no role when the surface is coated. In the PBS solution, there is a decrease

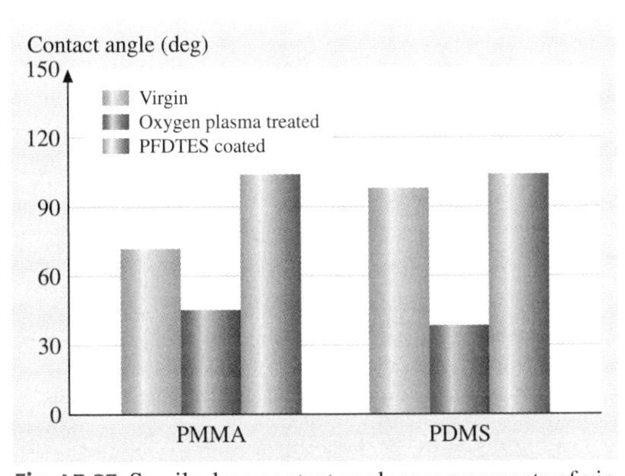

Fig. 47.37 Sessile drop contact angle measurements of virgin, oxygen-plasma-treated, and PFDTES-coated PMMA and PDMS surfaces. The maximum error in the data is ±2° (after [47.200])

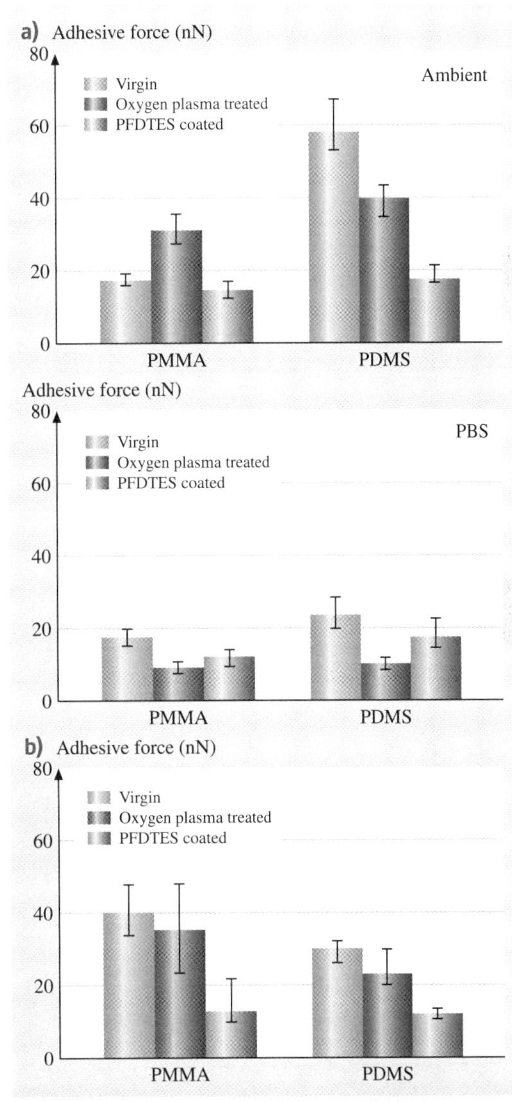

Fig. 47.38a,b Adhesion measurement of virgin, oxygen-plasma-treated, and PFDTES-coated PMMA and PDMS surfaces (**a**) with bare silicon nitride AFM tip in ambient, and in PBS environment, and (**b**) dip-coated tip with FBS in PBS environment (after [47.200])

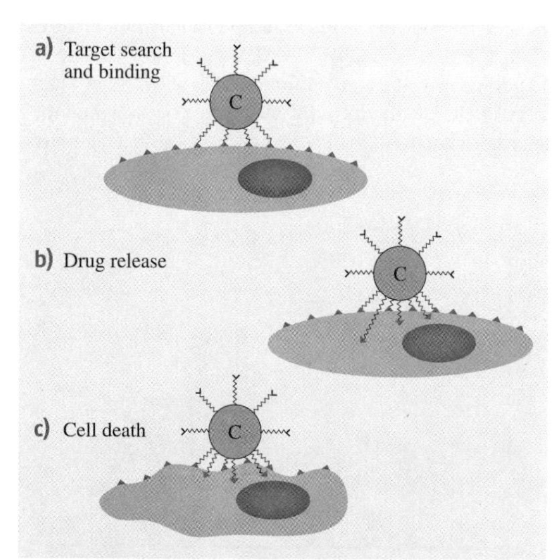

Fig. 47.39a–c The life cycle of a particle injected intravenously for drug delivery: (a) target search and binding, (b) drug release, and (c) cell death (after [47.142])

in adhesion values because of the lack of a meniscus contribution. The adhesion values for the FBS-coated tip in PBS are generally lower than for an uncoated tip in PBS.

In summary, the adhesion values of SAM-coated surfaces are lower than bare surfaces in various environments.

47.4.3 Submicron Particles for Therapeutics and Diagnostics

Submicron particles can be injected into the blood stream in human capillaries (as small as 4–5 μm) and employed to deliver drugs to diseased cells, to locate diseased cells or tumoral masses and estimate the state of disease, and to carry diagnostic agents (fluorescent molecules) to diseased cells in order to enhance imaging [47.142]. Particles exhibiting one of these characteristics can be considered as smart systems, which can function as purely therapeutic agents, purely diagnostic agents, or a combination of both. Small delivery particles include nanocrystals, synthetic vesicles, liposomes, and fabricated silicon.

Particles are reservoirs containing drug or diagnostic agents. These are covered with a layer of adhesive molecules (ligands) for attaching to selected target sites. When a particle is sufficiently close to select cell surfaces (a few nm), its ligands can interact with the cellular counterpart (receptors), which leads to firm attachment. Once the particle is arrested to its target, drug or diagnostic agent is delivered. A schematic of the lifecycle of a particle injected intravenously for drug delivery is shown in Fig. 47.39. The speed and the time needed for a particle circulating in the blood stream to reach the endothelium lining of the human capillary is dependent upon the distance and interactive forces. The particle can interact with the endothelium lining through buoyancy, van der Waals (vdW), electrostatic, and steric

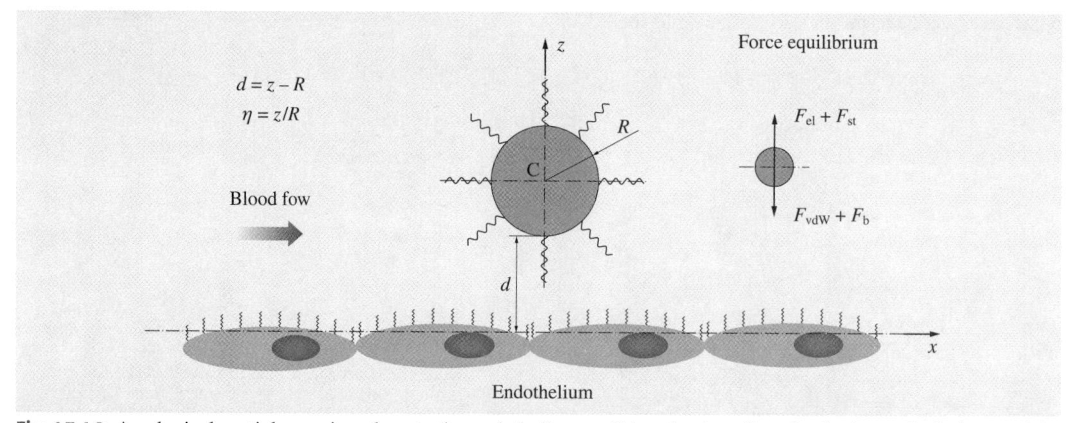

Fig. 47.40 A spherical particle moving close to the endothelium wall in a laminar flow. In the *inset*, the balance of the forces acting on the particle are sketched: F_{el}, F_{st}, F_{vdW}, and F_b correspond to the electrostatic force, steric force, van der Waals force, and buoyancy force, respectively (after [47.142])

forces. These interactions are a function of the material properties of the particles and any coating on it, in particular its relative density (particle density relative to blood), the electrostatic potential, and the dielectric constant as well as the particle radius. These properties can be optimized.

These interacting forces are weak, so the particle trajectory should be close to the endothelium lining, otherwise the particle may not get attracted to the lining to perform its intended function. *Decuzzi* et al. [47.142] developed an analytical model to predict the trajectory of a particle freely circulating in the blood stream and associated interaction forces. The model can be used to optimize the particle radii and material properties. Figure 47.40 shows a schematic of a spherical particle freely circulating in the blood stream with its center at a distance z from the endothelium wall. The particle has a radius R, and its trajectory is governed by the forces exerted by the blood stream and gravitational and electromagnetic interactions. We assume that the particle is sufficiently far from the endothelium wall that specific interactions (such as ligand–receptor interactions) can be ignored. At short range (below 1–3 nm), solvation and other steric forces dominate. However, the most important long-range forces (5–10 nm) between particles and wall surfaces in the presence of a liquid are buoyancy, van der Waals, electrostatic, and steric forces. The buoyancy force is related to the radius of the particle R and the relative density of the particle relative to blood. The van der Waals interaction (generally attractive) is related to R, its relative position with respect to the endothelium wall z, and the Hamaker constant A, which depends upon the dielectric constants of the media involved. The electrostatic double-layer (EDL) interaction (repulsive and attractive) is related to R, z, the ionic concentration, and the characteristic Debye length of the solution. Finally, the steric repulsive interaction is related to the unperturbed radius of gyration of polymer chains grafted onto the particle surface, R, and z. The value of various forces as a function of the particle radius R

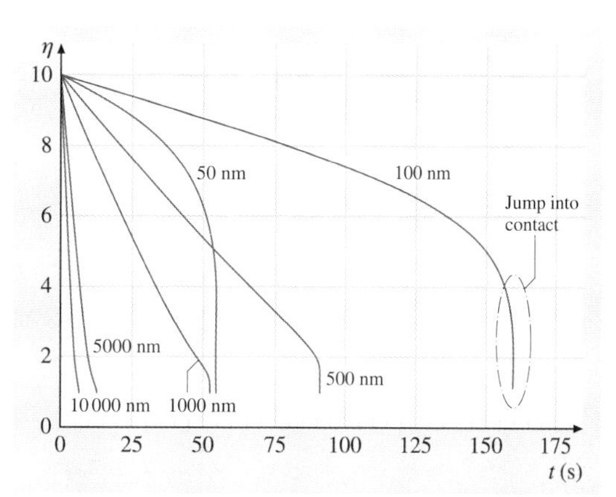

Fig. 47.41 The dimensionless particle position $\eta(=z/R)$ as a function of time t for different values of particle radius ($R = 10\,000$, 5000, 1000, 500, 100, and 50 nm). There exists a critical radius R_c at which the travel time is maximum (after [47.142])

for a fixed distance from the wall z is presented in Table 47.6. The dominating force is buoyancy when the particle radius is sufficiently large, and van der Waals when the particle is sufficiently small. Electrostatic and steric forces are negligible as long as the distance is larger than the 50 nm considered in this example.

The travel time needed to reach the wall depends upon the particle size, as shown in Fig. 47.41. In this figure, the dimensionless position of the particle center with respect to the endothelium wall η $(=z/R)$ is plotted as a function of time t for different particle radii, ranging from $R = 10\,\mu m$ to 50 nm. It was assumed that particles were initially at a distance $d_0(z_0 - R)$ equal to $9R$ from the endothelium wall (i.e., the center of the particle is at $z_0 = 10R$, $\eta_0 = z_0/R = 10$). Thus the distance traveled scales with the size of the particle. The data show that, as the radius of the particle decreases, the time needed to reach the wall increases up to a maximum beyond which it decreases as the ra-

Table 47.6 Values of the buoyancy force F_b, van der Waals force F_{vdW}, electrostatic force F_{el}, and steric force F_{sr} as functions of the particle radius for a fixed distance from the wall ($\eta = z/R = 2$ or $d = R$) [47.142]

R (nm)	F_b (pN)	F_{vdW} (pN)	F_{el} (pN)	F_{sr} (pN)
10 000	41.092	0.0005	≈ 0	≈ 0
1000	41.092×10^{-3}	0.005	≈ 0	10^{-33}
100	41.092×10^{-6}	0.05	10^{-52}	1.2×10^{-3}
50	51.3×10^{-7}	0.1	10^{-25}	3×10^{-2}

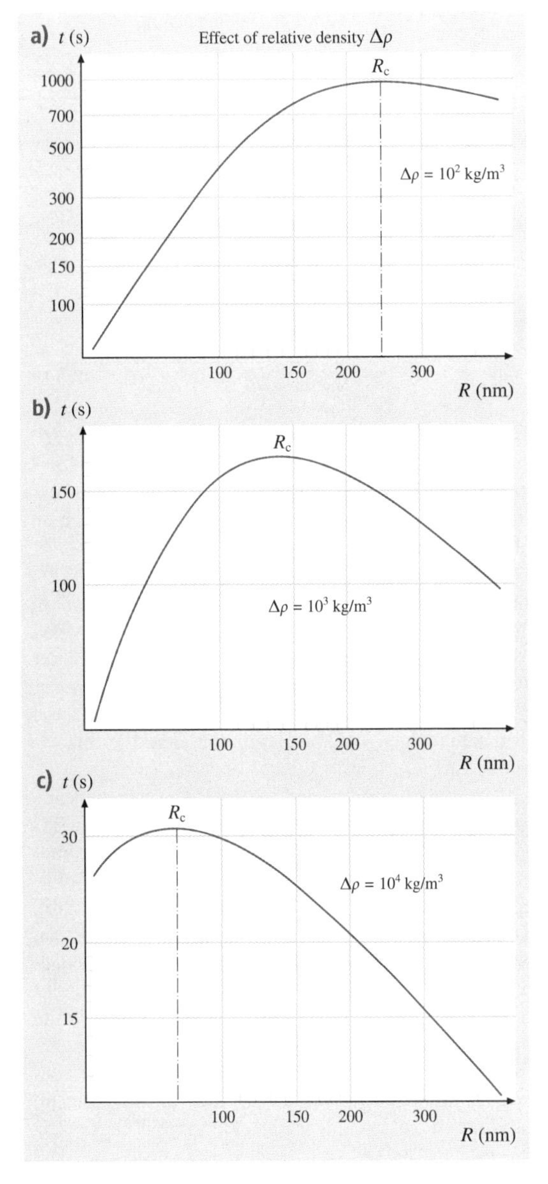

Fig. 47.42a–c Travel time as a function of particle radius R for different values of relative density: **(a)** $\Delta\rho = 10^2$, **(b)** $\Delta\rho = 10^3$, and **(c)** $\Delta\rho = 10^4\,\mathrm{kg/mm^3}$. The critical radius R_c depends upon $\Delta\rho$ (after [47.142]) ◀

dius is further reduced. Larger particles are initially far from the endothelium wall; a 10 μm particle is initially 100 μm away from the endothelium ($z = 10R$), where electrostatic and van der Waals interactions are negligible and particle motion is governed by buoyancy and hemodynamic resistance. As the particle approaches the endothelium wall, van der Waals attraction dominates, and a rapid increase in speed is observed with a *jump to contact* like behavior. Such a behavior is more clearly shown by particles with submicron radius, which are closer to the endothelium wall, so the van der Waals force dominates from the beginning. We note that there is a critical radius for which the time taken for the particle to travel to the wall is maximum. Selected radius should be smaller or larger than the critical radius (possibly smaller so that they can circulate freely even in smaller capillaries).

The effect of the relative density of the particle $\Delta\rho$ on the critical radius R_c at which travel time is maximum is plotted in Fig. 47.42. We note that, as $\Delta\rho$ reduces, R_c increases. As $\Delta\rho$ decreases, the effect of buoyancy becomes less important, and van der Waals attractive forces exert a greater influence as the particle radius decreases.

In summary, the interacting forces are weak, so the particle trajectory should be close to the endothelium lining. The trajectory and interaction forces depend upon the particle radii and material properties. These results suggest that particles for therapeutic and diagnostics should have a radius larger or smaller than a critical value (possibly smaller so that they can circulate freely even in smaller capillaries). The material properties of the particle and the polymer chains grafted onto it, such as the relative density of the particle, can be tuned specifically to the type of malignant tissue and the state of disease, improving the particle affinity with the diseased cells.

47.5 Surfaces with Roughness-Induced Superhydrophobicity, Self-Cleaning, and Low Adhesion

Various MEMS/NEMS and bioMEMS/bioNEMS require hydrophobic and self-cleaning surfaces and interfaces with low adhesion and friction. Hydrophobicity of a surface (wettability) is characterized by the static contact angle between a water droplet and the surface. If the liquid does not wet the surface, the value of the contact

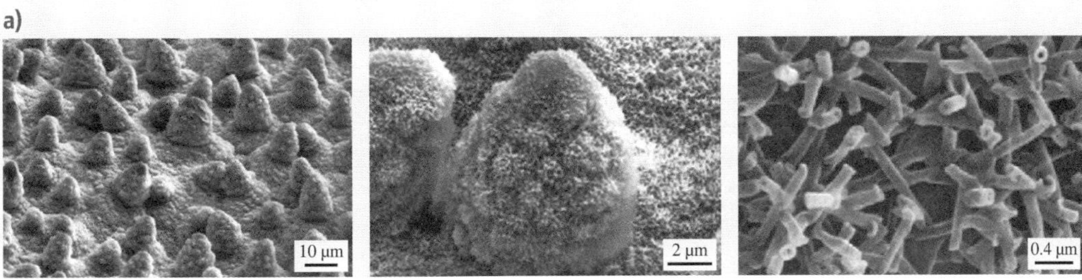

Fig. 47.43 (a) SEM micrographs (shown at three magnifications) of *Nelumbo nucifera* (lotus) leaf surface, which consists of microstructure formed by papillose epidermal cells covered with 3-D epicuticular wax tubules on the surface, which create nanostructure [47.204]

angle is $90 < \theta \leq 180°$. Surfaces with a contact angle between 150° and 180° are called superhydrophobic. In addition to the high contact angle, for self-cleaning, superhydrophobic surfaces should also have very low water contact angle hysteresis θ_H. Water droplets roll off (with some slip) on these surfaces and take contaminants with them, providing a self-cleaning ability known as the *lotus effect*. Contact angle hysteresis is the difference between advancing and receding contact angles, which are two stable values. It occurs due to surface roughness and surface heterogeneity. Contact angle hysteresis (CAH) reflects the irreversibility of the wetting/dewetting cycle. It is a measure of the energy dissipation during the flow of a droplet along a solid surface. At a low value of CAH, the droplet may roll in addition to slide, which facilitates the removal of contaminant particles. A surface with CAH of $< 10°$ is generally referred to as a self-cleaning surfaces. Surfaces with low CAH have a low water roll-off (tilt) angle, which denotes the angle to which a surface must be tilted for roll-off of water droplets. Self-cleaning surfaces are of interest in various applications, including self-cleaning windows, windshields, exterior paints for buildings and navigation ships, utensils, roof tiles, textiles, solar panels, and applications requiring antifouling and a reduction of drag in fluid flow, e.g., in micro-nanochannels in micro-nanofluidics. Superhydrophobic surfaces can also be used for energy conservation and conversion. Selection of a proper superhydrophobic surface allows the reduction of energy dissipation. Secondly, superhydrophobic and superoleophobic surfaces can be used for fuel economy. Third, the recently discovered effect of reversible superhydrophobicity provides the potential for new ways of energy conversion such as the microscale capillary engine.

Fig. 47.43 (b) Image of a water droplet sitting on the lotus leaf

Wetting may lead to the formation of menisci at the interface between solid bodies during sliding contact, which increases adhesion and friction. In some cases, the wet friction force can be greater than the dry friction force, which is usually undesirable [47.82–84, 126]. On the other hand, high adhesion is desirable in some applications, such as adhesive tapes and adhesion of cells to biomaterial surfaces; therefore, enhanced wetting would be desirable in these applications. Numerous applications, such as magnetic storage devices and micro-nanoelectromechanical systems (MEMS/NEMS), require surfaces with low adhesion and stiction [47.13, 34, 83, 152, 205].

Some natural surfaces, including leaves of water-repellent plants such as lotus, are known to be superhydrophobic and self-cleaning due to hierarchical roughness and the presence of a wax coating [47.206–208]. Figure 47.43 shows SEM micrographs (shown at three magnifications) of a superhydrophobic leaf of lotus (*Nelubo nucifera*). Lotus is characterized by papillose epidermal cells responsible for the creation of papillae or microbumps on the surfaces, covered with three-dimensional epicuticular wax tubules which are a mixture of very long-chain fatty acids molecules (compounds with chains > 20 carbon atoms) and create a nanostructure on the entire surface. The contact angle and contact angle hysteresis of the lotus leaf are ≈ 164 and $3°$, respectively [47.204, 208, 209].

Superhydrophobic surfaces can be achieved by either selecting low-surface-energy materials/coating, or by introducing roughness. In this section, we discuss design, fabrication, and characterization of roughness-induced superhydrophobic and self-cleaning surfaces by mimicking the lotus effect [47.204–206, 209–214].

47.5.1 Modeling of Contact Angle for a Liquid Droplet in Contact with a Rough Surface

If a droplet of liquid is placed on a smooth surface, the liquid and solid surfaces come together under equilibrium at a characteristic angle called the static contact angle θ_0 (Fig. 47.44a). The contact angle can be determined from the condition of the total energy of the system being minimized. Next, consider a rough solid surface with a typical size of roughness details smaller than the size of the droplet (of the order of a few hundred microns or larger) (Fig. 47.44a). For a droplet in contact with a rough surface without air pockets, referred to as a homogeneous interface, based on the minimization of the total surface energy of the system, the contact angle is given as by the Wenzel equation [47.215]

$$\cos\theta = R_f \cos\theta_0 , \quad (47.1)$$

where θ is the contact angle for rough surfaces and R_f is a roughness factor defined as the ratio of the solid–liquid area A_{SL} divided by its projection onto a flat plane A_F

$$R_f = \frac{A_{SL}}{A_F} . \quad (47.2)$$

The dependence of the contact angle on the roughness factor is presented in Fig. 47.44b for different values of θ_0, based on (47.1). It should be noted that (47.1) is valid only for moderate roughness, when $R_f \cos\theta_0 < 1$. The graph shows that, with an increase in the roughness factor, a hydrophobic surface becomes more hydrophobic, whereas a hydrophilic surface becomes even more hydrophilic. As an example, Fig. 47.44c shows a geometry with pyramidal asperities with rounded tops, which has complete packing. The size and shape of the asperities can be optimized for a desired roughness factor.

For higher roughness, air pockets may be formed between the asperities on the surface, which results in a composite interface consisting of a solid–liquid and a liquid–air fraction (Fig. 47.45a). In the case of such a composite interface, the contact angle is given as by the Cassie–Baxter equation [47.216]

$$\cos\theta = R_f \cos\theta_0 - f_{LA}(R_f \cos\theta_0 + 1) , \quad (47.3)$$

where f_{LA} is the fractional liquid–air contact area of the liquid–air interfaces under the droplet. In reality, some valleys will be filled with liquid and others with air, and the value of the contact angle is between the values predicted by (47.1) and (47.3). Examination of (47.3) shows that the contact angle increases with increasing R_f and f_{LA}. Even for a hydrophilic surface,

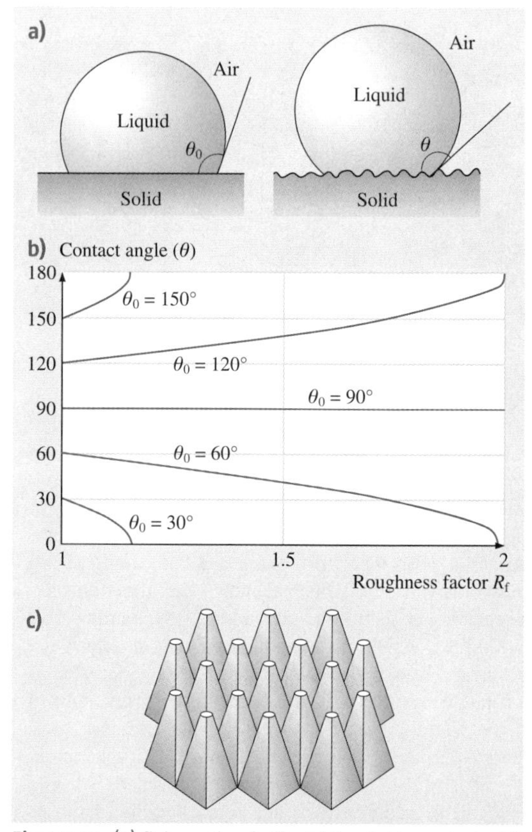

Fig. 47.44 (a) Schematic of a liquid droplet in contact with a smooth solid surface (contact angle θ_0) and a rough solid surface (contact angle θ), (b) contact angle for rough surface (θ) as a function of roughness factor (R_f) for various contact angles of the smooth surface (θ_0), and (c) schematic of round-topped pyramidal asperities with complete packing (after [47.210])

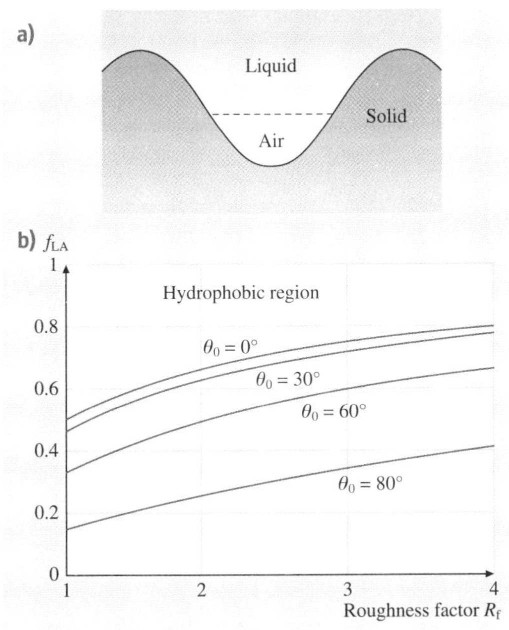

Fig. 47.45 (a) Schematic of formation of a composite solid–liquid–air interface for a rough surface; (b) f_{LA} requirement for a hydrophilic surface to be hydrophobic as a function of the roughness factor (R_f) and θ_0 [47.206]

the contact angle increases with increasing f_{LA}. At a high enough value of f_{LA}, a hydrophilic surface can become hydrophobic. The value of f_{LA} at which a hydrophilic surface could turn into a hydrophobic one is given as [47.206]

$$f_{LA} > \frac{R_f \cos\theta_0}{R_f \cos\theta_0 + 1} \quad \text{for } \theta < 90° \,. \tag{47.4}$$

Figure 47.45b shows the value of f_{LA} as a function of R_f required for different contact angles θ_0 for a hydrophilic surface to become hydrophobic. The graph shows that, unlike the so-called Wenzel regime, in the so-called Cassie–Baxter regime, even a hydrophilic surface can be made hydrophobic at a certain value of f_{LA} for a given θ_0.

As stated earlier, low contact angle hysteresis is desirable for self-cleaning. If a droplet sits over a tilted surface (Fig. 47.46) the contact angle at the front and back of the droplet corresponds to the advancing and receding contact angle, respectively. The advancing angle is greater than the receding angle, which results in the contact angle hysteresis. *Nosonovsky* and *Bhushan* [47.210] derived a relationship for contact an-

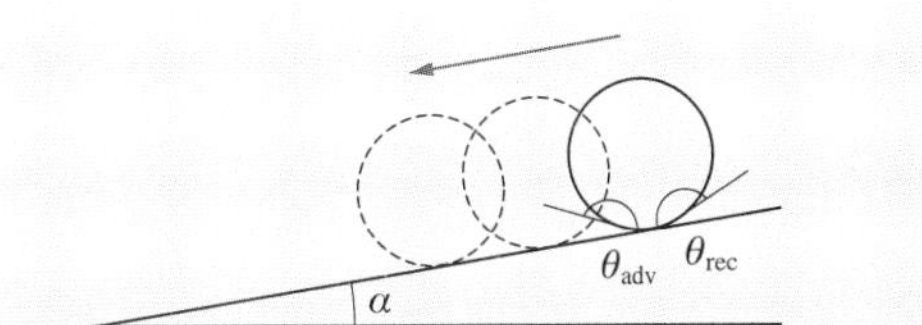

Fig. 47.46 Tilted surface profile (tilt angle α) with a liquid droplet; advancing and receding contact angles are θ_{adv} and θ_{rec}, respectively

gle hysteresis as a function of roughness, given as

$$\theta_{adv} - \theta_{rec} = \left(\sqrt{1 - f_{LA}}\right) R_f \frac{\cos\theta_{r0} - \cos\theta_{a0}}{\sqrt{2(R_f \cos\theta_0 + 1)}} \,. \tag{47.5}$$

Equation (47.5) shows that increasing roughness (high R_f) and decreasing fractional liquid–air contact area f_{LA} lead to an increase in contact angle hysteresis θ_H. Increasing f_{LA} is more efficient for decreasing θ_H; therefore, a composite interface is desirable for self-cleaning.

Formation of a composite interface is a multiscale phenomenon which depends upon the relative sizes of the liquid droplet and roughness details. Stability of the composite interface is an important issue. Even though it may be geometrically possible for the system to become composite, it may not be energetically profitable for the liquid to penetrate into valleys between asperities to form a homogenous interface. The destabilizing factors include capillary waves, nanodroplet condensation, surface inhomogeneity, and liquid pressure. *Nosonovsky* and *Bhushan* [47.210] have reported that convex surfaces lead to a stable interface. Microstructure resists capillary waves present on the liquid–air surface, and nanostructures prevent nanodroplets from filling the valleys between the asperities and pin the droplet. Therefore, hierarchical structure is required to resist these scale-dependent mechanisms resulting in high contact angle and low contact angle hysteresis.

47.5.2 Fabrication and Characterization of Microstructures, Nanostructures, and Hierarchical Structures

Various structures have been fabricated, and characterization of contact angles and adhesion and friction has been carried out, to validate modeling predictions and provide design guidelines [47.204, 206, 209, 211, 214].

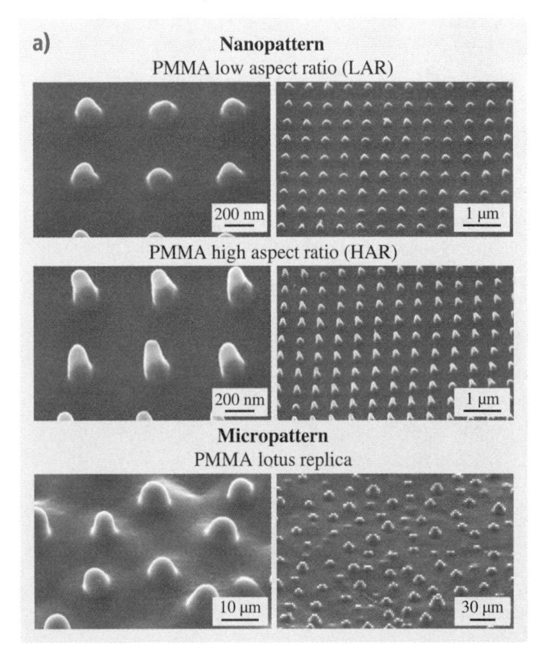

Fig. 47.47 (a) SEM micrographs of the two nanopatterned polymer surfaces (shown using two magnifications to show both the asperity shape and the asperity pattern on the surface) and the micropatterned polymer surface (lotus pattern, which has only microstructures on the surface)

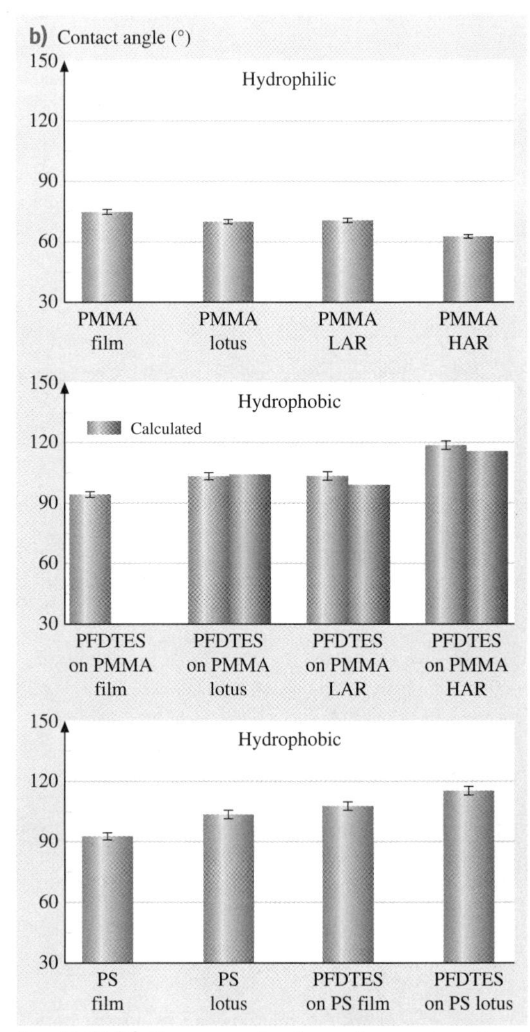

Fig. 47.47 (b) Contact angles for various patterned surfaces on PMMA and PS polymers, and calculated values using the Wenzel equation (after [47.206])

Micro- and Nanopatterned Polymer Surfaces

Nanopatterned poly(methyl methacrylate) (PMMA) surfaces were fabricated using soft lithography. To realize a micropatterned sample, a low-resolution replica of a lotus leaf was made and samples were fabricated using PMMA and polystyrene (PS) [47.206]. Figure 47.47a shows SEM images of the samples with nanopatterns with two aspect ratios using PMMA and with micropatterns of lotus replicas using PMMA. The PMMA surfaces were hydrophilic, and were made hydrophobic by coating them with a self-assembled monolayer of perfluorodecyltriethoxysilane (PFDTES). The static contact angles of various samples are presented in Fig. 47.47b. For hydrophilic surfaces, the contact angle decreases with roughness, while for hydrophobic surfaces it increases. Using the Wenzel equation, the contact angles of the hydrophobic nanopatterned and micropatterned surfaces were calculated using the contact angle of the flat surfaces and R_f. The measured contact angles of both nanopatterned samples are higher than the calculated values, whereas for the lotus pattern these are comparable. This suggests that nanopatterns benefit from air pocket formation. Furthermore, pinning at the top of the nanopatterns stabilizes the droplet. For the PS material, the contact angle of the lotus pattern also increased with increasing roughness factor.

Micropatterned Si Surfaces

Micropatterned surfaces with a square grid of cylindrical pillars were produced from single-crystal silicon (Si) using photolithography and coated with a self-assembled monolayer, making them hydrophobic [47.206]. Micropatterns were fabricated with a given diameter and height of pillars and with a range of pitch values. Optical profiler surface height maps of a representative sample are shown in Fig. 47.48a. Presence of Wenzel or Cassie–Baxter regime depends upon various factors, including the radius of the droplet and the roughness geometry. A transition criteria has been proposed by *Bhushan* and *Jung* [47.206], which provides the relationship for droplet radius and roughness geometry for the micropattern discussed here. For a small droplet suspended on a patterned surface, the local deformation is governed by surface effects. The curvature of a droplet is governed by the Laplace equation. For the patterned surface, the maximum droop of the droplet occurs in the center of the square formed by the four pillars, as shown in Fig. 47.48a. The maximum droop of the droplet can be found in the middle of two diagonally separated pillars, as shown in the figure, i.e., $(\sqrt{2}P-D)^2/8R$. If the droop is greater than the depth of the cavity, then the droplet will just contact the bottom of the cavities between pillars. If it is much greater, transition occurs from the Cassie–Baxter to the Wenzel regime for

$$\frac{\left(\sqrt{2}P-D\right)^2}{R} \geq H\,. \tag{47.6}$$

Figure 47.48b shows the static contact angle, contact angle hysteresis, and tilt angle as a function of pitch for a droplet with 1 mm radius (5 μl volume). The con-

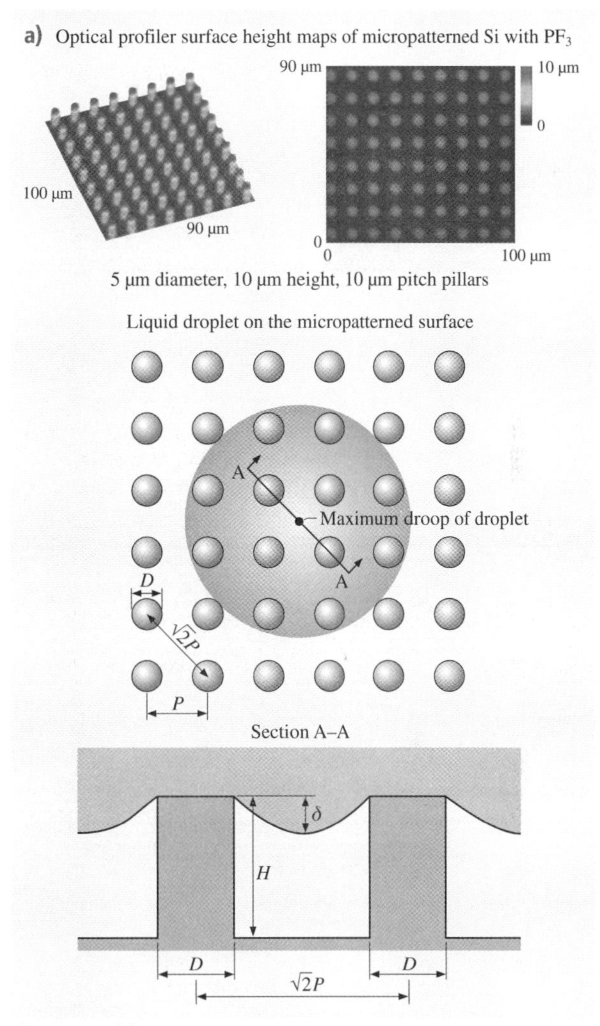

Fig. 47.48 (a) Surface height maps of a micropatterned surface using an optical profiler and a liquid droplet on the micropatterned surface, shown to obtain the transition criteria

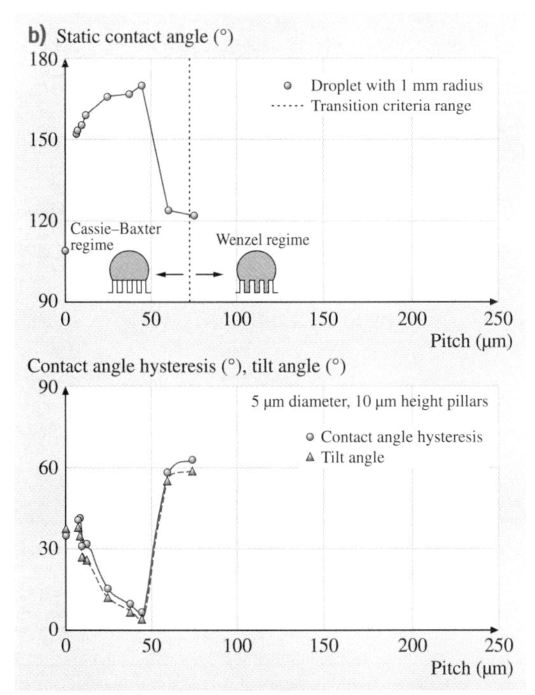

Fig. 47.48 (b) Static contact angle [*dotted line* represents the transition criteria range obtained using (47.6)] and contact angle hysteresis and tilt angles as a function of pitch value for a droplet with 1 mm radius (5 μl volume). Data at zero pitch correspond to a flat sample

tact angle of selected patterned surfaces is much higher than that of the flat surface. It first increases with increasing pitch values, then drops rapidly to a value slightly higher than that for the flat surface. In the first portion of the curve, it jumps to a high value corresponding to a superhydrophobic surface and continues to increase because open air space increases with increasing pitch, responsible for the greater propensity for air pocket formation. The sudden drop at a pitch value of $\approx 50\,\mu m$ corresponds to the transition from the Cassie–Baxter to the Wenzel regime. The dotted line corresponds to the value predicted from the transition criteria presented in (47.6); the measured and predicted values are close.

Figure 47.48b shows contact angle hysteresis and tilt angle as a function of pitch. Both angles are comparable. The angle first increases with increasing pitch, which has to do with pinning of the droplet at the sharp edges of the micropillars. As the pitch increases, there is greater propensity for air pocket formation and fewer sharp edges per unit area, which is responsible for the sudden drop in angle. Above a pitch value of $50\,\mu m$, the angle increases very rapidly because of transition to the Wenzel regime.

Droplet evaporation experiments have been conducted to investigate how droplet size influences this transition [47.206]. Figure 47.48c shows the radius of a droplet at which transition occurs as a function of pitch values. The experimental results (circles) are compared with the transition criterion (solid line, 47.6). It is found that the critical radius of impalement is in good agreement with our predictions. The critical radius of the droplet increases linearly with the pitch value. For surfaces with small pitch, the critical radius of the droplet can become quite small. This figure also shows optical micrographs of a water droplet before and just

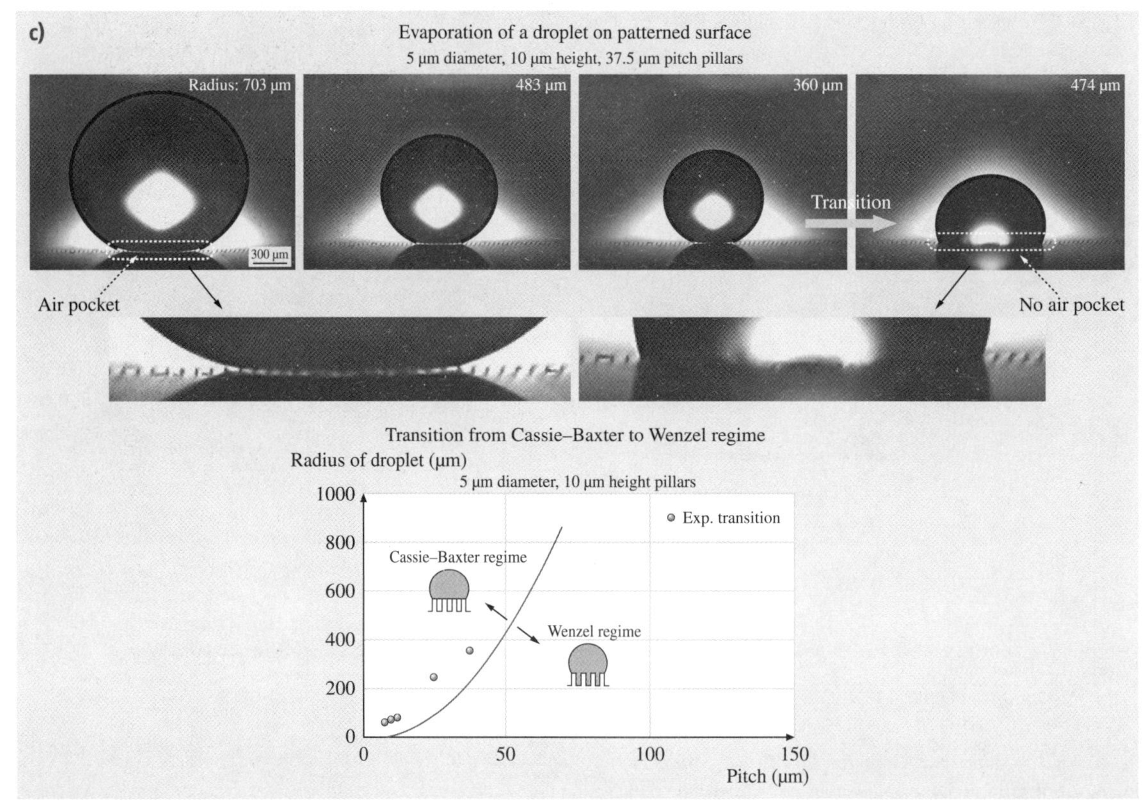

Fig. 47.48 (c) Radius of droplet for the regime transition as a function of pitch values. The experimental results (*circles*) are compared with the transition criterion (*solid line*, (47.6)) for the patterned surfaces with different pitch values. This figure also shows optical micrographs of a water droplet before and just after the transition (after [47.206])

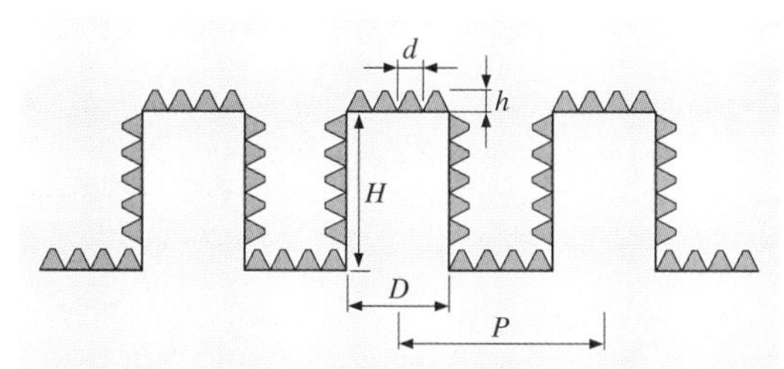

Fig. 47.49 Schematic of an ideal hierarchical surface. Microasperities consist of circular pillars with diameter D, height H, and pitch P. Nanoasperities consist of pyramidal nanoasperities of height h and diameter d with rounded tops (after [47.206])

after transition. Before the transition, air pockets are clearly visible at the bottom area of the droplet, but after the transition, air pockets are not found at the bottom of the droplet.

Based on the data above, one can achieve a very high contact angle ($\approx 170°$) and very low contact angle hysteresis ($\approx 2°$) at the critical pitch value.

Surfaces with Microstructure, Nanostructure, and Hierarchical Structure

It has been reported earlier that a hierarchical surface is needed to a develop a composite interface with high stability. The structure of an ideal hierarchical surface is shown in Fig. 47.49. The asperities should be high enough that the droplet does not touch the valleys. As an example, for a structure with circular pillars, the following relationship should hold for a composite interface: $(\sqrt{2}P - D)^2/R < H$ (47.6). As an example, for a droplet with a radius on the order of 1 mm or larger, a value of H of the order of 30 μm, and D of the order of 15 μm, a P of the order of 130 μm is optimum. Nanoasperities can pin the liquid–air interface and thus prevent liquid from filling the valleys between asperities. They are also required to support nanodroplets, which may condense in the valleys between large asperities. Therefore, nanoasperities should have a small pitch to handle nanodroplets, with radius < 1 mm down to a few nm. Structures with values of h of the order of 10 nm and d of the order of 100 nm can be easily fabricated.

Bhushan et al. [47.204, 209, 211] and *Koch* et al. [47.214] fabricated surfaces with microstructure, nanostructure, and hierarchical structure. A two-step molding process was used to fabricate microstructures by creating identical copies of a micropatterned Si surface and lotus leaves. Nanostructures were created by self-assembly of evaporating synthetic and plant waxes. Alkanes of varying chain length are common hydrophobic compounds of plant waxes. The alkane n-hexatriacontane ($C_{36}H_{74}$) was used for the development of platelet nanostructures. Tubule-forming waxes,

Fig. 47.50 (a) SEM micrographs of hierarchical structure using lotus and micropatterned Si replicas

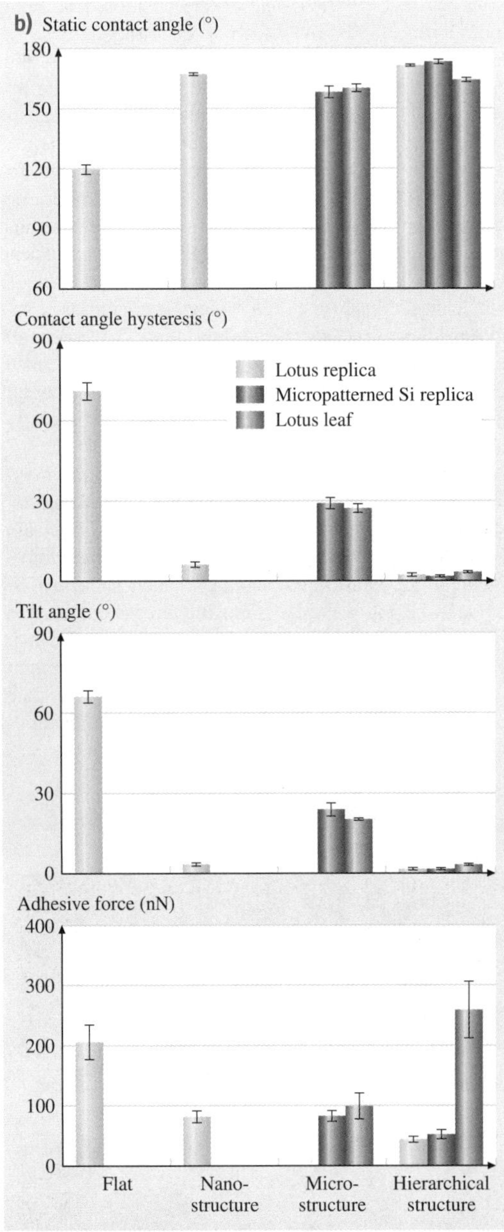

Fig. 47.50 (b) Bar chart showing the measured static contact angle, contact angle hysteresis, and tilt angle on various structures. The bar chart also shows adhesive forces for various structures. The *error bars* represents ±1 standard deviation [47.214]

isolated from leaves of *Tropaeolum majus* (L.) and *Nelumbo nucifera* (lotus), were used to create tubule structures. Figure 47.50a shows SEM micrographs of a hierarchical structure surface using lotus wax on lotus replica and micropatterned Si replica. The amount of wax used for evaporation was 0.8 $\mu g/mm^2$. In order to grow tubules, the specimens were placed in a chamber saturated with ethanol vapor at 50 °C. The static contact angle, contact angle hysteresis, tilt angle, and adhesive forces for various samples are shown in Fig. 47.50b. Figure 47.50b shows that the highest static contact angles of 173°, lowest contact angle hysteresis of 1°, and tilting angle varying between 1 and 2° were found for the hierarchical structured Si replica. The hierarchical structured lotus leaf replica shows a static contact angle of 171°, the same contact angle hysteresis (2°), and tilt angles of 1–2°, similar to that of the hierarchical Si replica. Fresh lotus leaf surface was reported to have a static contact angle of 164°, contact angle hysteresis of 3°, and a tilt angle of 3°. Therefore, the artificial hierarchical surfaces showed higher static contact angle and lower contact angle hysteresis. Structural differences between the original lotus leaf and the artificial lotus leaf de-

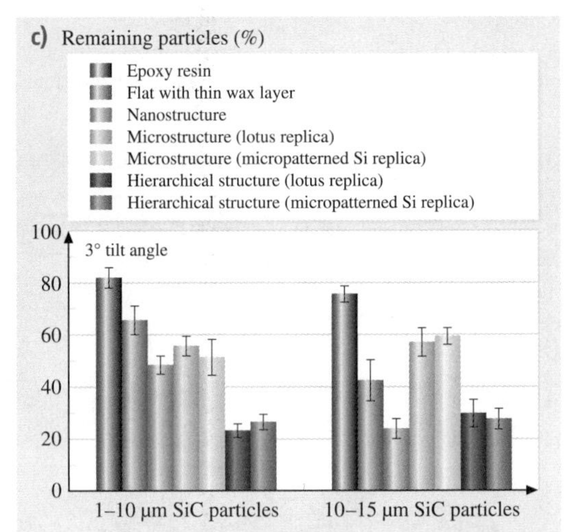

Fig. 47.50 (c) Bar charts showing remaining particles after applying droplets with nearly zero kinetic energy on various structures fabricated using lotus wax using 1–10 μm and 10–15 μm SiC particles. The experiments on the surfaces with lotus wax were carried out on stages tilted at 3°. The *error bars* represent ±1 standard deviation (after [47.204])

scribed here are limited to a difference in wax tubule length, which is 0.5–1 μm longer in the artificial lotus leaf.

Self-cleaning efficiency tests have also been carried out by *Bhushan* et al. [47.204]. The samples were exposed to contaminants in a contamination glass chamber and then cleaned with water droplets. Figure 47.50c shows that none of the investigated surfaces was fully cleaned by water rinsing. Most particles (70–80%) remained on smooth surfaces, and 50–70% of particles were found on microstructured surfaces. Most particles were removed from the hierarchical structured surfaces, but $\approx 30\%$ of the particles remained. A clear difference in particle removal, independent of particle size, was only found for flat and nanostructured surfaces, where larger particles were removed with higher efficiency. Observations of droplet behavior during movement on the surfaces showed that droplets rolled only on the hierarchical structured surfaces. On flat, microstructured, and nanostructured surfaces, the first droplets applied did not move, but continuous application of water droplets increased their volumes and led to sliding of these larger droplets. During this, some of the particles were removed from the surfaces. However, the droplets rolling on hierarchical structures did not collect dirt particles trapped in the cavities of the microstructures. The data clearly show that hierarchical structures have superior cleaning efficiency.

47.5.3 Summary

In the Wenzel regime, an increase in roughness on a hydrophilic surface decreases the contact angle, whereas on a hydrophobic surface it increases contact angle. However, in the Cassie–Baxter regime, air pocket formation can change a hydrophilic surface to a hydrophobic surface. Based on studies to explore the effect of droplet size and roughness geometry, the transition from the Cassie–Baxter regime to the Wenzel regime occurs below a certain radius of droplet and/or above a certain pitch value.

For fluid flow applications, for drag reduction, a surface should have high contact angle and low contact angle hysteresis. This condition should be achieved by a high value of fractional liquid–air contact area, f_{LA}, and relatively low value of roughness factor, R_f.

The fabricated hierarchical surface shows a high static contact angle of $\approx 170°$ and low contact angle hysteresis of $\approx 2°$, which provide superior superhydrophobic and self-cleaning surfaces.

47.6 Component-Level Studies

47.6.1 Surface Roughness Studies of Micromotor Components

Most of the friction forces resisting motion in a micromotor are concentrated near the rotor–hub interface, where continuous physical contact occurs. Surface roughness usually has a strong influence on the friction characteristics on the micro-nanoscale. A catalog of roughness measurements on various components of a MEMS device does not exist in the literature. Using an AFM, measurements on various component surfaces were made for the first time by *Sundararajan* and *Bhushan* [47.217]. Table 47.7 shows various surface roughness parameters obtained from 5 μm × 5 μm

Table 47.7 Surface roughness parameters and microscale coefficient of friction for various micromotor component surfaces measured using an AFM. Mean and $\pm 1\sigma$ values are given

	RMS roughness[a] (nm)	P–V distance[a] (nm)	Skewness[a], Sk	Kurtosis[a] K	Coefficient of microscale friction[b] (μ)
Rotor topside	21 ± 0.6	225 ± 23	1.4 ± 0.30	6.1 ± 1.7	0.07 ± 0.02
Rotor underside	14 ± 2.4	80 ± 11	-1.0 ± 0.22	3.5 ± 0.50	0.11 ± 0.03
Stator topside	19 ± 1	246 ± 21	1.4 ± 0.50	6.6 ± 1.5	0.08 ± 0.01

[a] Measured from a tapping-mode AFM scan of size 5 μm × 5 μm using a standard Si tip scanning at 5 μm/s in a direction orthogonal to the long axis of the cantilever

[b] Measured using an AFM in contact mode at 5 μm × 5 μm scan size using a standard Si_3N_4 tip scanning at 10 μm/s in a direction parallel to the long axis of the cantilever

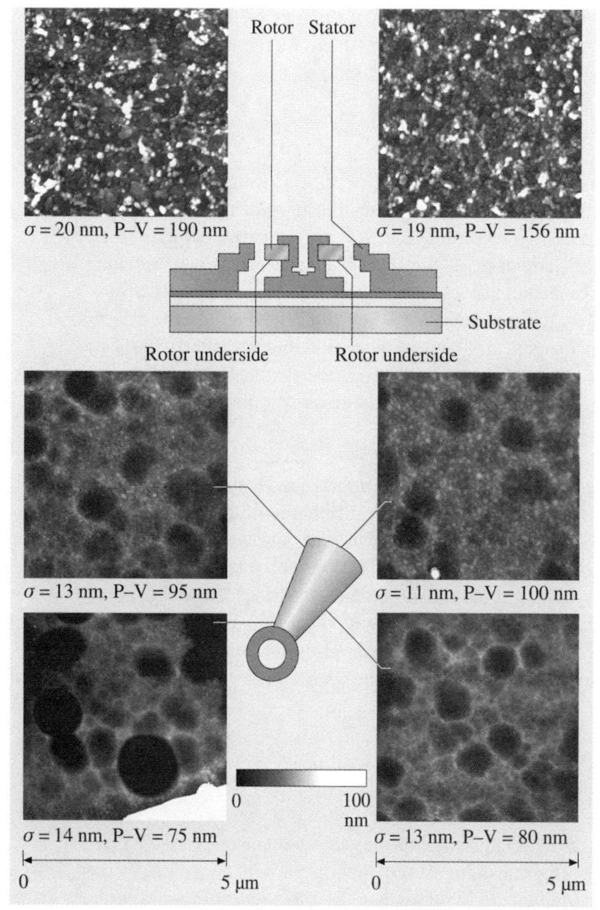

Fig. 47.51 Representative AFM surface height images obtained in tapping mode (5 μm × 5 μm scan size) of various component surfaces of a micromotor; root-mean-square (RMS) roughness and peak-to-valley values of the surfaces are given. The underside of the rotor exhibits drastically different topography from the topside (after [47.217])

scans of the various component surfaces of several unlubricated micromotors using the AFM in tapping mode. A surface with a Gaussian height distribution should have a skewness of 0 and kurtosis of 3. Although the rotor and stator top surfaces exhibit comparable roughness parameters, the underside of the rotors exhibits lower root-mean-square (RMS) roughness and peak-to-valley distance values. More importantly, the rotor underside shows negative skewness and lower kurtosis than the topsides, both of which are conducive to high real area of contact and hence high friction [47.80, 82].

The rotor underside also exhibits a higher coefficient of microscale friction than the rotor topside and stator, as shown in Table 47.7. Figure 47.51 shows representative surface height maps of the various surfaces of a micromotor measured using the AFM in tapping mode. The rotor underside exhibits varying topography from the outer edge to the middle and inner edge. At the outer edges, the topography shows smaller circular asperities, similar to the topside. The middle and inner regions show deep pits with fine edges that may have been created by the etchants used for etching of the sacrificial layer. It is known that etching can affect the roughness of surfaces in surface micromachining. The residence time of the etchant near the inner region is high, which is responsible for the larger pits. Figure 47.52 shows the roughness of the surface directly beneath the rotors (the base polysilicon layer). There appears to be a difference in the roughness between the portion of this surface that was initially underneath the rotor (region B) during fabrication and the portion that was away from the rotor and hence always exposed (region A). The former re-

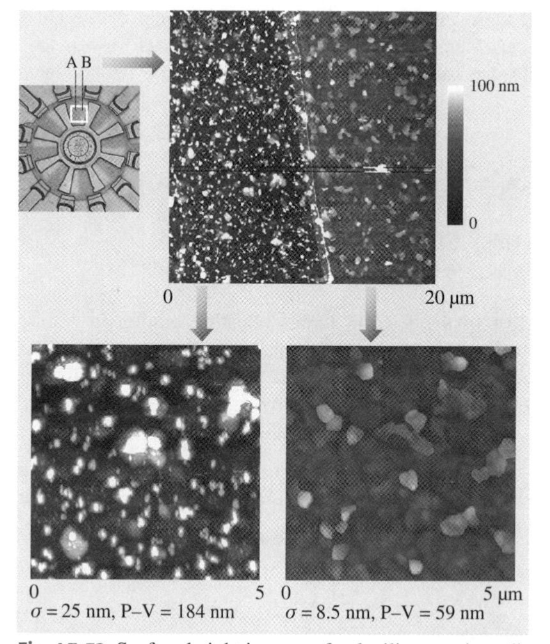

Fig. 47.52 Surface height images of polysilicon regions directly below the rotor. Region A is away from the rotor while region B was initially covered by the rotor prior to the release etch of the rotor. During this step, slight movement of the rotor caused region B to be exposed (after [47.217])

gion shows lower roughness than the latter region. This suggests that the surfaces at the rotor–hub interface that come into contact at the end of the fabrication process exhibit large real areas of contact that result in high friction.

47.6.2 Adhesion Measurements of Microstructures

Surface force apparatus (SFA) and AFMs are used to measure adhesion on micro- to nanoscales between two surfaces. In the SFA, adhesion of liquid films sandwiched between two curved and smooth surfaces is measured. In an AFM, as discussed earlier, adhesion between a sharp tip and the surface of interest is measured. The propensity for adhesion between two surfaces can be evaluated by studying the tendency of microstructures with well-defined contact areas, covering a wide spectrum of suspension compliances, to stick to the underlying substrate. The test structures which have been used include the cantilever beam array (CBA) technique with different lengths [47.218–221] and stand-off multiple dimples mounted on microstructures with a range of compliances, standing above a substrate [47.222]. The CBA technique, which is more commonly used, utilizes an array of micromachined polysilicon beams (for Si MEMS applications) on the mesoscopic length scale, anchored to the substrate at one end and with different lengths parallel to the surface. It relies on peeling and detachment of cantilever beams. The change in free energy or the reversible work done to separate unit areas

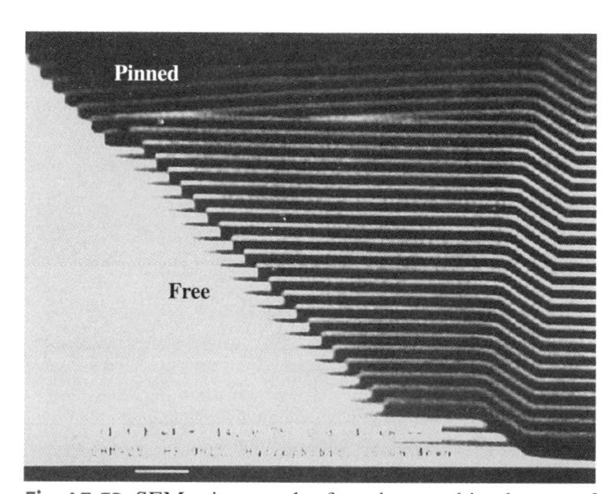

Fig. 47.53 SEM micrograph of a micromachined array of polysilicon cantilever beams of increasing length. The micrograph shows the onset of pinning for beams longer than 34 μm (after [47.218])

of two surfaces from contact is called the work of adhesion. To measure the work of adhesion, electrostatic actuation is used to bring all beams into contact with the substrate (Fig. 47.53) [47.218, 220]. Once the actuation force is removed, the beams begin to peel themselves off the substrate, which can be observed with an optical interference microscope (e.g., a Wyko surface profiler). For beams shorter than a characteristic length, the so-called detachment length, their stiffness is sufficient to free them completely from the substrate underneath. Beams larger than the detachment length remain adhered. The beams at the transition region start to detach and remain attached to the substrate just at the tips. For this case, by equating the elastic energy stored within the beam and the beam–substrate interfacial energy, the work of adhesion W_{ad} can be calculated by [47.218]

$$W_{ad} = \frac{3Ed^2t^3}{8\ell_d^4}, \tag{47.7}$$

where E is the Young's modulus of the beam, d is the spacing between the undeflected beam and the substrate, t is the beam thickness, and ℓ_d is the detachment length. The technique has been used to screen methods for adhesion reduction in polysilicon microstructures.

47.6.3 Microtriboapparatus for Adhesion, Friction, and Wear of Microcomponents

To measure adhesion, friction, and wear between two microcomponents, a microtriboapparatus has been used. Figure 47.54 shows a schematic of a microtriboapparatus capable of using MEMS components for tests [47.223]. In this apparatus, an upper specimen, mounted on a soft cantilever beam, comes into contact with a lower specimen mounted on a lower specimen holder. The apparatus consists of two piezos (x- and z-piezos) and four fiber-optic sensors (x- and z-displacement sensors, and x- and z-force sensors). For adhesion and friction studies, z- and x-piezos are used to bring the upper specimen and lower specimen into contact and to apply a relative motion in the lateral direction, respectively. The x- and z-displacement sensors are used to measure the lateral position of the lower specimen and vertical position of the upper specimen, respectively. The x- and z-force sensors are used to measure friction force and the normal load/adhesive force between these two specimens, respectively, by monitoring the deflection of the cantilever.

As most MEMS/NEMS devices are fabricated from silicon, study of silicon-on-silicon contacts is important. This contact was simulated by a flat single-crystal

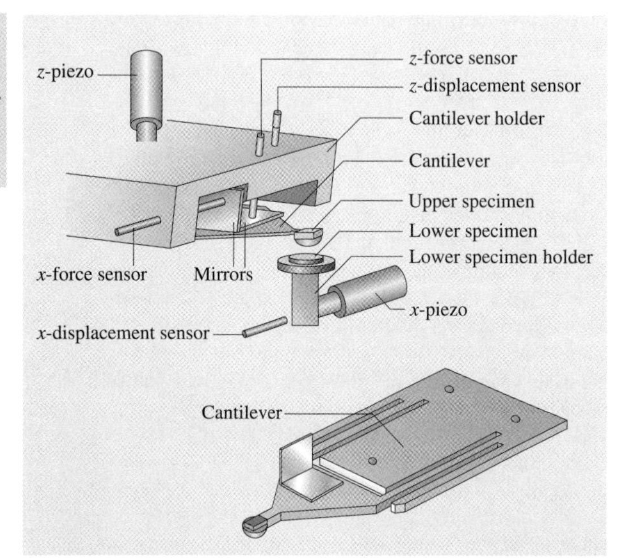

Fig. 47.54 Schematic of the microtriboapparatus including specially designed cantilever (with two perpendicular mirrors attached to the end), lower specimen holder, two piezos (x- and z-piezos), and four fiber-optic sensors (x- and z-displacement sensors and x- and z-force sensors) (after [47.223])

Si(100) wafer (phosphorus-doped) specimen sliding against a single-crystal Si(100) ball (1 mm in diameter, 5×10^{17} atoms/cm^3 boron doped) mounted on a stainless-steel cantilever [47.223, 224]. Both of them have a native oxide layer on their surfaces. The other materials studied were 10 nm-thick DLC deposited by filtered cathodic arc deposition on Si(100), 2.3 nm-thick chemically bonded PFPE (Z-DOL, BW) on Si(100), and hexadecane thiol (HDT) monolayer on evaporated Au(111) film to investigate their anti-adhesion performance.

It is well known that, in computer rigid disk drives, the adhesive force between a magnetic head and a magnetic disk increases rapidly with increasing rest time [47.34]. Considering that adhesion and friction are the major issues that lead to the failure of MEMS/NEMS devices, the effect of rest time on the microscale on Si(100), DLC, PFPE, and HDT was studied; the results are summarized in Fig. 47.55a. It is found that the adhesive force of Si(100) increases logarithmically with rest time up to a certain equilibrium time ($t = 1000$ s), after which it remains constant. Figure 47.55a also shows that the adhesive force of DLC, PFPE, and HDT does not change with rest time.

Single-asperity contact modeling of the dependence of meniscus force on rest time has been carried out by *Chilamakuri* and *Bhushan* [47.225], and the modeling results (Fig. 47.55b) verify experimental observations. Due to the presence of a thin film of adsorbed water on Si(100), a meniscus forms around the contacting asperities and grows with time until equilibrium occurs, which causes the effect of rest time on its adhesive force. The adhesive forces of DLC, PFPE, and HDT do not change with rest time, which suggests that the water meniscus is not present on their surfaces.

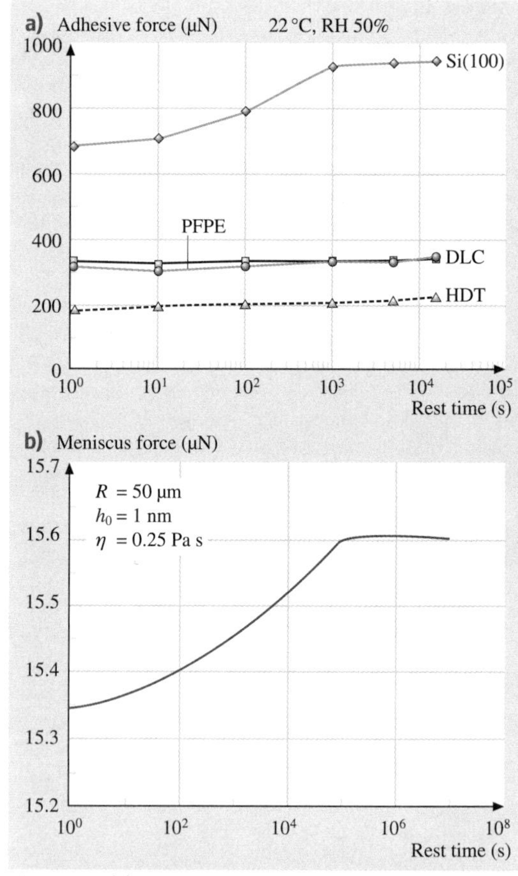

Fig. 47.55 **(a)** Influence of rest time on the adhesive force of Si(100), DLC, chemically bonded PFPE, and HDT. **(b)** Single-asperity contact modeling results of the effect of rest time on the meniscus force for an asperity of radius R in contact with a flat surface with a water film of thickness of h_0 and absolute viscosity of η_0 (after [47.225])

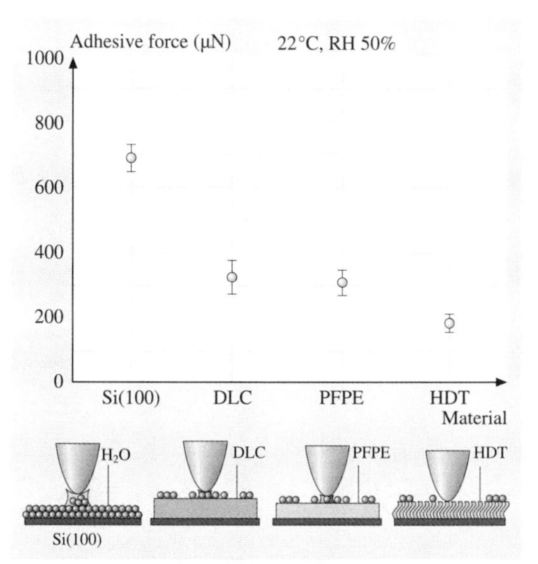

Fig. 47.56 Adhesive forces of Si(100), DLC, chemically bonded PFPE, and HDT at ambient condition, and a schematic showing the relative size of the water meniscus on different specimens

The measured adhesive forces of Si(100), DLC, PFPE, and HDT at rest time of 1 s are summarized in Fig. 47.56, which shows that the presence of solid films of DLC, PFPE, and HDT greatly reduces the adhesive force of Si(100), whereas HDT film has the lowest adhesive force. It is well known that the native oxide layer (SiO_2) on top of the Si(100) wafer exhibits hydrophilic properties, and water molecules, produced by capillary condensation of water vapor from the environment, can easily be adsorbed on this surface. The condensed water will form a meniscus as the upper specimen approaches the lower specimen surface. The meniscus force is a major contributor to the adhesive force. In the case of DLC, PFPE, and HDT, the films are found to be hydrophobic based on contact angle measurements, and the amount of condensed water vapor is low as compared with that on Si(100). It should be noted that the measured adhesive force is generally higher than that measured in AFM, because the larger radius of Si(100) ball as compared with that of an AFM tip induces larger meniscus and van der Waals forces.

To investigate the effect of velocity on friction, the friction force was measured as a function of velocity, as summarized in Fig. 47.57a. This indicates that, for Si(100), the friction force initially decreases with increasing velocity until equilibrium occurs. Figure 47.57a also indicates that the velocity has almost no effect on the friction properties of DLC, PFPE, and HDT. This implies that the friction mechanisms of DLC, PFPE, and HDT do not change with velocity. For Si(100), at high velocity, the meniscus is broken and does not have enough time to rebuild. In addition, it is also believed that tribochemical reaction plays an important role. High velocity leads to tribochemical reactions of Si(100) (which has SiO_2 native oxide) with water molecules to form $Si(OH)_4$ film. This film is removed and continuously replenished during sliding. The $Si(OH)_4$ layer at the sliding surface is known to be of low shear strength. The breaking of the water meniscus and the formation of the $Si(OH)_4$ layer results in a decrease in the friction force of Si(100). The DLC, PFPE, and HDT surfaces exhibit hydrophobic properties, and can adsorb few water molecules under ambient conditions. The aforementioned meniscus breaking and tribochemical reaction mechanisms do not exist for these films. Therefore, their friction force does not change with velocity.

The influence of relative humidity was studied in an environmentally controlled chamber. The adhesive force and friction force were measured by making measurements at increasing relative humidity; the results are summarized in Fig. 47.57b, which shows that, for Si(100), the adhesive force increases with relative humidity, but the adhesive force of DLC and PFPE only shows a slight increase when humidity is $> 45\%$, while the adhesive force of HDT does not change with humidity. Figure 47.57b also shows that, for Si(100), the friction force increases with an increase in relative humidity up to 45%, and then it shows a slight decrease with further increase in the relative humidity. For PFPE, there is an increase in the friction force when humidity is $> 45\%$. In the whole testing range, relative humidity does not have any apparent influence on the friction properties of DLC and HDT. In the case of Si(100), the initial increase of relative humidity up to 45% causes more adsorbed water molecules, which form a larger water meniscus that leads to an increase of friction force. However, at very high humidity of 65%, large quantities of adsorbed water can form a continuous water layer that separates the tip and sample surfaces, and acts as a kind of lubricant, which causes a decrease in the friction force. For PFPE, dewetting of lubricant film at humidity $> 45\%$ results in an increase in adhesive and friction forces. The DLC and HDT surfaces show hydrophobic properties, and increasing relative humidity does not play much of a role in their friction force.

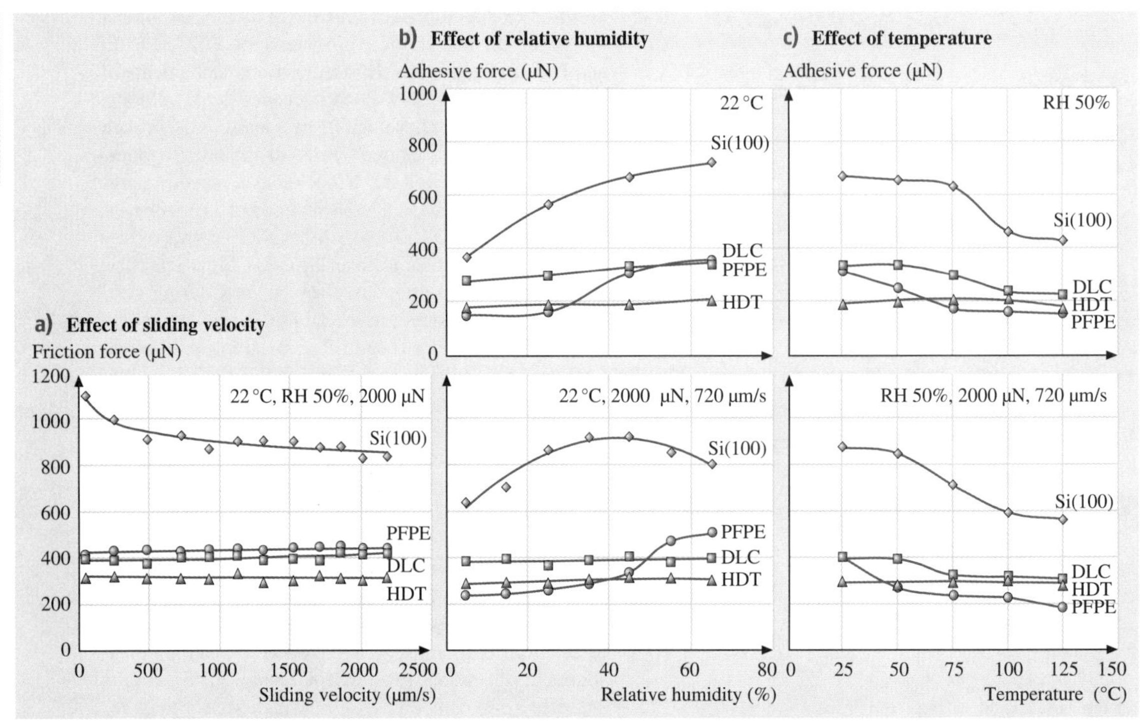

Fig. 47.57a–c The influence of (a) sliding velocity on the friction forces, (b) relative humidity on the adhesive and friction forces, and (c) temperature on the adhesive and friction forces of Si(100), DLC, chemically bonded PFPE, and HDT

The influence of temperature was studied using a heated stage. The adhesive force and friction force were measured by making measurements at increasing temperatures from 22 °C to 125 °C. The results are presented in Fig. 47.57c, which shows that, once the temperature is > 50 °C, increasing temperature causes a significant decrease of adhesive and friction forces of Si(100) and a slight decrease in the case of DLC and PFPE. However, the adhesion and friction forces of HDT do not show any apparent change with test temperature. At high temperature, desorption of water and reduction of the surface tension of water lead to the decrease of adhesive and friction forces of Si(100), DLC, and PFPE. However, in the case of HDT film, as only a few water molecules are adsorbed on the surface, the aforementioned mechanisms do not play a large role. Therefore, the adhesive and friction forces of HDT do not show any apparent change with temperature. Figure 47.57 shows that, in the whole velocity, relative humidity, and temperature test range, the adhesive force and friction force of DLC, PFPE, and HDT are always smaller than that of Si(100), whereas HDT has the smallest value.

To summarize, several methods can be used to reduce adhesion in microstructures. MEMS/NEMS surfaces can be coated with hydrophobic coatings such as PFPEs, SAMs, and passivated DLC coatings. It should be noted that other methods to reduce adhesion include the formation of dimples on the contact surfaces to reduce contact area [47.13, 80, 82, 84, 154, 155, 220]. Furthermore, an increase in the hydrophobicity of solid surfaces (high contact angle, approaching 180°) can be achieved by using surfaces with suitable roughness, in addition to lowering their surface energy [47.205, 206, 210]. The hydrophobicity of surfaces is dependent upon a subtle interplay between surface chemistry and mesoscopic topography. The self-cleaning mechanism or so-called *lotus effect* is closely related to the superhydrophobic properties of the biological surfaces, which usually show microsculptures on specific scales.

47.6.4 Static Friction Force (Stiction) Measurements in MEMS

In MEMS devices involving parts in relative motion to each other, such as micromotors, large friction forces become the factor limiting the successful operation and reliability of the device. It is generally known that most micromotors cannot be rotated as manufactured and require some form of lubrication. It is therefore critical to determine the friction forces present in such MEMS devices. To measure the static friction of a rotor-bearing interface in a micromotor in situ, *Tai* and *Muller* [47.226] measured the starting torque (voltage) and pausing position for different starting positions under a constant bias voltage. A friction–torque model was used to obtain the coefficient of static friction. To measure the in situ kinetic friction of the turbine and gear structures, *Gabriel* et al. [47.227] used a laser-based measurement system to monitor the steady-state spins and decelerations. *Lim* et al. [47.228] designed and fabricated a polysilicon microstructure to measure the static friction of various films in situ. The microstructure consisted of a shuttle suspended above an underlying electrode by a folded beam suspension. A known normal force was applied, and the lateral force was measured to obtain the coefficient of static friction. *Beerschwinger* et al. [47.229] developed a cantilever-deflection rig to measure the friction in LIGA-processed micromotors [47.230]. These techniques employ indirect methods to determine the friction forces, or involve fabrication of complex structures.

A novel technique to measure the static friction force (stiction) encountered in surface-micromachined polysilicon micromotors using an AFM has been developed by *Sundararajan* and *Bhushan* [47.217]. Continuous physical contact occurs during rotor movement (rotation) in micromotors between the rotor and lower hub flange. In addition, contact occurs at other locations between the rotor and the hub surfaces and between the rotor and the stator. Friction forces will be present at these contact regions during motor operation. Although the actual distribution of these forces is not known, they can be expected to be concentrated near the hub, where there is continuous contact. If we therefore represent the static friction force of the micromotor as a single force F_S acting at point P_1 (as shown in Fig. 47.58a), then the magnitude of the frictional torque about the center of the motor (O) that must be overcome before rotor movement can be initiated is

$$T_S = F_S \ell_1 , \tag{47.8}$$

where ℓ_1 is the distance OP_1, which is assumed to be the average distance from the center at which the friction force F_S occurs. Now consider an AFM tip moving against a rotor arm in a direction perpendicular to the long axis of the cantilever beam (the rotor arm edge closest to the tip is parallel to the long axis of the cantilever beam), as shown in Fig. 47.58a. When the tip encounters the rotor at point P_2, the tip will twist, generating a lateral force between the tip and the rotor (event A in Fig. 47.58b). This reaction force will generate a torque about the center of the motor. Since the tip is trying to move further in the direction shown, the tip will continue to twist to a maximum value, at which the lateral force between the tip and the rotor becomes high enough such that the resultant torque T_f about the center of the motor equals the static friction torque T_S. At this point, the rotor will begin to rotate, and the twist of the cantilever decreases sharply (event B in Fig. 47.58b). The twist of the cantilever is measured in the AFM as a change in the lateral deflection signal (in volts), which is the underlying concept of friction force microscopy (FFM). The change in the lateral deflection signal corresponding to the above-mentioned events as the tip approaches the rotor is shown schematically in Fig. 47.58c. The value of the peak V_f is a measure of the force exerted on the rotor by the tip just before the

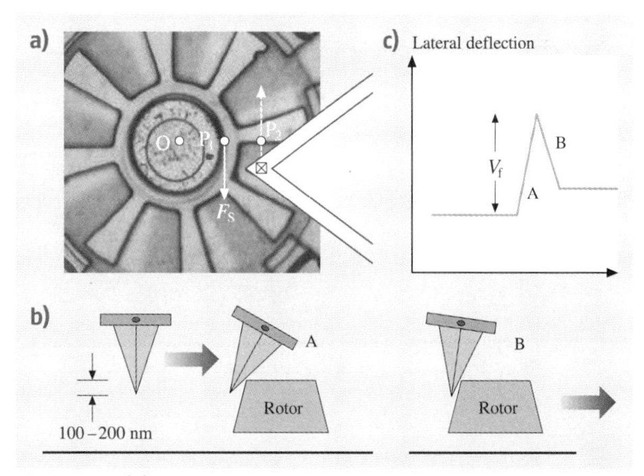

Fig. 47.58 (a) Schematic of the technique used to measure the force F_S required to initiate rotor movement using an AFM/FFM. (b) As the tip is pushed against the rotor, the lateral deflection experienced by the rotor due to the twisting of the tip prior to rotor movement is a measure of the static friction force F_S of the rotors. (c) Schematic of the lateral deflection expected during the aforementioned experiment. The peak V_f is related to the state of the rotor (after [47.217])

static friction torque is matched and the rotor begins to rotate.

Using this technique, the viability of PFPE lubricants for micromotors has been investigated, and the effect of humidity on the friction forces of unlubricated and lubricated devices was studied as well. Figure 47.59 shows static friction forces, normalized by the weight of the rotor, of unlubricated and lubricated micromotors as a function of rest time and relative humidity. Rest time here is defined as the time elapsed between the first experiment conducted on a given motor (solid symbol at time zero) and subsequent experiments (open symbols). Each open symbol data point is an average of six measurements. It can be seen that, for the unlubricated motor and the motor lubricated with a bonded layer of Z-DOL(BW), the static friction force is highest for the first experiment and then drops to an almost constant level. In the case of the motor with an as-is mobile layer of Z-DOL, the values remain very high up to 10 days after lubrication. In all cases, there is negligible difference in the static friction force at 0% and 45% RH. At 70% RH, the unlubricated motor exhibits a substantial increase in the static friction force, while the motor with bonded Z-DOL shows no increase in static friction force due to the hydrophobicity of the lubricant layer. The motor with an as-is mobile layer of the lubricant shows consistently high values of static friction force that vary little with humidity.

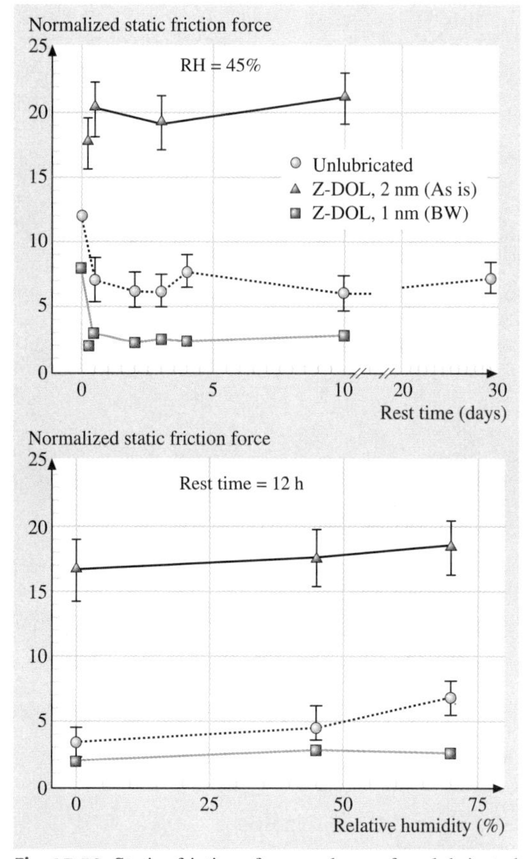

Fig. 47.59 Static friction force values of unlubricated motors and motors lubricated using PFPE lubricants, normalized by the rotor weight, as a function of rest time and relative humidity. The rest time is defined as the time elapsed between a given experiment and the first experiment in which motor movement was recorded (time 0). The motors were allowed to sit at a particular humidity for 12 h prior to measurements (after [47.217])

Figure 47.60 summarizes static friction force data for two motors, M1 and M2, along with schematics of the meniscus effects for the unlubricated and lubricated surfaces. Capillary condensation of water vapor from the environment results in the formation of meniscus bridges between the contacting and near-contacting asperities of two surfaces in close proximity to each other, as shown in Fig. 47.60. For unlubricated surfaces, more menisci are formed at higher humidity, resulting in higher friction force between the surfaces. The formation of meniscus bridges is supported by the fact that the static friction force for unlubricated motors increases at high humidity (Fig. 47.60). Solid bridging may occur near the rotor–hub interface due to silica residues after the first etching process. In addition, the drying process after the final etch can result in liquid bridging formed by the drying liquid due to meniscus force at these areas [47.80, 82, 218, 219]. The initial static friction force will therefore be quite high, as evidenced by the solid data points in Fig. 47.60. Once the first movement of the rotor permanently breaks these solid and liquid bridges, the static friction force of the motors will drop (as seen in Fig. 47.60) to a value dictated predominantly by the adhesive energies of the rotor and hub surfaces, the real area of contact between these surfaces, and meniscus forces due to water vapor in the air, at which point, the effect of lubricant films can be observed. Lubrication with a mobile layer, even a thin one, results in very high static friction forces due to meniscus effects of the lubricant liquid itself at and near the contact regions. It should be noted that a motor submerged in a liquid lubricant would result in a fully flooded lubrication regime. In this case there is no meniscus contribution, and only the viscous contribution to the friction forces would be relevant. However, submerging the device in a lubricant may not be a practical method. A solid-like hydrophobic lubricant layer (such as bonded Z-DOL) results in favorable friction characteristics of the motor. The hydrophobic nature of the lubricant inhibits meniscus formation between the contact surfaces and maintains low friction even at high humidity (Fig. 47.60). This suggests that solid-like hydrophobic lubricants are ideal for lubrication of MEMS, while mobile lubricants result in increased values of static friction force.

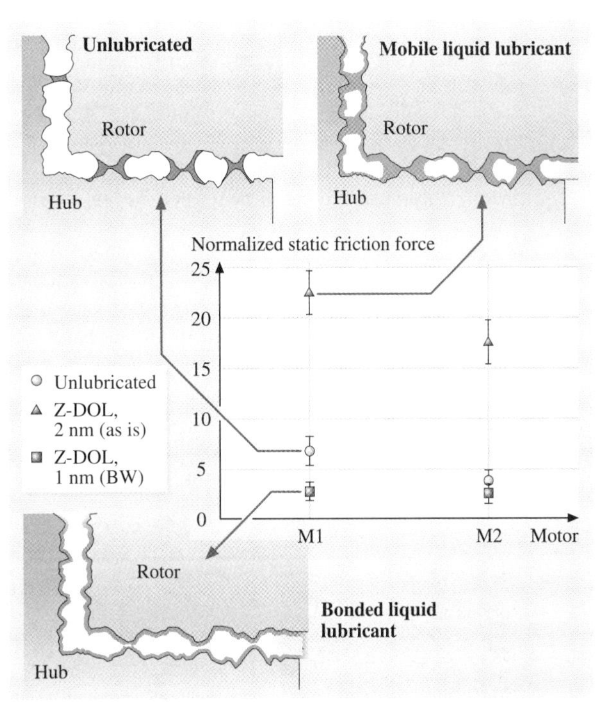

Fig. 47.60 Summary of the effects of liquid and solid lubricants on the static friction force of micromotors. Despite the hydrophobicity of the lubricant used (Z-DOL), a mobile liquid lubricant (Z-DOL, as-is) leads to very high static friction force due to increased meniscus forces, whereas a solid-like lubricant (bonded Z-DOL, BW) appears to provide some amount of reduction in static friction force

47.6.5 Mechanisms Associated with Observed Stiction Phenomena in Digital Micromirror Devices (DMD) and Nanomechanical Characterization

DMDs are used in digital projection displays, as described earlier. The DMD has a layered structure, consisting of an aluminum alloy micromirror layer, a yoke and hinge layer, and a metal layer on a CMOS memory array [47.27–29]. A blown-up view of the DMD and the corresponding AFM surface height images are presented in Fig. 47.61 [47.108]. Single-layered aluminum alloy films are used for the construction of micromirrors; these are also sometimes used for the construction of hinges, spring tips, and landing sites. The aluminum-alloy films are overwhelmingly comprised of aluminum; trace elements (including Ti and Si) are present to suppress contact spiking and electromigration, which may occur if current densities become high during electrostatic operation. Multilayered sputtered SiO_2/TiN/Al alloy films are now generally used for the landing-site structure to minimize refraction

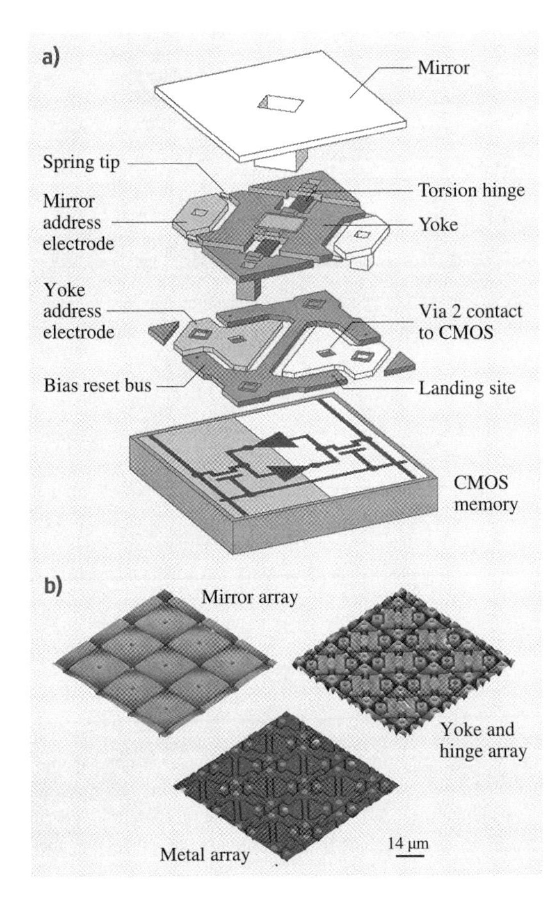

Fig. 47.61 (a) Exploded view of a DMD pixel. (b) AFM surface height images of various arrays. The DMD layers were removed by ultrasonic method (after [47.108]) ◄

throughout the visible region of the electromagnetic spectrum in order to increase the contrast ratio in projection display systems [47.231, 232]. These multilayered films are also generally used for hinges and spring tips. A low-surface-energy SAM is maintained on the surfaces of the DMD, which is packaged in a hermetic environment to minimize stiction during contact between the spring tip and the landing site. A SAM of perfluorinated n-alkanoic acid ($C_nF_{2n-1}O_2H$) (e.g., perfluorodecanoic acid or PFDA, $CF_3(CF_2)_8COOH$) applied by the vapor-phase deposition process is used. A getter strip of PFDA is included inside the hermetically sealed enclosure containing the chip, which acts as a reservoir in order to maintain a PFDA vapor within the package.

In order to identify a stuck mirror and characterize its nanotribological properties, the chip was scanned using an AFM [47.108]. It was found that it is hard to tilt the stuck micromirror back to its normal position by adding a normal load at the rotatable corner of the micromirror; this is thus called a *hard* stuck micromirror. An example of a stuck micromirror is shown in Fig. 47.62a. Once the stuck micromirror was found, the region was repeatedly scanned at a large normal load, up to 300 nN. After several scans, the stuck micromirror was removed. Once the stuck micromirror was removed, the surrounding micromirrors could also be removed by continuous scanning under a large normal load (Fig. 47.62a, bottom row). The adhesive force of the landing site underneath the stuck micromirror and the normal micromirror are presented in Fig. 47.62b, which clearly indicates that the landing site underneath the stuck micromirror has much larger adhesion. A 1 μm × 1 μm view of landing sites under stuck and

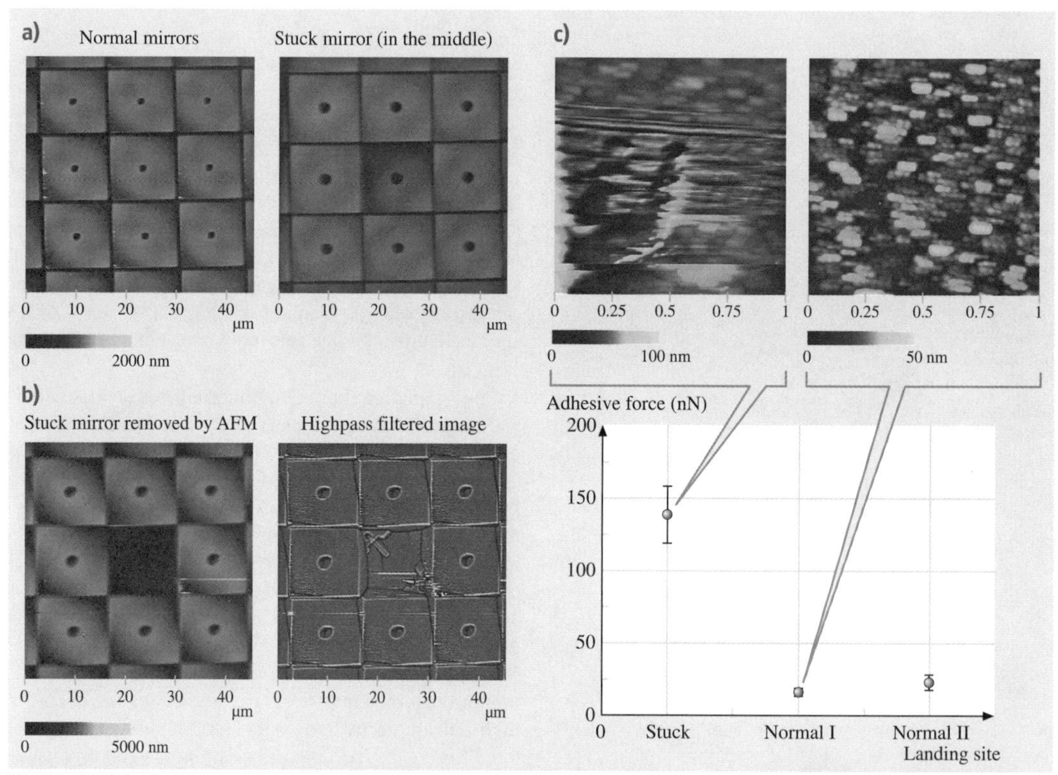

Fig. 47.62 **(a)** The *top row* shows AFM surface height images of a stuck micromirror surrounded by eight normal micromirrors. **(b)** *Left image* shows the stuck micromirror which was removed by an AFM tip after repeated scanning at high normal load. The *right image* in the bottom row presents a high-pass-filtered image, in which the residual hinge that sits underneath the removed micromirror is clearly observed. **(c)** AFM surface height images and adhesive forces of the landing sites underneath the two normal micromirrors and the stuck micromirror (after [47.108])

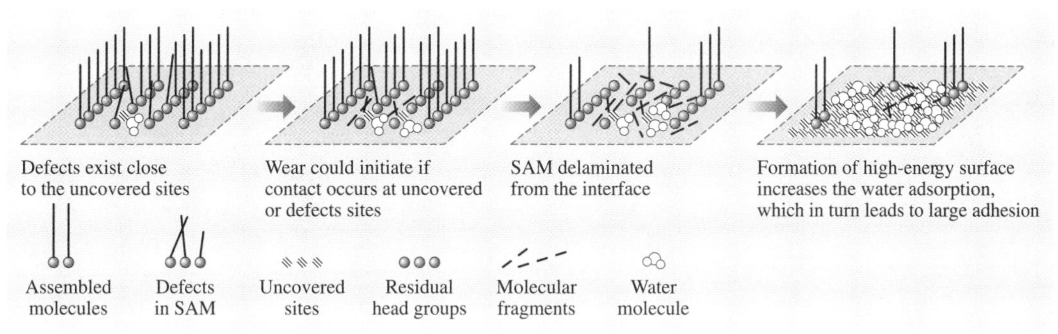

Fig. 47.63 Suggested mechanisms for wear and stiction (after [47.108])

normal micromirrors are also shown in Fig. 47.62b. The landing site under the stuck micromirror has an apparent U-shaped wear mark, which is surrounded by a smeared area.

Liu and *Bhushan* [47.108] calculated contact stresses to examine if the stresses were high enough to cause wear at the spring tip–landing site interface. The calculated contact stress value was ≈ 33 MPa, which is substantially lower than the hardness, therefore much plastic deformation and consequently wear was not expected. Wear mark was only found on a very few landing sites on the DMD, which means that the SAM coating can generally endure such high contact stresses. Based on data reported in the literature, coverage of vapor-deposited SAMs is expected to be $\approx 97\%$. The bond strength of the molecules close to the boundary of the uncovered sites is expected to be weak. Thus, the uncovered sites and the adjacent molecules are referred to as defects in the SAM coating. Occasionally, if contact occurs at the defect sites, the large cyclic stress may be close to the critical load, and lead to initial delamination of the SAM coating at the interface. The continuous contact leads to the formation of a high-surface-energy surface by exposure of the fresh substrate and formation of SAM fragments. This eventually leads to an increase in stiction by the formation of large menisci. Once this happens, the stress at the contact area is increased, which would accelerate the wear. Based on this hypothesis, suggested mechanisms for the wear and stiction of the landing site are summarized in Fig. 47.63. Wear initiates at the defect sites, and consequent high stiction can result in high wear. Improving the coverage and wear resistance of SAM coatings could enhance the yield of DMD.

In some cases, the micromirrors are not fully stuck and can be moved by applying a load at the rotatable corner of the micromirror with a discontinuous motion, which is called *soft* stiction. Soft-stuck micromirrors studied by *Liu* and *Bhushan* [47.109] were identified in quality inspection. These micromirrors encountered slow transition from one end to the other $(+1/-1)$. Figure 47.64 shows AFM surface height images of a location showing a stuck mirror (S) and surrounding normal micromirrors N_i $(i = 1, 2, 3)$. Surprisingly, the images of the stuck and normal micromirror array are almost the same. On the micromirrors of interest, tilting test was performed at the corner of the micromirrors; the rotatable direction of the microarray is indicated by an arrow in Fig. 47.64. A load–displacement curve for the stuck micromirror is presented in Fig. 47.65; it is not smooth and appears serrated. This clearly indicates that, although the micromirror S can be rotated, it rotates with hesitation. In regimes 1 and 2, as marked in Fig. 47.65, the slopes are much higher. In order to understand the stiction mechanisms, stiction of the landing sites of normal and stuck mirrors were measured. Unlike a hard-stuck mirror, adhesive

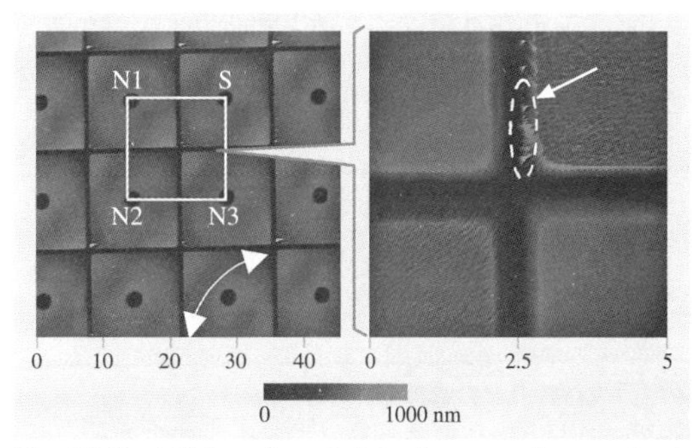

Fig. 47.64 AFM surface height images of normal micromirrors and a soft-stuck micromirror. The soft-stuck micromirror is labeled S, and the normal micromirrors studied are labeled N1, N2, and N3 (after [47.109])

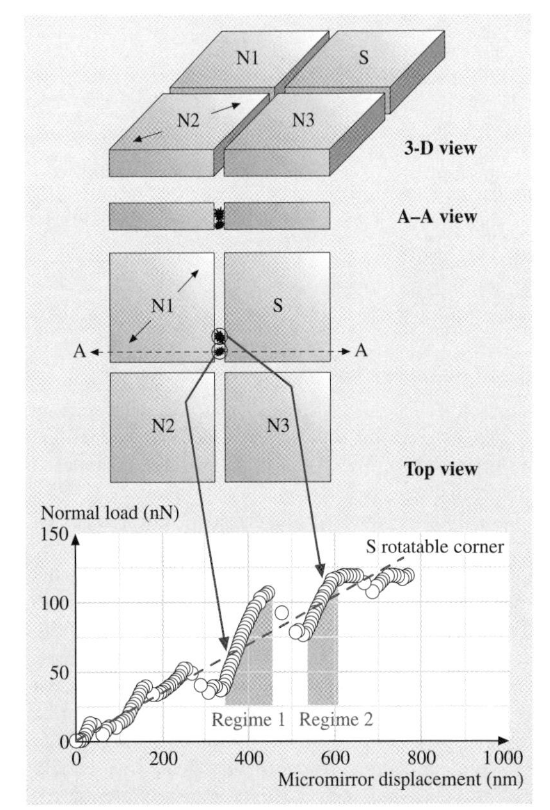

Fig. 47.65 Load–displacement curve obtained on the rotatable corner of micromirror S, and schematic to illustrate the suggested mechanism for occurrence of soft stiction

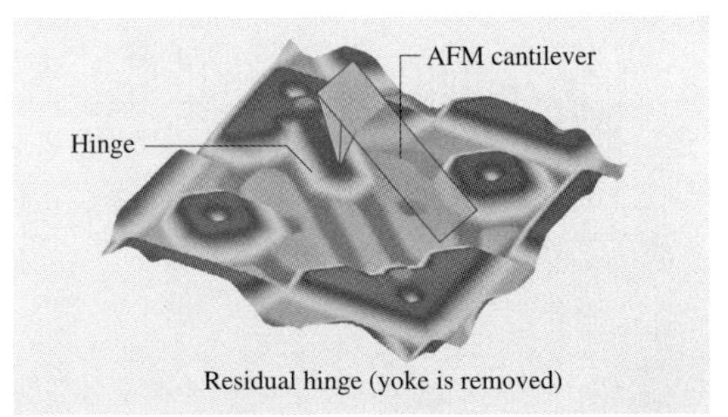

Fig. 47.66 AFM surface height image of the residual hinge and schematic diagram of the relative position of the hinge and AFM tip during the nanoscale bending and fatigue tests. The tip is located at the free end of the hinge (after [47.233])

forces of soft-stuck and normal mirrors were comparable, which suggests that the SAM coating is intact for the soft-stuck mirror. It was found that a high normal load (≈ 900 nN) and of the order of a couple of hundred scans were required to remove the soft-stuck micromirrors by the AFM, whereas, only about 300 nN and about ten scans were required to remove a hard-stuck mirror. After careful examination of the AFM images of the micromirror sidewalls in Fig. 47.64 (bottom left), it was noted that there were contaminant particles attached to the sidewalls of the micromirror S. It is therefore believed that, during the tilting test, for the micromirror S (see schematic in Fig. 47.65) a regime with a sharper slope will occur in the displacement curve. Extra force is required to overcome the resistance induced by these sidewall contamination particles. This is believed to be the reason for the slow transition of the micromirror during quality inspection.

Finally, nanomechanical characterization of various layers used in the construction of landing sites, hinge, and micromirror materials have been measured by *Wei* et al. [47.231, 232]. Bending and fatigue studies of hinges have been carried out by *Liu* and *Bhushan* [47.233] and *Bhushan* and *Liu* [47.234] to measure their stiffness and fatigue properties. For these studies, the micromirror was removed. During removal, the micromirror–yoke structure was removed simultaneously, leaving the hinge mounted on one end of the array (Fig. 47.66). The stiffness of the Al hinge was reported to be comparable to that of bulk Al. The Al hinge exhibited higher modulus than the SiO_2 hinge. The fatigue properties depended upon the preparation of the hinge for testing.

47.7 Conclusions

The field of MEMS/NEMS and bioMEMS/bioNEMS has expanded considerably over the last decade. The large surface-to-volume ratio of these devices results in very high surface forces, such as adhesion and friction/stiction, that seriously undermine the performance and reliability of devices. There is a need for fundamental understanding of adhesion, friction, stiction, wear, and lubrication, and the role of contamination and environment, all on nanoscale. Most mechanical properties are known to be scale dependent, therefore the properties of nanostructures need to be measured. Using AFM-based techniques, researchers have conducted nanotribology and nanomechanics studies of materials and devices. In addition, component-level testing has been carried out to improve understanding of the nanotribological phenomena observed in MEMS/NEMS.

Macroscale and microscale tribological studies of silicon and polysilicon films have been performed. The effect of doping, oxide films, and environment on the tribological properties of these popular MEMS/NEMS materials has also been studied. SiC film is found to be a good candidate material for use in high-temperature MEMS/NEMS devices. Perfluoroalkyl self-assembled monolayers and bonded perfluoropolyether lubricants appear to be well suited for lubrication of micro-nanodevices under a range of environmental conditions. DLC coatings can also be used for low friction and wear.

For bioMEMS/bioNEMS, adhesion between biological molecular layers and the substrate, and friction and wear of biological layers can be important. Adhesion of biomolecules on Si substrate surfaces for various bioMEMS applications can be improved by nanopatterning and the chemical linker method. Friction and wear mechanisms of protein layers have been studied. The trajectory in the blood stream of submicron particles used for therapeutic and diagnostics purposes needs to be optimized in order for them to bond to target sites on the endothelium wall, for which an analytical model has been developed. Roughness-induced hierarchical surfaces have been designed and fabricated for superhydrophobicity, self-cleaning, and low adhesion and friction.

Surface roughness measurements of micromachined polysilicon surfaces have been carried out using an AFM. The roughness distribution on surfaces is strongly dependent upon the fabrication process. Adhesion and friction of microstructures can be measured using a novel microtriboapparatus. Adhesion and friction measurements on silicon-on-silicon confirm AFM measurements that hexadecane thiol and bonded perfluoropolyether films exhibit superior adhesion and friction properties. Static friction force measurements of micromotors have been performed using an AFM. The forces are found to vary considerably with humidity. A bonded layer of perfluoropolyether lubricant is found to satisfactorily reduce the friction forces in the micromotor. Tribological failure modes of digital micromirror devices are either *hard* or *soft* stiction. In hard stiction, the tip on the yoke remains stuck to the landing site underneath. The mechanism responsible for this hard stiction is localized damage of the SAM on the landing site, whereas in soft stiction the mirror–yoke assembly rotates with hesitation. The mechanism responsible for soft stiction is contaminant particles present at the mirror sidewalls.

AFM/FFM-based techniques show the capability to study and evaluate nanotribology and nanomechanics related to MEMS/NEMS and bioMEMS/bioNEMS devices.

47.A Micro-Nanofabrication Techniques

Micro-nanofabrication techniques include top-down methods, in which one builds down from the large to the small, and bottom-up methods, in which one builds up from the small to the large.

47.A.1 Top-Down Techniques

The top-down fabrication methods used in the construction of MEMS/NEMS include lithographic and nonlithographic techniques to produce micro- and nanostructures. The lithographic techniques fall into three basic categories: bulk micromachining, surface micromachining, and LIGA (a German acronym for *Lithographie Galvanoformung Abformung*, i.e., lithography, electroplating, and molding). The first two approaches, bulk and surface micromachining, mostly use planar photolithographic fabrication processes developed for semiconductor devices in producing two-dimensional (2-D) structures [47.13, 19, 235–238]. The various steps involved in these two fabrication processes are shown schematically in Fig. 47.67. Bulk micromachining employs anisotropic etching to remove sections through the thickness of a single-crystal silicon wafer, typically 250–500 μm thick. Bulk micromachining is a proven high-volume production process and is routinely used to fabricate microstructures such as accelerometers, pressure sensors, and flow sensors. In surface micromachining, structural and sacrificial films are alternatively deposited, patterned, and etched to produce a freestanding structure. These films are typically made

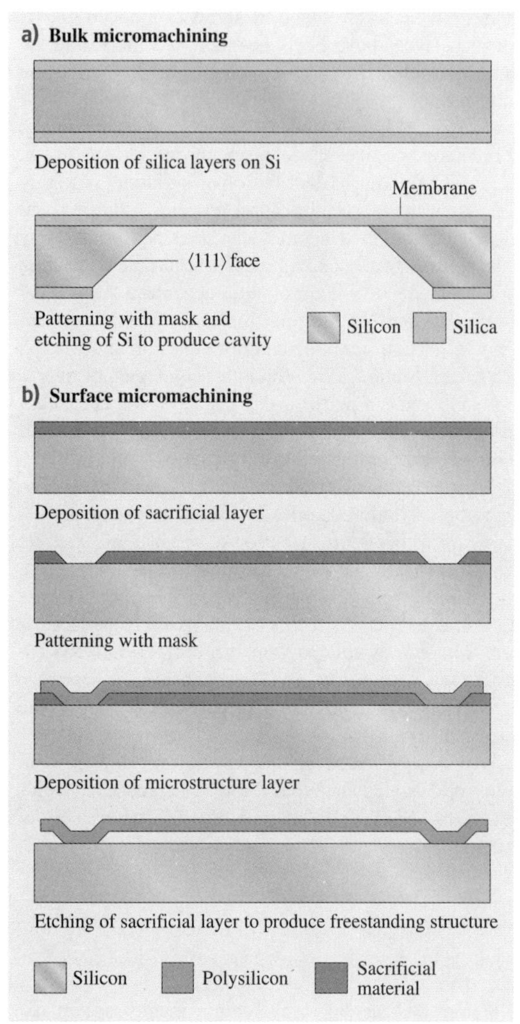

Fig. 47.67a,b Schematic of process steps involved in (**a**) bulk micromachining and (**b**) surface micromachining for fabrication of MEMS

of low-pressure chemical vapor deposition (LPCVD) polysilicon film with 2–20 μm thickness. Surface micromachining is used to produce sensors, actuators, micromirror arrays, motors, gears, and grippers. The resolution in photolithography is dependent upon the wavelength of light. A commonly used light source is an argon fluoride excimer laser with 193 nm wavelength (ultraviolet, UV), used in patterning 90 nm lines and spaces. Deep-UV wavelengths, x-ray lithography, e-beam lithography, focused ion beam lithography, maskless lithography, liquid-immersion lithography, and STM writing by removing material atom by atom are some of the recent developments for sub-100 nm patterning.

The fabrication of nanostructures such as nanochannels with sub-10 nm resolution can be accomplished through several routes: electron beam (e-beam) lithography and sacrificial layer lithography (SLL). The process for e-beam lithographic technique is a finely focused electron beam that is exposed onto a resist surface; the exposure duration and location is controlled with the use of a computer [47.239, 240]. When the resist is exposed to the electron beam, the electrons either break or join the molecules in the resist, so the local characteristics are changed in such a way that further processes can either remove the exposed part (positive resist) or the unexposed part (negative resist). The resist material determines whether the molecules will break or join together and thus whether a positive or negative image is produced. E-beam lithography can be used either to create photolithographic masks for replication or to create devices directly. The masks that are created can be used for either optical or x-ray lithography. One limitation of e-beam lithography is that throughput is drastically reduced since a single electron beam is used to create the entire exposure pattern on the resist. While this technique is slower than conventional lithographic techniques, it is ideal for prototype fabrication because no masks are required.

In the SLL process, the use of a sacrificial layer allows direct control of nanochannel dimensions as long as there exists a method for removing the sacrificial layer from the structural layers with absolute selectivity. A materials system with such selectivity is the silicon/silicon oxide system, which is used widely in the microfabrication of MEMS devices. The use of sidewall deposition of the sacrificial layer and subsequent etching allows for fabrication of high-density nanochannels for biomedical applications, based on surface micromachining [47.78]. Figure 47.68 shows a schematic of the process steps in the sacrificial layer lithography based on *Hansford* et al.'s [47.78] work on fabrication of polysilicon membranes with nanochannels. As with all membrane protocols, the first step in the fabrication is etching of the support ridge structure into the bulk silicon substrate. A low-stress silicon nitride (LSN, or simply nitride), which functions as an etch-stop layer, is then deposited using LPCVD. The base structural polysilicon layer (base layer) is deposited on top of the etch-stop layer. The plasma etching of holes in the base layer is what defines the shape of the pores.

The buried nitride etch stop acts as an etch stop for the plasma etching of a polysilicon base layer. After the pore holes are etched through the base layer, the pore sacrificial thermal oxide layer is grown on the base layer. The basic requirement for the sacrificial layer is the ability to control its thickness with high precision across the entire wafer. Anchor points are defined in the sacrificial oxide layer to connect the base layer mechanically to the plug layer (necessary to maintain the pore spacing between layers). This is accomplished by using the same mask shifted from the pore holes. This produces anchors in one or two corners of each pore hole, which provides the desired connection between the structural layers while opening as much pore area as possible. After the anchor points are etched through the sacrificial oxide, the plug polysilicon layer is deposited (using LPCVD) to fill in the holes. To open the pores at the surface, the plug layer is planarized using chemical mechanical polishing (CMP) down to the base layer, leaving the final structure with the plug layer only in the pore hole openings. When the silicon wafer is ready for release, a protective nitride layer is deposited on the wafer (completely covering both sides of the wafer). The backside etch windows are etched in the protective layer, exposing the silicon wafer in the desired areas, and the wafer is placed in a KOH bath to etch. After the silicon wafer is completely removed up to the membrane (as evidenced by the smooth buried etch-stop layer), the protective, sacrificial, and etch-stop layers are removed by etching in concentrated HF. Etching of the sacrificial layer in polysilicon film defines nanochannels.

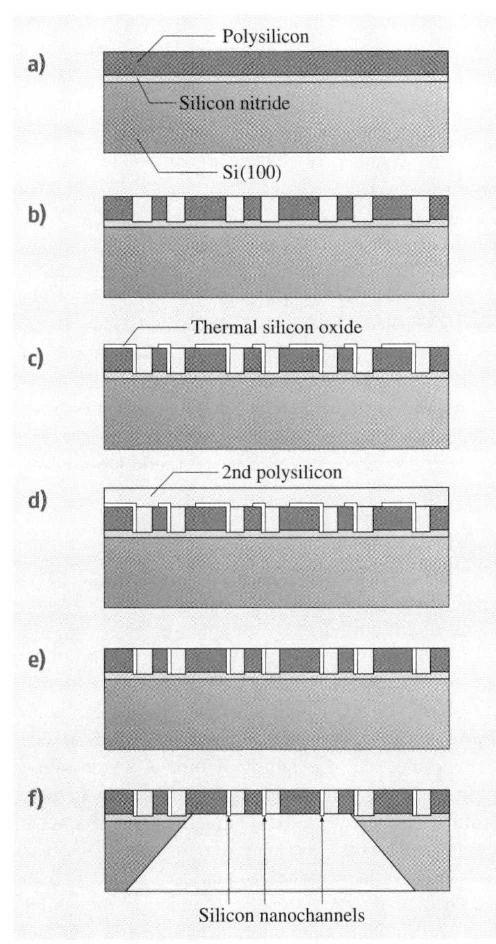

Fig. 47.68a–f Schematic of process steps involved in sacrificial layer lithography. (**a**) Growth of silicon nitride layer (etch stop) and base polysilicon deposition, (**b**) hole definition in base, (**c**) growth of thin sacrificial oxide and patterning of anchor points, (**d**) deposition of plug polysilicon, (**e**) planarization of plug layer, and (**f**) deposition and patterning of protective nitride layer through etch, followed by etching of protective, sacrificial, and etch layers for final release of the structure in HF (after [47.78])

The LIGA process is based on combined use of x-ray lithography, electroplating, and molding processes. X-rays produced by synchroton radiation are used to prepare the mold. The steps involved in the LIGA process are shown schematically in Fig. 47.69. LIGA is used to produce high-aspect-ratio MEMS (HARMEMS) devices that are up to 1 mm in height and only a few microns in width or length [47.241]. The LIGA process yields very sturdy 3-D structures due to their increased thickness. One of the limitations of silicon microfabrication processes originally used for fabrication of MEMS devices is the lack of suitable materials that can be processed. However, with LIGA, a variety of nonsilicon materials such as metals, ceramics, and polymers can be processed.

Nonlithographic micromachining processes, primarily in Europe and Japan, are also being used for fabrication of millimeter-scale devices using direct material microcutting or micromechanical machining (such as microturning, micromilling, and microdrilling) or removal by energy beams (such as microspark erosion, focused ion beam, laser ablation, and laser polymerization) [47.19, 242]. Hybrid technologies including LIGA and high-precision micromachining techniques have been used to produce miniaturized motors, gears, actuators, and connectors [47.87, 91, 92, 243].

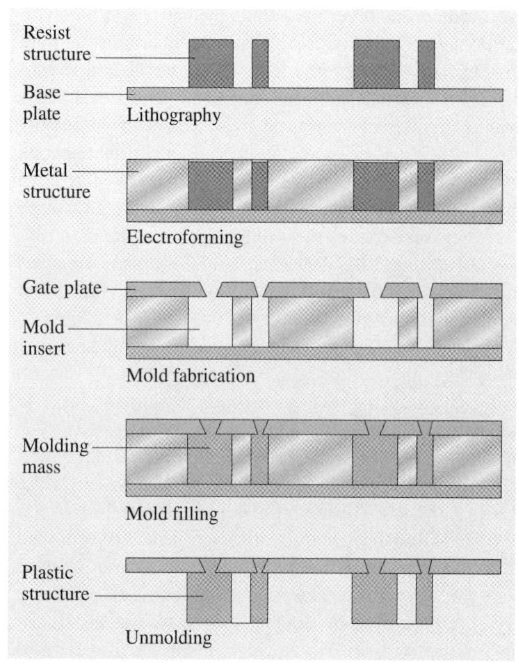

Fig. 47.69 Schematic of process steps involved in LIGA fabrication of MEMS

These millimeter-scale devices may find more immediate applications.

The micro-nanofabrication technique known as *soft lithography* is a nonlithographic technique [47.54, 60, 244, 245] in which a master or mold is used to generate patterns, defined by the relief on its surface, on polymers by replica molding [47.246], hot embossing (nanoimprint lithography) [47.247], or by contact printing (known as microcontact printing, μCP) [47.248]. Soft lithography is faster, less expensive, and more suitable for most biological applications than glass or silicon micromachining. Polymers have established an important role in bioMEMS/bioNEMS because of their reduced cost. The use of polymers also offers a wide range of material properties to allow tailoring of biological interactions for improved biocompatibility. Polymer fabrication is believed to be about an order of magnitude cheaper than silicon fabrication.

Replica molding is the transfer of a topographic pattern by curing or solidifying a liquid precursor against the original patterned mold. The mold or stamp is generally made of a two-part polymer (elastomer and curing agent), such as poly(dimethylsiloxane) (PDMS) from photolithographically generated photoresist master. Solvent-based embossing, or imprinting, uses a solvent to restructure a polymer film. Hot embossing, also called nanoimprint lithography, usually refers to the transfer of a pattern from a micromachined quartz or metal master to a pliable plastic sheet. Heat and high pressure allow the plastic sheet to become imprinted. These sheets can then be bonded to various plastics such as poly(methyl methacrylate) (PMMA). Nanoimprint lithography can produce patterns on a surface with 10 nm resolution. Contact printing uses a patterned stamp to transfer ink (mostly self-assembled monolayer) onto a surface in a pattern defined by the raised regions of a stamp. These techniques can be used to pattern line widths as small as 60 nm.

Replica molding is commonly used for mass-produced, disposable plastic micro-nanocomponents, for example micro-nanofluidic chips, generally made of PDMS and PMMA [47.245, 249]; it is more flexible in terms of materials choice for construction than conventional photolithography.

To assemble microsystems, microrobots are used. Microrobotics include building blocks such as steering links, microgrippers, conveyor system, and locomotive robots [47.17].

47.A.2 Bottom–Up Fabrication (Nanochemistry)

The bottom–up approach (from small to large) largely relies on nanochemistry [47.39,40,42–47]. The bottom–up approach includes chemical synthesis, spontaneous *self-assembly* of molecular clusters (molecular self-assembly) from simple reagents in solution or biological molecules as building blocks to produce three-dimensional nanostructures as done by nature, quantum dots (nanocrystals) of arbitrary diameter (about 10–10^5 atoms), molecular-beam epitaxy (MBE) and organometallic vapor-phase epitaxy (OMVPE) to create specialized crystals one atomic or molecular layer at a time, and manipulation of individual atoms by scanning tunneling microscope, atomic force microscope, or atom optics. The self-assembly must be encoded; that is, one must be able to precisely assemble one object next to another to form a designed pattern. A variety of nonequilibrium plasma chemistry techniques are also used to produce layered nanocomposites, nanotubes, and nanoparticles. Nanostructures can also be fabricated using mechanosynthesis with proximal probes.

References

47.1 Anonymous: *Microelectromechanical Systems: Advanced Materials and Fabrication Methods, NMAB-483* (National Academy Press, Washington 1997)

47.2 M. Roukes: Nanoelectromechanical systems face the future, Phys. World **2**, 25–31 (2001)

47.3 Anonymous: Small Tech 101 – An Introduction to Micro- and Nanotechnology, Small Times (2003)

47.4 M. Schulenburg: *Nanotechnology – Innovation for Tomorrow's World* (European Commission, Research DG, Brussels 2004)

47.5 J.C. Eloy: Status of the MEMS Industry 2005, Report of Yole Developpement, France, presented at SPIE Photonic West, San Jose (2005)

47.6 S. Lawrence: Nanotech grows up, Technol. Rev. **108**(6), 31 (2005)

47.7 R.J. Chen, H.C. Choi, S. Bangsaruntip, E. Yenilmez, X. Tang, Q. Wang, Y.L. Chang, H. Dai: An investigation of the mechanisms of electrode sensing of protein adsorption on carbon nanotube devices, J. Am. Chem. Soc. **126**, 1563–1568 (2004)

47.8 D. Srivastava: Computational nanotechnology of carbon nanotubes. In: *Carbon Nanotubes: Science and Applications*, ed. by M. Meyyappan (CRC Press, Boca Raton 2004) pp. 25–63

47.9 W.G. van der Wiel, S. De Franceschi, J.M. Elzerman, T. Fujisawa, S. Tarucha, L.P. Kouwenhoven: Electron transport through double quantum dots, Rev. Mod. Phys. **75**, 1–22 (2003)

47.10 R.S. Muller, R.T. Howe, S.D. Senturia, R.L. Smith, R.M. White: *Microsensors* (IEEE Press, New York 1990)

47.11 I. Fujimasa: *Micromachines: A New Era in Mechanical Engineering* (Oxford Univ. Press, Oxford 1996)

47.12 W.S. Trimmer (Ed.): *Micromachines and MEMS, Classic and Seminal Papers to 1990* (IEEE Press, New York 1997)

47.13 B. Bhushan: *Tribology Issues and Opportunities in MEMS* (Kluwer, Dordrecht 1998)

47.14 G.T.A. Kovacs: *Micromachined Transducers Sourcebook* (McGraw-Hill, Boston 1998)

47.15 S.D. Senturia: *Microsystem Design* (Kluwer, Boston 2000)

47.16 M. Elwenspoek, R. Wiegerink: *Mechanical Microsensors* (Springer, Berlin Heidelberg 2001)

47.17 M. Gad-el-Hak: *The MEMS Handbook* (CRC Press, Boca Raton 2002)

47.18 T.R. Hsu: *MEMS and Microsystems: Design and Manufacture* (McGraw-Hill, Boston 2002)

47.19 M. Madou: *Fundamentals of Microfabrication: The Science of Miniaturization*, 2nd edn. (CRC Press, Boca Raton 2002)

47.20 A. Hierlemann: *Integrated Chemical Microsensor Systems in CMOS Technology* (Springer, Berlin Heidelberg 2005)

47.21 T.A. Core, W.K. Tsang, S.J. Sherman: Fabrication technology for an integrated surface-micromachined sensor, Solid State Technol. **36**(10), 39–47 (1993)

47.22 J. Bryzek, K. Peterson, W. McCulley: Micromachines on the March, IEEE Spectrum **5**, 20–31 (1994)

47.23 J.S. Aden, J.H. Bohorquez, D.M. Collins, M.D. Crook, A. Gacia, U.E. Hess: The third-generation HP thermal inkjet printhead, HP Journal **45**(1), 41–45 (1994)

47.24 H. Le: Progress and trends in ink-jet printing technology, J. Imaging Sci. Technol. **42**, 49–62 (1998)

47.25 R. Baydo, A. Groscup: Getting to the heart of ink jet: Printheads, Beyond Recharg. **10**(May), 10–12 (2001)

47.26 E.R. Lee: *Microdrop Generation* (CRC Press, Boca Raton 2003)

47.27 L.J. Hornbeck, W.E. Nelson: Bistable deformable mirror device, OSA Tech. Dig. Ser. **8**, 107–110 (1988)

47.28 L.J. Hornbeck: Digital Light Processing update – status and future applications, SPIE **3634**, 158–170 (1999)

47.29 L.J. Hornbeck: The DMD^{TM} projection display chip: A MEMS-based technology, MRS Bulletin **26**, 325–327 (2001)

47.30 K. Suzuki: Micro electro mechanical systems (MEMS) micro-switches for use in DC, RF, and optical applications, Jpn. J. Appl. Phys. **41**, 4335–4339 (2002)

47.31 V.M. Lubecke, J.C. Chiao: MEMS Technologies for Enabling High Frequency Communication Cicuits, Proc. IEEE 4th Int. Conf. Telecom. in Modern Satellite, Cable and Broadcasting Services, Nis (IEEE, New York 1999) pp. 1–8

47.32 C.R. Giles, D. Bishop, V. Aksyuk: MEMS for lightwave networks, MRS Bulletin **4**, 328–329 (2001)

47.33 A. Hierlemann, O. Brand, C. Hagleitner, H. Baltes: Microfabrication techniques for chemical/biosensors, Proc. IEEE **91**, 839–863 (2003)

47.34 B. Bhushan: *Tribology and Mechanics of Magnetic Storage Devices*, 2nd edn. (Springer, New York 1996)

47.35 H. Hamilton: Contact recording on perpendicular rigid media, J. Mag. Soc. Jpn. **15**(S2), 481–483 (1991)

47.36 D.A. Horsley, M.B. Cohn, A. Singh, R. Horowitz, A.P. Pisano: Design and fabrication of an angular microactuator for magnetic disk drives, J. Microelectromech. Syst. **7**, 141–148 (1998)

47.37 T. Hirano, L.S. Fan, D. Kercher, S. Pattanaik, T.S. Pan: HDD tracking microactuator and its integration issues, Proc. ASME Int. Mech. Eng. Congr. Expo., Vol. 2, ed. by A.P. Lee, J. Simon, F.K. Foster, R.S. Keynton (2000) pp. 449–452

47.38 T. Fukuda, F. Arai, L. Dong: Assembly of nanodevices with carbon nanotubes through nanorobotic manipulations, Proc. IEEE **91**, 1803–1818 (2003)

47.39 K.E. Drexler: *Nanosystems: Molecular Machinery, Manufacturing and Computation* (Wiley, New York 1992)

47.40 G. Timp (Ed.): *Nanotechnology* (Springer, New York 1999)
47.41 M.S. Dresselhaus, G. Dresselhaus, P. Avouris (Eds.): *Carbon Nanotubes*, Top. Appl. Phys., Vol. 80 (Springer, Berlin Heidelberg 2001)
47.42 E.A. Rietman: *Molecular Engineering of Nanosystems* (Springer, New York 2001)
47.43 W.A. Goddard, D.W. Brenner, S.E. Lyshevski, G.J. Iafrate (Eds.): *Handbook of Nanoscience, Engineering, and Technology* (CRC Press, Boca Raton 2002)
47.44 H.S. Nalwa (Ed.): *Nanostructured Materials and Nanotechnology* (Academic, San Diego 2002)
47.45 C.P. Poole, F.J. Owens: *Introduction to Nanotechnology* (Wiley, Hoboken 2003)
47.46 B. Bhushan: *Handbook of Micro/Nanotribology*, 2nd edn. (CRC, Boca Raton 1999)
47.47 B. Bhushan: Nanotribology and nanomechanics, Wear **259**, 1507–1531 (2005)
47.48 J.A. Stroscio, D.M. Eigler: Atomic and molecular manipulation with a scanning tunneling microscope, Science **254**, 1319 (1991)
47.49 P. Vettiger, J. Brugger, M. Despont, U. Drechsler, U. Dürig, W. Häberle, M. Lutwyche, H. Rothuizen, R. Stutz, R. Widmer, G. Binning: Ultrahigh density, high data-rate NEMS based AFM data storage system, Microelectron. Eng. **46**, 11–27 (1999)
47.50 C. Stampfer, A. Jungen, C. Hierold: Fabrication of Discrete Carbon Nanotube Based Nanoscaled Force Sensor, Proc. IEEE Sens. 2004, Vienna (IEEE, New York 2004) pp. 1056–1059
47.51 C. Hierold: *Carbon Nanotube Devices* (Wiley-VCH, Weinheim 2008)
47.52 B. Bhushan: *Mechanics and Reliability of Flexible Magnetic Media*, 2nd edn. (Springer, New York 2000)
47.53 Anonymous: *International Technology Roadmap for Semiconductors* (2004) http://public.itrs.net/
47.54 A. Manz, H. Becker (Eds.): *Microsystem Technology in Chemistry and Life Sciences*, Top. Curr. Chem., Vol. 194 (Springer, Berlin Heidelberg 1998)
47.55 J. Cheng, L.J. Kricka (Eds.): *Biochip Technology* (Harwood Academic Publishers, Philadelphia 2001)
47.56 M.J. Heller, A. Guttman (Eds.): *Integrated Microfabricated Biodevices* (Marcel Dekker, New York 2001)
47.57 C.L.P San, E.P.H. Yap (Eds.): *Frontiers in Human Genetics* (World Scientific, Singapore 2001)
47.58 C.H. Mastrangelo, H. Becker (Eds.): *Microfluidics and BioMEMS*, Proc. SPIE, Vol. 4560 (SPIE, Bellingham 2001)
47.59 H. Becker, L.E. Locascio: Polymer microfluidic devices, Talanta **56**, 267–287 (2002)
47.60 D.J. Beebe, G.A. Mensing, G.M. Walker: Physcis and applications of microfluidics in biology, Annu. Rev. Biomed. Eng. **4**, 261–286 (2002)
47.61 A. van der Berg (Ed.): *Lab-on-a-Chip: Chemistry in Miniaturized Synthesis and Analysis Systems* (Elsevier, Amsterdam 2003)
47.62 P. Gravesen, J. Branebjerg, O. Jensen: Microfluidics – A review, J. Micromech. Microeng. **3**, 168–182 (1993)
47.63 S.N. Bhatia, C.S. Chen: Tissue engineering at the micro-scale, Biomed. Microdevices **2**, 131–144 (1999)
47.64 R.P. Lanza, R. Langer, J. Vacanti (Eds.): *Principles of Tissue Engineering*, 2nd edn. (Academic, San Diego 2000)
47.65 E. Leclerc, K.S. Furukawa, F. Miyata, T. Sakai, T. Ushida, T. Fujii: Fabrication of microstructures in photosensitive biodegradable polymers for tissue engineering applications, Biomaterials **25**, 4683–4690 (2004)
47.66 K. Park (Ed.): *Controlled Drug Delivery: Challenges and Strategies* (American Chemical Society, Washington 1997)
47.67 R.S. Shawgo, A.C.R. Grayson, Y. Li, M.J. Cima: BioMEMS for drug delivery, Curr. Opin. Solid State Mater. Sci. **6**, 329–334 (2002)
47.68 P.A. Oeberg, T. Togawa, F.A. Spelman: *Sensors in Medicine and Health Care* (Wiley, New York 2004)
47.69 J.V. Zoval, M.J. Madou: Centrifuge-based fluidic platforms, Proc. IEEE **92**, 140–153 (2000)
47.70 W.C. Tang, A.P. Lee: Defense applications of MEMS, MRS Bulletin **26**, 318–319 (2001), Also see www.darpa.mil/mto/mems
47.71 M.R. Taylor, P. Nguyen, J. Ching, K.E. Peterson: Simulation of microfluidic pumping in a genomic DNA blood-processing cassette, J. Micromech. Microeng. **13**, 201–208 (2003)
47.72 R. Raiteri, M. Grattarola, M. Butt, P. Skladal: Micromechanical Cantilever-Based Biosensor, Sens. Actuators B: Chemical **79**, 115–126 (2001)
47.73 B. Bhushan, D.R. Tokachichu, M.T. Keener, S.C. Lee: Morphology and adhesion of biomolecules on silicon based surfaces, Acta Biomater. **1**, 327–341 (2005)
47.74 H.P. Lang, M. Hegner, C. Gerber: Cantilever array sensors, Mater. Today **4**, 30–36 (2005)
47.75 F. Patolsky, C. Lieber: Nanowire nanosensors, Mater. Today **8**(4), 20–28 (2005)
47.76 M. Scott: MEMS and MOEMS for National Security Applications. In: Reliability, Testing, and Characterization of MEMS/MOEMS II, Proc Proc. SPIE Vol. 4980 (SPIE, Bellingham, Washington 2003) pp. xxxvii–xliv
47.77 T.A. Desai, D.J. Hansford, L. Kulinsky, A.H. Nashat, G. Rasi, J. Tu, Y. Wang, M. Zhang, M. Ferrari: Nanopore technology for biomedical applications, Biomed. Devices **2**, 11–40 (1999)
47.78 D. Hansford, T. Desai, M. Ferrari: Nano-scale size-based biomolecular separation technology. In: *Biochip Technology*, ed. by J. Cheng, L.J. Kricka (Harwood Academic Publishers, New York 2001) pp. 341–361

47.79 F.J. Martin, C. Grove: Microfabricated drug delivery systems: Concepts to improve clinical benefits, Biomed. Microdevices **3**, 97–108 (2001)

47.80 B. Bhushan: *Principles and Applications of Tribology* (Wiley, New York 1999)

47.81 B. Bhushan: *Modern Tribology Handbook*, Vol. 1/2, ed. by B. Bhushan (CRC Press, Boca Raton 2001)

47.82 B. Bhushan: *Introduction to Tribology* (Wiley, New York 2002)

47.83 B. Bhushan: Adhesion and stiction: Mechanisms, measurement techniques, and methods for reduction, J. Vac. Sci. Technol. B **21**, 2262–2296 (2003)

47.84 B. Bhushan: *Nanotribology and Nanomechanics – An Introduction*, 2nd edn. (Springer, Berlin Heidelberg 2008)

47.85 Y.C. Tai, L.S. Fan, R.S. Muller: IC-processed micromotors: Design, technology and testing, Proc. IEEE Micro Electro Mech. Syst. (1989) pp. 1–6

47.86 S.M. Spearing, K.S. Chen: Micro-gas turbine engine materials and structures, Ceram. Eng. Sci. Proc. **18**, 11–18 (2001)

47.87 H. Lehr, S. Abel, J. Doppler, W. Ehrfeld, B. Hagemann, K.P. Kamper, F. Michel, C. Schulz, C. Thurigen: Microactuators as driving units for microrobotic systems, Proc. Microrobotics: Compon Appl. **2906**, 202–210 (1996)

47.88 L.G. Fréchette, S.A. Jacobson, K.S. Breuer, F.F. Ehrich, R. Ghodssi, R. Khanna, C.W. Wong, X. Zhang, M.A. Schmidt, A.H. Epstein: High-speed microfabricated silicon turbomachinery and fluid film bearings, J. MEMS **14**, 141–152 (2005)

47.89 L.X. Liu, Z.S. Spakovszky: Effect of Bearing Stiffness Anisotropy on Hydrostatic Micro Gas Journal Bearing Dynamic Behavior, Proc. ASME Turbo Expo 2005 (Reno 2005), Paper No. GT-2005-68199

47.90 M. Mehregany, K.J. Gabriel, W.S.N. Trimmer: Integrated fabrication of polysilicon mechanisms, IEEE Trans. Electron Devices **35**, 719–723 (1988)

47.91 H. Lehr, W. Ehrfeld, B. Hagemann, K.P. Kamper, F. Michel, C. Schulz, C. Thurigen: Development of micro-millimotors, Minim. Invasive Ther. Allied Technol. **6**, 191–194 (1997)

47.92 F. Michel, W. Ehrfeld: Microfabrication technologies for high performance microactuators. In: *Tribology Issues and Opportunities in MEMS*, ed. by B. Bhushan (Kluwer, Dordrecht 1998) pp. 53–72

47.93 E.J. Garcia, J.J. Sniegowski: Surface micromachined microengine, Sens. Actuators A **48**, 203–214 (1995)

47.94 D.M. Tanner, N.F. Smith, L.W. Irwin, W.P. Eaton, K.S. Helgesen, J.J. Clement, W.M. Miller, J.A. Walraven, K.A. Peterson, P. Tangyunyong, M.T. Dugger, S.L. Miller: *MEMS Reliability: Infrastructure, Test Structures, Experiments, and Failure Modes, SAND2000-0091* (Sandia National Laboratories, Albuquerque, New Mexico 2000), Download from www.prod.sandia.gov

47.95 S.S. Mani, J.G. Fleming, J.A. Walraven, J.J. Sniegowski, M.P. de Beer, L.W. Irwin, D.M. Tanner, D.A. LaVan, M.T. Dugger, H. Jakubczak, W.M. Miller: Effect of W Coating on Microengine Performance, Proc. 38th Annu. Int. Reliab. Phys. Symp. (IEEE, New York 2000) pp. 146–151

47.96 M.G. Hankins, P.J. Resnick, P.J. Clews, T.M. Mayer, D.R. Wheeler, D.M. Tanner, R.A. Plass: Vapor deposition of amino-functionalized self-assembled monolayers on MEMS, Proc. SPIE **4980**, 238–247 (2003)

47.97 J.K. Robertson, K.D. Wise: An electrostatically actuated integrated microflow controller, Sens. Actuators A **71**, 98–106 (1998)

47.98 B. Bhushan: Nanotribology and Nanomechanics of MEMS Devices, Proc. 9th Annu. Workshop MEMS (IEEE, New York 1996) pp. 91–98

47.99 R.E. Sulouff: MEMS opportunities in accelerometers and gyros and the microtribology problems limiting commercialization. In: *Tribology Issues and Opportunities in MEMS*, ed. by B. Bhushan (Kluwer, Dordrecht 1998) pp. 109–120

47.100 J.R. Martin, Y. Zhao: Micromachined Device Packaged to Reduce Stiction, US Patent 5694740 (1997)

47.101 G. Smith: The application of microtechnology to sensors for the automotive industry, Microelectron. J. **28**, 371–379 (1997)

47.102 M. Parsons: Design and manufacture of automotive pressure sensors, Sensors **18**, 32–46 (2001)

47.103 L.S. Chang, P.L. Gendler, J.H. Jou: Thermal mechanical and chemical effects in the degradation of the plasma-deposited α-SiC:H passivation layer in a multlayer thin-film device, J. Mater Sci. **26**, 1882–1890 (1991)

47.104 V.A. Aksyuk, F. Pardo, D. Carr, D. Greywall, H.B. Chan, M.E. Simon, A. Gasparyan, H. Shea, V. Lifton, C. Bolle, S. Arney, R. Frahm, M. Paczkowski, M. Haueis, R. Ryf, D.T. Neilson, J. Kim, R. Giles, D. Bishop: Beam-steering micromirrors for large optical cross-connects, J. Lightwave Technol. **21**, 634–642 (2003)

47.105 S.A. Henck: Lubrication of digital micromirror devices, Tribol. Lett. **3**, 239–247 (1997)

47.106 M.R. Douglass: Lifetime estimates and unique failure mechanisms of the digital micromirror devices (DMD), Proc. 36th Annu. Int. Reliab. Phys. Symp. (IEEE, New York 1998) pp. 9–16

47.107 M.R. Douglass: DMD reliability: A MEMS success story, Proc. SPIE **4980**, 1–11 (2003)

47.108 H. Liu, B. Bhushan: Nanotribological characterization of digital micromirror devices using an atomic force microscope, Ultramicroscopy **100**, 391–412 (2004)

47.109 H. Liu, B. Bhushan: Investigation of nanotribological and nanomechanical properties of the digital micromirror device by atomic force microscope, J. Vac. Sci. Technol. A **22**, 1388–1396 (2004)

47.110 L.J. Hornbeck: Low Surface Energy Passivation Layer for Micromechanical Devices, US Patent 5602671 (1997)

47.111 R.A. Robbins, S.J. Jacobs: Lubricant Delivery for Micromechanical Devices, US Patent No. 6300294 B1 (2001)

47.112 I. DeWolf, W.M. van Spengen: Techniques to study the reliability of metal RF MEMS capacitive switches, Microelectron. Reliab. **42**, 1789–1794 (2002)

47.113 B. Bhushan, K.J. Kwak: Platinum-coated probes sliding at up to 100 mm s^{-1} against coated silicon wafers for AFM probe-based recording technology, Nanotechnology **18**, 345504 (2007)

47.114 B. Bhushan, K.J. Kwak: Noble metal-coated probes sliding at up to 100 mm s^{-1} against PZT films for AFM probe-based ferroelectric recording technology, J. Phys. D **20**, 225013 (2008), invited

47.115 K.J. Kwak, B. Bhushan: Platinum-coated probes sliding at up to 100 mm/s against lead zirconate titanate films for atomic force microscopy probe-based ferroelectric recording technology, J. Vac. Sci. Technol. A **26**, 783–793 (2008)

47.116 B. Bhushan, K. Kwak, M. Palacio: Nanotribology and nanomechanics of AFM probe-based data recording technology, J. Phys. D **20**, 365207 (2008)

47.117 B. Bhushan, M. Palacio, B. Kinzig: AFM-based nanotribological and electrical characterization of ultrathin wear-resistant ionic liquid films, J. Colloid Interface Sci. **317**, 275–287 (2008)

47.118 M. Palacio, B. Bhushan: Ultrathin wear-resistant ionic liquid films for novel MEMS/NEMS applications, Adv. Mater. **20**, 1194–1198 (2008)

47.119 M. Palacio, B. Bhushan: Molecularly thick dicationic liquid films for nanolubrication, J. Vac. Sci. Technol. A **27**(4), 986–995 (2009)

47.120 T. Ohnstein, T. Fukiura, J. Ridley, U. Bonne: Micromachined silicon microvalve, Proc. IEEE-MEMS Workshop (IEEE, New York 1990) pp. 95–98

47.121 S. Shoji, M. Esashi: Microflow devices and systems, J. Micromech. Microeng. **4**, 157–171 (1994)

47.122 M. Stehr, S. Messner, H. Sandmaier, R. Zenergle: The VAMP – A new device for handing liquids or gases, Sens. Actuators A **57**, 153–157 (1996)

47.123 P. Woias: Micropumps – Summarizing the first two decades. In: *Proc. of SPIE – Microfluidics and BioMEMS*, Vol. 4560, ed. by C.H. Mastrangelo, H. Becker (SPIE, Bellingham, Washington 2001) pp. 39–52

47.124 N.T. Nguyen, X. Huang, T.K. Chuan: MEMS-micropumps: A review, ASME J. Fluids Eng. **124**, 384–392 (2002)

47.125 B. Bustgens, W. Bacher, W. Menz, W.K. Schomburg: Micropump manufactured by thermoplastic molding, Proc. IEEE-MEMS Workshop (IEEE, Piscataway 1994) pp. 18–21

47.126 C.H. Ahn, M.G. Allen: Fluid micropumps based on rotary magnetic actuators, MEMS '95: IEEE 8th Int. Workshop on MEMS (IEEE, Piscataway 1995) pp. 408–412

47.127 J. Doepper, M. Clemens, W. Ehrfeld, S. Jung, K.P. Kaemper, H. Lehr: Micro gear pumps for dosing of viscous fluids, J. Micromech. Microeng. **7**, 230–232 (1997)

47.128 J. Henniker: Triboelectricity in polymers, Nature **196**, 474 (1962)

47.129 M. Sakaguchi, H. Kashiwabara: A generation mechanism of triboelectricity due to the reaction of mechaniradicals with mechanoions which are produced by mechanical fracture of solid polymer, Colloid Poly. Sci. **270**, 621–626 (1992)

47.130 G.R. Freeman, N.H. March: Triboelectricity and some associated phenomena, Mater Sci. Technol. **15**, 1454–1458 (1999)

47.131 B. Bhushan, D. Tokachichu, M.T. Keener, S.C. Lee: Nanoscale adhesion, friction, and wear studies of biomolecules on silicon based surfaces, Acta Biomater. **2**, 39–49 (2006)

47.132 G. Wei, B. Bhushan, N. Ferrell, D. Hansford: Microfabrication and nanomechanical characterization of polymer microelectromechanical systems for biological applications, J. Vac. Sci. Technol. A **23**, 811–819 (2005)

47.133 S.K. Cho, H. Moon, C.-J. Kim: Creating, transporting, cutting, and merging liquid droplets by electrowetting-based actuation for digital microfluidic circuits, J. Microelectromech. Syst. **12**, 70–80 (2003)

47.134 A.R. Wheeler, H. Moon, C.A. Bird, R.R.O. Loo, C.-J. Kim, J.A. Loo, R.L. Garrell: Digital microfluidics with in-line sample purification for proteomics analysis with MALDI-MS, Anal. Chem. **77**, 534–540 (2005)

47.135 S.C. Lee, M.T. Keener, D.R. Tokachichu, B. Bhushan, P.D. Barnes, B.R. Cipriany, M. Gao, L.J. Brillson: Protein binding on thermally grown silicon dioxide, J. Vac. Sci. Technol. B **23**, 1856–1865 (2005)

47.136 J. Black: *Biological Performance of Materials: Fundamentals of Biocompatibility* (Marcel Dekker, New York 1999)

47.137 B. Bhushan, K. Kwak, S. Gupta, S.C. Lee: Nanoscale adhesion, friction and wear studies of biomolecules on SAM-coated silica and alumina based surfaces, J. R. Soc. Interface **6**(37), 719–733 (2009)

47.138 K.K. Lee, B. Bhushan, D. Hansford: Nanotribological characterization of perfluoropolymer thin films for BioMEMS applications, J. Vac. Sci. Technol. A **23**, 804–810 (2005)

47.139 B. Bhushan, T. Kasai, G. Kulik, L. Barbieri, P. Hoffmann: AFM study of perfluorosilane and alkylsilane self-assembled monolayers for anti-stiction in MEMS/NEMS, Ultramicroscopy **105**, 176–188 (2005)

47.140 B. Bhushan, D. Hansford, K.K. Lee: Surface modification of silicon and PDMS surfaces with vapor phase deposited ultrathin fluorosilane films for biomedical nanodevices, J. Vac. Sci. Technol. A **24**, 1197–1202 (2006)

47.141 T. Kasai, B. Bhushan, G. Kulik, L. Barbieri, P. Hoffmann: Nanotribological study of perfluorosilane SAMs for anti-stiction and low wear, J. Vac. Sci. Technol. B **23**, 995–1003 (2005)
47.142 P. Decuzzi, S. Lee, B. Bhushan, M. Ferrari: A theoretical model for the margination of particles with blood vessels, Ann. Biomed. Eng. **33**, 179–190 (2005)
47.143 B. Bhushan: Nanotribology of carbon nanotubes, J. Phys. D **20**, 365214 (2008)
47.144 H. Guckel, D.W. Burns: Fabrication of micromechanical devices from polysilicon films with smooth surfaces, Sens. Actuators **20**, 117–122 (1989)
47.145 G.T. Mulhern, D.S. Soane, R.T. Howe: Supercritical Carbon Dioxide Drying of Microstructures, Proc. Int. Conf. on Solid-State Sens. Actuators (IEEE, New York 1993) pp. 296–299
47.146 K.F. Man, B.H. Stark, R. Ramesham: *A Resource Handbook for MEMS Reliability, Rev. A* (JPL Press, Jet Propulsion Laboratory, California Institute of Technology, Pasadena 1998)
47.147 S. Kayali, R. Lawton, B.H. Stark: MEMS reliability assurance activities at JPL, EEE Links **5**, 10–13 (1999)
47.148 S. Arney: Designing for MEMS Reliability, MRS Bulletin **26**, 296–299 (2001)
47.149 K.F. Man: MEMS reliability for space applications by elimination of potential failure modes through testing and analysis (NASA 2001)
47.150 B. Bhushan, A.V. Kulkarni, W. Bonin, J.T. Wyrobek: Nano/picoindentation measurement using a capacitance transducer system in atomic force microscopy, Philos. Mag. **74**, 1117–1128 (1996)
47.151 S. Sundararajan, B. Bhushan: Development of AFM-based techniques to measure mechanical properties of nanoscale structures, Sens. Actuators A **101**, 338–351 (2002)
47.152 B. Bhushan, J.N. Israelachvili, U. Landman: Nanotribology: Friction, wear and lubrication at the atomic scale, Nature **374**, 607–616 (1995)
47.153 B. Bhushan: Nanotribology and nanomechanics of MEMS/NEMS and BioMEMS/BioNEMS materials and devices, Microelectron. Eng. **84**, 387–412 (2007)
47.154 B. Bhushan: Nanotribology, nanomechanics and nanomaterials characterization, Philos. Trans. R. Soc. A **366**, 1351–1381 (2008)
47.155 B. Bhushan: Nanotribology and nanomechanics in nano/biotechnology, Philos. Trans. R. Soc. A **366**, 1499–1537 (2008)
47.156 B. Bhushan: Biomimetics: Lessons from nature – An overview, Philos. Trans. R. Soc. A **367**, 1445–1486 (2009)
47.157 M. Mehregany, C.A. Zorman, N. Rajan, C.H. Wu: Silicon carbide MEMS for harsh environments, Proc. IEEE **86**, 1594–1610 (1998)
47.158 J.S. Shor, D. Goldstein, A.D. Kurtz: Characterization of n-type β-SiC as a Piezoresistor, IEEE Trans. Electron Devices **40**, 1093–1099 (1993)
47.159 C.A. Zorman, A.J. Fleischmann, A.S. Dewa, M. Mehregany, C. Jacob, S. Nishino, P. Pirouz: Epitaxial growth of 3C-SiC films on 4 in. diam Si(100) silicon wafers by atmospheric pressure chemical vapor deposition, J. Appl. Phys. **78**, 5136–5138 (1995)
47.160 C.A. Zorman, S. Roy, C.H. Wu, A.J. Fleischman, M. Mehregany: Characterization of polycrystalline silicon carbide films grown by atmospheric pressure chemical vapor deposition on polycrystalline silicon, J. Mater. Res. **13**, 406–412 (1998)
47.161 C.H. Wu, S. Stefanescu, H.I. Kuo, C.A. Zorman, M. Mehregany: Fabrication and Testing of Single Crystalline 3C-SiC Piezoresistive Pressure Sensors, Technical Digest – 11th Int. Conf. Solid State Sensors and Actuators – Eurosensors **XV** (Munich 2001) pp. 514–517
47.162 A.A. Yasseen, C.H. Wu, C.A. Zorman, M. Mehregany: Fabrication and testing of surface micromachined polycrystalline SiC micromotors, IEEE Electron Device Lett. **21**, 164–166 (2000)
47.163 B. Bhushan, B.K. Gupta: *Handbook of Tribology: Materials, Coatings and Surface Treatments* (Krieger, Malabar 1997), Reprint edition
47.164 J.F. Shackelford, W. Alexander, J.S. Park (Eds.): *CRC Material Science and Engineering Handbook*, 2nd edn. (CRC Press, Boca Raton 1994)
47.165 B.K. Gupta, J. Chevallier, B. Bhushan: Tribology of ion bombarded silicon for micromechanical applications, ASME J. Tribol. **115**, 392–399 (1993)
47.166 B.K. Gupta, B. Bhushan, J. Chevallier: Modification of tribological properties of silicon by boron ion implantation, Tribol. Trans. **37**, 601–607 (1994)
47.167 B.K. Gupta, B. Bhushan: Nanoindentation studies of ion implanted silicon, Surf. Coat. Technol. **68/69**, 564–570 (1994)
47.168 B. Bhushan, V.N. Koinkar: Tribological studies of silicon for magnetic recording applications, J. Appl. Phys. **75**, 5741–5746 (1994)
47.169 G.M. Pharr: The anomalous behavior of silicon during nanoindentation, Thin Films **239**, 301–312 (1991)
47.170 D.L. Callahan, J.C. Morris: The extent of phase transformation in silicon hardness indentation, J. Mater. Res. **7**, 1612–1617 (1992)
47.171 N.A. Fleck, G.M. Muller, M.F. Ashby, J.W. Hutchinson: Strain gradient plasticity: Theory and experiment, Acta Metall. Mater. **42**, 475–487 (1994)
47.172 B. Bhushan, S. Venkatesan: Friction and wear studies of silicon in sliding contact with thin-film magnetic rigid disks, J. Mater. Res. **8**, 1611–1628 (1993)
47.173 S. Venkatesan, B. Bhushan: The role of environment in the friction and wear of single-crystal silicon in sliding contact with thin-film magnetic rigid disks, Adv. Info Storage Syst. **5**, 241–257 (1993)
47.174 S. Venkatesan, B. Bhushan: The sliding friction and wear behavior of single-crystal, polycrystalline and oxidized silicon, Wear **171**, 25–32 (1994)

47.175 B. Bhushan: Chemical, mechanical and tribological characterization of ultra-thin and hard amorphous carbon coatings as thin as 3.5 nm: Recent developments, Diam. Rel. Mater. **8**, 1985–2015 (1999)

47.176 B. Bhushan, S. Sundararajan, X. Li, C.A. Zorman, M. Mehregany: Micro/nanotribological studies of single-crystal silicon and polysilicon and SiC films for use in MEMS devices. In: *Tribology Issues and Opportunities in MEMS*, ed. by B. Bhushan (Kluwer, Dordrecht 1998) pp. 407–430

47.177 S. Sundararajan, B. Bhushan: Micro/nanotribological studies of polysilicon and SiC films for MEMS applications, Wear **217**, 251–261 (1998)

47.178 X. Li, B. Bhushan: Micro/nanomechanical characterization of ceramic films for microdevices, Thin Solid Films **340**, 210–217 (1999)

47.179 H. Liu, B. Bhushan: Nanotribological characterization of molecularly-thick lubricant films for applications to MEMS/NEMS by AFM, Ultramicroscopy **97**, 321–340 (2003)

47.180 V.N. Koinkar, B. Bhushan: Micro/nanoscale studies of boundary layers of liquid lubricants for magnetic disks, J. Appl. Phys. **79**, 8071–8075 (1996)

47.181 V.N. Koinkar, B. Bhushan: Microtribological studies of unlubricated and lubricated surfaces using atomic force/friction force microscopy, J. Vac. Sci. Technol. A **14**, 2378–2391 (1996)

47.182 Z. Tao, B. Bhushan: Bonding, degradation, and environmental effects on novel perfluoropolyether lubrications, Wear **259**, 1352–1361 (2005)

47.183 B. Bhushan, M. Cichomski, Z. Tao, N.T. Tran, T. Ethen, C. Merton, R.E. Jewett: Nanotribological characterization and lubricant degradation studies of metal-film magnetic tapes using novel lubricants, ASME J. Tribol. **129**, 621–627 (2007)

47.184 M. Palacio, B. Bhushan: Surface potential and resistance measurements for detecting wear of chemically-bonded and unbonded molecularly-thick perfluoropolyether lubricant films using atomic force microscopy, J. Colloid Interface Sci. **315**, 261–269 (2007)

47.185 M. Palacio, B. Bhushan: Wear detection of candidate MEMS/NEMS lubricant films using atomic force microscopy-based surface potential measurements, Scr. Mater. **57**, 821–824 (2007)

47.186 T. Stifter, O. Marti, B. Bhushan: Theoretical investigation of the distance dependence of capillary and van der Waals forces in scanning force microscopy, Phys. Rev. B **62**, 13667–13673 (2000)

47.187 B. Bhushan, A.V. Kulkarni, V.N. Koinkar, M. Boehm, L. Odoni, C. Martelet, M. Belin: Microtribological characterization of self-assembled and langmuir-blodgett monolayers by atomic force and friction force microscopy, Langmuir **11**, 3189–3198 (1995)

47.188 B. Bhushan, H. Liu: Nanotribological properties and mechanisms of alkylthiol and biphenyl thiol self-assembled monolayers studied by AFM, Phys. Rev. B **63**, 245412:1–11 (2001)

47.189 H. Liu, B. Bhushan, W. Eck, V. Stadler: Investigation of the adhesion, friction, and wear properties of biphenyl thiol self-assembled monolayers by atomic force microscopy, J. Vac. Sci. Technol. A **19**, 1234–1240 (2001)

47.190 H. Liu, B. Bhushan: Investigation of nanotribological properties of self-assembled monolayers with alkyl and biphenyl spacer chains, Ultramicroscopy **91**, 185–202 (2002)

47.191 B. Bhushan, M. Cichomski, E. Hoque, J.A. DeRose, P. Hoffmann, H.J. Mathieu: Nanotribological characterization of perfluoroalkylphosphonate self-assembled monolayers deposited on aluminum-coated silicon substrates, Microsyst. Technol. **12**, 588–596 (2006)

47.192 N.S. Tambe, B. Bhushan: Nanotribological characterization of self assembled monolayers deposited on silicon and aluminum substrates, Nanotechnology **16**, 1549–1558 (2005)

47.193 Z. Tao, B. Bhushan: Degradation mechanisms and environmental effects on perfluoropolyether, self assembled monolayers, and diamondlike carbon films, Langmuir **21**, 2391–2399 (2005)

47.194 Anonymous: *Properties of Silicon*, EMIS Data Rev., Vol. 4 (INSPEC, London 1988)

47.195 Anonymous: *MEMS Materials Database* (2002) http://www.memsnet.org/material/

47.196 E. Eteshola, M.T. Keener, M. Elias, J. Shapiro, L.J. Brillson, B. Bhushan, S.C. Lee: Engineering functional protein interfaces for immunologically modified field effect transistor (immunoFET) by molecular genetics means, J. R. Soc. Interface **5**, 123–127 (2008)

47.197 K. Kallury, P.M. MacDonald, M. Thompson: Effect of surface water and base catalysis on the silanization of silica by (aminopropyl)alkoxysilanes studied by x-ray photoelectron spectroscopy and 13C cross-polarization/magic angle spinning nuclear magnetic resonance, Langmuir **10**, 492–499 (1994)

47.198 J.H. Moon, J.W. Shin, S.Y. Kim, J.W. Park: Formation of uniform aminosilane thin layers: an imine formation to measure relative surface density of the amine group, Langmuir **12**, 4621–4624 (1996)

47.199 Y. Han, D. Mayer, A. Offenhausser, S. Ingebrandt: Surface activation of thin silicon-oxides by wet cleaning and silanization, Thin Solid Films **510**, 175–180 (2006)

47.200 D.R. Tokachichu, B. Bhushan: Bioadhesion of polymers for BioMEMS, IEEE Trans. Nanotechnol. **5**, 228–231 (2005)

47.201 N.S. Tambe, B. Bhushan: Identifying materials with low friction and adhesion for nanotechnology applications, Appl. Phys. Lett. **86**, 061906-1–061906-3 (2005)

47.202 N.S. Tambe, B. Bhushan: Micro/nanotribological characterization of PDMS and PMMA used for

BioMEMS/NEMS applications, Ultramicroscopy **105**, 238–247 (2005)
47.203 B. Bhushan, Z. Burton: Adhesion and friction properties of polymers in microfluidic devices, Nanotechnology **16**, 467–478 (2005)
47.204 B. Bhushan, Y.C. Jung, K. Koch: Micro-, nano-, and hierarchical structures for superhydrophobicity, self-cleaning and low adhesion, Philos. Trans. R. Soc. A **367**, 1631–1672 (2009)
47.205 M. Nosonovsky, B. Bhushan: *Multiscale Dissipative Mechanisms and Hierarchical Surfaces* (Springer, Berlin Heidelberg 2008)
47.206 B. Bhushan, Y.C. Jung: Wetting, adhesion and friction of superhydrophobic and hydrophilic leaves and fabricated micro/nanopatterned surfaces, J. Phys. D **20**, 225010 (2008)
47.207 K. Koch, B. Bhushan, W. Barthlott: Diversity of structure, morphology, and wetting of plant surfaces, Soft Matter **4**, 1943–1963 (2008), invited
47.208 K. Koch, B. Bhushan, W. Barthlott: Multifunctional surfaces structures of plants: An inspiration for biomimetics, Prog. Mater. Sci. **18**, 843–855 (2009), invited
47.209 B. Bhushan, K. Koch, Y.C. Jung: Fabrication and characterization of the hierarchical structure for superhydrophobicity, Ultramicroscopy **109**(8), 1029–1034 (2009)
47.210 M. Nosonovsky, B. Bhushan: Roughness-induced superhydrophobicity: A way to design non-adhesive surfaces, J. Phys. D **20**, 225009 (2008)
47.211 B. Bhushan, K. Koch, Y.C. Jung: Nanostructures for superhydrophobicity and low adhesion, Soft Matter **4**, 1799–1804 (2008)
47.212 M. Nosonovsky, B. Bhushan: Multiscale friction mechanisms and hierarchical surfaces in nano- and bio-tribology, Mater. Sci. Eng. R **58**, 162–193 (2007)
47.213 M. Nosonovsky, B. Bhushan: Biologically-inspired surfaces: broadening the scope of roughness, Adv. Funct. Mater. **18**, 843–855 (2008)
47.214 K. Koch, B. Bhushan, Y.C. Jung, W. Barthlott: Fabrication of artificial lotus leaves and significance of hierarchical structure for superhydrophobicity and low adhesion, Soft Matter **5**, 1386–1393 (2009)
47.215 R.N. Wenzel: Resistance of solid surfaces to wetting by water, Ind. Eng. Chem. **28**, 988–994 (1936)
47.216 A. Cassie, S. Baxter: Wetting of porous surfaces, Trans. Faraday Soc. **40**, 546–551 (1944)
47.217 S. Sundararajan, B. Bhushan: Static friction and surface roughness studies of surface micromachined electrostatic micromotors using an atomic force/friction force microscope, J. Vac. Sci. Technol. A **19**, 1777–1785 (2001)
47.218 C.H. Mastrangelo, C.H. Hsu: Mechanical stability and adhesion of microstructures under capillary forces – Part II: Experiments, J. Microelectromech. Syst. **2**, 44–55 (1993)
47.219 R. Maboudian, R.T. Howe: Critical review: Adhesion in surface micromechanical structures, J. Vac. Sci. Technol. B **15**, 1–20 (1997)
47.220 C.H. Mastrangelo: Surface force induced failures in microelectromechanical systems. In: *Tribology Issues and Opportunities in MEMS*, ed. by B. Bhushan (Kluwer, Dordrecht 1998) pp. 367–395
47.221 M.P. De Boer, T.A. Michalske: Accurate method for determining adhesion of cantilever beams, J. Appl. Phys. **86**, 817 (1999)
47.222 R.L. Alley, G.J. Cuan, R.T. Howe, K. Komvopoulos: The effect of release-etch processing on surface microstructure stiction, Proc. Solid State Sensor and Actuator Workshop, ed. by C.H. Mastrangelo, C.H. Hsu (IEEE, New York 1992) pp. 202–207
47.223 H. Liu, B. Bhushan: Adhesion and friction studies of microelectromechanical systems/nanoelectromechanical systems materials using a novel microtriboapparatus, J. Vac. Sci. Technol. A **21**, 1528–1538 (2003)
47.224 B. Bhushan, H. Liu, S.M. Hsu: Adhesion and friction studies of silicon and hydrophobic and low friction films and investigation of scale effects, ASME J. Tribol. **126**, 583–590 (2004)
47.225 S.K. Chilamakuri, B. Bhushan: A comprehensive kinetic meniscus model for prediction of long-term static friction, J. Appl. Phys. **15**, 4649–4656 (1999)
47.226 Y.C. Tai, R.S. Muller: Frictional study of IC processed micromotors, Sens. Actuators A **21–23**, 180–183 (1990)
47.227 K.J. Gabriel, F. Behi, R. Mahadevan, M. Mehregany: In situ friction and wear measurement in integrated polysilicon mechanisms, Sens. Actuators A **21–23**, 184–188 (1990)
47.228 M.G. Lim, J.C. Chang, D.P. Schultz, R.T. Howe, R.M. White: Polysilicon Microstructures to Characterize Static Friction, Proc. IEEE Micro Electro Mechanical Systems (IEEE, New York 1990) pp. 82–88
47.229 U. Beerschwinger, S.J. Yang, R.L. Reuben, M.R. Taghizadeh, U. Wallrabe: Friction measurements on LIGA-processed microstructures, J. Micromech. Microeng. **4**, 14–24 (1994)
47.230 D. Mathason, U. Beerschwinger, S.J. Yuong, R.L. Reuben, M. Taghizadeh, S. Eckert, U. Wallrabe: Effect of progressive wear on the friction characteristics of nickel LIGA processed rotors, Wear **192**, 199–207 (1996)
47.231 G. Wei, B. Bhushan, S.J. Jacobs: Nanomechanical characterization of digital multilayered thin film structures for digital micromirror devices, Ultramicroscopy **100**, 375–389 (2004)
47.232 G. Wei, B. Bhushan, S.J. Jacobs: Nanoscale indentation fatigue and fracture toughness measurements of multilayered thin film structures for digital micromirror devices, J. Vac. Sci. Technol. A **22**, 1397–1405 (2004)

47.233 H. Liu, B. Bhushan: Bending and fatigue study on a nanoscale hinge by an atomic force microscope, Nanotechnology **15**, 1246–1251 (2004)
47.234 B. Bhushan, H. Liu: Characterization of nanomechanical and nanotribological properties of digital micromirror devices, Nanotechnology **15**, 1785–1791 (2004)
47.235 R.C. Jaeger: *Introduction to Microelectronic Fabrication*, Vol. 5 (Addison-Wesley, Reading 1988)
47.236 J. Voldman, M.L. Gray, M.A. Schmidt: Microfabrication in biology and medicine, Annu. Rev. Biomed. Eng. **1**, 401–425 (1999)
47.237 J.W. Judy: Microelectromechanical systems (MEMS): Fabrication, design, and applications, Smart Mater. Struct. **10**, 1115–1134 (2001)
47.238 C. Liu: *Foundations of MEMS* (Pearson Prentice Hall, Upper Saddle River 2006)
47.239 G. Brewer: *Electron-Bean Technology in Microelectronic Fabrication* (Academic, New York 1980)
47.240 K. Valiev: *The Physics of Submicron Lithography* (Plenum, New York 1992)
47.241 E.W. Becker, W. Ehrfeld, P. Hagmann, A. Maner, D. Munchmeyer: Fabrication of microstructures with high aspect ratios and great structural heights by synchrotron radiation lithography, galvanoforming, and plastic moulding (LIGA process), Microelectron. Eng. **4**, 35–56 (1986)
47.242 C.R. Friedrich, R.O. Warrington: Surface characterization of non-lithographic micromachining. In: *Tribology Issues and Opportunities in MEMS*, ed. by B. Bhushan (Kluwer, Dordrecht 1998) pp. 73–84
47.243 M. Tanaka: Development of desktop machining microfactory, Riken Review **34**, 46–49 (2001)
47.244 Y. Xia, G.M. Whitesides: Soft lithography, Angew. Chem. Int. Ed. **37**, 550–575 (1998)
47.245 H. Becker, C. Gaertner: Polymer microfabrication methods for microfluidic analytical applications, Electrophoresis **21**, 12–26 (2000)
47.246 Y. Xia, E. Kim, X.M. Zhao, J.A. Rogers, M. Prentiss, G.M. Whitesides: Complex optical surfaces formed by replica molding against elastomeric masters, Science **273**, 347–349 (1996)
47.247 S.Y. Chou, P.R. Krauss, P.J. Renstrom: Imprint lithography with 25-nanometer resolution, Science **272**, 85–87 (1996)
47.248 A. Kumar, G.M. Whitesides: Features of gold having micrometer to centimeter dimensions can be formed through a combination of stamping with an elastomeric stamp and an alkanethiol ink followed by chemical etching, Appl. Phy. Lett. **63**, 2002–2004 (1993)
47.249 J.C. McDonald, D.C. Duffy, J.R. Anderson, D.T. Chiu, H. Wu, O.J.A. Schueller, G.M. Whitesides: Fabrication of microfluidic systems in poly(dimethylsiloxane), Electrophoresis **21**, 27–40 (2000)

48. Friction and Wear in Micro- and Nanomachines

Maarten P. de Boer, Alex D. Corwin, Frank W. DelRio, W. Robert Ashurst

The prediction and characterization of multi-length-scale tribological phenomena is challenging, yet essential for the advancement of micro- and nanomachine technology. Here, we consider theoretical underpinnings of multiasperity friction, review various approaches to measure micro- and nanoscale friction, and discuss the effect of monolayer coatings to reduce friction. We then focus on test results from a friction-based actuator called a nanotractor. The experimental procedures and data analysis used to measure friction, adhesion force, and wear are detailed. We observe and discuss a variety of phenomena including nanoscale slip with an associated bifurcation in the transition to motion, contact aging and deaging, a stick–slip/steady sliding bifurcation behavior, and wear. We anticipate great progress towards reliable, contacting micro- and nanomachines by linking theory and experiment to nano- and microscale tribological phenomena and by improving the testing, materials, and processing methods used to characterize these phenomena.

48.1 **From Single- to Multiple-Asperity Friction** 1743
48.1.1 Micromachined Test Structures 1744
48.1.2 Monolayer Lubrication in Nano- and MEMS Tribology 1746

48.2 **Nanotractor Device Description** 1747
48.2.1 Nanotractor Dynamic Friction Measurement 1748
48.2.2 Nanotractor Static Friction Measurements 1749
48.2.3 Nanotractor Release Time Measurement 1752
48.2.4 Stick–Slip Testing Using the Nanotractor 1754
48.2.5 Wear Testing Using the Nanotractor 1755

48.3 **Concluding Remarks** 1755

References 1756

Friction and wear phenomena present challenges and opportunities for micro- and nanosystems. In many possible applications, counterfaces rub against each other. Examples of how this can occur are illustrated in Fig. 48.1, where a close-up of features in a complex locking mechanism [48.1] are shown. Here, a large gear ratio is employed to amplify torque. The gear teeth mesh and rotate about hubs that are made by a five-structural-level polycrystalline silicon (polysilicon) surface micromachining technology at Sandia National Laboratories, known as the SUMMiT V process [48.2]. The gears are equipped with features known as dimples to minimize contact and hence adhesion with the substrate. The dimples also prevent the gears from tilting excessively, as does the guide shown at the top right of Fig. 48.1. The output gear meshes with a linear rack that is guided via long rails, and rubbing contact is also made there. To move from right to left, a pin must

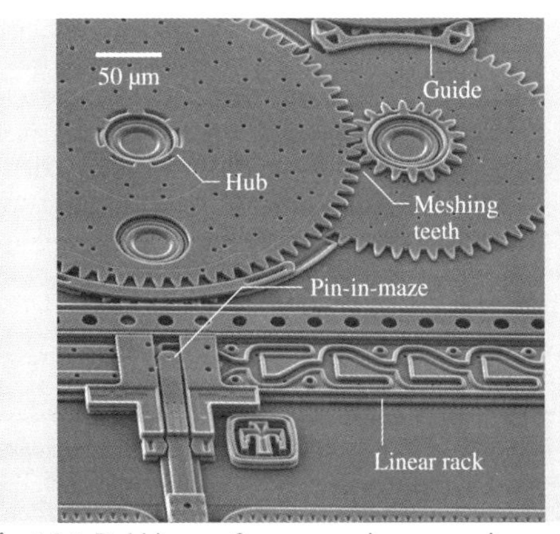

Fig. 48.1 Rubbing surfaces occur in many micromachine applications, and pose a reliability challenge. This figure shows an example of complex mechanical logic

correctly traverse a maze inside the linear rack. If an incorrect decision is made, a ratchet and pawl mechanism (not shown), which also requires rubbing contact, prevents resetting.

In the example shown in Fig. 48.1, friction is generally a parasitic effect and efforts are directed towards minimizing it. Eliminating friction in micromachines has proven difficult, yet therein lies opportunity. A number of researchers have chosen to take advantage of friction, achieving impressive micro- and nanoscale actuation performance. The actuators are microscale stepper motors with step sizes on the order of 10–100 nm, and with long travel ranges up to hundreds of μm. Also, their actuation forces can reach the mN scale. These characteristics compare favorably with the more commonly used comb drives, which typically move 10–20 μm and deliver forces in the low μN range. Such features make such stepper motors attractive, including for positioning for optical [48.5], data storage [48.6, 7], and medical [48.8] applications. We also envision that they may be very useful for testing properties of micro- and nanoscale specimens as well as for manipulating nanoscale instruments.

Several representative implementations of microscale linear actuators, in which actuation is based on frictional effects, are illustrated in Fig. 48.2. Each of these devices has unique design features, performance tradeoffs, and processing requirements. An asymmetric bushing geometry enables the motion of the scratch drive actuator (SDA) in Fig. 48.2a [48.3]. The layout and the drive signal for the SDA are simple, the device area is small ($\approx 30\,\mu m \times 100\,\mu m$), the applied voltage is moderate (30–100 V, depending on layer thicknesses), it travels unidirectionally a distance of 500 μm or more [48.9], and it generates reasonably large forces (tens of μN) [48.3]. A model that predicts operating voltages has been developed, but it does not treat frictional forces [48.10]. The shuffle motor [48.11] and nanotractor [48.12] ($150\,\mu m \times 600\,\mu m$) designs place voltage-controlled clamps at the ends of an actuator plate, enabling bidirectional motion. These actuators can be modeled well and the nanotractor, pictured in Fig. 48.2b, achieves a theoretical maximum force value of $F_{max} = 0.5$ mN [48.12]. A subsequent design known as the μWalker with two degrees of freedom has been demonstrated [48.4, 13] as shown in Fig. 48.2c. Both the nanotractor and the two-dimensional (2-D) μWalker achieve other theoretical expectations including step size on the order of tens of nm and velocity proportional to step frequency up to the mm/s range. The 2-D μWalker is realized in a four-mask process with two layers of structural polysilicon. It takes advantage of a Si_3N_4 coating for in-plane isolation of voltage applied to the clamps, and also as a mechanical layer to enhance wear characteristics. The nanotractor is fabricated in the SUMMiT V process and is also a model friction test structure.

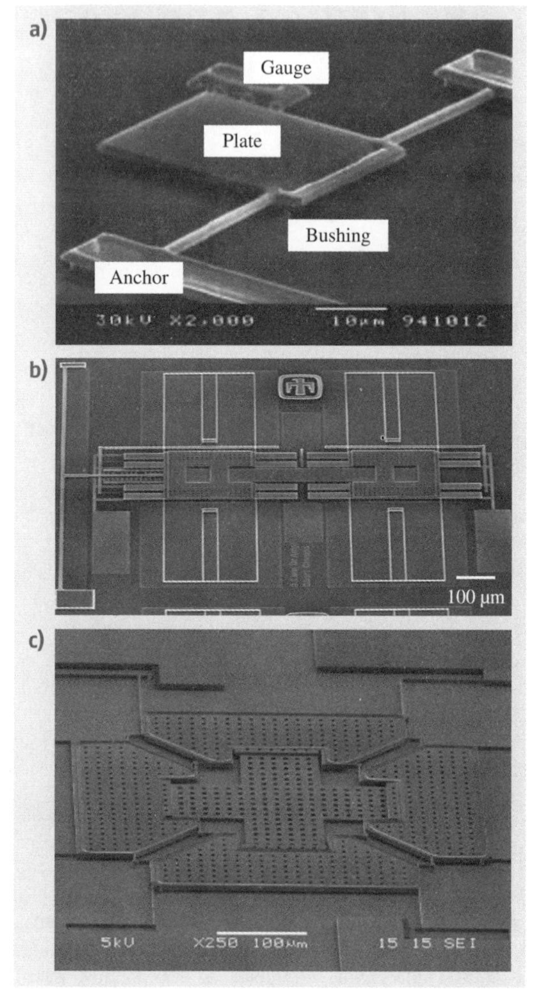

Fig. 48.2a–c Rubbing surfaces can also be used to create powerful actuators such as (a) scratch drive actuator (after [48.3], © IEEE by permission). (b) Nanotractor. (c) 2-D μWalker (after [48.4], © Koninklijke Bibliotheek, Den Haag)

In the following sections, we shall describe theoretical underpinnings of multiasperity friction, micromachined test structures to measure friction, and monolayer coatings that have been used to reduce friction in microelectromechanical systems (MEMS). The major focus of this chapter will then be on testing, metrology, friction, and wear in experiments with nanotractors.

48.1 From Single- to Multiple-Asperity Friction

According to *Amontons*' empirically deduced law [48.14], the friction force F_f is directly proportional to the applied normal load F_n and independent of the apparent contact area A_a as described by

$$F_f = \mu F_n , \quad (48.1)$$

where μ is the coefficient of friction. For surfaces at rest, the coefficient of static friction is denoted by μ_s, while for surfaces in relative motion, the coefficient of dynamic friction is denoted by μ_d.

By considering two surfaces in mechanical contact, it becomes clear that real surfaces are rough on the microscale, consisting of many peaks and valleys. Thus, the *real* contact area A_r between two surfaces is actually defined by the highest peaks, also known as asperities, and is much less than the *apparent* contact area A_a. *Bowden* and *Tabor* [48.15] presented the idea that friction is related to the real (as opposed to apparent) area of contact via two basic mechanisms: shearing and plowing. Shearing refers to the force required to break the junctions at contacting asperities, while plowing represents the force needed for a hard asperity to displace softer material. In the absence of plowing (i. e., for two materials with roughly the same hardness, as is usually the case in micromachined materials), friction is directly proportional to the real contact area via

$$F_f = \tau A_r , \quad (48.2)$$

where τ is the shear strength of the contact. Hence Amontons' law is recovered if A_r is proportional to normal load.

The proportionality between real contact area and applied load, however, seems to disappear for two asperities in *elastic* contact. *Hertz* [48.16] provided the first analysis of the stress distribution and displacement at the interface of two elastic spheres, which can be simplified to a sphere in contact with a flat. The total load F_n compressing a sphere of radius R into a flat surface can be related to the contact area A_r by

$$A_r = \pi \left(\frac{3F_n R}{4E^*} \right)^{\frac{2}{3}} , \quad (48.3)$$

where the combined elastic modulus of the two contacting surfaces $E^* = [(1-\nu_1^2)/E_1 + (1-\nu_2^2)/E_2]^{-1}$. *Johnson*, *Kendall*, and *Roberts* (JKR) later observed that the contact area between two rubber spheres was larger than that predicted by the Hertz theory and developed a model to account for the surface forces at the interface [48.17]. The model was based on a balance between the stored elastic energy, the mechanical energy in the applied load, and the surface energy. *Derjaguin*, *Muller*, and *Toporov* (DMT) used a thermodynamic approach to consider the molecular forces in and around the contact zone [48.18]. Unlike the JKR approach, though, the profile of the sphere outside of the contact area was assumed to be Hertzian (i. e., the surface forces are small enough that their effect on the deformation of the sphere can be neglected). Initially, the models seemed to contradict each other. *Tabor* [48.19] determined that the models represent the extreme ends of a spectrum, and developed a parameter μ_T to span the spectrum. Accordingly, the JKR theory is suitable

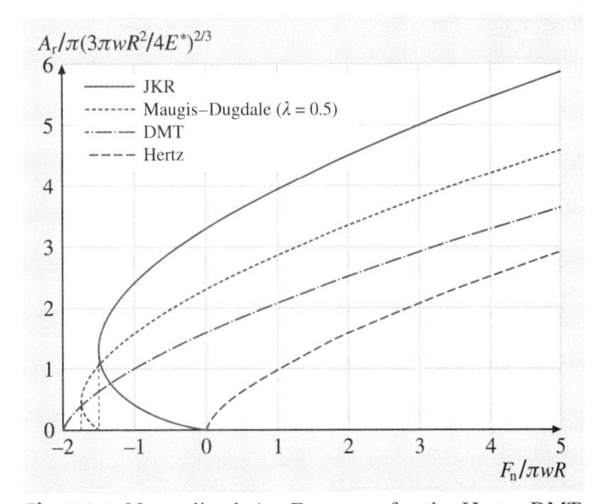

Fig. 48.3 Normalized A_r–F_n curves for the Hertz, DMT, Maugis–Dugdale, and JKR contact models. As the transition parameters μ_T and λ increase from zero to infinity, there is a continuous transition from the DMT to the JKR regime. In the absence of adhesion, all of the models approach the Hertzian case. The work of adhesion is w

for elastically compliant materials with a large radius and surface energy, while the DMT theory is appropriate for elastically stiff materials with a small radius and surface energy. *Maugis* [48.20] later developed a semi-analytical solution using a Dugdale approximation; the transition from DMT to JKR was described in terms of the parameter λ, which is related to μ_T via $\lambda = 1.16\mu_T$. Figure 48.3 illustrates the relationship between A_r and F_n for all of these models. In the presence of attractive forces, the contact area increases at a given load, which increases the friction force, as depicted by (48.2). More importantly, all of the models exhibit a nonlinear relationship between real contact area and applied load, contradicting expectations from Amontons' law.

A number of researchers [48.22–24] have successfully used the aforementioned single-asperity relationships to model friction measurements taken via friction force microscopy (FFM). However, micro- and nanodevices fabricated using surface micromachining techniques often have contacting surfaces with nanometer-scale surface roughness. As a result, it becomes important to develop *multiple*-asperity contact models to elucidate the impact of surface topography on the relationship between contact area and applied load. In the simplest case, the rough surface consists of a series of asperities all at the same height, as shown in Fig. 48.4a. Here, the applied load is divided evenly among all of the contacting asperities, which results in $A_r \propto F_n^{2/3}$. *Archard* [48.21, 25] improved on this idea by considering a *uniform* distribution of spherical asperities (radii of curvature R_1) in contact with a rigid flat as shown in Fig. 48.4b. For elastic deformation, $A_r \propto F_n^{4/5}$. To examine multiscale roughness, a second set of protuberances (radii of curvature R_2) can be evenly distributed over the surface of the existing spheres, such that $R_2 \ll R_1$, as shown in Fig. 48.4c. In this case, $A_r \propto F_n^{14/15}$. With even smaller protuberances of radius R_3, as shown in Fig. 48.4d, the relationship becomes $A_r \propto F_n^{44/45}$. This indicates that proportionality between area and load is not necessarily the result of plastic flow in the contact zones, as originally proposed, but can be due to elastic contact between rough surfaces.

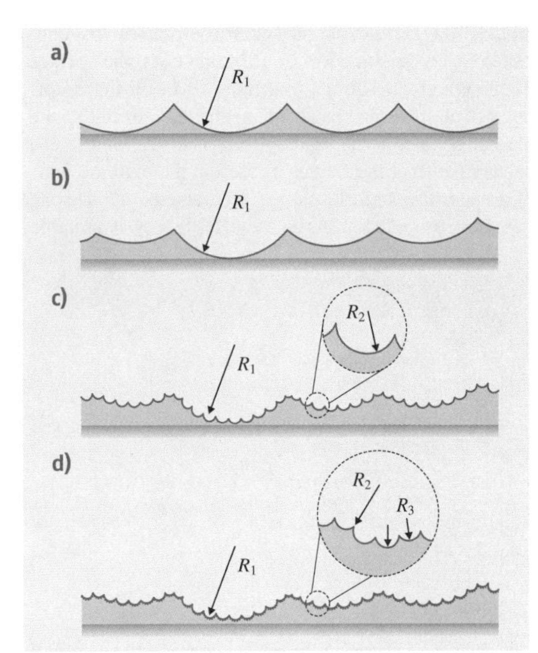

Fig. 48.4a–d Multiple-asperity models from [48.21]. Assuming elastic deformation, the relationship between the contact area A_r and the applied load F_n are (**a**) $A_r \propto F_n^{2/3}$, (**b**) $A_r \propto F_n^{4/5}$, (**c**) $A_r \propto F_n^{14/15}$, and (**d**) $A_r \propto F_n^{44/45}$. In general, the index n in the relationship $A_r \propto F_n^n$ ranges from $\frac{2}{3}$ for a series of asperities at the same height to ≈ 1 for a multiple-asperity surface with several sets of protuberances

Greenwood and *Williamson* (GW) considered a *statistical* distribution of N asperities with constant radius of curvature R in contact with a flat surface [48.26]. For a Gaussian distribution of asperity heights, A_r is almost exactly proportional to F_n. *Fuller* and *Tabor* [48.27] extended the GW model for JKR contacts to study rough surface adhesion and defined the adhesion parameter θ, which represents the competition between the compressive forces exerted by the higher asperities and the adhesive forces acting between the lower asperities (analogous to the Tabor parameter). In a similar manner, *Maugis* [48.28] extended the GW model for DMT contacts. For infinite θ (work of adhesion $w = 0$), the Maugis theory is equivalent to the GW model and $A_r \propto F_n$. On the other hand, as θ decreases (w increases), the relationship between contact area and applied load becomes more nonlinear.

48.1.1 Micromachined Test Structures

From this survey of multiple asperity contact models, we see that (1) adhesive forces and (2) asperity distribution can significantly impact the relationships between friction force, contact area, and applied load. Therefore, although widely observed, Amontons' law is not necessarily expected. While there are techniques in place to convert topographic data from real surfaces into Gaussian asperity distribution data [48.29], it is unclear as to whether these statistical models accurately represent the surface topography [48.30]. Therefore, it becomes necessary to fabricate micromachined test structures that can measure friction directly. Moreover, micromachined structures are appropriate test vehicles for studying friction and wear because their performance will reflect any effects that are due to details of the fabrication process. These include technologically relevant issues such as the effect of sidewall slope on vertical surfaces, surface roughness, and the ability to coat micromachined structures with friction-reducing coatings in spite of sometimes tortuous access paths.

One of the earliest systematic studies of microscale friction was carried out by *Howe* and coworkers at UC Berkeley [48.33]. Since then, a number of researchers have investigated this critical issue using a variety of techniques and devices. A common approach is to design a moveable beam that can be brought into contact and slid against a counter surface (see, e.g., [48.31, 34–39]). An example from [48.31] is shown in Fig. 48.5a,b. The measurements from this study showed that statistical contact models as described above may not apply to lightly loaded, small apparent contact areas in MEMS. This is because such models would predict less than one contact within the contact area. The researchers also found that adhesion between the surfaces must be recognized as an important contribution to the applied normal load. Most microdevice friction work has involved so-called *sidewall* devices such as this one, where the contacting surfaces are the result of an etching procedure and may have vertical striations and significant directional anisotropies. The topography of sidewalls is influenced by lithography, etching, *and* grain boundaries as seen in Fig. 48.5c [48.32], and often exhibits significantly higher roughness than in-plane surfaces where topography is primarily determined by grain growth and etching phenomena.

Komvopoulos and coworkers [48.38] developed a sidewall microscale friction device that involved a push drive and a shear drive. Testing of the device consisted of first loading the contact and then shearing the newly formed interface while monitoring the position of the shear drive with a charge-coupled device (CCD) camera, with a spatial resolution of about 0.3 μm. It was shown that the engineering coefficient of friction exhibited a nonlinear dependence on contact pressure that was unique to the testing environment; higher friction was observed at higher humidity levels. Further, the adhesion forces, significant at the microscale, added substantially to the friction when the external load was comparable to the adhesion force [48.38].

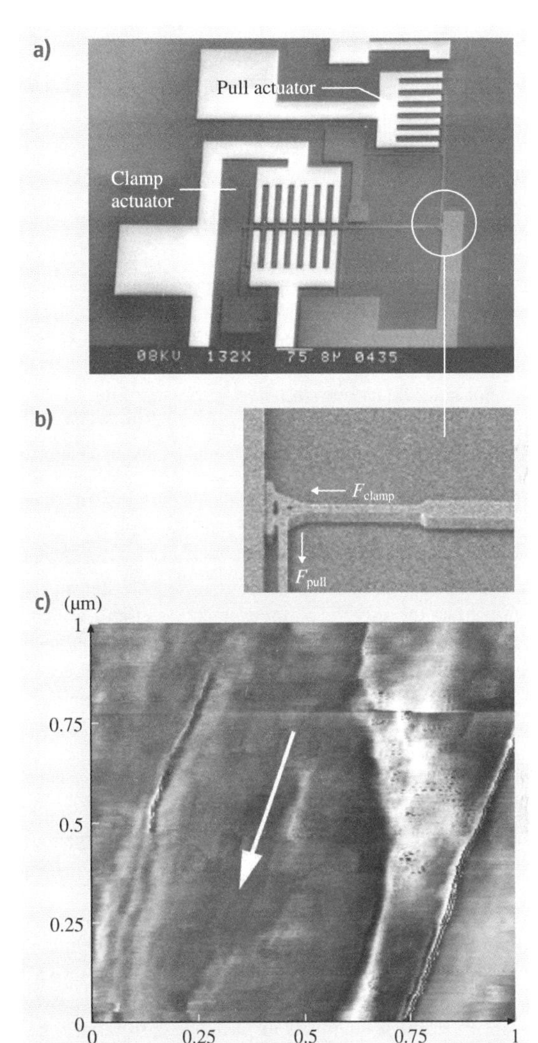

Fig. 48.5 (a) A MEMS test structure (b) designed to measure sidewall friction (after [48.31], © VSP). (c) Topographical image of a sidewall surface, where the root-mean-square (RMS) roughness is 40 nm (after [48.32], © Xavier by permission)

Frenken and coworkers recently developed a similar sidewall device, but included an on-device, real-time, electrical capacitance readout mechanism for high-resolution displacement measurement (about 4 nm peak-to-peak noise sensitivity). Their device, called the *Leiden MEMS tribometer*, consists of two orthogonally oriented comb drives that are used to position a test slider against a fixed surface. The comb drives are used to generate loading and shearing forces on the contact surface. They indicated two regimes in the friction behavior depending on the load. For the low-loading regime, a wearless stick–slip phenomenon is observed which is repeatable over many cycles but stochastic in position and presumably related to the details of the topography. In the high-loading regime, wear was observed, and although the stick–slip behavior exhibited in the low-load regime disappeared, the motion was not smooth, as ongoing wear processes altered the surface topography and changed the contact mechanics [48.37].

48.1.2 Monolayer Lubrication in Nano- and MEMS Tribology

While the results in the previous section are certainly interesting, the effects of monolayer coatings, which significantly reduce friction, were not studied. It has been well established that monolayer lubricants are effective at reducing friction at the nanoscale. Much of this knowledge has been achieved through nanotribology experiments involving atomic force microscopy (AFM) [48.40, 41] or a surface forces apparatus (SFA) [48.42] where ideal systems (i. e., pristine environments, single crystal substrates, perfect monolayers, etc.) are employed to probe fundamental aspects of friction phenomena using single-asperity contacts.

Nanotribology studies have shown significant (and sometimes conflicting) impact of monolayer thickness (i. e., precursor chain length) and terminal group on friction [48.43–46]. Similar to rough surfaces, there are two basic phenomena involved in single-asperity friction: plowing and shearing [48.41]. With regard to monolayer coatings, plowing refers to the action of the asperity tip deforming the monolayer, thereby dissipating energy by introducing chain defects that may elastically relax after the tip has passed, while shearing refers to the localized disruption of intermolecular forces as a result of the shear interaction of the film surface with the tip. Some researchers report an increase in friction as chain length increases [48.47], while others report a decrease in friction [48.45]. These conflicting findings suggest that additional experimental factors contribute to the frictional behavior of monolayer films, and the operating conditions must be considered in the friction study.

As discussed above, the friction force may be proportional to the true contact area, leading to a nonlinear dependence of friction with load. This has been supported by experimental data for inorganic materials. However, in many studies of organic monolayers, a linear relationship is found [48.43, 45, 48–50]. This could be attributed to plastic deformation, or to multiple contacting nanoasperities on the single AFM tip, or to viscoelastic deformation during the sliding. It is important to note that many of the fundamental mechanistic studies of nanoscale friction have involved high-modulus inorganic materials with similar bulk and surface properties. Since monolayers are comparatively compliant materials, usually deposited on a stiff material, the validity of applying findings from such fundamental studies to monolayer systems can be questioned. Another aspect to consider is that deviation from linearity as in the JKR model [48.17] may be small and experimentally obscured for low loadings, typical of AFM studies [48.41]. Recently, a continuum thin coating theory has been developed to address these issues [48.51, 52].

Some researchers have successfully applied monolayer films to micromachines for the purpose of studying their effects on friction [48.12, 53–56]. The most convenient monolayers to apply to polysilicon microdevices involve silane-based linking reactions with the oxide layer present on the surface. From the silane class of materials, the most commonly studied monolayer film is that produced by octadecyltrichlorosilane (OTS). This particular film has received much attention in the literature due to its self-assembly characteristics [48.57]. Recent work using AFM methods has shown that the friction behavior of the film is dependent on the local 2-D phase (liquid expanded or liquid condensed) at the asperity contact [48.58] and that liquid condensed phases are likely to be associated with grain-boundary areas for polysilicon devices [48.59]. OTS films have been utilized as antifriction layers for MEMS devices and studied in that capacity. It is generally accepted by the MEMS community that OTS and other monolayer films reduce the friction of microdevices when properly integrated and applied to the device.

Clearly, it would be interesting to link single-asperity measurements made by nanotribologists to multiasperity measurements made by MEMS tribologists. This would involve a detailed understanding of the surface properties, loading characteristics, and topography. Indeed, asperity radii of polycrystalline silicon are on the order of 20–200 nm, in the same range as AFM tips. However, the richness of MEMS tribology measurements, such as the nonlinear effect of pressure [48.38] and wearless regime [48.37] mentioned above, are only beginning to be explored. At the macroscale, friction measurements have revealed complex behavior including aging, velocity dependence, and stick–slip bifurcations, and a theory known as *rate-and-state* friction has been advanced [48.60–64]. This development comes from detailed measurements with different combinations of springs and different puller velocities, and it applies to a wide range of materials from rock [48.61] to cardboard [48.62] to plastic [48.65]. This theory applies the notion of contact rejuvenation and solves coupled differential equations of rate (instantaneous velocity) and state (contact age) variables [48.66,67]. Quite possibly, microscale friction will have related dependencies. As an initial effort to explore some of these potential dependencies, the loading protocol has been varied in a systematic fashion using the nanotractor. As we shall see, strong effects that may be related to rate-and-state friction are observed.

48.2 Nanotractor Device Description

The findings assembled from uncoated and coated sidewall devices have significant value to the MEMS community and provide guidance on how to more reliably design and operate microdevices. However, statistical contact mechanics models, which rely on numerous contacts, do not apply because of the small contact areas and nonvertical nature of sidewall etching. On the other hand, the nanotractor achieves uniform loading over a large area of in-plane surfaces during friction testing so that such models will apply.

Figure 48.6a is a scanning electron microscopy (SEM) cross section showing the details of how the nominally flat model frictional interface is created in the clamp regions. Normal force is symmetrically applied via the electrodes to the right and left of the counterfaces. The counterfaces are each electrically grounded and make mechanical contact. The air gap breaks down at 300 V; the device is actuated at 200 V or less to avoid any issues with electrostatic damage. The inset shows that the upper counterface is nominally flat, meaning that calculations of real contact area from statistical theories [48.26,29,30,68,69] can be used without having to make a correction for surface tilt or curvature. This counterface is 3 μm wide, and there are two such counterfaces in each clamp. The length of the clamps in the results reported below is 600 μm. Hence the total apparent contact area for both friction clamps is

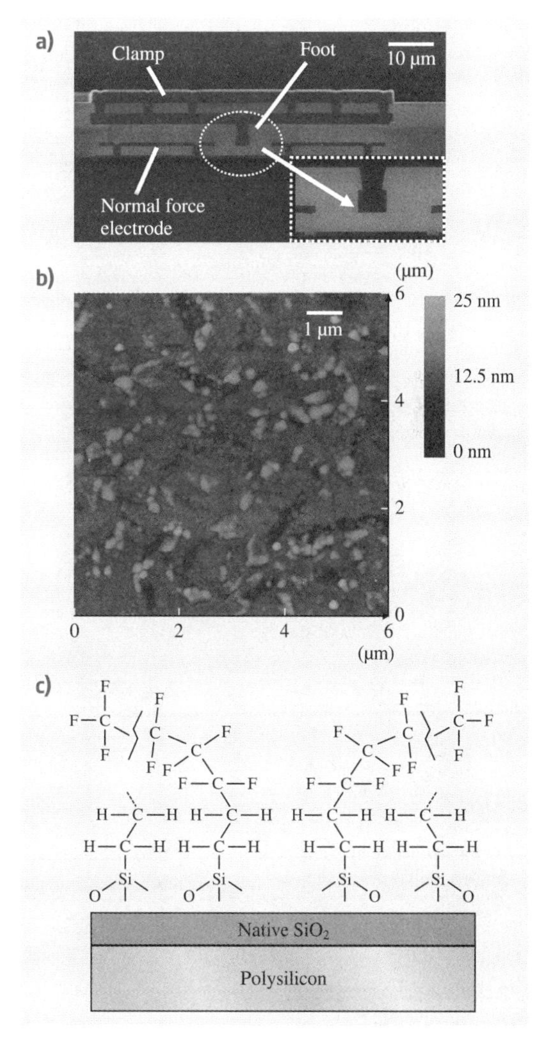

Fig. 48.6 (a) Nanotractor SEM cross section through a clamp. The *inset* shows the detail of the frictional counterfaces. The oxide material surrounding the polysilicon is removed by HF acid etching prior to monolayer lubricant deposition. (b) AFM micrograph of a typical polysilicon surface after coating with a FOTAS organic monolayer. (c) Schematic representation of the disordered FOTAS monolayer, with some chains bending out of the plane of the paper ▶

$A_a = 7200\,\mu m^2$. The electrode area enables continuous control of normal load F_n up to $\approx 10\,mN$, where F_n is proportional to V_c^2. Here, V_c is the clamping voltage.

Figure 48.6b shows the surface roughness, which is mainly due to the grain structure on these surfaces (i. e., not due to plasma etching), with a typical root-mean-square value of 5 nm. Although the contact geometry is much improved, it remains a challenging task to calculate the real contact area accurately. In one approach, *DelRio* et al. [48.70] imported data from AFM surface topography maps and conducted pixel-by-pixel analysis to show that adhesion in MEMS is dominated by van der Waals forces across noncontacting surfaces. Agreement within a factor of two with the experimental data was obtained. A voxel calculation based on directly imported data has also been used to better estimate the areas of real contact between polycrystalline gold MEMS surfaces [48.71]. It would also be of interest to apply a theory that incorporates skew and kurtosis [48.69], or a theory that calculates real contact area by solving a diffusion equation of an autocorrelated surface power spectrum [48.30].

The experimental data that follows were measured using nanotractors coated by a 1.2 nm-thick monolayer coating (tridecafluoro-1,1,2,2-tetrahydrooctyl-*tris* (dimethylamino)silane, $CF_3C_5F_{10}C_2H_4Si(N(CH_3)_2)_3$, FOTAS) deposited from the vapor phase [48.72]. This monolayer is represented schematically in Fig. 48.6c. The eight-carbon-chain molecule has a sufficiently high vapor pressure to allow vapor-phase deposition, which makes it highly reproducible in a manufacturing setting. However, it is too short to form a self-assembled monolayer; rather the chains attached to the surface are believed to be somewhat disordered. This lack of order may be responsible for the memory effects reported below.

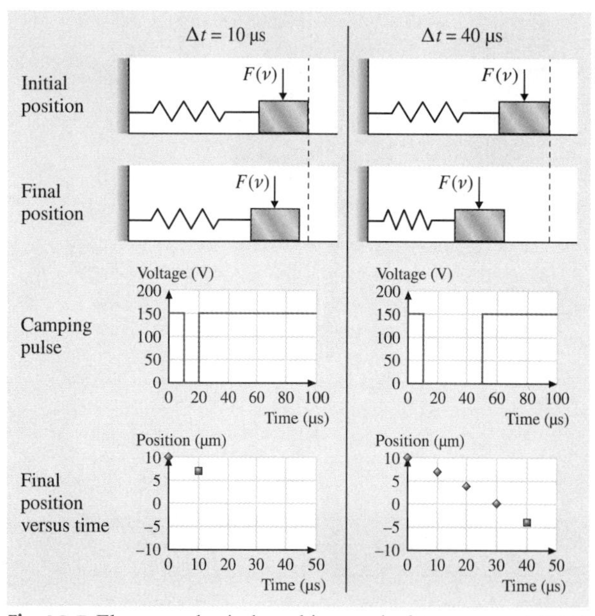

Fig. 48.7 Electromechanical strobing method to capture MEMS friction dynamics

48.2.1 Nanotractor Dynamic Friction Measurement

In studying its dynamic motion, the nanotractor can be simply thought of as a mass and a spring, riding atop a planar surface, with a normal force generated by electrostatics. By walking the nanotractor (as described in [48.12]), the mass can be pulled out to some distance, and then released. The resulting motion will be influenced by both air damping (dependent on the pressure) and Coulomb friction damping (depending on the normal force exerted on the mass), as well as adhesive force between the mass and the plane. For suitably small friction, the mass can be expected to oscillate several times about its zero position with decaying amplitude before finally coming to rest.

Corwin and *de Boer* have developed a mechanical stop motion technique for studying this motion [48.54] through the use of pulsed clamping, shown diagrammatically in Fig. 48.7. They walk the nanotractor out to a specified position ($x_0 = 10\,\mu m$ for the example shown). The position is determined using a high-accuracy, subpixel interpolated pattern-matching technique (from National Instruments Vision Toolkit as utilized in scripting software developed at Sandia National Laboratories called MEMScript). With a 50x objective, this technique achieves approximately 5 nm in-plane resolution. Once the nanotractor is walked to x_0 (within $\approx 50\,nm$ as limited by its step size), it is held in place solely by the electrostatic clamp between two polysilicon surfaces. The clamp is then released for $10\,\mu s$, and then clamped again (clamping time is about $1\,\mu s$). The position of the clamp is recorded using pattern matching, and the nanotractor is again walked back to the specified position. The clamp is now released for $20\,\mu s$ and its position is recorded. The same procedure is repeated for $30\,\mu s$, $40\,\mu s$, etc., continuing until the resting position of the nanotractor no longer changes with release duration. The position of the nanotractor

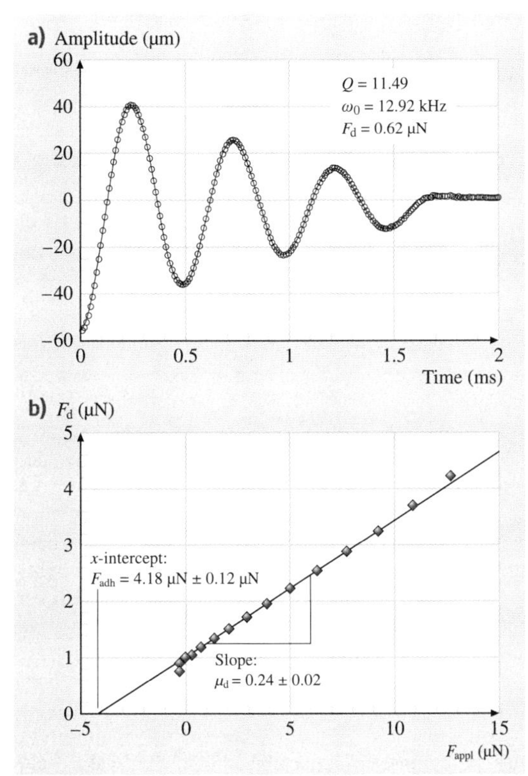

Fig. 48.8a,b Dynamic friction test results. (**a**) Position versus time data. (**b**) Friction force versus applied load data, which is used to find F_{adh} (after [48.54], © AIP by permission)

as a function of release duration can then be assembled into a position versus time plot, as is seen in Fig. 48.8 ($x_0 = -58\,\mu m$ for these results).

The position data in Fig. 48.8 is modeled with a second-order nonlinear differential equation that takes into account both air damping and dry friction damping

$$m\ddot{x} + b\dot{x} + F_d \operatorname{sgn}(\dot{x}) + kx = 0 . \quad (48.4)$$

Here, the contribution from the air damping is written as a velocity-dependent force, $b\dot{x}$ (as the gap between the nanotractor and the substrate is small [48.73]), and the friction is written as $F_d \operatorname{sgn}(\dot{x})$ where the sgn function simply returns 1 or -1 depending on the sign of the argument. Also, k is the spring constant of the folded beam suspension (shown schematically in Fig. 48.7) and m is the mass of the nanotractor.

Solving (48.4) piecewise analytically, the undamped resonant frequency ω_0, F_d, and the quality factor Q can be determined. The measured value of ω_0 is consistent with the expected resonant frequency given the nanotractor mass and the suspension spring constant. The experimentally determined quality factor, $Q = m\omega_0/b = 11.5$, is in good agreement with a first-order Couette calculation of $Q = 14.5$ [48.73]. The friction force F_d is nonzero even though the suspension spring provides a slight out-of-plane force (with spring constant k_z), indicating the presence of an adhesive force.

Following Amontons' law, the total frictional force is found by summing all normal forces as

$$F_d = \mu_d(F_c + mg - k_z z) + \mu_d F_{adh} , \quad (48.5)$$

including the restoring force from the suspension spring ($k_z z$), the gravitational mass (mg), the electrostatic force on the clamp (F_c), and a surface attraction adhesive term (F_{adh}). Here the applied force is $F_{appl} = F_c + mg - k_z z$.

The value of μ_d is determined by repeating the procedure illustrated in Fig. 48.7, systematically increasing F_c and measuring the resulting F_d. The results are shown in Fig. 48.8b. Following (48.5), a line is fitted to extract both dynamic friction and adhesion. For the data shown, $\mu_d = 0.24 \pm 0.02$ and $F_{adh}/A_a = 0.6\,nN/\mu m^2$. The latter value is somewhat smaller than the value of $6\,nN/\mu m^2$ extracted from adhesion models [48.70]. However, in this device the nanotractor foot was not entirely flat, an empirical difficulty that was corrected in later experiments.

With a nominally flat foot, *Corwin* and *de Boer* have recently demonstrated an improved version of this technique that minimizes the number of measurements required to determine the adhesion and dynamic friction of the nanotractor. The technique is detailed in [48.74], and reduces the problem to fitting linear curves over the first half-cycle of motion. The adhesion value found was $F_{adh}/A_a = 3\,nN/\mu m^2$, in reasonable agreement with the adhesion model calculation [48.70]. In subsequent measurements reported below, the foot is nominally flat, as shown in Fig. 48.6a.

48.2.2 Nanotractor Static Friction Measurements

To make a static friction measurement, the nanotractor is walked to an initial position and held in place with a prescribed normal force, as shown in Fig. 48.9a. The tangential force experienced during the hold is kx_0. The normal force F_n is then ramped down by lowering the clamp voltage. (Here it is assumed that $F_n = F_c$ because the other terms in (48.5) are small for large F_n).

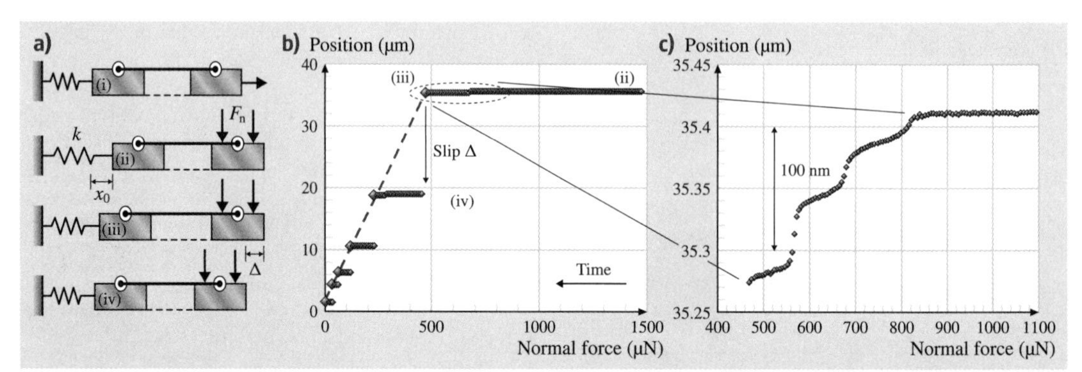

Fig. 48.9 (**a**) Schematic for the friction test. (**b**) Position versus normal force data at the micron scale. (**c**) Nanoscale slip before the transition to gross slip as measured by moiré interferometry

The block position was measured seven times per second to ±5 nm accuracy by the optical pattern-matching method described above.

The results of a typical measurement are shown in Fig. 48.9b. At each large jump (denoted by diamonds), an equilibrium between tangential force and friction force exists, and following Amontons' law a coefficient of static friction can be determined according to

$$\mu_s = \frac{F_s}{F_n}, \quad (48.6)$$

where F_s (the static friction force opposing the spring force supplied by the linear suspension spring) and F_n (the electrostatic normal force) are the values associated with the static friction event. A more accurate coefficient of friction is found by taking the slope of the dashed line in Fig. 48.9b. In that case, adhesion is taken into account [48.54].

Taking advantage of the ±5 nm displacement resolution, in Fig. 48.9c the region just before a large jump can be expanded. Motion on the order of 100 nm is observed before the large jump. This nanometer-scale slip has been observed reproducibly over many measurements and is of concern for high-precision positioning applications. To understand this interesting characteristic in more detail, *Corwin* and *de Boer* have investigated the effect of three parameters:

- t_h – the time the nanotractor is held before the normal force ramp down begins
- F_h – the normal load at which the clamps are held before the ramp down
- $\dot{F}_n$ – the rate at which the normal force is ramped down [48.75].

Their effects are shown in Figs. 48.10–48.12 and described below.

In the data presented in Fig. 48.10a, the nanotractor is held at position $x = x_0$ and the normal load is held at F_h (designated by a vertical dashed line on the right) for hold time t_h. The normal load is then ramped down until an off-scale jump is observed. The experiment is repeated for a different hold time after the nanotractor is repositioned at $x = x_0$. The data for different hold times in Fig. 48.10a is offset by 0.1 μm in order to make it distinguishable. The circle for each hold time represents the onset of detectable slip. If the load is held for only a few seconds ($t_h \leq 32$ s) as it was in Fig. 48.9 ($t_h = 1$ s), frictional creep is observed before the transition to gross slip, as designated by the label "creep" in Fig. 48.10a. On the other hand, if it is held in place for a longer time ($t_h \geq 64$ s), the test structure reveals no motion before the transition to gross slip, as indicated by the label "off-scale jump" in the figure. There is also a critical normal force F_{nc}, as designated by a vertical dashed line. If detectable motion occurs before F_n reaches F_{nc}, the system exhibits frictional creep before the transition to gross slip. If F_n descends below F_{nc} before detectable slip, the transition to gross slip is abrupt. Hence, there is a bifurcation in the transition to motion – it either occurs as an inertial jump or as frictional creep. These measurements are highly reproducible (the same effects were observed over multiple fabrication lots and also if the hold time order was randomized) and show that the loading protocol matters in MEMS friction measurements. In Fig. 48.10b, the μ_s values, calculated from the circles in Fig. 48.10a, are plotted. A characteristic logarithmic aging coefficient β is found.

Fig. 48.10 (**a**) Complex aging behavior observed by varying the hold time t_h; see text for description. (**b**) The logarithmic aging behavior is characterized by the slope β ▶

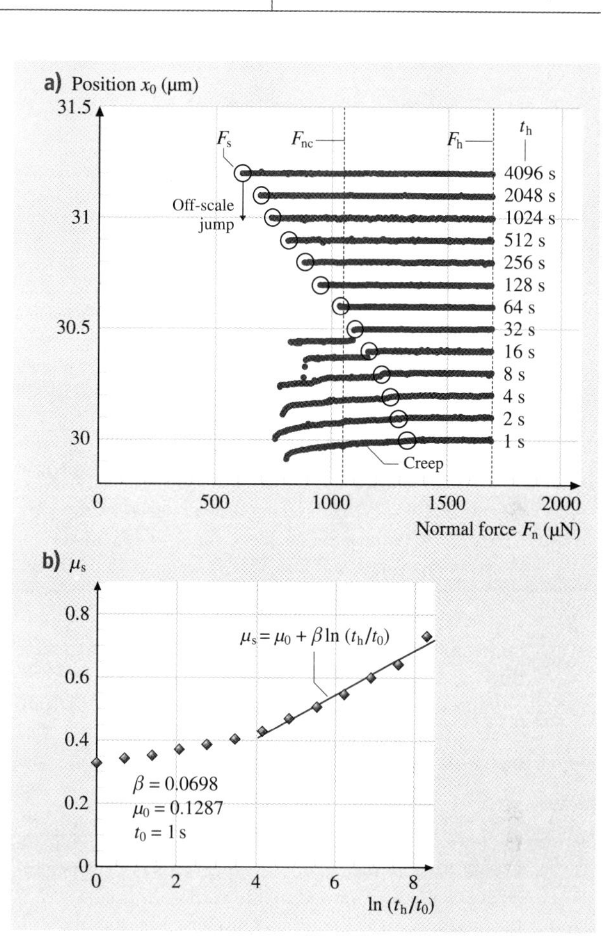

In Fig. 48.11, the effect of changing the hold load F_h before the normal load ramp down is shown. Here it is seen that, as F_h is increased, the aging rate β decreases and approaches zero for sufficiently large F_h. Logarithmic frictional aging is a characteristic signature of rate–state behavior, and is generally attributed to material creep; that is, the contact area increases with time due to the heavy loading at the contacts [48.65], similar to in indentation creep. This has been directly observed in soda lime glass and acrylic plastic [48.76]. Hence, the result of Fig. 48.11 is counterintuitive. This effect can better be thought of as an increase in aging rate as the ratio of shear to normal force increases. A similar result has been reported for a rough glassy polymer on silanized glass [48.77] and can be ascribed to the associated biasing of a glassy material energy landscape [48.78, 79].

Although not shown, each F_h series in Fig. 48.11 is derived from a plot similar to Fig. 48.10a. From such plots a bifurcation in the transition to sliding is again observed, and F_{nc} remains the same, independent of F_h. A qualitative explanation for the existence of F_{nc} is as follows. If the interface ages significantly during the hold, then μ_s becomes large. As the normal load ramps down, F_n reaches a low value before motion initiates. Once this occurs, the contacts are rejuvenated, and μ_s becomes much lower. This large change in μ_s, coupled with the low value of F_n, results in an immediate transition to inertial sliding. Conversely, if t_h is small, the contacts do not age much. Therefore, F_n does not reach a very low value before the onset of motion. The small change in μ_s, coupled with the high value of F_n at which sliding initiates, results in only a very small change in position, and hence the slip on the surface is stable. Detailed measurements reported in [48.75] indeed show that this apparent frictional creep is a convolution of true frictional creep and the continuing normal force ramp down.

Although the dependence of μ_s on t_h in Figs. 48.10 and 48.11 demonstrates that there is an aging effect, the protocol for measuring this is not necessarily fully specified. The normal force ramp-down rate $\dot{F}_n$ may also affect the μ_s values because the state of the interface, in terms of the number of contacts per unit area, decreases while the normal load is ramped down. When the ramp-down time t_{rd} (on the order of tens of seconds for the data shown thus far) is on the order of t_h, an interaction

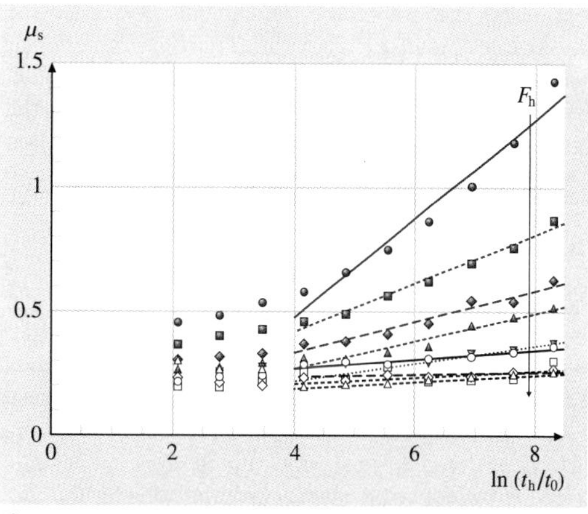

Fig. 48.11 Increasing the hold force F_h (from 1500 to 4800 μN here) decreases the aging rate

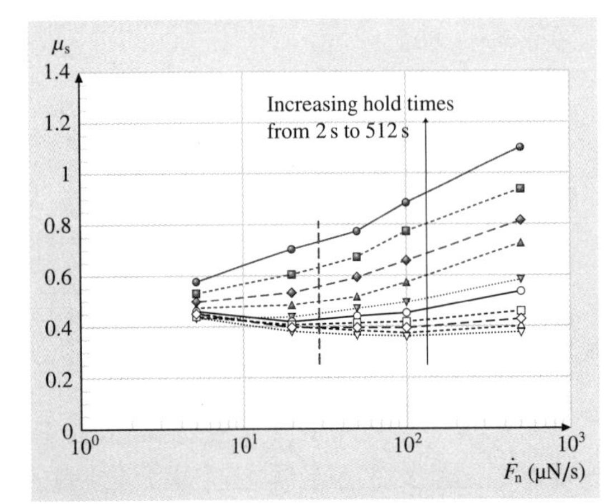

Fig. 48.12 By varying the ramp-down rate, μ_s changes by a factor of 3. In the previous plots (Figs. 48.10 and 48.11), the ramp-down rate was 20–35 μN/s, as indicated by the *dashed vertical line*

between t_{rd} and t_h is expected. When it is much greater than t_{rd}, t_h will possibly control the μ_s value.

The data presented in Fig. 48.12 show the strong effect of $\dot{F}_n$ on μ_s. Here, the ramp-down rate was increased by increasing the voltage increments during the ramp down, while the camera frame rate was fixed at seven frames per second. The ramp-down rate for Figs. 48.10 and 48.11 was 20–35 μN/s, while in Fig. 48.12 it is varied from 5 to 500 μN/s. Further increase of $\dot{F}_n$ was limited by the camera frame rate. For the largest $t_h = 512$ s, μ_s increases monotonically with $\dot{F}_n$. For the smallest t_h and $\dot{F}_n$ values, some aging occurs during the ramp down, which tends to increase the μ_s value for small $\dot{F}_n$. To obtain Fig. 48.12, a series of data similar to Fig. 48.10a was taken for each $\dot{F}_n$ value. Each value for μ_s in Fig. 48.12 was again taken from the first observable motion. Each plot revealed a bifurcation in the transition to slip, again with $F_{nc} \approx 1000$ μN.

48.2.3 Nanotractor Release Time Measurement

The data in Fig. 48.12 show that, even at the highest $\dot{F}_n$ value (where, the ramp-down time t_{rd} is much less than the hold time t_h), the μ_s value does not saturate. Given the strong dependencies observed, *Corwin* and *de Boer* considered the possibility that the concept of static friction may not be the most appropriate one for the normal force ramp-down test protocol [48.80]. Recently, the phenomenon of interface *deaging* has been directly measured on PMMA/PMMA interfaces [48.81]. There, history-dependent unloading and a true contact area that decreases with time in a deaging process whose time depends on prior aging were observed. The real contact area was directly observed using a high-speed camera with a total internal reflection method. Perhaps an important parameter characterizing the interface shear strength in these experiments is a release (or deaging) time.

In order to examine this idea, a series of experiments was carried out utilizing the maximum possible ramp-down rate (i. e., release on the order of 1 μs), and measurements of the release time were made. Details are provided in [48.80], and are briefly described here. The setup that was implemented is shown in Fig. 48.13. The nanotractor was walked to the initial position x_0 as shown in Fig. 48.13a. It was held at that position with a normal force F_h for a time t_h. Then, the load was *instantaneously* dropped to F_r. At this point, a high-speed camera (Phantom V 5.0) operating at 10 000 frames per second was triggered as shown in Fig. 48.13b, and frames were stored in the camera memory. Using simple dynamics calculations assuming $\mu_s = 0.3$ and a displacement resolution of 1 μm, the expected response time is 6 μs, significantly less than the time for one frame. If this calculation applies, the motion will always occur in the first frame.

Figure 48.14, however, shows that release times from less than 100 μs to almost 50 s are measured, spanning nearly six orders of magnitude. For fixed values of F_n and F_r, an increase in hold time t_h leads to an increase in release time t_r. Thus aging between

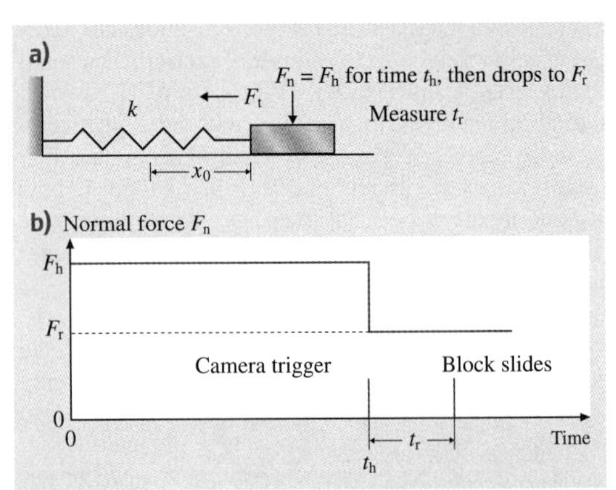

Fig. 48.13a,b Schematic of the release time test. **(a)** Nanotractor block against stretched spring, **(b)** timing diagram to measure release time t_r

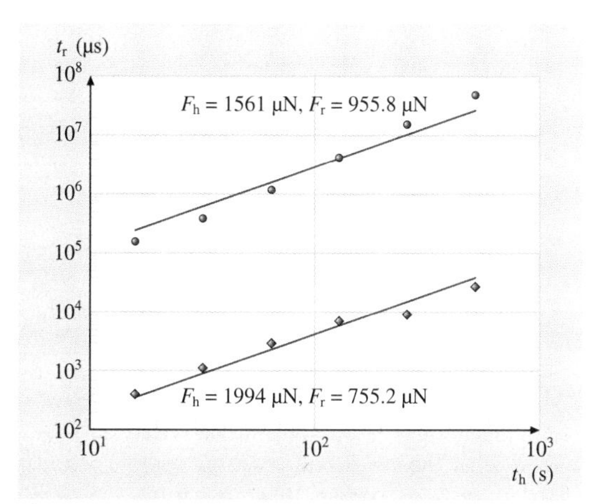

Fig. 48.14 Dependence of release time t_r on t_h

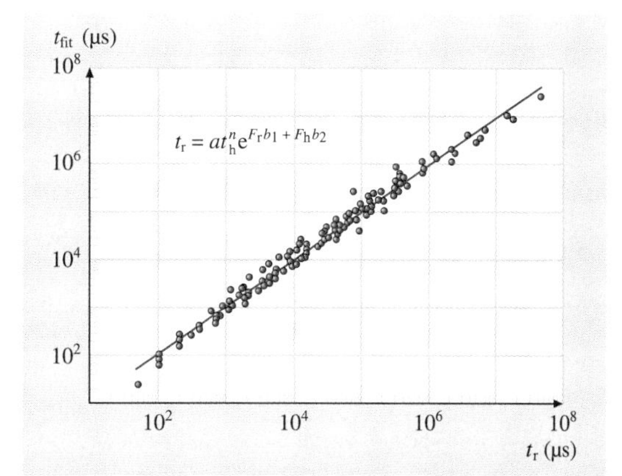

Fig. 48.15 Combining the fits from three different measurements, the data collapse onto a single fit ranging over six orders of magnitude of time

the two contacting surfaces is also evidenced in these measurements (assuming that deaging is directly related to aging [48.81]). The two data sets in Fig. 48.14 were taken for different pairs of the parameters F_h and F_r. The linear behavior suggests a power-law dependence of release time on hold time of the form $t_r \propto t_h^n$. The line through both sets of data are plotted with the same value of the exponent n, demonstrating that the scaling of t_r with t_h is independent of both F_h and F_r.

A direct demonstration that aging occurs was shown by resetting it after a hold. By moving the block by as little as 50 nm (a single step of the nanotractor) after aging but before dropping the release force, the release time enhancement due to aging was entirely removed.

Similar to Fig. 48.14, measurements of t_r as a function of F_r and F_h also revealed characteristic dependencies. A full functional form is a combination of the three dependencies, and can be written as

$$\log(t_r) = \log(a) + n\log(t_h) + b_1 F_r + b_2 F_h\ . \qquad (48.7)$$

This data can be approximated by a least-squares fit with the four parameters, a, n, b_1, and b_2, extracted from Fig. 48.14 and the other measurements just mentioned. Figure 48.15 displays a plot of fitted release times t_{fit} versus measured release times and demonstrates that (48.7) fits the data well over six orders of magnitude of time.

These results show that release time in this system depends on the full load history through the hold time t_h, the hold force F_h, and also on the release force F_r. This memory effect is similar to observations by *Rubinstein* et al. [48.81]. Through measurements of the true contact area of PMMA blocks, they observed history-dependent unloading and a true contact area that decreased with time in a deaging process whose time depended on prior aging, similar to the measurement of release time. Viewing deaging time as a measure of prior aging, this empirical model can be thought of as revealing the functional dependencies of aging, and because unloading takes place almost instantaneously, this model does not depend on setting a particular ramp-down rate.

This observed deaging time behavior can qualitatively explain the previously observed dependencies of measured static friction coefficient on the ramp-down rate. If ramping down the normal force slowly, the contacting surfaces have more time to pull apart and measurable slip occurs at a higher normal force, leading to a smaller measured μ_S. When ramping down rapidly, the surfaces have less time to separate and measurable slip occurs at a much lower normal force, leading to a larger measured μ_S.

The aging effects reported above were observed on the FOTAS monolayer. It can be expected that the data will qualitatively change for different monolayers as well as different environmental conditions. Therefore, to fully assess the friction of a given monolayer for use in MEMS or nanoelectromechanical systems (NEMS) application, much more testing will be necessary.

Preliminary data indicates that, if a better ordered monolayer, e.g., OTS [48.82] is applied, results are substantially different. In particular, OTS is seen to exhibit significantly more creep. In the release time measurements, instead of observing a clearly identifiable release time after which motion ensued, an

OTS-coated nanotractor instead displayed continuous creep, moving at a much faster rate after the normal force was dropped. Similar behavior has been observed for the OTS-coated nanotractor in static friction measurements [48.55], where compared with the FOTAS-coated devices, OTS-coated devices showed much more creep. Also, measurements of μ_s and μ_d are nearly indistinguishable with OTS.

48.2.4 Stick–Slip Testing Using the Nanotractor

Stick–slip is a widely observed phenomenon, usually occurring at low spring constants and low velocities, as shown in Fig. 48.16a. At higher velocities or spring constants, steady sliding occurs. Very likely, the transition from stick–slip to steady sliding will be an important phenomenon to control and understand in microsystems. The friction test protocol used above is natural for microdevice testing schemes considering that the nanotractor is connected to the substrate via a spring, but is different from that of most friction measurements. More commonly in macroscale friction testing, normal load is held constant while tangential force is raised until slip is measured. Unlike conventional macroscale experiments, in the microscale measurements shown above, the apparent rate of aging β is not fundamental, but rather depends on the normal load ramp-down rate. To make more meaningful comparisons across the length scales, it is necessary to build a microfriction apparatus that is analogous to a conventional stick–slip apparatus.

A prototype system to measure stick–slip is shown in Fig. 48.16b. Here the nanotractor serves as a stepper motor providing a *constant-velocity* puller (stepper motors are also used as pullers in macroscale experiments; see, e.g., [48.62]). A folded-beam spring connects the nanotractor to a friction block. The spring separation is monitored using a vernier scale that extends *inside* the spring. The design of the friction block on the left of Fig. 48.16b is similar to that of a nanotractor clamp in which normal force is provided by electrostatic actuation. In this design, the spring constant is only changed by varying its line width, and hence the spring constant is fixed for a given structure. However, normal force can be easily varied on the same device.

A total travel distance of 300 μm is possible with this apparatus. Hence, a relatively large surface region can be sampled. Figure 48.16b schematically shows a device that has been moved 150 μm to the left in preparation for a test. To make measurements, a high-speed camera is again used. The nanotractor velocity is controlled by the step frequency and average velocities v up to 1 mm/s are possible.

Figure 48.17 shows that a stick–slip to steady sliding bifurcation can be induced by changing the spring constant. In Fig. 48.17a, $k = 0.2$ N/m and stick–slip is observed. In Fig. 48.17b $k = 2.0$ N/m and steady sliding

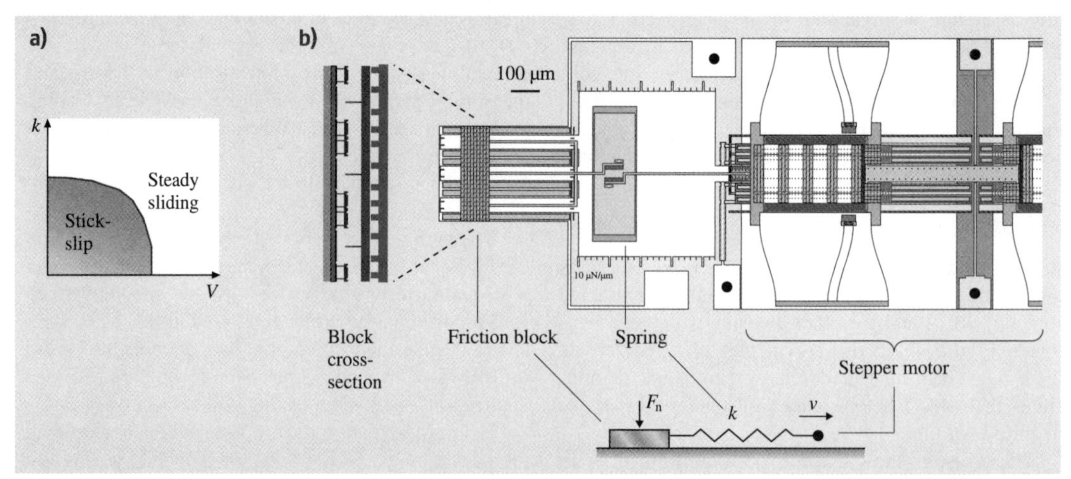

Fig. 48.16 (a) General form of kinetic phase diagram. **(b)** Surface-micromachined stick–slip pull system. The optical image shows a device that has been moved 150 μm to the left to maximize travel distance during the test. Here, the nanotractor becomes the puller. A folded beam with an internal vernier forms the spring of constant k and a structure similar to the nanotractor clamp becomes the friction block

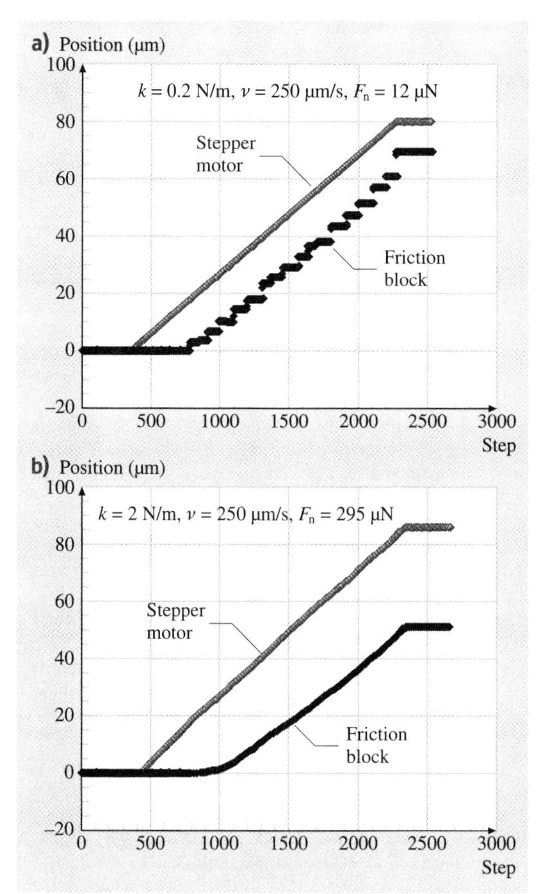

Fig. 48.17a,b Stick–slip results with an FOTAS coating. **(a)** $k = 0.2\,\text{N/m}$, stick–slip; **(b)** $k = 2\,\text{N/m}$, steady sliding

is observed. The same qualitative behavior occurred for each k, independent of velocity from 25 to $250\,\mu\text{m/s}$. These preliminary experiments show that stick–slip exists and can be characterized with this apparatus. Much more work to understand these phenomena at this scale will be needed.

48.2.5 Wear Testing Using the Nanotractor

As a final example of friction phenomena that can be explored with the nanotractor, we discuss results of wear experiments. Normally in the walking process, the clamp voltage is changed from a large value (150 V) to zero before the clamp is moved. Here instead *Flater* et al. [48.56] reduced the clamp voltage to a nonzero value, typically 20 V. In these experiments, the nanotractor walked against a nonlinear load spring, developing a tangential force of $100\,\mu\text{N}$ at each end of the travel. A nanotractor coated by a monolayer of FOTAS [48.72] consistently walked $27 \pm 2\,\mu\text{m}$ for a fixed number of steps. Failure abruptly occurred at 7000 cycles. On the other hand, an oxide-coated nanotractor prepared by critical-point drying walked consistently over only 500 cycles. While it continued to operate, the distance traveled was highly sporadic, ranging from 0 to $25\,\mu\text{m}$. Failure to walk occurred at 5000 cycles. Wear of the exposed polysilicon counterface at the end of the clamp can be imaged by AFM or SEM, as shown in [48.56].

These nanotractor measurements, including the dynamic friction measurement and the static friction measurements, required numerous experimental trials in order to gain the data. An assumption made is that the nanotractor friction does not change over the course of these measurements. The observation that the data is consistent from one experimental trial to the next supports this assumption. The observation that the FOTAS-coated nanotractor consistently walks for up to several thousand cycles even when loaded further supports these methods for taking data. Eventually, however, wear will affect these measurements, and this is an important issue being addressed by researchers [48.83, 84].

48.3 Concluding Remarks

These results on a nominally flat multiasperity polysilicon interface coated by a FOTAS monolayer show that Amontons' law (modified for the adhesion force due to van der Waals forces over the vast noncontacting areas) applies for dynamic friction, but not for static friction. This latter quantity rather depends strongly on loading protocol. We have described nanotribology measurements, microscale measurements, and various contact mechanics models. The *holy grail* in this field is to link these three areas using appropriate physical models. It is also important to apply data from friction measurements directly to actuator performance, and initial progress in this area has been reported [48.85]. Although the results presented here are mainly at the microscale, we imagine that similar test structures and actuator designs will also be of interest for NEMS. Surface roughness, controlled by grain growth and etching processes, may also be similar. Therefore, we expect that much of what is learned from microscale friction may apply to the nanoscale as well. Technologically, progress is continuing on many fronts. This includes the development of new materials such as SiC [48.86] and diamond-like materials [48.87], as well as methods to achieve improved hydrophobicity [48.88], and to reduce wear [48.83]. Taken together, significant technological impact can be expected from this burgeoning area of MEMS and NEMS interface science.

References

48.1 M.S. Rodgers, J.J. Sniegowski: Five-level polysilicon surface micromachine technology: application to complex mechanical systems, Solid-State Sens. Actuator Workshop, Hilton Head Island (1998) pp. 144–149

48.2 J.J. Sniegowski, M.P. de Boer: IC-compatible polysilicon surface micromachining, Annu. Rev. Mater. Sci. **30**, 299–333 (2000)

48.3 T. Akiyama, D. Collard, H. Fujita: Scratch drive actuator with mechanical links for self-assembly of three-dimensional MEMS, J. Microelectromech. Syst. **6**(1), 10–17 (1997)

48.4 E. Sarajlic: Electrostatic microactuators fabricated by vertical trench isolation technology. Ph.D. Thesis (University of Twente, Twente 2005)

48.5 L. Eldada: Advances in telecom and datacom optical components, Opt. Eng. **40**(7), 1165–1178 (2001)

48.6 E. Sarajlic, E. Berenschot, N. Tas, H. Fujita, G. Krijnen, M. Elwenspoek: High performance bidirectional electrostatic inchworm motor fabricated by trench isolation technology, TRANSDUCERS '05, 13th Int. Conf. Solid-State Sens. Actuators Microsyst. (2005) pp. 53–56

48.7 P. Vettiger, G. Cross, M. Despont, U. Drechsler, U. Dürig, B. Gotsmann, W. Häberle, M.A. Lantz, H.E. Rothuizen, R. Stutz, G.K. Binnig: The "Millipede" – nanotechnology entering data storage, IEEE Trans. Nanotechnol. **1**(1), 39–55 (2002)

48.8 J.M. Zara, S.M. Bobbio, S. Goodwin-Johansson, S.W. Smith: Intracardiac ultrasound scanner using a micromachine (MEMS) actuator, IEEE Trans. Ultrason. Ferroelectr. Freq. Control **47**(4), 984–993 (2000)

48.9 T. Akiyama, K. Shono: Controlled stepwise motion in polysilicon microstructures, J. Microelectromech. Syst. **2**(3), 106–110 (1993)

48.10 R.J. Linderman, V.M. Bright: Nanometer precision positioning robots utilizing optimized scratch drive actuators, Sens. Actuators A **91**, 292–300 (2001)

48.11 N. Tas, J. Wissink, L. Sander, T. Lammerink, M. Elwenspoek: Modeling, design and testing of the electrostatic shuffle motor, Sens. Actuators A **70**, 171–178 (1998)

48.12 M.P. de Boer, D.L. Luck, W.R. Ashurst, A.D. Corwin, J.A. Walraven, J.M. Redmond: High-performance surface-micromachined inchworm actuator, J. Microelectromech. Syst. **13**(1), 63–74 (2004)

48.13 E. Sarajlic, E. Berenschot, H. Fujita, G. Krijnen, M. Elwenspoek: Bidirectional electrostatic linear shuffle motor with two degrees of freedom, IEEE Int. Conf. MEMS (2005) pp. 391–394

48.14 G. Amontons: On the resistance originating in machines, Mém. Acad. R. A, 206–222 (1699), in French

48.15 F.P. Bowden, D. Tabor: *Friction and Lubrication of Solids: Part I* (Oxford Univ. Press, Oxford 1950)

48.16 H. Hertz: On the contact of elastic solids, J. Reine Angew. Math. **92**, 156–171 (1882), in German

48.17 K.L. Johnson, K. Kendall, A.D. Roberts: Surface energy and the contact of elastic solids, Proc. R. Soc. Lond. Ser. A **324**, 301–313 (1971)

48.18 B.V. Derjaguin, V.M. Muller, Y.P. Toporov: Effect of contact deformations on the adhesion of particles, J. Colloid Interface Sci. **53**(2), 314–326 (1975)

48.19 D. Tabor: Surface forces and surface interactions, J. Colloid Interface Sci. **58**, 2–13 (1977)

48.20 D. Maugis: Adhesion of spheres: The JKR-DMT transition using a Dugdale model, J. Colloid Interface Sci. **150**, 243–269 (1992)

48.21 J.F. Archard: Elastic deformation and the laws of friction, Proc. R. Soc. Lond. Ser. A **243**, 190–205 (1957)

48.22 R.W. Carpick, N. Agrait, D.F. Ogletree, M. Salmeron: Measurement of interfacial shear (friction) with an ultrahigh vacuum atomic force microscope, J. Vac. Sci. Technol. B **14**, 1289–1295 (1996)

48.23 M. Enachescu, R.J.A. van den Oetelaar, R.W. Carpick, D.F. Ogletree, C.F.J. Flipse, M. Salmeron: Observation of proportionality between friction and contact area at the nanometer scale, Tribol. Lett. **7**, 73–78 (1999)

48.24 M.A. Lantz, S.J. O'Shea, M.E. Welland, K.L. Johnson: Atomic force microscope study of contact area and friction on $NbSe_2$, Phys. Rev. B **55**, 10776–10785 (1997)

48.25 J.F. Archard: Contact and rubbing of flat surfaces, J. Appl. Phys. **24**, 981–988 (1953)

48.26 J.A. Greenwood, J.B.P. Williamson: Contact of nominally flat surfaces, Proc. R. Soc. Lond. Ser. A **295**, 300–319 (1966)

48.27 K.N.G. Fuller, D. Tabor: The effect of surface roughness on the adhesion of elastic solids, Proc. R. Soc. Lond. Ser. A **345**, 327–342 (1975)

48.28 D. Maugis: On the contact and adhesion of rough surfaces, J. Adhes. Sci. Technol. **10**(2), 161–175 (1996)

48.29 J.I. McCool: Comparison of models for the contact of rough surfaces, Wear **107**, 37–60 (1986)

48.30 B.N.J. Persson: On the nature of surface roughness with application to contact mechanics, sealing, rubber friction and adhesion, J. Phys. Condens. Matter **17**(44), 1071–1142 (2005)

48.31 N.R. Tas, C. Gui, M. Elwenspoek: Static friction in elastic adhesion contacts in MEMS, J. Adhes. Sci. Technol. **17**(4), 547–561 (2003)

48.32 W.R. Ashurst, M.P. de Boer, C. Carraro, R. Maboudian: An investigation of sidewall adhesion in MEMS, Appl. Surf. Sci. **212/213**, 735 (2003)

48.33 M.G. Lim, J.C. Chang, D.P. Schultz, R.T. Howe, R.M. White: Polysilicon microstructures to characterize static friction, Proc. IEEE MEMS Workshop, Napa Valley (1990) pp. 82–88

48.34 W.R. Ashurst, Y.J. Jang, L. Magagnin, C. Carraro, M.M. Sung, R. Maboudian: Nanometer-thin titania films with SAM-level stiction and superior wear resistance for reliable MEMS performance, Proc. 17th IEEE Int. Conf. MEMS (2004) pp. 153–156

48.35 W.R. Ashurst, C. Yau, C. Carraro, C. Lee, R.T. Howe, R. Maboudian: Alkene-based monolayer films as anti-stiction coatings for polysilicon MEMS, Sens. Actuators A **9**(3), 239–248 (2001)

48.36 D.C. Senft, M.T. Dugger: Friction and wear in surface-micromachined tribological test devices, Proc. SPIE **3224**, 31–38 (1997)

48.37 W.M. van Spengen, J.W.M. Frenken: The Leiden tribometer: Real time dynamic friction loop measurements with an on-chip tribometer, Tribol. Lett. **28**, 149–156 (2007)

48.38 S.J. Timpe, K. Komvopoulos: The effect of adhesion on the static friction properties of sidewall contact interfaces of microelectromechanical devices, J. Microelectromech. Syst. **15**(6), 1612–1621 (2006)

48.39 S.J. Timpe, K. Komvopoulos: Microdevice for measuring friction and adhesion properties of sidewall contact interfaces of microelectromechanical systems, Rev. Sci. Instrum. **78**, 065106-1–065106-9 (2007)

48.40 R.W. Carpick, D.F. Ogletree, M. Salmeron: A general equation for fitting contact area and friction versus load measurements, J. Colloid Interface Sci. **211**(2), 395–400 (1997)

48.41 G.J. Leggett: Friction force microscopy of self-assembled monolayers: Probing molecular organisation at the nanometre scale, Anal. Chim. Acta **479**, 17–38 (2003)

48.42 Y.L. Chen, J.N. Israelachvili: Effects of ambient conditions on adsorbed surfactant and polymer monolayers, J. Phys. Chem. **96**, 7752–7760 (1992)

48.43 J.B.D. Green, M.T. McDermott, M.D. Porter, L.M. Siperko: Nanometer-scale mapping of chemically distinct domains at well-defined organic interfaces using frictional force microscopy, J. Phys. Chem. B **99**(27), 10960–10965 (1995)

48.44 G.Y. Liu, P. Fenter, C.E.D. Chidsey, D.F. Ogletree, P. Eisenberger, M. Salmeron: An unexpected packing of fluorinated n-alkane thiols on Au(111): A combined atomic-force microscopy and x-ray diffraction study, J. Chem. Phys. **101**(5), 4301–4306 (1994)

48.45 M.T. McDermott, J.B.D. Green, M.D. Porter: Scanning force microscopic exploration of the lubrication capabilities of n-alkanethiolate monolayers chemisorbed at gold: Structural basis of microscopic friction and wear, Langmuir **13**, 2504–2510 (1997)

48.46 E.W. van der Vegte, G. Hadziioannou: Scanning force microscopy with chemical specificity: An extensive study of chemically specific tip-surface interactions and the chemical imaging of surface functional groups, Langmuir **13**(16), 4357–4368 (1997)

48.47 Y. Liu, D.F. Evans, Q. Song, D.W. Grainger: Structure and frictional properties of self-assembled surfactant monolayers, Langmuir **12**(5), 1235–1244 (1996)

48.48 B.D. Beake, G.J. Leggett: Friction and adhesion of mixed self-assembled monolayers studied by chemical force microscopy, Phys. Chem. Chem. Phys. **1**, 3345–3350 (1999)

48.49 H.I. Kim, T. Koini, T.R. Koini, S.S. Perry: Systematic studies of the frictional properties of fluorinated monolayers with atomic force microscopy: Comparison of CF_3- and CH_3-terminated films, Langmuir **13**(26), 7192–7196 (1997)

48.50 E.W. van der Vegte, A. Subbotin, G. Hadziioannou, P.R. Ashton, J.A. Preece: Nanotribological properties of unsymmetrical n-dialkyl sulfide monolayers on gold: Effect of chain length on adhesion, friction, and imaging, Langmuir **16**, 3249–3256 (2000)

48.51 E.D. Reedy: Thin-coating contact mechanics with adhesion, J. Mater. Res. **21**(10), 2660–2668 (2006)

48.52 E.D. Reedy: Contact mechanics for coated spheres that includes the transition from weak to strong adhesion, J. Mater. Res. **22**(9), 2617–2622 (2007)

48.53 W.R. Ashurst, C. Yau, C. Carraro, R. Maboudian, M.T. Dugger: Dichlorodimethylsilane as an anti-stiction monolayer for MEMS: A comparison to the octadecyltrichlosilane self-assembled monolayer, J. Microelectromech. Syst. **10**(1), 41–49 (2001)

48.54 A.D. Corwin, M.P. de Boer: Effect of adhesion on dynamic and static friction in surface micromachining, Appl. Phys. Lett. **84**(13), 2451–2453 (2004)

48.55 A.D. Corwin, M.D. Street, R.W. Carpick, W.R. Ashurst, M.P. de Boer: Pre-sliding tangential deflections can govern the friction of MEMS devices, 2004 ASME/STLE Jt. Int. Tribol. Conf., Long Beach (2004) pp. TRIB2004-64630

48.56 E.E. Flater, A.D. Corwin, M.P. de Boer, M.J. Shaw, R.W. Carpick: In-situ wear studies of surface micromachined interfaces subject to controlled loading, Wear **260**(6), 580–593 (2006)

48.57 C. Carraro, O.W. Yauw, M.M. Sung, R. Maboudian: Observation of three growth mechanisms in self-assembled monolayers, J. Phys. Chem. B **102**(23), 4441–4445 (1998)

48.58 E.E. Flater, W.R. Ashurst, R.W. Carpick: Nanotribology of octadecyltrichlorosilane monolayers and silicon: Self-mated versus unmated interfaces and local packing density effects, Langmuir **23**(18), 9242–9252 (2007)

48.59 B.G. Bush, F.W. DelRio, J. Opatkiewiez, R. Maboudian, C. Carraro: Effect of formation temperature and roughness on surface potential of octadecyltrichlorosilane self-assembled monolayer on silicon surfaces, J. Phys. Chem. A **111**, 12339–12343 (2007)

48.60 J.H. Dieterich: Time-dependent friction in rocks, J. Geophys. Res. **77**(20), 3690–3697 (1972)

48.61 J.H. Dieterich, B. Kilgore: Implications of fault constitutive properties for earthquake prediction, Proc. Natl. Acad. Sci. USA **93**, 3787 (1996)

48.62 F. Heslot, T. Baumberger, B. Perrin, B. Caroli, C. Caroli: Creep, stick-slip and dry-friction dynamics: experiments and a heuristic model, Phys. Rev. E **49**(6), 4973–4988 (1994)

48.63 J.R. Rice, A.L. Ruina: Stability of frictional slipping, J. Appl. Mech. **50**(2), 343–349 (1983)

48.64 A.L. Ruina: Slip instability and state variable friction laws, J. Geophys. Res. **88**(NB12), 359–370 (1983)

48.65 P. Berthoud, T. Baumberger, C. G'Sell, J.-M. Hiver: Physical analysis of the state- and rate-dependent friction law: static friction, Phys. Rev. B **59**(22), 14313–14327 (1999)

48.66 T. Baumberger, C. Caroli, B. Perrin, O. Ronsin: Nonlinear analysis of the stick-slip bifurcation in the creep-controlled regime of dry friction, Phys. Rev. E **51**(5), 4005–4010 (1995)

48.67 Y.F. Lim, K. Chen: Dynamics of dry friction: A numerical investigation, Phys. Rev. E **58**(5), 5637–5642 (1998)

48.68 W.R. Chang, I. Etsion, D.B. Bogy: An elastic-plastic model for the contact of rough surfaces, J. Tribol. Trans. ASME **109**, 257–263 (1987)

48.69 Y. Ning, A.A. Polycarpou: Extracting summit roughness parameters from random Gaussian surfaces accounting for asymmetry of the summit heights, J. Tribol. Trans. ASME **126**(4), 761–766 (2004)

48.70 F.W. DelRio, M.P. de Boer, J.A. Knapp, E.D. Reedy, P.J. Clews, M.L. Dunn: The role of van der Waals forces in adhesion of micromachined surfaces, Nat. Mater. **4**(8), 629–634 (2005)

48.71 D.J. Dickrell, M.T. Dugger, M.A. Hamilton, W.G. Sawyer: Direct contact-area computation for MEMS using real topographic surface data, J. Microelectromech. Syst. **16**(5), 1263–1268 (2007)

48.72 M.G. Hankins, P.J. Resnick, P.J. Clews, T.M. Mayer, D.R. Wheeler, D.M. Tanner, R.A. Plass: Vapor deposition of amino-functionalized self-assembled monolayers on MEMS, Proc. SPIE **4980**, 238–247 (2003)

48.73 Y.H. Cho, A.P. Pisano, R.T. Howe: Viscous damping model for laterally oscillating microstructures, J. Microelectromech. Syst. **3**(2), 81–87 (1994)

48.74 A.D. Corwin, M.P. de Boer: Frictional aging and sliding bifurcation in monolayer-coated micromachines, J. Microelectromech. Syst. **18**(2), 250–262 (2009)

48.75 A.D. Corwin, M.P. de Boer: A linearized method to measure dynamic friction of microdevices, Exp. Mech. **49**(3), 395–401 (2009)

48.76 J.H. Dieterich, B.D. Kilgore: Direct observations of frictional contacts – New insights for state-dependent properties, Pure Appl. Geophys. **143**(1–3), 283–302 (1994)

48.77 L. Bureau, T. Baumberger, C. Caroli: Rheological aging and rejuvenation in solid friction contacts, Eur. Phys. J. E **8**(3), 331–337 (2002)

48.78 J. Rottler, M.O. Robbins: Unified description of aging and rate effects in yield of glassy solids, Phys. Rev. Lett. **95**(22), 225504-1–225504-4 (2005)

48.79 P. Sollich: Rheological constitutive equation for a model of soft glassy materials, Phys. Rev. E **58**(1), 738–759 (1998)

48.80 A.D. Corwin, M.P. de Boer: Frictional deaging in micromachined surfaces, J. Appl. Phys. (2009), submitted

48.81 S.M. Rubinstein, G. Cohen, J. Fineberg: Contact area measurements reveal loading-history dependence of static friction, Phys. Rev. Lett. **96**, 256103 (2006)

48.82 U. Srinivasan, M.R. Houston, R.T. Howe, R. Maboudian: Alkyltrichlorosilane-based self-assembled monolayer films for stiction reduction in silicon micromachines, J. Microelectromech. Syst. **7**(2), 252–260 (1998)

48.83 D.B. Asay, M.T. Dugger, J.A. Ohlhausen, S.H. Kim: Macro- to nanoscale wear prevention via molecular adsorption, Langmuir **24**(1), 155–159 (2008)

48.84 S.S. Mani, J.G. Fleming, J.J. Sniegowski, M.P. de Boer, L.W. Irwin, J.A. Walraven, D.M. Tanner, D.A. LaVan: Selective W for coating and releasing MEMS devices, Mater. Res. Soc. Symp. Proc. **605**, 135–140 (2000)

48.85 M. Patrascu, S. Stramigioli: Stick-slip actuation of electrostatic stepper micropositioners for data storage – The microWalker, Proc. 2005 Int. Conf. MEMS NANO Smart Syst. (2005) pp. 81–86

48.86 M. Mehregany, C.A. Zorman: SiC MEMS: Opportunities and challenges for applications in harsh environments, Thin Solid Films **356**, 518–524 (1999)
48.87 S. Cho, I. Chasiotis, T.A. Friedmann, J.P. Sullivan: Young's modulus, Poisson's ratio and failure properties of tetrahedral amorphous diamond-like carbon for MEMS devices, J. Micromech. Microeng. **15**(4), 728–735 (2005)
48.88 B. Bhushan, M. Nosonovsky, Y.C. Jung: Towards optimization of patterned superhydrophobic surfaces, J. R. Soc. Lond. Interface **4**(15), 643–648 (2007)

49. Failure Mechanisms in MEMS/NEMS Devices

W. Merlijn van Spengen, Robert Modliñski, Robert Puers, Anne Jourdain

The commercialization of MEMS/NEMS devices is proceeding slower than expected, because the reliability problems of microscopic components differ from macroscopically known behavior. In this chapter, we provide an overview of the state of the art in MEMS/NEMS reliability. We discuss the specific, MEMS-related problems caused by stiction due to surface forces and electric charge. Materials issues such as creep and fatigue are treated as well. Nanoscale wear is covered briefly. MEMS packaging is also discussed, because the reliability of MEMS/NEMS components critically depends on the available protection from the environment.

49.1 **Failure Modes and Failure Mechanisms** ... 1762
49.2 **Stiction and Charge-Related Failure Mechanisms** ... 1763
49.2.1 Stiction Due to Surface Forces ... 1763
49.2.2 Stiction Due to Electrostatic Attraction ... 1766
49.3 **Creep, Fatigue, Wear, and Packaging-Related Failures** ... 1769
49.3.1 Creep ... 1769
49.3.2 Fatigue ... 1773
49.3.3 Wear ... 1775
49.3.4 Packaging ... 1776
49.4 **Conclusions** ... 1779
References ... 1779

With the rapid spread of micro-electromechanical systems (MEMS) technology in the 1990s, it has become clear that the reliability of microscopic mechanical devices is a large concern [49.1]. It is one of the main reasons that much of the micromechanical technology available all over the world in laboratories has not been commercialized yet. The route from a working prototype to a commercial MEMS component should not be underestimated: the pioneering effort of Texas Instruments to bring the digital micromirror device (DMD) MEMS chip to the market took more than 10 years after its invention [49.2]. The same holds for gyroscopes and accelerometers: many years of hard work were spent on the conversion of laboratory prototypes to commercial products [49.3].

The failure mechanisms of MEMS/NEMS (in the rest of this chapter we will use MEMS as a generic name that also includes devices with nanometer-scale features, which are increasingly being referred to as nanoelectromechanical systems (NEMS)) components are widely differing from those of their macroscopic counterparts, and this is the main reason that this trajectory takes so long. Many MEMS designers still do not take into account the implications that the scale difference has on reliability when designing new devices, although the paradigmatic approach of *design for reliability*, i. e. taking into account reliability issues from the earliest design stage, is gaining more acceptance. A second point is that, even if people are willing to take into account all the available information on the failure modes and mechanisms of MEMS, they soon discover that not so much is known at all. Still, over the last few years, the information available on MEMS failure mechanisms has been steadily growing [49.4]. At present, we are not yet in a position to predict quantitatively when or where most of the different failure mechanisms will occur, but we can describe them in some detail and give rules of thumb to avoid many of them.

In this chapter we will introduce the basic MEMS failure mechanisms and point out when they are important and in which respects they differ from macroscopic failure mechanisms. We will study the most important of them in some detail and give recommendations for procedures that are known to counteract them.

49.1 Failure Modes and Failure Mechanisms

We have to start with a clear distinction between failure modes and failure mechanisms. The failure mode is the apparent failure. It might be that e.g. a MEMS beam is stuck to the substrate. If this causes a device to malfunction, the failure mode of the device is a stuck beam. This failure mode may be caused by different failure mechanisms, such as stiction of the beam (stiction is a generic term for surfaces being stuck due to surface forces), it may be a fatigue fracture failure or caused by an overload condition, and so on. The failure mechanism is hence the physical cause of the failure mode. In Fig. 49.1 the failure mode is broken beams, but the failure mechanism is rough handling, causing a mechanical overload condition.

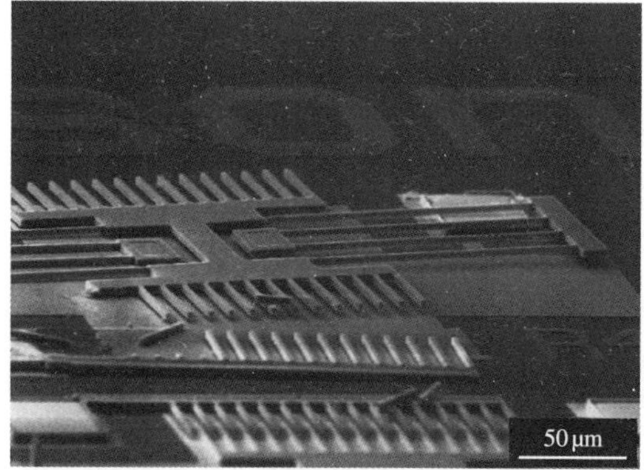

Fig. 49.1 The failure mode of this resonator is broken beams; the failure mechanism is rough handling, causing an overload stress in the beams and support

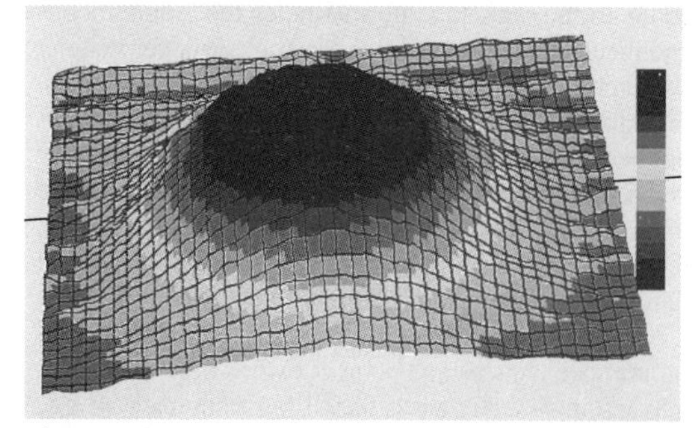

Fig. 49.2 Example of optical assessment of dynamic MEMS behavior: laser television (TV) holography measurement image of the first resonance mode of a MEMS pressure sensor membrane (after [49.5])

This distinction is important because often the failure mode is reported when something is wrong, while the problem can only be accurately addressed by investigating the failure mechanism. In Table 49.1, the most commonly encountered failure mechanisms are given (most of them taken from [49.6]).

Dedicated equipment to assess failure mechanisms is definitely needed, but only a few set ups have been reported to date. Sandia National Laboratories is famous for their Sandia high-volume measurement of micromachine reliability (SHiMMeR) system, where they can test a significant number of rotating micromachines simultaneously under various environmental conditions [49.7]. The operation of the micromotors is monitored with advanced software tools [49.8]. Inter-University Microelectronics Center (IMEC) has a dedicated system to assess radio-frequency (RF)

Table 49.1 Common MEMS failure mechanisms

Fracture	Overload fracture Fatigue fracture
Creep	Applied stress Intrinsic stress Thermal stress
Stiction	Capillary forces Van der Waals molecular forces Electrostatic forces Solid bridging
Electromigration	
Wear	Adhesive Abrasive Corrosive
Degradation of dielectrics	Leakage Charging Breakdown
Delamination	
Contamination	
Pitting of contacting surfaces	
Electrostatic discharge (ESD)	

MEMS switch reliability, where environmental conditions can be altered with in situ optical access and various stimuli can be applied [49.5]. Interferometry- and vibrometry-based optical systems of various kinds are in use to monitor the dynamic behavior of MEMS devices [49.9–12] (Fig. 49.2). Some commercial laboratories also have their dedicated systems but do not disclose their exact set ups.

Other equipment that is used for failure mechanism assessment has not been developed exclusively for MEMS, but the MEMS reliability community is relying heavily on them: scanning electron microscopy (SEM), focused ion beam (FIB) cross-sectioning, atomic force microscopy (AFM), optical microscopy, and various electronic actuation and detection systems are the most widespread among them.

49.2 Stiction and Charge-Related Failure Mechanisms

49.2.1 Stiction Due to Surface Forces

One of the most important problems in MEMS/NEMS devices is stiction, i. e. the unintentional sticking of contacting surfaces (Fig. 49.3). Because forces intentionally generated in MEMS by e.g. electrostatic actuation are generally very small, surface forces can dominate the device behavior. The fact that MEMS surfaces are so smooth that a large part of the surfaces is within the range of the atomic forces of the other surface aggravates the matter further. The most important surfaces forces commonly listed are: forces due to capillary condensation, van der Waals molecular forces, and chemical and hydrogen bonds between the surfaces.

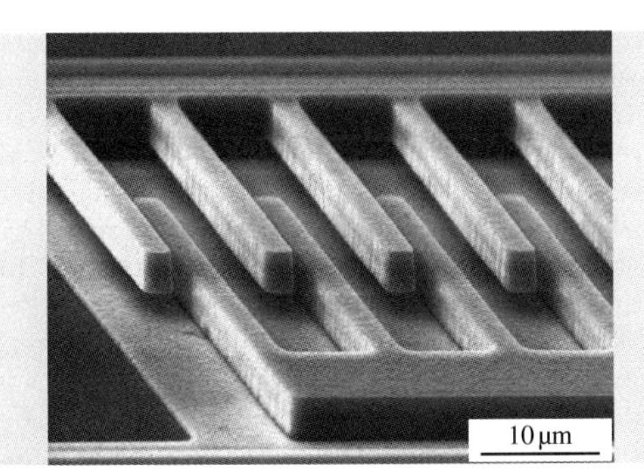

Fig. 49.3 Stiction failure of a comb drive

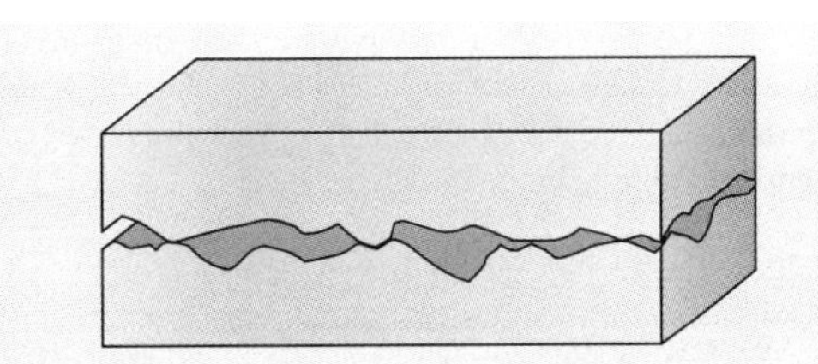

Fig. 49.4 Two rough surfaces in contact

Surfaces tend to stick together when they are dried after the release etch, and a good overview of what happens is given by *Mastrangelo* and *Hsu* [49.14, 15]. From a reliability point of view, in-use stiction is more important.

The capillary condensation, van der Waals forces, and bonding force effects mentioned have in common that they are true interface forces. The force depends in this case on the situation at the interface, and not on effects far away from it. All these forces have in common that they are strong for surfaces very close together but fall off relatively fast when the distance is larger than a few nanometers.

As a result, surface roughness is an important parameter governing the stiction force, because the surface roughness of MEMS devices is commonly found to be of the same order of magnitude as the range of the surface forces (Fig. 49.4). This was demonstrated for the first time by *De Boer* et al. [49.13]. They monitored one-

Table 49.2 Deflection and surface interaction energy of S- and arc-shaped beams (after [49.13])

	S-shaped beam	Arc-shaped beam
Deflection $u(x)$	$u(x)=h\left(\frac{x}{s}\right)^2\left(3-2\frac{x}{s}\right)$	$u(x)=h\left(\frac{x}{s}\right)^2\left(3-\frac{x}{s}\right)$
Surface interaction energy Γ [a]	$\Gamma=\frac{3}{2}E\frac{t^3h^2}{s^4}$	$\Gamma=\frac{3}{8}E\frac{t^3h^2}{s^4}$

[a] E = Young's modulus of the beam material, the other parameters are given in Fig. 49.6

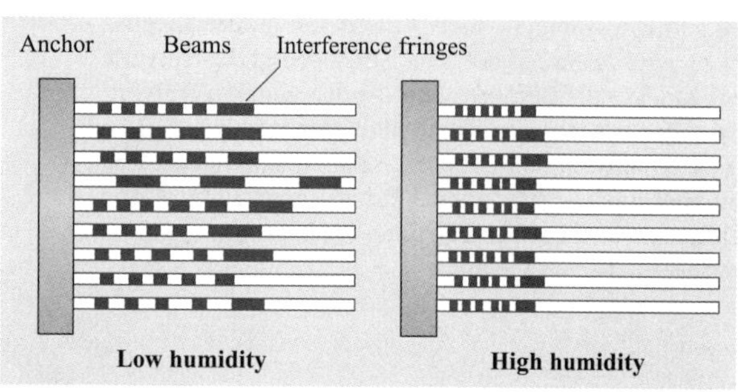

Fig. 49.5 Schematic top view of the measurement of adhesion as a function of relative humidity using interference imaging by *De Boer* et al. [49.13]. The *dark parts* are fringes created with an interferometric objective. The fringe pattern reveals where the beams are bending, and where they are flat, because fringes are only observed where the beams are not running perfectly in plane

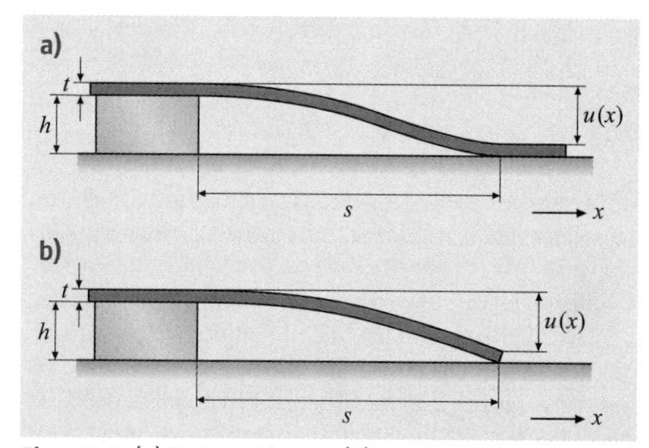

Fig. 49.6 (**a**) S-shaped beam; (**b**) arc-shaped beam

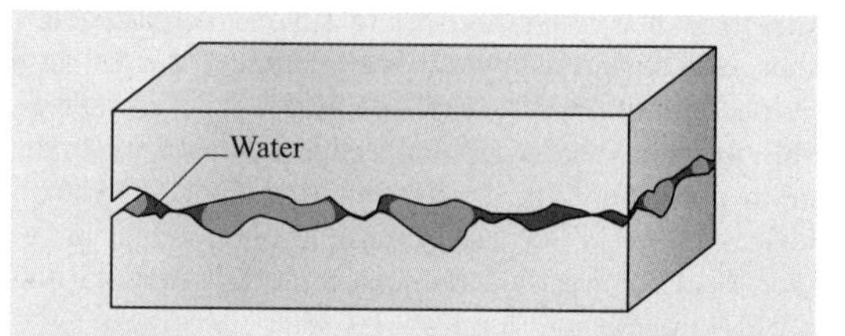

Fig. 49.7 Capillary condensation occurs on all places where the two contacting surfaces are closer together than a characteristic distance

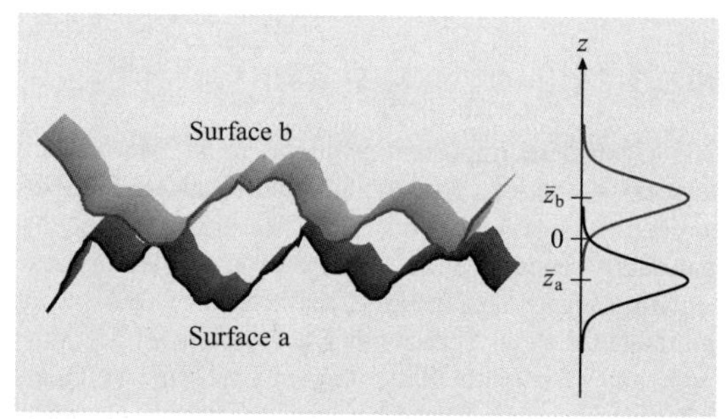

Fig. 49.8 The two surface-height distribution functions give rise to a distance distribution function $h(z)$ describing how large the part of the surface is that has an interaction with the other surface at a certain distance

side-clamped free-standing cantilever MEMS structures that were stuck over a considerable part of their length, using an interferometer (Fig. 49.5). Fringes in the image indicated which part of the beams was free-standing and which part was stuck. Because the restoring spring force of the beams was known, the length over which the beams were not stuck gave a measure for the surface interaction energy (Fig. 49.6 and Table 49.2).

We have to distinguish for this calculation between beams that adhere only at the tip (*arc-shaped*) and beams that adhere over a considerable length (*S-shaped*). It was argued by De Boer and coworkers that the S-shaped beam gives a more precise measure for the surface interaction energy because the arc-shaped beam introduces a large statistical error; only the properties of the surface at the very small contacting tip produce the measured result. Arc-shaped beams can very easily be used to find the surface interaction energy to first order however, because we only have to look at which beams are stuck and which beams are free-standing.

De Boer et al. [49.13] found that the surface interaction energy varied considerably with the relative humidity and attributed this effect to capillary condensation. Small liquid bridges are supposed to form when two hydrophilic surfaces are in close contact (Fig. 49.7). This effect had already been reported by *Bowden* and *Tabor* in 1950 [49.16] with a macroscopic measuring apparatus, but only with the development of MEMS technology has capillary condensation become a big technological issue. Because only the surface area that is closer to the other surface than a characteristic distance will be bridged by capillary condensed water, generally only part of the surface area is in contact in this sense.

Maboudian and *Howe* [49.17], *Legtenberg* et al. [49.18] and *Tas* et al. [49.19] provided the first quantitative models of stiction, making use the work of *Israelachvili* [49.20] on more generic systems. They described the surface forces as a function of the distance between two flat surfaces. The relation between surface interaction energy and distance for capillary forces (a function of humidity) and of van der Waals molecular forces (a function of the Hamaker constant of the molecules at the surface) was presented with very general applicability.

The roughness of the surfaces (some parts are closer together and contribute more to the stiction force than others) was discussed briefly by *Komvopoulos* et al. [49.21] and more quantitatively by *van Spengen* et al. [49.22]. The procedure boils down to the determination of a distance distribution function $h(z)$, giving the amount of surface at every distance z from the other surface (Fig. 49.8). The total surface interaction energy Γ_i due to a certain i-th force $e_i(z)$, e.g. the van der Waals force, is then given by

$$\Gamma_i = \int_0^\infty e_i(z)h(z)\,\mathrm{d}z\,. \tag{49.1}$$

This principle is generally applicable. As long as the forces do not influence each other, they will be reasonably well described by (49.1) and can be added together.

It is not easy to predict the distance distribution function $h(z)$, as the two surfaces may deform each other at the contacting points, giving rise to both plastic and elastic deformation. If we know the height distribution of the contacting surfaces independently, a rough estimate can be obtained by assuming only plastic deformation. This gives rise to the *bearing area* representation [49.22], from which the distance distribution can be obtained. Unfortunately, the surface interaction energy value obtained in this way is expected to be too low, as elastic deformation is expected to bring larger parts of the surfaces closer together than in the fully plastic case. Purely elastic deformation can be treated by the Greenwood–Williamson model [49.23], but has been found to be useful from an experimental point of view only for very soft, elastic materials like rubber. The technologically more interesting cases (silicon on silicon, metals etc.) are best described with a mixed plastic/elastic model, but unfortunately only numerical simulations can be employed to describe this [49.24, 25].

A more experimental approach to characterize the rough contact between two MEMS surfaces is followed by *Kogut* et al. [49.26], who use electrical current measurements through the contacting points to estimate the real contact area.

Whether the resulting surface interaction energies from the different force contributions result in a stiction failure or not depends on the restoring force. The peel number N_p (first coined by *Mastrangelo* and *Hsu* [49.15] and very well explained by *Zhao* [49.27]) can be used to find out whether a structure will suffer from stiction. For $N_\mathrm{p} < 1$ the surfaces will adhere, while for $N_\mathrm{p} > 1$ the surfaces will separate again if they are brought into contact.

The most cumbersome stiction force is that caused by capillary condensation. Water will only capillary condense onto hydrophilic surfaces, so a hydrophobic coating can be applied to prevent this. These layers are not unlike the boundary lubricants that have been used for decades to lower friction. Different ways of creating hydrophobic coatings have been applied but there is no industry standard yet. Antistiction coatings are generally made by either plasma deposition or self-assembled monolayer (SAM) deposition. The standard film quality test method is to bring a small water droplet to the surface and monitor the contact angle.

The gas used for the plasma deposition is usually a low-molecular-weight fluorinated organic compound, such as CHF_3 [49.28] or C_4F_{10} [49.29]. The MEMS device is bombarded with all kinds of free radicals and other reactive species generated from these molecules and forms a well-adhered water-repelling coating on the surfaces of the device. A big problem with plasma deposition is obtaining conformal coatings. Tricks such as having a Faraday cage inside the chamber are advantageous [49.29], but even then the conformity is usually not better than 2 : 1 for the top and bottom surfaces of a MEMS device.

The durability of plasma-deposited coatings is relatively good, although they tend to come off in boiling water. *Man* et al. [49.29] tested their films for water contact-angle changes at elevated temperatures. By determining the lifetime at different temperatures and using an Arrhenius relation for the mean time to failure (MTTF)

$$\mathrm{MTTF} \propto \exp\left(\frac{E_\mathrm{a}}{k_\mathrm{B}T}\right), \tag{49.2}$$

they calculated a lifetime of more than 10 years at 150 °C. Knowing the Boltzmann constant k_B, one can obtain the activation energy E_a by performing experiments at different temperatures T. Impact wear was also tested.

To overcome the conformity problems associated with plasma deposition, and also because very thin layers are possible in this way, SAM coatings have been investigated since the middle of the 1990s. *Srinivasan* et al. [49.30] reported extensively on the subject. Silicon microstructures are released in HF, and while still in the liquid phase, the HF solution is replaced by a trichlorosilane solu-

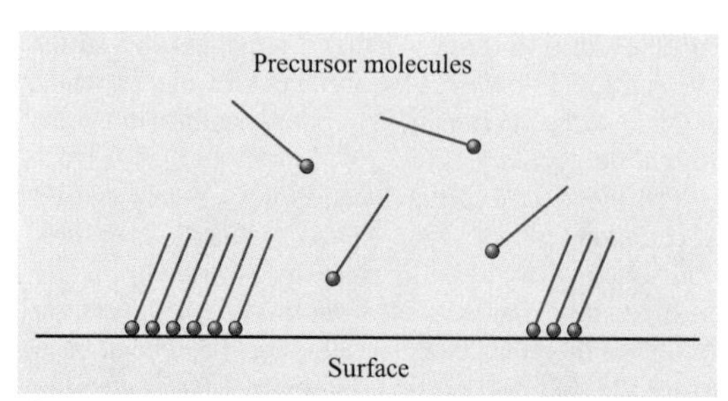

Fig. 49.9 SAM formation on a MEMS surface

tion. The most common are octadecyltrichlorosilane (OTS, $CH_3(CH_2)_{17}SiCl_3$) and 1H,1H,2H,2H-perfluorodecyltrichlorosilane (FDTS, $CF_3(CF_2)_7(CH_2)$ $SiCl_3$). Figure 49.9 shows how such a film is formed on the surface of a MEMS device. A processing problem is that the reaction depends on trace amounts of water in the reaction environment. If the processing conditions are not extremely well controlled, micelles are formed as well as highly interconnected surface films that are easily peeled off [49.31].

Thermal stability tests show that, in air, the fluorocarbon films are able to stand 400 °C while the hydrocarbon films will degrade above 100 °C. In nitrogen no difference was observed [49.31]. *Kim* et al. [49.32] have suggested that the use of dimethyldichlorosilane (DDMS, $(CH_3)_2SiCl_2$) results in a process that is less sensitive to temperature, humidity and coating time. The films thus created withstood temperatures of above 450 °C without degradation.

De Boer et al. [49.13] found that FDTS films were susceptible to *water island formation* in a 90% relative humidity environment after only a few hours, while OTS films did not show such a dramatic effect. Whether this was a real difference or merely an effect of a difference in surface film quality has not been elucidated.

Another method besides antistiction coatings to reduce stiction problems is roughening the surfaces so that only small parts of the surfaces are in intimate contact. *Maboudian* et al. [49.22] show how this can be accomplished with different etchants (this also decreases van der Waals contributions). An absolutely water-free environment will also prevent capillary condensation problems altogether. The hermeticity of packages to accomplish this is discussed in the packaging section of this chapter (Sect. 49.3.4). The best solution to stiction due to surface interaction energies is the use of very stiff structures and high forces, which will separate the surfaces even under high adhesive contact forces. A typical higher limit for the surface interaction energy due to capillary forces is $0.1\,J/m^2$ [49.17], which corresponds to water everywhere between the surfaces. Surface interaction energies in rough devices and with antistiction coatings are typically 10–1000 times lower. If these numbers are taken into account, finite element modeling of the structures before they are produced can be used to predict whether stiction due to surface forces will be a reliability threat or not.

49.2.2 Stiction Due to Electrostatic Attraction

Stiction due to dielectric charging is different from the previously discussed forces in the sense that it is not a true interface force. Charges may be present at the interface but may equally well reside in the bulk of a dielectric material. Electrostatic forces are very important on the microscale and hence an in-depth investigation of charge effects is important.

Charges may be generated in microsystems in different ways. It has been found that in devices with rubbing surfaces tribocharging may occur [49.33]. A rotating silicon micromotor was tested until it remained stuck due to the accumulated charge. Irradiation with a focused ion beam (FIB) neutralized the charges, after which the device resumed operation. Another situation where charge may be generated is where sparks

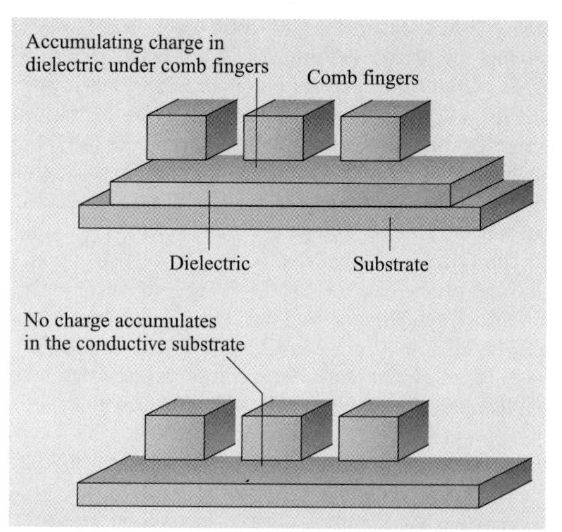

Fig. 49.10 Ionizing radiation can cause charging of the dielectric under the cantilevers (that are part of the moving mass) of an accelerometer and hence cause drift and in extreme cases stiction of the moving part

in a small gap spray charge onto a surface [49.34]. Contact potentials of materials are usually $< 0.5\,V$ and are not expected to give any trouble in normal applications.

Charges generated by irradiation with ionizing radiation constitute a considerable challenge in MEMS devices. Accelerometers subjected to nuclear radiation were found to drift with increased dose because the dielectric under the moving mass of such a device was charged [49.35, 36]. Accelerometers without a dielectric under the moving mass were found to be much less susceptible, because no charge can accumulate (Fig. 49.10). One can imagine that excessive radiation would cause the moving mass to snap in and constitute a stiction failure, but this has not yet been reported.

The largest problems with stiction due to electrostatic attraction are found in devices where a large voltage is applied across a thin dielectric, such as in RF MEMS switches [49.37]. The high field causes charge injection into the dielectric by Schottky- or Poole–Frenkel-type conduction and is experimentally found to depend exponentially on the actuation voltage [49.38]. The situation resembles that of the high-k-dielectric gate oxides of the newest generations of metal–oxide–semiconductor (MOS) transistors, where similar charging effects threaten the stability of the threshold voltage [49.39].

The exact charging properties of a dielectric thin film depend on the deposition conditions. A *leaky* dielectric with a low resistivity may have many trapping sites, but can still be good from a technological point of view if the charge can flow away easily [49.40]. The films have to be empirically optimized because no consistent theory to describe good films has been developed yet.

Much more is known about how parasitic charges, when they are present, influence device behavior. For a free-sanding structure with small-scale deflections, *Wibbeler* et al. [49.34] have presented a theory that describes the deflection as a function of the accumulated charge. *Reid* [49.41] has shown finite element modeling (FEM) simulations for large deflections, pull-in and pull-out effects and the effect of parasitic charge on them.

The most comprehensive efforts, which are very useful from the point of view of failure due to stiction, are the treatises of *van Spengen* et al. [49.42] for uniform charge densities in the dielectric and *Rottenberg* et al. [49.43] for inhomogeneous charge densities. These two papers consider RF MEMS switches, but the principles are generic.

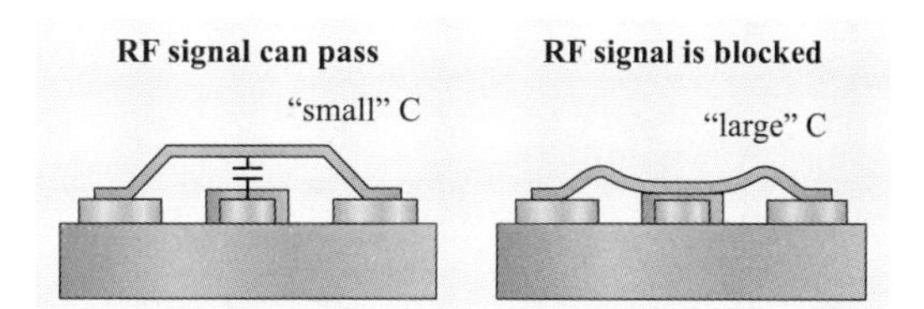

Fig. 49.11 RF MEMS switch cross section

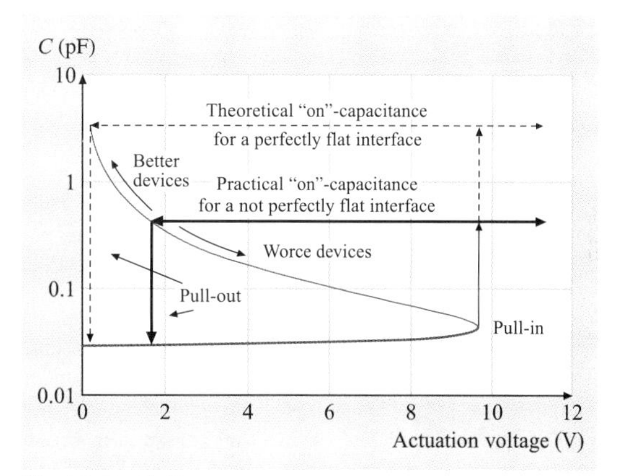

Fig. 49.12 C–V curve

In these models, elementary electrostatics is used to describe the peculiar dynamic effects in MEMS with a structure like that in Fig. 49.11. A free-standing structure (the *bridge*) is located above a dielectric with a lower conductor. The whole assembly behaves as a capacitor. When a voltage is applied between the bridge conductor and the lower conductor, the bridge is attracted. If the bridge is moved over more than one third of the gap, the system becomes unstable and the bridge lands on the dielectric (pull-in). At this point the capacitance of the system becomes much higher and this is the reason that it can be used to affect RF electronic signals. A much lower voltage is required to keep it there, so we have to lower the actuation voltage considerably to let it *pull out* again. This hysteresis is well known from RF MEMS C–V curves (Fig. 49.12).

Both positive and negative actuation voltages can cause the bridge to pull in (the C–V curve is symmetric for positive and negative voltages). For a uniform parasitic charge, the curve starts to shift. Negative charge causes the curve to shift to the right, while positive charge shifts the whole curve to the left on the voltage axis.

The effect of positive charge is easily understood: when the curve shifts so much to the left that the pull-

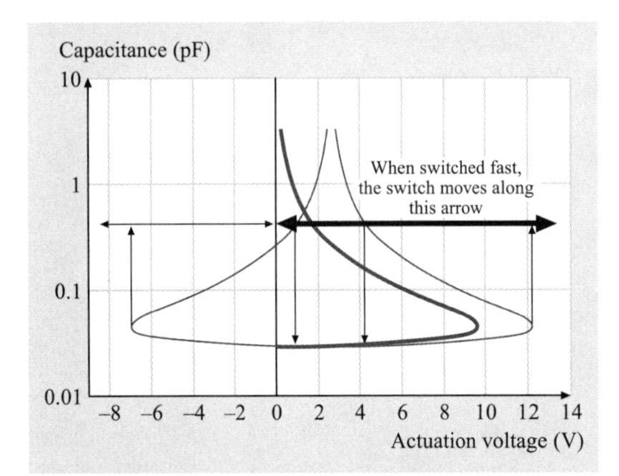

Fig. 49.13 With square-wave actuation, after the negative pull-out voltage has crossed 0 V due to a parasitic charge, the device moves only along the *bold arrow* and remains stuck ◀

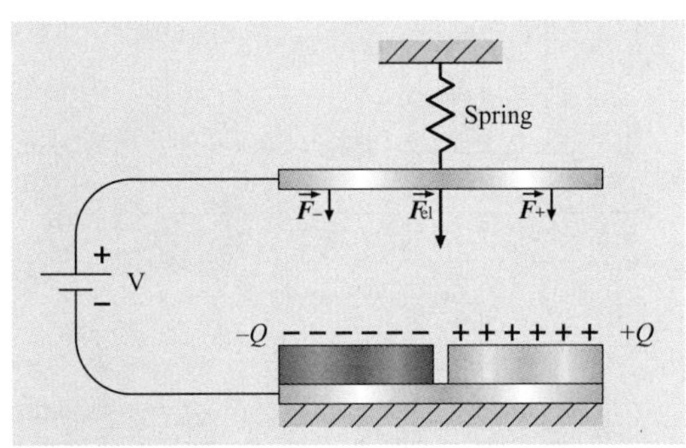

Fig. 49.14 A nonuniform charge can generate attraction of the upper conductor even when the net charge is zero (after [49.43])

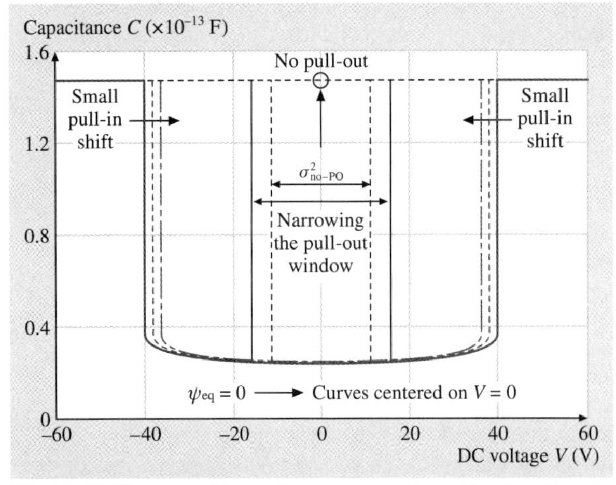

Fig. 49.15 The nonuniform charge causes the pull-out voltage to move to 0 V; in extreme cases the pull-out voltage disappears (after [49.43])

out voltage crosses the 0 V line, the bridge will remain stuck even if the actuation voltage is removed. A negative charge results in a more tricky behavior. For small charge values, the pull-in and pull-out voltages will shift slightly upward so no stiction is expected. If too much charge accumulates, the pull out of the negative-going actuation will shift through 0 V (Fig. 49.13). For slow, positive-going voltages, this makes no difference. When we increase the voltage on an initially free-standing bridge, it will pull in at the shifted positive pull-in voltage, and return to its upward position when the voltage is again slowly decreased to below the pull-out voltage. Only when we go negative in actuation voltage will the bridge be pulled in again at the shifted negative pull-in voltage.

But when we change the actuation voltage quickly, as is often the case when devices are actuated, e.g. by a square wave, the behavior is completely different. We start in the same condition as before, with the bridge up, and increase the voltage quickly to above the pull-in voltage, the bridge will be pulled in after a characteristic time that is related to the mechanical resonance frequency of the device. When we quickly reduce the voltage to 0 V, the bridge stays behind again and is still down when the voltage is 0 V. However, now we are in a completely different situation. Because the pull-out voltage of the negative branch of the curve is positive, the bridge will remain stuck. The bridge does not know whether it came from a negative voltage or from a positive voltage. For a negative voltage it would only pull out above 0 V at the negative-branch pull-out voltage, so it will remain stuck. In this way both positive and negative charges can cause a MEMS device to remain stuck. It is even possible to calculate the approximate magnitude of the charge that is required to cause stiction in either case. The important case in which the negative pull-out voltage crosses zero and causes a device failure has also been verified experimentally [49.42].

Not only charge injection due to high voltages across the dielectric may cause this kind of reliability problems to emerge; ionizing radiation can also play a similar role. It has to be kept in mind however that, if the bridge or other moving structure does not intentionally or accidentally hit the surface below, stiction can only take place when the charge causes a shift that is so large that the much higher pull-in voltage crosses 0 V from either side.

The previous explanation is valid only for uniform charge distributions (either bulk or surface charge). *Rottenberg* et al. [49.43] have extended the theory to deal with nonuniform charge distributions that give rise to effects that have also been experimentally observed. The nonuniform distribution can cause a nonzero force even when the net charge on the dielectric is zero (Fig. 49.14). This causes a narrowing of the C–V curve so that the positive and negative pull-out and pull-in are closer together (Fig. 49.15).

In conclusion, even relatively small charge values of both polarities that are present in or on a dielectric may affect the device reliability when the moving part intentionally or accidentally hits the lower surface. If this is not the case, stiction can only occur when the C–V curve is shifted so much that the pull-in voltage crosses 0 V.

Reduction of the sensitivity for charging phenomena in MEMS devices can be accomplished in different ways. The most obvious way is the reduction of charging by optimizing the dielectrics for low charging. This can be done either by making a very-high-quality dielectric with few trapping sites, or by increasing the trap density but making sure that the average time a charge remains in a trap is very short (the *leaky dielectric*). Another option is to design the dielectrics that are prone to charging out of the device concept. In RF MEMS this can be done with separate actuation pads where no dielectric is located so that no voltage acts across the high-k dielectric. In the accelerometer example of Fig. 49.10, this was done by removing the dielectric altogether. The third option, but probably the least satisfactory because of its added complexity and device cost, is to use advanced actuation schemes to reduce charging. *Goldsmith* et al. [49.38] use an advanced actuation waveform with a high initial voltage to pull the bridge down and a lower *holding* voltage to keep it down, so that charging is reduced. Polarity reversal is also used to sweep out the charges that are accumulating [49.44]. Both of these are useful with actuation signals that are known in advance and do not depend on the particular application in mind.

49.3 Creep, Fatigue, Wear, and Packaging-Related Failures

49.3.1 Creep

Creep is the permanent and irreversible deformation of a ductile material occurring below the yield strength. Its importance depends on temperature, mechanical stress, time, and material composition. It is easy to think of a solid metal as having a well-defined yield strength below which it does not flow and above which flow is rapid. This is unfortunately true only at absolute zero. Above this temperature, plastic flow is a dynamic process, where the strength of solids depends on strain and temperature. The irreversible time-dependent part of a strain ε is called creep.

The strain rate $\partial\varepsilon/\partial t$ (elongation speed) is extremely temperature-sensitive. It can often be approximated by a rate equation

$$\frac{\partial\varepsilon}{\partial t} = A\exp\left(-\frac{q}{k_\mathrm{B}T}\right), \qquad (49.3)$$

where A and q are the fit parameters, k_B is Boltzmann's constant and T is the absolute temperature.

Thermal vibrations aid the applied stress in overcoming specific barriers (e.g. obstacles) to induce plastic flow in the metal. Slip, climb, shear on grain boundaries, and vacancy diffusion inside the grain (the so-called Nabarro–Herring effect) are the mechanisms responsible for deformation in metals [49.46]. The last of these is significant only at very high temperatures and under very small stresses. The normal operating temperatures in most practical engineering applications are too low and usually the operating stresses are too

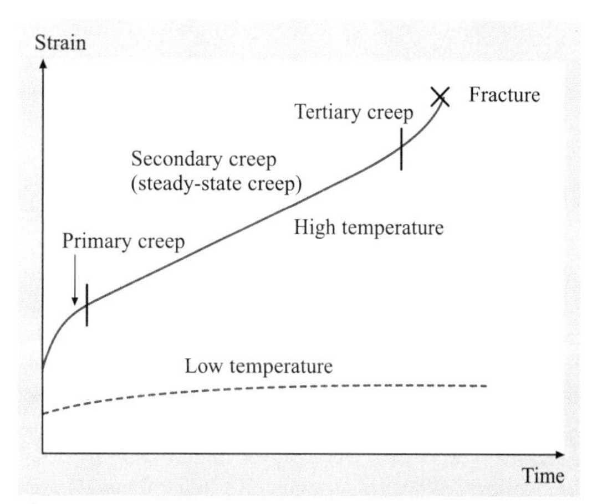

Fig. 49.16 Low-temperature and high-temperature creep under a constant stress (after [49.45])

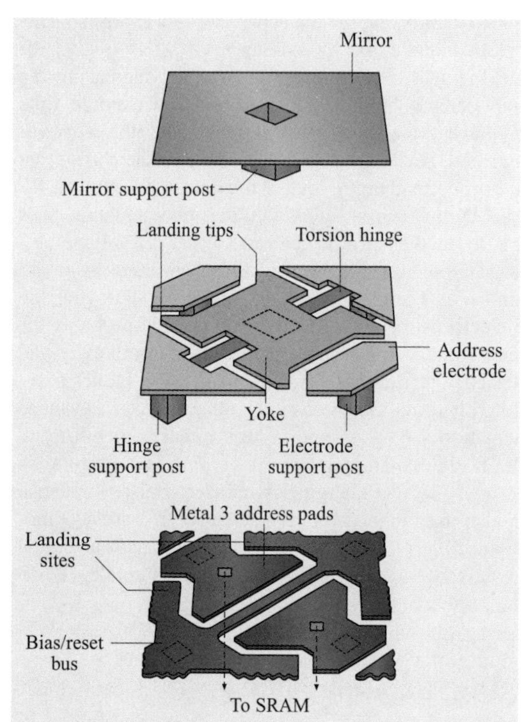

Fig. 49.17 Digital micromirror device for projection displays (© Texas Instruments)

high for the Nabarro–Herring mechanism to occur, so we can conclude that creep deformation in metals depends primarily on dislocation movement (slip, climb and grain-boundary shearing).

Three creep stages can be distinguished: primary, secondary (so-called steady-state creep) and tertiary creep, which ends up with fracture (Fig. 49.16). Especially devices made out of low-melting-point metals are sensitive to creep. This is the case because the melting temperature of a material has a major influence on its creep behavior. There exists a threshold temperature, which divides the deformation of a material by creep into two parts: high-temperature and low-temperature creep (Fig. 49.17). In the regime of high temperature, creep is a rather fast deformation and depends strongly on temperature, whereas in the low-temperature regime, the creep rate decreases logarithmically with time. It is commonly accepted that this threshold temperature T_c is given by $T_c = 0.5T_m$, where T_m is the absolute melting temperature, although each material must be evaluated separately [49.45]. For instance, T_c is around 200 °C for aluminum and its alloys and 315 °C for titanium alloys.

Shape changes caused by creep can be undesirable and represent a limiting factor in the lifetime of a device. For instance, blades on the spinning rotors in turbine engines are endangered by creep. They slowly grow in length during operation and must be replaced before they touch the housing.

Some MEMS, like for instance RF MEMS switches can be affected by creep, and the creep properties of thin metal films are not necessarily the same as those of the bulk material.

Creep in MEMS

Creep is well characterized at the macroscopic level (usually under constant-stress/constant-temperature conditions), but has not been studied extensively in microsystems and therefore it can represent a reliability problem in MEMS. Many microdevices are fabricated by adopting low-melting-point metals (in particular aluminum). In most recent research work, long-term or accelerated reliability tests (i. e. creep tests) seldom attract much attention. In fact, feasibility studies are rarely conducted when starting research work (e.g. academic research). However, reliability became an issue in applications, where durability and reproducibility are a key element for commercial success. Creep can form a danger for commercial devices fabricated from low-melting-point metals, especially when operating at higher temperatures and stresses. Therefore, at an early stage of the design of metal MEMS, creep should be taken into account.

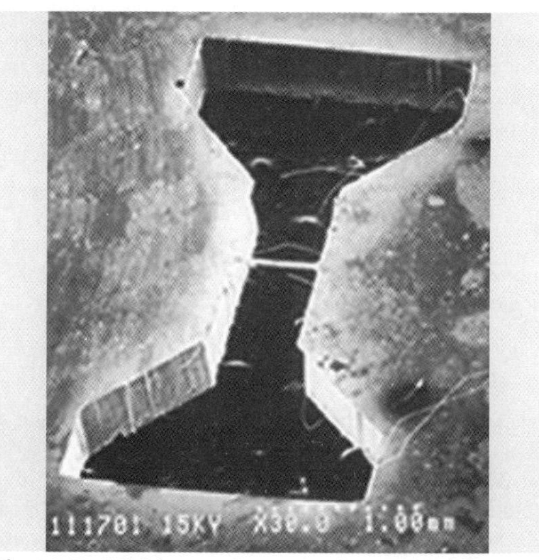

Fig. 49.18 SEM image of a test specimen (after [49.47])

An important source of information on creep in MEMS is the work done by Texas Instruments on the digital micromirror device (DMD). *Van Kassel* et al. [49.48] have shown that creep was a problem in their micromirrors. This reliability problem was termed the hinge memory effect. Figure 49.18 depicts the sketch of a DMD. A single mirror is mounted on hinges and electrostatic attraction between the mirror structure and a semiconductor memory cell below is responsible for a ±10° tilt. Hinge memory occurs when a mirror is operated in the same direction for a long period of time (high duty cycle). A residual tilt when the voltage is removed appears over time. The lifetime of the mirror is reached when the residual tilt becomes excessive (dead pixels). *Van Kassel* et al. [49.48] related this effect to creep of the hinge material. The lifetime was increased by selecting more creep-resistant materials and by improving the fabrication process. The hinge memory effect was thermally activated. At 65 °C and high duty cycle, mirrors could operate for about 5000 h continuously, while at 45 °C they remained functional for more than 100 000 h even at high duty cycles.

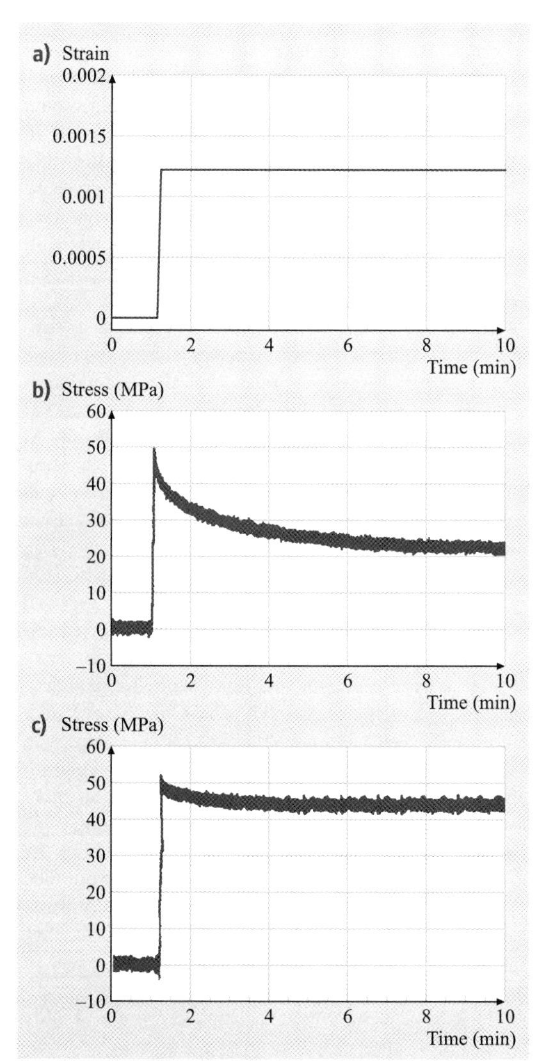

Fig. 49.19a–c Strain change with time (**a**) and corresponding stress change with time for Al beams (**b**) and AlTi beams (**c**) (after [49.47])

Lee et al. [49.47] presented a test specimen for measuring fatigue and creep in microscale structures. They patterned wafers in the shape of common (macroscopic) tensile test specimens with 50 μm wide 500 μm long metal beams (Fig. 49.19). Al and Al-1.5 at. %Ti beams were tested in a piezoactuator-driven test apparatus. In stress relaxation tests (the samples were quickly loaded to a certain stress and then held at this strain while monitoring the change of stress), *Lee* et al. [49.47] observed a load drop of 56 and 15% for Al and AlTi beams, respectively. The measurements were carried out for 10 min at room temperature. Figure 49.20 shows these results. *Lee* et al. believe that the difference in the amount of relaxation was from Al_3Ti precipitates that are formed at the Al grain boundaries in the alloy samples. They showed that the microstructure accounted for the relaxation difference in the investigated films.

Cho et al. [49.49] presented creep data at elevated temperatures of electrodeposited lithography, electro-

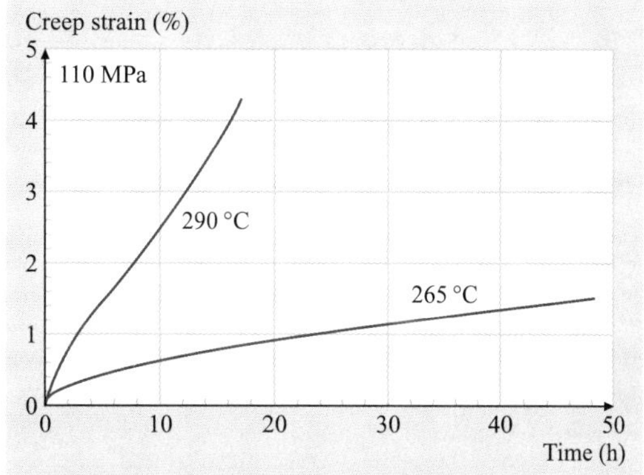

Fig. 49.20 Creep curves at 265 and 290 °C under the stress of 110 MPa (after [49.49])

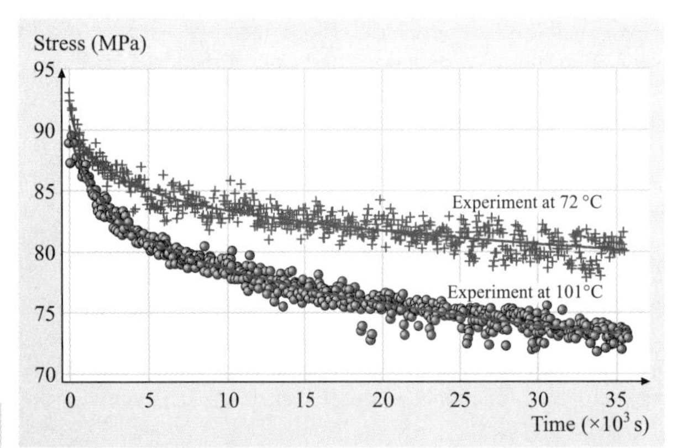

Fig. 49.21 Stress relaxations for Al-0.4 at. %Cu 5 μm thick film at 72 and 101 °C. *Solid lines* represent the fitted creep mechanism: dislocation glide limited by obstacles (after [49.50])

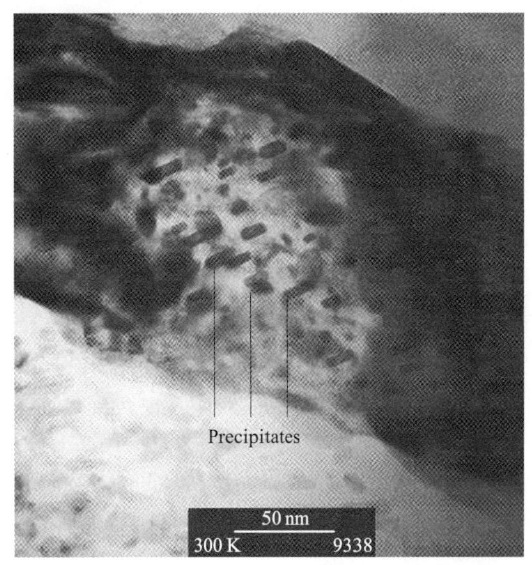

Fig. 49.22 TEM image of AlCuMgMn. Strong and highly dense precipitates hinder the dislocation movement in the grain interior

forming and molding (LIGA, from the German initials) nickel dog-bone-shaped microsamples. Figure 49.21 presents creep tests conducted at 265 and 290 °C at stresses significantly below the yield strength. The results clearly show the primary-creep regime, where the creep rate decreases with time, and a steady-state regime, where strain rate is constant in time. *Cho* et al. showed that the time needed to attain 0.5% creep strain (considered as being a significant deformation) was more that 6 h at 265 °C, 1 h at 290 °C and only about 4 min at 400 °C.

It is clear that the creep phenomenon is thermally activated and much more harmful to devices at elevated temperatures. However, *Larson* et al. [49.51] and *Yin* et al. [49.52] reported that some creep activity of nanocrystalline Ni was observed even at room and low temperatures. This means that for those materials care must be taken under almost any temperature conditions.

Hardened Pt alloys for MEMS have been presented by *Brazzle* et al. [49.53] to serve as creep-resistant materials. Solution-hardened Pt alloys for improved reliability of electromagnetically actuated micromirror devices exhibited precision angular displacement without observable fatigue and creep when used in flexures. *Brazzle* et al. [49.53] found that alloy 851 (79%Pt-15%Rh-6%Ru) and Pt(10%Ir) were very hard flexure materials. The alloying components (rhodium, ruthenium and iridium) impede dislocation movement, thereby increasing the yield strength and reducing fatigue and creep.

Modlinski et al. [49.50] have presented a simple way of selecting creep-resistant aluminum alloys for use in RF MEMS switches, but the technique is generically applicable. They propose substrate curvature measurements to study creep properties of Al alloy films using isothermal tensile stress relaxation (Fig. 49.22). The tensile stress relaxation data fitted well to a known deformation mechanism (dislocation glide limited by obstacles) in the temperature range 60–110 °C. The Al alloys studied were characterized by two creep parameters: the activation energy ΔF and the athermal flow stress τ. The higher these parameters are, the more creep-resistant the alloy is. They selected a very promising Al alloy with the composition Al93.5Cu4.4Mg1.5Mn0.6 which has a very high ΔF (6.4 ± 2.5 eV) and τ (920 ± 420 MPa). It was also shown that there is a relation between the creep parameters and the microstructure of the Al alloys in more detail. The activation energy describes the strength of the precipitates, and also gives information about the coherency of the second-phase particles as well. The athermal flow stress, on the other hand, gives information on the density and arrangement of the obstacles. A transmission electron microscope (TEM) image of the AlCuMgMn alloy revealed very small, strong, coherent and/or semi-coherent (high-ΔF) and highly dense (high-τ) precipitates.

Creep is caused by the motion of dislocations inside grains in this temperature range. Impeding the dislocation motion inside the grains is a very effective way of increasing creep resistivity of an Al99.6Cu0.4 alloy thin film. It was done by a macroscopically well-known hardening mechanism: precipitation hardening. An increase of both ΔF (by 50%) and τ was observed.

RF MEMS switches are starting to be commercialized, but reliability is still a problem. Hot switching and the impact of RF power on the reliability of the MEMS switches has been extensively investigated, mainly by simulations. *Jensen* et al. [49.54], *Rottenberg* et al. [49.55] and many others reported that temperatures due to RF power present in some parts of a movable metal bridge are expected to be as high as 250 °C [49.54] and can easily exceed 100 °C. The high residual stresses (50–200 MPa) present in an RF MEMS bridge structure and the high temperatures caused by RF power may lead to creep deformation, deteriorating the functionality of a switch over time.

Creep Conclusion

At the macroscale level, creep tests are performed using tensile specimens subjected to constant stress and temperature. These tests are rather easy to prepare and perform. In microscale experiments, the situation is very different. There are no standardized specimens, and there is no standard way of investigating creep in microdevices. Some perform the tests on real MEMS devices (DMD); some try to mimic macroscopic tensile test specimens and use them in the microworld. A very good idea, although not often used, is to implement well-known macroscopic techniques to render a material more creep-resistant. For example, hardening mechanisms such as solution hardening or precipitation hardening are interesting in this respect. These treatments produce stronger obstacles and hence decrease dislocation motion.

The microstructure (grain sizes, precipitates in case of alloys) of deposited films is very process-dependent. There are no standard deposited films. For instance, Al alloy films will exhibit different creep behavior when sputtered under different conditions. Their properties depend on the deposition methods as well as conditions. This variation is a considerable challenge for microscopic creep research.

A human factor is certainly an issue in understanding the state of the art of creep in MEMS. Most engineers and scientists dealing with MEMS are silicon- or, in general semiconductor-oriented. Creep is not really an issue in monocrystalline silicon, polysilicon or silicon-germanium. There is probably a lack of know-how on plastic deformation in the MEMS society.

Research work in a field of metal plasticity in MEMS has been started. Increasing numbers of publications on understanding and investigating creep phenomenon on the microscale are being issued. MEMS engineers and scientists are facing new problems and challenges in metal MEMS, but fundamental research programs, discussions and exchange of views and knowledge of the plastic deformation of metals between specialists will speed up research in this field.

49.3.2 Fatigue

Fatigue of Metals

The only large source of information on fatigue of metal MEMS devices is again the research carried out by Texas Instruments of the digital micromirror device (DMD). Macroscopically, the lifetime of many metal systems is limited by fatigue. A large, repetitive mechanical stress causes the accumulation of dislocation defects at the surface, resulting in fatigue cracks. In microscopic systems, the structural parts are only one, or a few grains thick, with very little dislocations. Therefore, not enough damage can accumulate at the surface to cause fatigue failures. It has been found indeed that MEMS devices suffer much less from fatigue than their macroscopic counterparts [49.56].

Fatigue behavior of materials is often shown in a so-called S-curve, in which the number of stress cycles to fracture is shown as a function of the applied maximum stress level. We usually find that at very high stresses only a small number of cycles is required to fracture the device, the low-cycle fatigue regime (Fig. 49.23). For lower stresses, the number of cycles increases. For very long low-stress cycling times, we find that different materials have different properties. Most steels e.g. have a fatigue limit (Fig. 49.23a). Below this limit, the structure survives forever. Other materials, like aluminum, do not have such a threshold. The curve

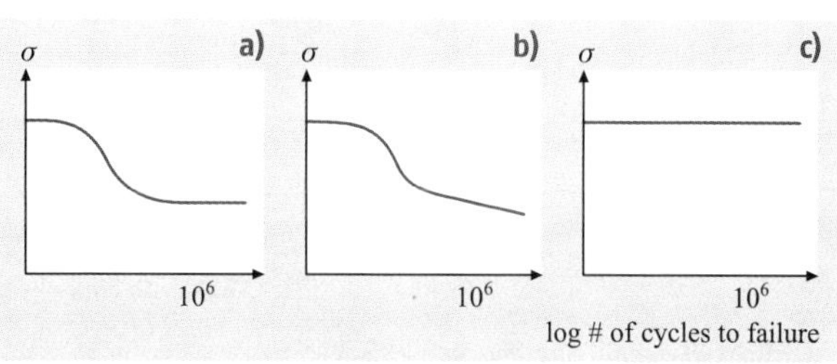

Fig. 49.23 **(a)** S-curve with threshold; **(b)** S-curve without threshold; **(c)** S-curve for ceramic

decreases slowly, even for very large numbers of cycles (Fig. 49.23b). Brittle materials such as ceramics do not have a cycling fatigue effect; below the fracture strength the lifetime is infinite, above it, they immediately break without any cyclic stress effect (Fig. 49.23c).

It has been found that high-cycle fatigue is thin films is very unlikely, as explained above. Aluminum RF MEMS switches have likewise never been reported to suffer from high-cycle fatigue [49.58]. An interesting feature of microscopic structures however is that low-cycle fatigue has been found in aluminum [49.59, 60]. This raises the question whether in thin films at very high cyclic stresses new defects are being formed, while at lower stresses only existing defects can move to the surface (and there are too few of them to do any harm). More research is definitely needed to figure out how the process really behaves, and where the turning point is between microscopic and macroscopic fatigue behavior. For the moment we can only say qualitatively that metals that macroscopically tend to have no threshold for fatigue do have such a threshold when used as thin films. Therefore the S-curve for thin-film aluminum resembles that of bulk steel (Fig. 49.23a) instead of bulk aluminum (Fig. 49.23b).

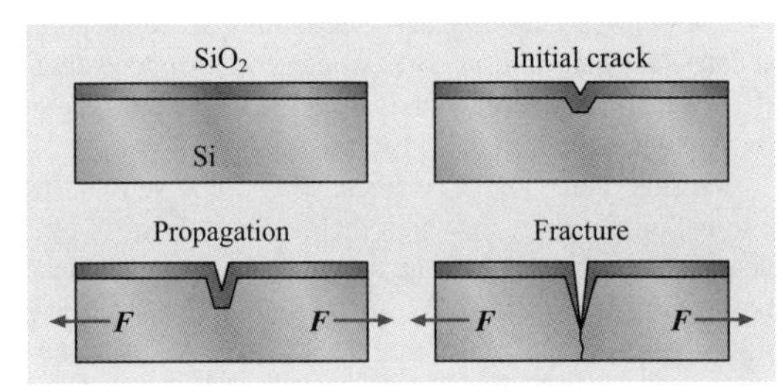

Fig. 49.24 SSC failure mechanism in silicon

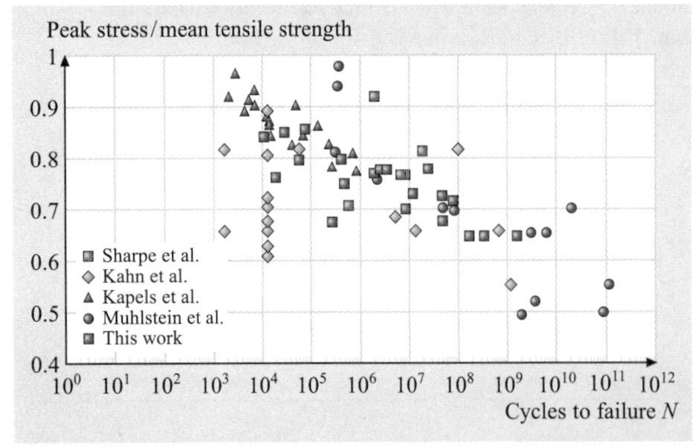

Fig. 49.25 *Bagdahn* and *Sharpe* [49.57] show that a definite trend appears if the applied stress is plotted as a function of the number of cycles to failure, contradicting the basic assumption of SCC

Fatigue of Silicon

From the early days of MEMS technology on, silicon in its mono- and polycrystalline form has been extensively used as a structural material. Silicon is a brittle material, and as such, should not intrinsically suffer degradation from fatigue (Fig. 49.23c). However, in air it is always covered with a native oxide, and concern was raised that this oxide might be susceptible to a stress corrosion cracking (SCC) phenomenon that is well known in glass technology [49.61]. SCC is the stress-assisted hydrolysis of the SiO_2 bonds at the crack tip under the influence of tensile stress and ambient humidity, causing crack propagation because atomic bonds are broken.

The phenomenon is thought to happen in the following way (Fig. 49.24). A small crack at the surface in the SiO_2 native oxide grows slightly due to SCC under the influence of a tensile stress, so that the silicon below the crack is slightly closer to the surface and oxidizes. Now the crack can propagate through the extended oxide layer, causing a new oxidation step in the silicon below. This continues until the whole structure fractures.

Connally et al. [49.62–66] initiated research into SCC in silicon microstructures, and indeed found a large humidity influence on the crack propagation rate. *Muhlstein* et al. [49.67] measured a more realistic configuration in which they did not only monitor crack propagation but also took into account crack initiation. Another advantage over the earlier measurements was that the stress intensity factor at the tip was not a required parameter to have a meaningful description. Similar measurements were also reported by *Kapels* et al. [49.68] and *Kahn* et al. [49.69].

It is likely that there is a critical stress in the silicon device, above which SCC may occur. The native oxide of silicon is normally under a compressive stress [49.6] and the applied external stress should be so large that this stress has to become tensile for SCC to occur at all. If the stress in the silicon is kept below this transition point, where the oxide surface layer turns from its intrinsic compressive stress to tensile, no problems due to SCC are expected.

Bagdahn and *Sharpe* [49.57] put all the data available, including their own, in one table and showed that SCC cannot explain all the properties of the behavior that has been found. Being a chemical, stress-assisted process, SCC should not have a pronounced cycling-speed dependency; only the absolute time that the stress is applied counts towards the reaction rate. *Bagdahn* and *Sharpe* [49.57] showed that the data available provide a definite trend of the number of cycles to failure as a function of the applied stress, which contradicts the SCC assumption. We see in Fig. 49.25 that these effects only show up at very high stresses, near the fracture strength of silicon. Therefore in practical MEMS devices the fatigue of silicon may not be a big issue because it is good design practice to keep some margin to the fracture strength for brittle materials anyway. However, in harsh environments it may turn out to be a problem.

49.3.3 Wear

Wear is one of the major limiting factors in MEMS technology. After its occurrence had been shown by *Gabriel* et al. [49.71] and *Mehregany* et al. [49.72] in silicon micromotors in 1990, it was quickly recognized that wear is a severe limiting factor for almost all MEMS devices incorporating rubbing/sliding surfaces. At Sandia National Laboratories, a large effort was spent on understanding this effect, with experiments on a dedicated micromotor test vehicle (Fig. 49.26). This resulted in a quantitative empirical model to predict wear in certain types of micromotors [49.73]. Beautiful focused ion beam (FIB) cross sections in this work show that wear is indeed severe (Fig. 49.27). The wearing mechanism proposed is a kind of *adhesive wear*, where the surfaces adhere at their contacting points and shear off parts of the other surface (Fig. 49.28). Abrasive and corrosive wear take place at higher contact forces, or when a chemical reaction assists the wear process. A comprehensive overview of their work can be found in their MEMS reliability report [49.74].

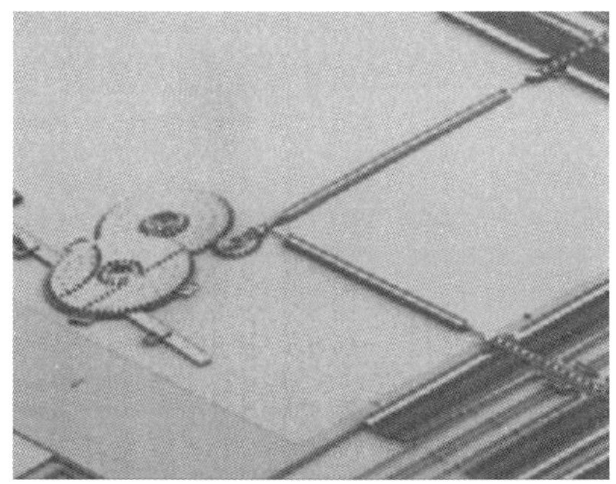

Fig. 49.26 Sandia's micromotor test vehicle to study wear

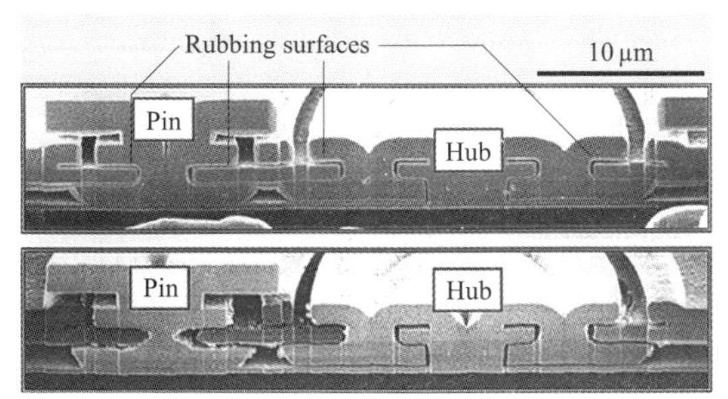

Fig. 49.27 FIB cross section of a new and a worn micromotor

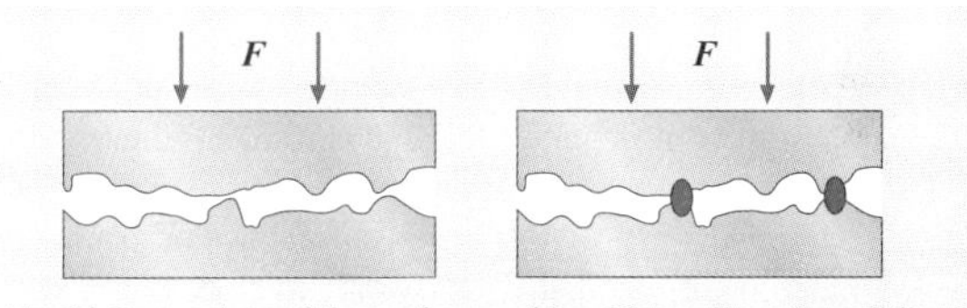

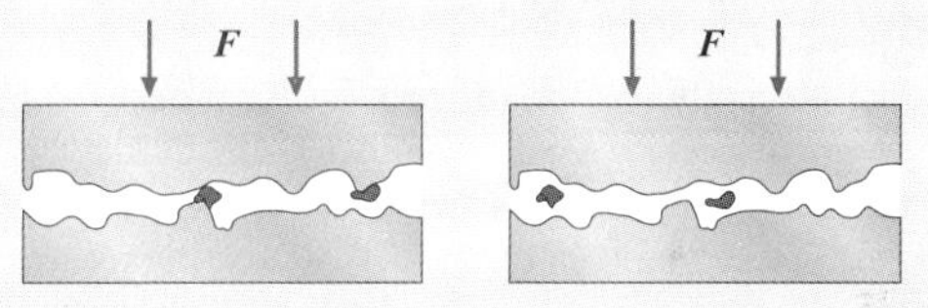

Fig. 49.28 Principle of adhesive wear

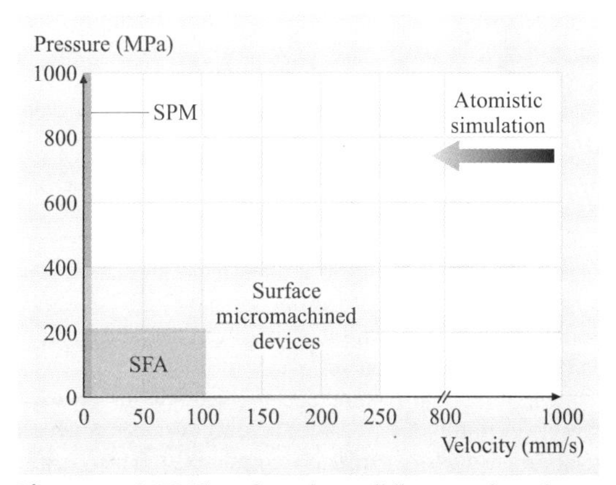

Fig. 49.29 MEMS surfaces have sliding speeds and contact forces not easily assessed by the available analysis methods: scanning probe microscopy (SPM) and surface force apparatus (SFA) (after [49.70])

A good overview from the middle of the 1990s is given by *Komvopoulos* [49.75]. Studying wear at a more fundamental level is difficult, because typical analytical measurement methods do not cover the whole range of speeds and contact forces in MEMS devices (Fig. 49.29) [49.70]. Atomic force microscopy/friction force microscopy (AFM/FFM) systems are very accurate in mimicking a single asperity, but only at, for MEMS devices, very low speeds [49.76]. A dedicated *tribo-apparatus* measurement system [49.77] fills part of the gap, but a large area remains outside the scope of the different (microscopic) measurement methods. Atomistic simulations can usually only be applied at much higher speeds, because otherwise computing time becomes prohibitively long. As a result, wear is often studied using MEMS devices themselves as the measurement vehicle [49.78–81], but, of course, this is generally not as accurate as one would like.

Rymuza [49.82, 83] proposed the use of graphitelike materials embedded in the wear layer of a MEMS device. As the device loses its surface layer due to wear, small amounts of graphite are released to lubricate the device.

Recently, a lot of activity has been going on with respect to coating silicon MEMS devices with a friction-reducing and/or wear-resistant coating after the device has been released. This coating usually takes the form of a hard, chemically inert layer with a thickness of nanometers. Experiments have been performed using tungsten [49.84], SiN_x [49.85], diamondlike carbon (DLC) [49.86–89] ultrananocrystalline diamond (UNCD) [49.90], and SiC [49.91–93], among others. SAM coatings (Sect. 49.2.1) have also been investigated for this purpose but tend to wear off during rubbing [49.94].

The results of these studies show that especially DLC, UNCD and SiC are very promising coating materials, but also that their properties severely depend on the exact composition, roughness and surface termination (passivation) of the film. At this moment, there is no quantitative theory that is able to predict which coating compositions/properties are the best. Therefore optimization of these films is performed mainly using an experimental approach.

One finds that commercial applications of devices that have continuously rubbing/sliding surfaces, such as rotary micromotors, currently do not exist. This means that wear problems in MEMS devices are still so large that designers of commercial products avoid designs where rubbing surfaces have to survive more than a couple of million cycles.

49.3.4 Packaging

Introduction

Many MEMS contain movable and fragile parts. A specific packaging approach is required for protection during fabrication as well as during operation. For instance, standard wafer dicing will, if not destroy the MEMS device, at least introduce some contamination and particles that can lead to failure. Hence packaging must be carried out on the wafer during wafer processing, prior to die singulation. This packaging step is referred to as wafer-level or 0-level packaging. It defines the first protective interface for the MEMS device, achieved through on-wafer encapsulation of the movable parts in a sealed cavity. The cavity must be strong, for many applications hermetic, and equipped with electrical signal feedthroughs. An example of a 0-level packaging scheme for RF MEMS switches, which relies on wafer or chip stacking techniques using benzocyclobutene (BCB) as the bonding and sealing material, is shown in Fig. 49.30 [49.95].

Failure Modes and Failure Mode Investigation

Important requirements of MEMS packaging are resistance (against shear stress, pull stress, mechanical, thermal loads, etc.), and hermeticity against moisture, liquids, particles or gas contaminants, but also hermeticity to keep a controlled ambient (gas and pressure) inside the package. The most common failure modes that can occur are mainly related to the sealing material itself. For instance, a package can lose its shear or pull resistance in the case of bad adhesion between the sealing material and the MEMS die, although this is rarely the case and is process-related. More important is the loss of hermeticity in the case of an hermetic package. This can happen with the appearance of cracks in the sealing layer or at the bonding interfaces when the package device is subject to thermal shocks, for instance. The difference between the thermal coefficients of expansion of the different materials can induce a very high stress at the bonding interfaces and create some cracks, thus inducing gross leaks.

The importance of this can be seen in the following example. The package of Fig. 49.31b is made of a Si capping chip flip-chip bonded on top of an RF tunable capacitor using a solder seal (SnPb). Figure 49.32 shows the tuning ratio of the capacitor as a function of the actuation voltage, right after packaging and laser dicing, after storage in air, and after drying in dry N_2 at 75 °C, respectively. Compared

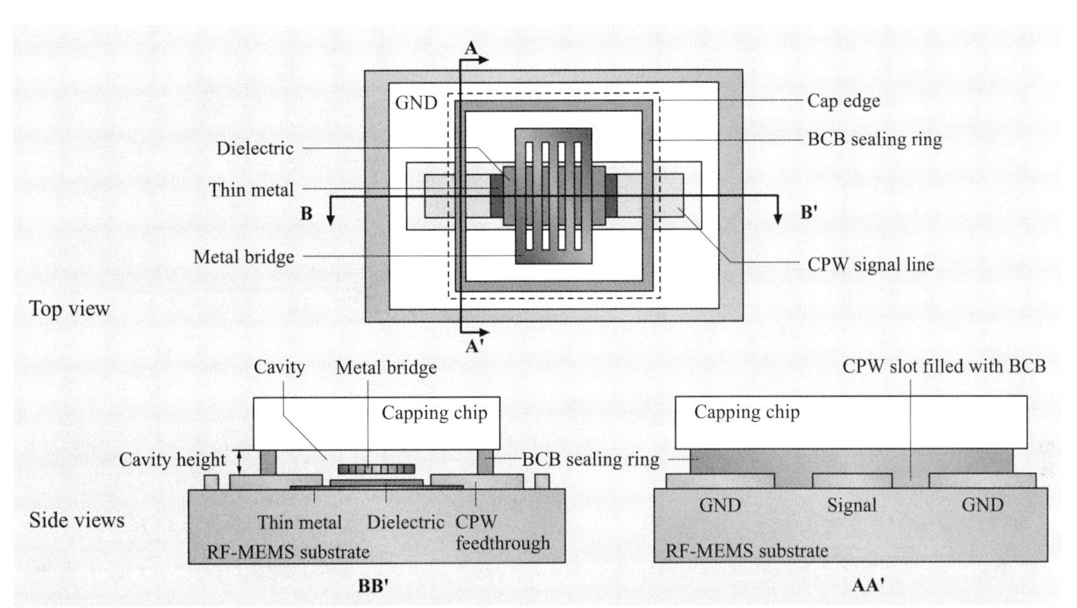

Fig. 49.30 Illustration of a wafer or 0-level packaged RF MEMS switch. The fragile MEMS part is housed in a cavity formed by a capping chip, bonded to the RF MEMS substrate using a layer of BCB

to the curve obtained right after dicing, the tuning curve after storage in air is shifted towards lower actuation voltages and shows nonreversible tuning. This phenomenon could be explained by a moisture-related charging effect of the RF capacitor when the package is stored in ambient air. This is confirmed by the tuning curve obtained after drying the packaged device in dry N_2 at 75 °C: the initial reversible tuning curve is restored. This clearly shows that, in this example, the package is not hermetic to moisture and that this influences the electrical performance of the MEMS device. Possible causes of packaging failure here could be microcracks in the solder sealing ring, bad adhesion between the solder and the MEMS die, particle contamination in the solder, or a local scarcity of solder.

For the same reasons, in the case of a (supposedly) hermetic package containing a controlled ambient (pressure and gas), the loss of hermeticity due to a leak in the sealing material will fill the package with air after a determined period of time, depending on the leak

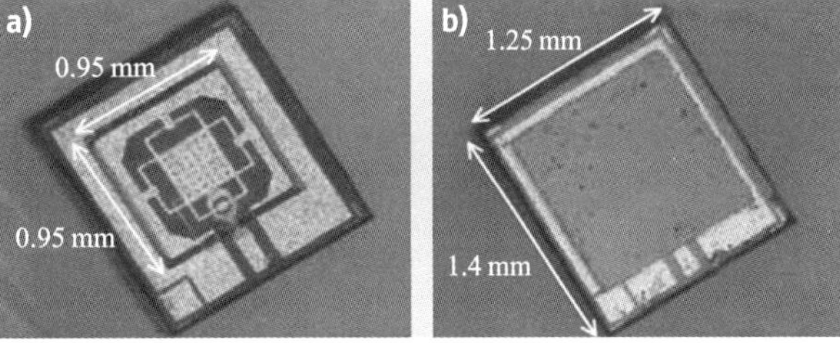

Fig. 49.31a,b Picture of a laser-diced *naked* (a) and packaged (b) RF tunable capacitor using solder (SnPb) flip-chip assembly

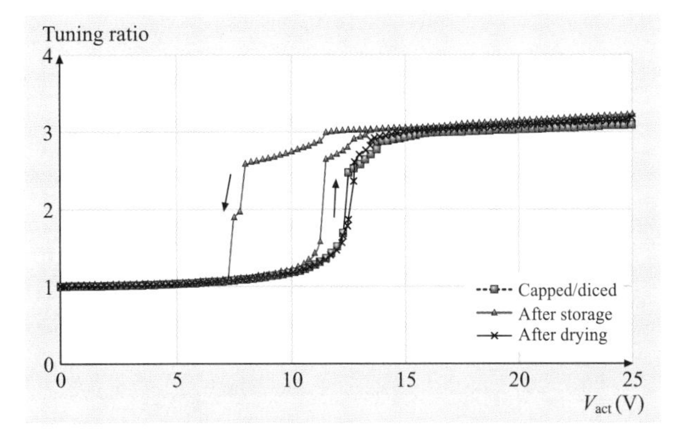

Fig. 49.32 Tuning ratio of the packaged tunable capacitor of Fig. 49.20 right after dicing (*squares*), after storage in ambient air (*triangles*) and after drying in dry N_2 at 75 °C (*crosses*) ▶

Fig. 49.33 UV inspection of a leaky package using the dye-penetrant method

Fig. 49.34 SAM picture of a Si cap bonded to a Si wafer using polymer. The *black areas* indicate positions where the bond failed (*arrows*) (after [49.96])

rate. For instance, a package of 100 nl (a few tens of nl are typically encountered in MEMS packaging) with an initial internal pressure of 1 mbar and displaying a leak rate as low as 10^{-10} mbar l/s will be filled with air after only 11 days.

Different testing methods exist or have been developed to first test the package (e.g. shear and pull strength, hermeticity, etc.) and second, to investigate the failure mechanisms. One can always refer to the MIL-STD-883 procedure to obtain the testing conditions and the rejection limits for the test [49.97]. This is the case for instance for the pull and shear testing (MIL-STD-883, method 2019.5). Hermeticity (fine leak) testing as specified by the MIL-STD-883 (method 1014.9) is still a problem because the fine-leak test con-

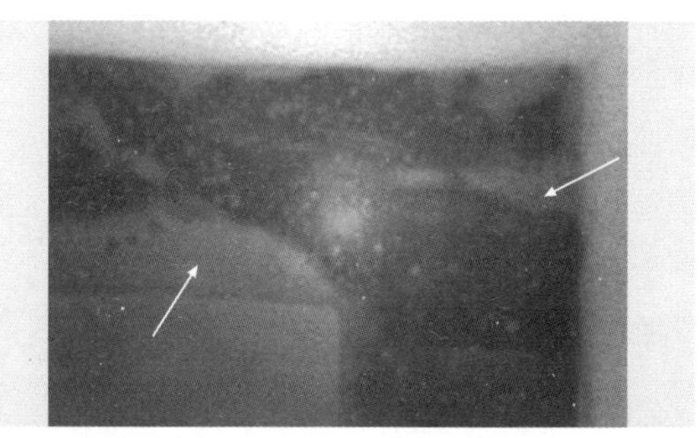

Fig. 49.35 X-ray image of a solder-sealed Si package. *White areas* are assumed to be solder oxide (*unbonded areas*)

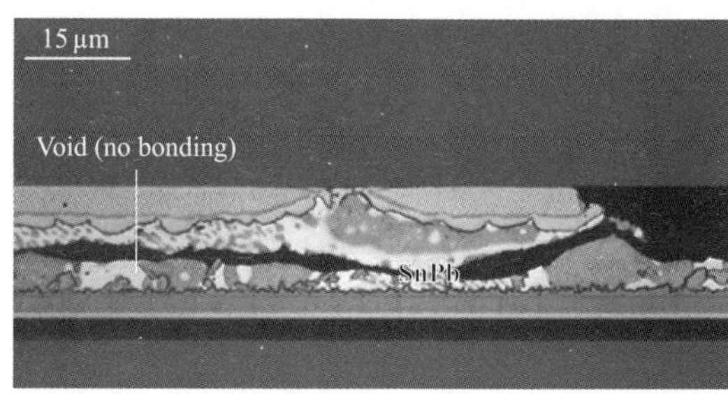

Fig. 49.36 Optical microscopy image of a cross section of a solder-sealed Si package

ditions and specifications are clearly not applicable to MEMS packages due to the small cavity volumes they deal with [49.98]. Alternative methods are needed to investigate the leak rate of MEMS packages. However, gross-leak testing using fluorocarbon fluids is still applicable and is a good method to localize a leak path (e.g. a crack or a bad bonding) in the sealing material or the package itself. Similar to gross-leak testing is the dye penetrant method, as illustrated in Fig. 49.33. The package is immersed in a pressurized yellow dye (type Zyglo ZL-56 from Magnaflux). Next, the sample is rinsed with water, dried, and sprayed with a developer (type Zyglo ZP-9F). This developer has the property of sucking out the dye from the package if a leak exists, and some yellow fringes are then clearly visible under UV light.

Another method to localize a gross leak in a package is scanning acoustic microscopy (SAM). A picture of a polymer-sealed leaky package is shown in Fig. 49.34 [49.96]. The black areas indicate that there is no adhesion between the sealing layer and the Si chip.

Other methods to look at the bonding interface are x-ray imaging and cross sectioning. Figure 49.35 is an x-ray image of a solder-sealed Si package. Irregularities in the solder ring are visible, and the lighter areas are assumed to be oxidized solder areas, and thus potential leaky paths.

Cross sectioning is of course destructive for the package, but it can also be used to investigate the bond quality, as shown in the example of Fig. 49.36. In this picture, voids in the sealing layer are clearly visible, and are potential source of gross leaks.

Conclusion

Investigating the failure mechanisms of the package is at least as important as for the MEMS itself, because without a good and robust package, most MEMS devices cannot survive operating or storage conditions. Understanding why a package failed will help to improve the next generation of MEMS. A lot of development still has to be done in packaging reliability testing, and this requires dedicated methods of investigation, which is now slowly emerging.

49.4 Conclusions

We have seen that a lot about the reliability of MEMS/NEMS is already known, and that in some cases even macroscopic principles can be applied to improve the reliability of micromachines over the state of the art. The fact that reliability is such a problem is not because MEMS/NEMS are intrinsically unreliable. They are not, and in fact – due to the low mass and high strength of materials on a microscale – the reliability that can be attained is even higher than in many macroscopic situations. The reason that reliability is hampering the commercialization of MEMS has already briefly been addressed in the creep section: to produce any MEMS at all is already a considerable feat, so that reliability issues are often not taken into account in the early stages of development. In the macroscopic world, instead, reliability is one of the key design issues of almost any product.

This sense still has to pervade the MEMS society, although things are changing. Especially in commercial laboratories, *design for reliability* is the new mantra, and independent research and development labs have set up dedicated groups to study fundamental issues in microscale reliability, such as stiction, charging, creep, fatigue, wear and packaging. If this trend continues over the coming years, MEMS may become just as pervasive as microelectronics is now.

References

49.1 W.M. Miller, D.M. Tanner, S.L. Miller, K.A. Peterson: MEMS reliability. The challenge and the promise, Proc. 4th Annu. Reliab. Chall. (Dublin 1998) pp. 4-1–4-7

49.2 M.R. Douglass: DMD reliability. A MEMS success story, Proc. SPIE **4980**, 1–11 (2003)

49.3 J. Bienstman: From product to production in automotive MEMS, Proc. MicroMech. Eur., ed. by R. Puers (Leuven 2004) pp. 107–108

49.4 W.M. van Spengen: MEMS reliability from a failure mechanisms perspective, Microelectron. Reliab. **43**, 1049–1060 (2003)

49.5 W.M. van Spengen, R. Puers, R. Mertens, I. De Wolf: High resolution optical investigation of small out-of-plane movements and fast vibrations; characterization and failure analysis of MEMS, Microsyst. Technol. **10**, 89–96 (2004)

49.6 B. Stark (Ed.): *MEMS Reliability Assurance Guidelines for Space Applications* (National Aeronautics Space Administration (NASA) and Jet Propulsion Laboratory (JPL), California Institute of Technology, Pasadena 1999)

49.7 D.M. Tanner, N.F. Smith, D.J. Bowman, W.P. Eaton, K.A. Peterson: First reliability test of a surface micromachined microengine using SHiMMeR, Proc. SPIE **3224**, 14–23 (1997)

49.8 G.F. LaVigne, S.L. Miller: A performance analysis system for MEMS using automated imaging methods, Proc. IEEE Int. Test Conf. (IEEE, Washington 1998) pp. 442–447

49.9 P. Krehl, S. Engemann, C. Rembe, E.P. Hofer: High-speed visualization, a powerful diagnostic tool for microactuators – Retrospect and prospect, Microsyst. Technol. **5**, 113–132 (1999)

49.10 C. Rembe, L. Muller, R.S. Muller, R.T. Howe: Full three-dimensional motion characterization of a gimballed electrostatic microactuator, Proc. IEEE Annu. Int. Reliab. Phys. Symp. (IRPS), ed. by E.S. Sneyder (IEEE, Orlando 2001) pp. 91–98

49.11 J.S. Burdess, A.J. Harris, D. Wood, R.J. Pitcher, D. Glennie: A system for the dynamic characterization of microstructures, J. MEMS **6**, 322–328 (1997)
49.12 M.R. Hart, R.A. Conant, K.Y. Lau, R.S. Muller: Stroboscopic interferometer system for dynamic MEMS characterization, J. MEMS **9**, 409–418 (2000)
49.13 M.P. de Boer, J.A. Knapp, T.M. Mayer, T.A. Michalske: The role of interfacial properties on MEMS performance and reliability, Proc. SPIE **3825**, 2–15 (1999)
49.14 C.H. Mastrangelo, C.H. Hsu: Mechanical stability and adhesion of microstructures under capillary forces – Part I: Basic theory, J. MEMS **2**, 33–43 (1993)
49.15 C.H. Mastrangelo, C.H. Hsu: Mechanical stability and adhesion of microstructures under capillary forces – Part II: Experiments, J. MEMS **2**, 44–55 (1993)
49.16 F.P. Bowden, D. Tabor: *The Friction and Lubrication of Solids* (Clarendon, Oxford 1950)
49.17 R. Maboudian, R.T. Howe: Critical review: Adhesion in surface micromechanical structures, J. Vac. Sci. Technol. B **15**, 1–20 (1997)
49.18 R. Legtenberg, H.A.C. Tilmans, J. Elders, M. Elwenspoek: Stiction of surface micromachined structures after rinsing and drying: Model and investigation of adhesion mechanisms, Sens. Actuators A **43**, 230–238 (1994)
49.19 N. Tas, T. Sonnenberg, H. Jansen, R. Legtenberg, M. Elwenspoek: Stiction in surface micromachining, J. Micromech. Microeng. **6**, 385–397 (1996)
49.20 J. Israelachvili: *Intermolecular and Surface Forces* (Academic, London 1991)
49.21 K. Komvopoulos: Surface engineering and microtribology for microelectro-mechanical systems, Wear **200**, 305–327 (1996)
49.22 W.M. van Spengen, W.M.R. Puers, I. De Wolf: A physical model to predict stiction in MEMS, J. Micromech. Microeng. **12**, 702–713 (2002)
49.23 J.A. Greenwood, J.B.P. Williamson: Contact of nominally flat surfaces, Proc. R. Soc. A **295**, 300–319 (1966)
49.24 B. Bhushan: Methodology for roughness measurement and contact analysis for optimization of interface roughness, IEEE Trans. Mag. **32**, 1819–1825 (1996)
49.25 B. Bhushan: *Principles and Applications of Tribology* (Wiley, New York 1999)
49.26 L. Kogut, K. Komvopoulos: Analysis of interfacial adhesion based on electrical contact resistance, J. Appl. Phys. **94**, 6386–6390 (2003)
49.27 Y.-P. Zhao, L.S. Wang, T.X. Yu: Mechanics of adhesion in MEMS – A review, J. Adhes. Sci. Technol. **17**, 519–546 (2003)
49.28 J. Elders, H.V. Jansen, M. Elwenspoek: Materials analysis of fluorocarbon films for MEMS applications, Proc. Investig. Micro Struct. Sens. Actuators Mach. Robot. Syst. (New York 1994) pp. 170–175
49.29 P.F. Man, B.P. Gogoi, H. Mastrangelo: Elimination of post-release adhesion in microstructures using conformal fluorocarbon coatings, J. MEMS **6**, 25–34 (1997)
49.30 U. Srinivasan, M.R. Houston, R.T. Howe, R. Maboudian: Alkyltrichlorosilane-based self-assembled monolayer films for stiction reduction in silicon micromachines, J. MEMS **7**, 252–260 (1998)
49.31 R. Maboudian, W.R. Ashurst, C. Carraro: Self-assembled monoayers as anti-stiction coatings for MEMS: Characteristics and recent developments, Sens. Actuators A **82**, 219–223 (2000)
49.32 B.-H. Kim, C.-H. Oh, K. Chun, T.-D. Chung, J.-W. Byun, Y.-S. Lee: A new class of surface modifiers for stiction reduction, Proc. 12th Int. Conf. Micro Electro Mech. Syst. (Piscataway 1999) pp. 189–193
49.33 K.A. Peterson, P. Tangyunyong, A.A. Pimentel: Failure analysis of surface micromachined microengines, Proc. SPIE **3512**, 190–200 (1998)
49.34 J. Wibbeler, G. Pfeifer, M. Hietschold: Parasitic charging of dielectric surfaces in capacitive microelectromechanical systems (MEMS), Sens. Actuators A **71**, 74–80 (1998)
49.35 A.R. Knudson, S. Buchner, P. McDonald, W.J. Stapor, A.B. Campbell, K.S. Grabowski, D.L. Knies: The effects of radiation on MEMS accelerometers, IEEE Trans. Nucl. Sci. **43**, 3122–3126 (1996)
49.36 C.I. Lee, A.H. Johnston, W.C. Tang, C.E. Barnes: Total dose effects on microelectromechanical systems (MEMS): Accelerometers, IEEE Trans. Nucl. Sci. **43**, 3127–3132 (1996)
49.37 G.M. Rebeiz: *RF MEMS: Theory, Design and Technology* (Wiley, Hoboken 2003)
49.38 C.L. Goldsmith, J. Ehmke, A. Malczewski, B. Pillans, S. Eshelman, Z. Yao, J. Brank, M. Eberly: Lifetime characterization of capacitive RF MEMS switches, Proc. IEEE MTT-S Int. Microw. Symp. (IEEE, New York 2001) pp. 227–230
49.39 S. Zafar, A. Callegari, E. Gusev, M.V.P. Fischetti: Charge trapping in high k gate dielectric stacks, Proc. Int. Electron Devices Meet. (IEDM), ed. by S. Ikeda (IEEE, San Francisco 2002) pp. 517–520
49.40 J.C. Ehmke, C.L. Goldsmith, Z.J. Yao, S.M. Eshelman: Method and apparatus for switching high frequency signals, US Patent 6391675 (1999)
49.41 J.R. Reid: Simulation and measurement of dielectric charging in electrostatically actuated capacitive microwave switches, Proc. Model. Simul. Microsyst. (MSM), ed. by M. Laudon, B. Romanowicz (NSTI, San Juan 2002) pp. 250–253
49.42 W.M. van Spengen, R. Puers, R. Mertens, I. De Wolf: A comprehensive model to predict the charging and reliability of capacitive RF MEMS switches, Micromech. Microeng. **14**, 514–521 (2004)
49.43 X. Rottenberg, B. Nauwelaers, W. De Raedt, H.A.C. Tilmans: Distributed dielectric charging and its impact on RF MEMS devices, Proc. 34th Eur.

Microw. Conf. (Artech House, Amsterdam 2004) pp. 77–80
49.44 W.M. van Spengen, R. Puers, I. De Wolf: RF MEMS reliability – The challenge, the physics, and the reward, Proc. MME, ed. by R. Puers (Leuven 2004) pp. 319–325
49.45 H.E. Boyer: *Atlas of Creep and Stress-Rupture Curves* (ASM International, Metals Park 1988)
49.46 F.N.R. Nabarro, H.L. de Villiers: *The Physics of Creep* (Taylor Francis, New York 1995)
49.47 H.-J. Lee, G. Cornella, J.C. Bravman: Stress relaxation of free-standing aluminum beams for microelectromechanical systems applications, Appl. Phys. Lett. **76**, 3415–3417 (2000)
49.48 P.F. Van Kessel, L.J. Hornbeck, R.E. Meier, M.R. Douglass: A MEMS-based projection display, Proc. IEEE **86**, 1687–1704 (1998)
49.49 H.S. Cho, K.J. Hemker, K. Lian, J. Goettert, G. Dirras: Measured mechanical properties of LIGA Ni structures, Sens. Actuators A **103**, 59–63 (2003)
49.50 R. Modlinski, A. Witvrouw, P. Ratchev, V. Simons, A. Jourdain, H.A.C. Tilmans, R. Puers, J. den Toonder, I. De Wolf: Creep as a reliability problem in MEMS, J. Microelectron. Reliab. **44**, 1733–1738 (2004)
49.51 K.P. Larsen, A.A. Rasmussen, J.T. Ravnkilde, M. Ginnerup, O. Hansen: MEMS devices for bending test: Measurements of fatigue and creep of electroplated nickel, Sens. Actuators A **103**, 156–164 (2003)
49.52 W.M. Yin, S.H. Whang, R. Morshams, C.H.P. Xiao: Creep behavior of nanocrystalline nickel at 290 and 373 K, Mater. Sci. Eng. A **2301**, 18–22 (2001)
49.53 J.D. Brazzle, W.P. Taylor, B. Ganesh, J.J. Price, J.J. Bernstein: Solution hardened platinum alloy flexure materials for improved performance and reliability of MEMS devices, J. Micromech. Microeng. **15**, 43–48 (2005)
49.54 B.D. Jensen, J.L. Volakis, K. Saitou, K. Kurabayashi: Impact of skin effect on thermal behavior of RF-MEMS switches, 6th ASME-JSME Conf., ed. by S. Nishio, A. Lavine (Kona, Hawaii 2003), TED-AJ03-420
49.55 X. Rottenberg, B. Nauwelaers, W. De Raedt, H.A.C. Tilmans: RF current and power handling of RF-MEMS shunt switches, Proc. MEMSWAVE (2004) pp. C1–C4
49.56 M.R. Douglass: Lifetime estimates and unique failure mechanisms of the digital micromirror device, 36th Int. Reliab. Phys. Symp. (IRPS), ed. by A.N. Campbell (IEEE, Reno 1998) pp. 9–16
49.57 J. Bagdahn, W.N. Sharpe Jr.: Fatigue of polysilicon silicon under long-term cyclic loading, Sens. Actuators A **103**, 9–15 (2003)
49.58 J.J. Yao: RF MEMS from a device perspective, J. Micromech. Microeng. **10**, R9–R38 (2000)
49.59 D.T. Read, J.W. Dally: Fatigue of microlithographically patterned free-standing aluminum thin film under axial stresses, J. Electron. Packag. **117**, 1–6 (1995)
49.60 G. Cornella, R.P. Vinci, R. Suryanarayanan Iyer, R.H. Dauskardt, J.C. Bravman: Observations of low-cycle fatigue of Al thin films for MEMS applications, Mater. Res. Soc. Symp. Proc. **518**, 81–86 (1998)
49.61 S.M. Wiederhorn, E.R. Fuller Jr., R. Thomson: Micromechanisms of crack growth in ceramics and glasses in corrosive environments, Met. Sci. **14**, 450–458 (1980)
49.62 W.W. van Arsdell, S.B. Brown: Subcritical crack growth in silicon MEMS, J. MEMS **8**, 319–327 (1999)
49.63 J.A. Connally, S.B. Brown: Micromechanical fatigue testing, TRANSDUCERS '91, Int. Conf. Solid-State Sens. Actuators Dig. (New York 1991) pp. 953–956
49.64 S.B. Brown, W. van Arsdell, C.L. Muhlstein: Materials reliability in MEMS devices, TRANSDUCERS '97, Int. Conf. Solid-State Sens. Actuators Dig., Vol. 1 (New York 1997) pp. 591–594
49.65 S.B. Brown, E. Jansen: Reliability and long term stability of MEMS, Summer Top. Meet. Dig. Opt. MEMS Their Appl. (New York 1996) pp. 9–10
49.66 S.B. Brown, G. Povirk, J. Connally: Measurement of slow crack growth in silicon and nickel micromechanical devices, Proc. Micro Electro Mech. Syst. Investig. Micro Struct. Sens. Actuators Mach. Syst. (IEEE, New York 1993) pp. 99–102
49.67 C.L. Muhlstein, S.B. Brown, R.O. Ritchie: High cycle fatigue and durability of polycrystalline silicon thin films in ambient air, Sens. Actuators A **94**, 177–188 (2001)
49.68 H. Kapels, R. Aigner, J. Binder: Fracture strength and fatigue of polysilicon determined by a novel thermal actuator, Trans. Electron. Dev. **47**, 1522–1528 (2000)
49.69 H. Kahn, N. Tayebi, R. Ballerini, R.L. Mullen, A.H. Heuer: Fracture toughness of polysilicon MEMS devices, Sens. Actuators A **82**, 274–280 (2000)
49.70 A.D. Romig Jr., M.T. Dugger, P.J. McWorther: Materials issues in microelectromechanical devices: Science, engineering, manufacturability and reliability, Acta Mater. **51**, 5837–5866 (2003)
49.71 K.J. Gabriel, F. Behi, R. Mahadevan: In-situ friction and wear measurements in integrated polysilicon mechanisms, Sens. Actuators A **A21–A23**, 184–188 (1990)
49.72 M. Mehregany, S.D. Senturia, J.H. Lang: Friction and wear in microfabricated harmonic side-drive motors, Tech. Dig. IEEE Solid-State Sens. Actuator Workshop (IEEE, 1990) pp. 17–22
49.73 D.M. Tanner, W.M. Miller, W.P. Eaton, L.W. Irwin, K.A. Peterson, M.T. Dugger, D.C. Senft, N.F. Smith, P. Tanyunyong, S.L. Miller: The effect of frequency on the lifetime of a surface micromachined microengine driving a load, Int. Reliab. Phys. Symp. Proc. (IRPS), ed. by A.N. Campbell (IEEE, Reno 1998) pp. 26–35

49.74 D.M. Tanner: *MEMS Reliability: Infrastructure, Test Structures, Experiments and Failure Modes*, Sandia Rep. (Sandia National Laboratories, Livermore 2000), available from National Technical Information Service, US Department of Commerce, Springfield or http://www.sandia.gov/mstc/technologies/micromachines/tech-info/bibliography/biblog_char.html
49.75 K. Komvopoulos: Surface engineering and microtribology for microelectro-mechanical systems, Wear **200**, 305–327 (1996)
49.76 J.A. Ruan, B. Bhushan: Atomic-scale friction measurement using friction force microscopy. 1. General – Principles and new measurement techniques, J. Tribol. **116**, 378–388 (1994)
49.77 H. Lui, B. Bhushan: Adhesion and friction studies of microelectromechanical systems/nanoelectromechanical systems materials using a novel triboapparatus, J. Vac. Sci. Technol. A **21**, 1528–1538 (2003)
49.78 M.P. de Boer: A hinged-pad test structure for sliding friction measurement in micromachining, Proc. SPIE **3512**, 241–250 (1998)
49.79 S.L. Miller, J.J. Sniegowski, G. LaVigne, P.J. McWorther: Friction in surface micromachined microengines, Proc. SPIE **2722**, 197–204 (1996)
49.80 D.C. Senft, M.T. Dugger: Friction and wear in surface micromachined tribological test devices, Proc. SPIE **3224**, 31–38 (1997)
49.81 M.P. de Boer, T.M. Mayer: Tribology of MEMS, MSR Bulletin **26**, 302–304 (2001)
49.82 Z. Rymuza: Control tribological and mechanical properties of MEMS surfaces. Part 1: Critical review, Microsyst. Technol. **5**, 173–180 (1999)
49.83 Z. Rymuza, M. Misiak, L. Kuhn, K. Schmidt-Szalowski, Z. Ranek-Boroch: Control tribological and mechanical properties of MEMS surfaces. Part 2: Nanomechanical behavior of self-lubricating ultrathin films, Microsyst. Technol. **5**, 181–188 (1999)
49.84 J.G. Fleming, S.S. Mani, J.J. Sniegowski, R.S. Blewer: Tungsten coating for improved wear resistance and reliability of microelectromechanical devices, US Patent 6290859 (2001)
49.85 U. Beerschwinger, D. Mathieson, R.L. Reuben, S.J. Yang: A study of wear on MEMS contact morphologies, J. Micromech. Microeng. **4**, 95–105 (1994)
49.86 R. Bandorf, H. Lüthje, T. Staedler: Influencing factors on microtribology of DLC films for MEMS and microactuators, Diam. Relat. Mater. **13**, 1491–1493 (2004)
49.87 A.P. Musinho, R.D. Mansano, M. Massi, J.M. Jaramillo: Micro-machine fabrication using diamond-like carbon films, Diam. Relat. Mater. **12**, 1041–1044 (2003)
49.88 A. Erdemir: Superlubricity and wearless sliding in diamondlike carbon films, Proc. Mater. Res. Soc. **697**, 391–403 (2002)
49.89 A. Erdemir: Design criteria for superlubricity in carbon films and related microstructures, Tribol. Int. **37**, 577–583 (2004)
49.90 A.R. Krauss, O. Auciello, D.M. Gruen, A. Jayatissa, A. Sumant, J. Tucek, D.C. Mancini, N. Moldovan, A. Erdemir, D. Ersoy, M.N. Gardos, H.G. Busmann, E.M. Meyer, M.Q. Ding: Ultrananocrystalline diamond thin films for MEMS and moving mechanical assembly devices, Diam. Relat. Mater. **10**, 1952–1962 (2001)
49.91 W.R. Ashurst, M.B.J. Wijesundra, C. Carraro, R. Maboudian: Tribological impact of SiC encapsulation of released polycrystalline silicon microstructures, Tribol. Lett. **71**, 195–198 (2004)
49.92 S. Sundararajan, B. Bhushan: Micro/nanotribological studies of polysilicon and SiC films for MEMS applications, Wear **217**, 251–261 (1998)
49.93 X. Li, B. Bhushan: Micro/nanotribological characterization of ceramic films for microdevices, Thin Solid Films **340**, 210–217 (1999)
49.94 R. Maboudian, W.R. Ashurst, C. Carraro: Tribological challenges in microelectromechanical systems, Tribol. Lett. **12**, 95–100 (2002)
49.95 H.A.C. Tilmans, H. Ziad, H. Jansen, O. Di Monaco, A. Jourdain, W. De Raedt, X. Rottenberg, E. De Backer, A. De Caussemaeker, K. Baert: Wafer-level packaged RF-MEMS switches fabricated in a CMOS fab, Proc. IEDM 2001 (IEEE, Washington 2001) pp. 921–924
49.96 I. De Wolf, W.M. van Spengen, R. Modlinski, A. Jourdain, A. Witvrouw, P. Fiorini, H.A.C. Tilmans: Reliability and failure analysis of RF MEMS switches, Proc. ISTFA 2002 (ASM, Phoenix 2002) pp. 275–281
49.97 Military Standard (MIL-STD-883): Test Methods and Procedures for Microelectronics
49.98 A. Jourdain, P. De Moor, S. Pamidighantam, H.A.C. Tilmans: Investigation of the hermeticity of BCB-sealed cavities for housing (RF-)MEMS devices, Proc. MEMS 2002 (IEEE, Las Vegas 2002) pp. 677–680

50. Mechanical Properties of Micromachined Structures

Harold Kahn

To be able to accurately design structures and make reliability predictions in any field, it is first necessary to know the mechanical properties of the materials that make up the structural components. The devices encountered in the fields of microelectromechanical systems (MEMS) and nanoelectromechanical systems (NEMS), are necessarily very small, and so the processing techniques and the microstructures of the materials used in these devices may differ significantly from bulk structures. Also, the surface-area-to-volume ratios in such structures are much higher than in bulk samples, and so surface properties become much more important. In short, it cannot be assumed that the mechanical properties measured for a bulk specimen of a material will apply when the same material is used in MEMS and NEMS. This chapter will review the techniques that have been used to determine the mechanical properties of micromachined structures, especially residual stress, strength and Young's modulus. The experimental measurements that have been performed will then be summarized, in particular the values obtained for polycrystalline silicon (polysilicon).

50.1 **Measuring Mechanical Properties of Films on Substrates** 1783
50.1.1 Residual Stress Measurements 1783
50.1.2 Mechanical Measurements Using Nanoindentation 1784

50.2 **Micromachined Structures for Measuring Mechanical Properties** 1785
50.2.1 Passive Structures 1785
50.2.2 Active Structures 1788

50.3 **Measurements of Mechanical Properties** . 1795
50.3.1 Mechanical Properties of Polysilicon 1795
50.3.2 Mechanical Properties of Other Materials 1798

References 1799

50.1 Measuring Mechanical Properties of Films on Substrates

In order to accurately determine the mechanical properties of very small structures, it is necessary to test specimens made from the same materials, processed in the same way, and of the same approximate size. Not surprisingly it is often difficult to handle specimens this small. One solution is to test the properties of films that remain on substrates. Micro- and nanomachined structures are typically fabricated from films that are initially deposited onto a substrate, are subsequently patterned and etched into the appropriate shapes, and are then finally released from the substrate. If the testing is performed on the continuous film, before patterning and release, the substrate can be used as an effective *handle* for the specimen (in this case, the film). Of course, since the films are attached to the substrate, the types of tests possible are severely limited.

50.1.1 Residual Stress Measurements

One common measurement easily performed on films attached to substrates is residual film stress. The curvature of the substrate is measured before and after film deposition. Curvature can be measured in a number of ways. The most common technique is to scan a laser across the surface (or scan the substrate beneath the laser) and detect the angle of the reflected signal. Alternatively, profilometry, optical interferometry or even atomic force microscopy can be used. As expected, tools that map a surface or perform multiple linear scans can give more accurate readings than tools that measure only a single scan.

Assuming that the film is thin compared to the substrate, the average residual stress in the film σ_f is given by the Stoney equation

$$\sigma_f = \frac{1}{6}\frac{E_s}{(1-\nu_s)}\frac{t_s^2}{t_f}\left(\frac{1}{R_1}-\frac{1}{R_2}\right), \quad (50.1)$$

where the subscripts f and s refer to the film and substrate, respectively; t is thickness, E is Young's modulus, ν is Poisson's ratio, and R is the radius of curvature before (R_1) and after (R_2) film deposition [50.2]. For a typical (100)-oriented silicon substrate, $E/(1-\nu)$ (also known as the biaxial modulus) is equal to 180.5 GPa, independent of in-plane rotation [50.3]. This investigation can be performed on the as-deposited film or after any subsequent annealing step, provided no changes occur to the substrate.

This measurement will reveal the average residual stress of the film. Typically, however, the residual stresses of deposited films will vary throughout the thickness of the film. One way to detect this, using substrate curvature techniques, is to etch away a fraction of the film and repeat the curvature measurement. This can be iterated any number of times to obtain a residual stress profile for the film [50.4]. Alternatively, tools have been designed that can measure the substrate curvature during the deposition process itself, in order to obtain information on how the stresses evolve [50.5].

An additional feature of some of these tools is the ability to heat the substrates while performing the stress measurement. An example of the results obtained in such an experiment is shown in Fig. 50.1 [50.1], for an aluminium film on a silicon substrate. The slope of the heating curve gives the difference in thermal expansion between the film and the substrate. When the heating curve changes slope and becomes nearly horizontal, the yield strength of the film has been reached.

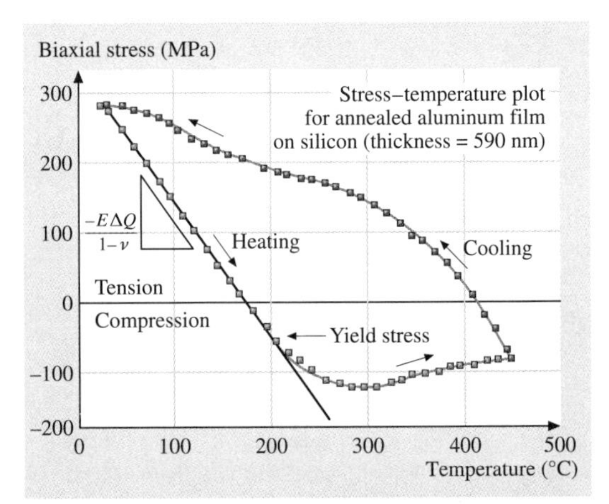

Fig. 50.1 Typical results for residual stress as a function of temperature for an aluminum film on a silicon substrate (after [50.1]). The stresses were determined by measuring the curvature of the substrate before and after film deposition, using the reflected signal of a laser scanned across the substrate surface

50.1.2 Mechanical Measurements Using Nanoindentation

Aside from residual stress, it is difficult to measure the mechanical properties of films attached to substrates without the measurement being affected by the presence of the substrate. Recent developments in nanoindentation equipment have allowed this technique to be used in some cases. With specially designed tools, indentation can be performed using very low loads. If the films being investigated are thick and rigid enough, measurements can be made that are not influenced by the presence of the substrate. Of course, this can be verified by depositing the same film onto different substrates. By continuously monitoring the displacement as well as the load during indentation, a variety of properties can be measured, including hardness and Young's modulus [50.6]. This area is covered in more detail in a separate chapter.

For brittle materials, cracks can be generated by indentation, and strength information can be gathered. But the exact stress fields created during the indentation process are not known exactly, and therefore quantitative values for strength are difficult to determine. Anisotropic etching of single-crystal silicon has been performed to create 30 μm tall structures that were then indented to examine fracture toughness [50.7], but this is not possible with most materials.

50.2 Micromachined Structures for Measuring Mechanical Properties

Certainly the most direct way to measure the mechanical properties of small structures is to fabricate structures that would be conducive to such tests. Fabrication techniques are sufficiently advanced that virtually any design can be realized, at least in two dimensions. Two basic types of devices are used for mechanical property testing: *passive* structures and *active* structures.

50.2.1 Passive Structures

As mentioned previously, the main difficulty encountered when testing very small specimens is handling. One way to circumvent this problem is to use passive structures. These structures are designed to act as soon as they are released from the substrate and to provide whatever information they are designed to supply without further manipulation. For all of these passive structures, the forces acting on them come from the internal residual stresses of the structural material. For devices on the micron scale or smaller, gravitational forces can be neglected, and therefore internal stresses are the only source of actuation force.

Stress Measurements

Since internal residual stresses act upon the passive devices when they are released, it is natural to design a device that can be used to measure residual stresses. One such device, a rotating microstrain gage, is shown in Fig. 50.2. There are many different microstrain gauge designs, but all operate via the same principle. In Fig. 50.2a, the large pads, labeled A, will remain anchored to the substrate when the rest of the device is released. Upon release, the device will expand or contract in order to relieve its internal residual stresses. A structure under tension will contract, and a structure under compression will expand. For the structure in Fig. 50.2, compressive stress will cause the legs to lengthen. Since the two opposing legs are not attached to the central beam at the same point – they are offset, they will cause the central beam to rotate when they expand. The device in Fig. 50.2 contains two independent gauges that point to one another. At the ends of the two central beams are two parts of a Vernier scale. By observing this scale, one can measure the rotation of the beams.

If the connections between the legs and the central beams were simple pin connections, the strain ε of the legs (the fraction of expansion or contraction) could be determined simply by the measured rotation and the geometry of the device, namely

$$\varepsilon = \frac{d_{\text{beam}} d_{\text{offset}}}{2 L_{\text{central}} L_{\text{leg}}} , \qquad (50.2)$$

where d_{beam} is the lateral deflection of the end of one central beam, d_{offset} is the distance between the connections of the opposing legs, L_{central} is the length of the central beam (measured to the center point between the leg connections), and L_{leg} is the length of the leg. However, since the entire device was fabricated from a single polysilicon film, this cannot be the case; some bending must occur at the connections. As a result, to get an accurate determination of the strain relieved upon release, finite element analysis (FEA) of the structure must be performed. This is a common situation for microdevices. FEA is a powerful tool for determining the displacements and stresses of nonideal geometries. One drawback is that the Young's modulus of the material must be known in order to do the FEA as well as to convert the measured strain into a stress value. But Young's moduli are known for many micromachined materials or they can be measured using other techniques.

Other devices besides rotating strain gauges have been designed that can measure residual stresses. One of the simplest is a doubly clamped beam, a long, narrow beam of constant width and thickness that is anchored to the substrate at both ends. If the beam contains a tensile stress it will remain straight, but if the beam contains a compressive stress it will buckle if its length exceeds a critical value l_{cr} according to the Euler buckling criterion [50.8]

$$\varepsilon_{\text{r}} = -\frac{\pi^2}{3} \left(\frac{h}{l_{\text{cr}}} \right)^2 , \qquad (50.3)$$

where ε_{r} is the residual strain in the beam and h is the width or thickness of the beam, whichever is less. To determine the residual strain, a series of doubly clamped beams of varying lengths are fabricated. In this way, the critical length for buckling l_{cr} can be deduced after release. One problem with this technique is that during the release process, any turbulence in the solution will lead to enhanced buckling of the beams, and a low value for l_{cr} will be obtained.

For films with tensile stresses, a similar analysis can be performed using ring-and-beam structures, also called Guckel rings after their inventor, Henry Guckel. A schematic of this design is shown in Fig. 50.3 [50.8].

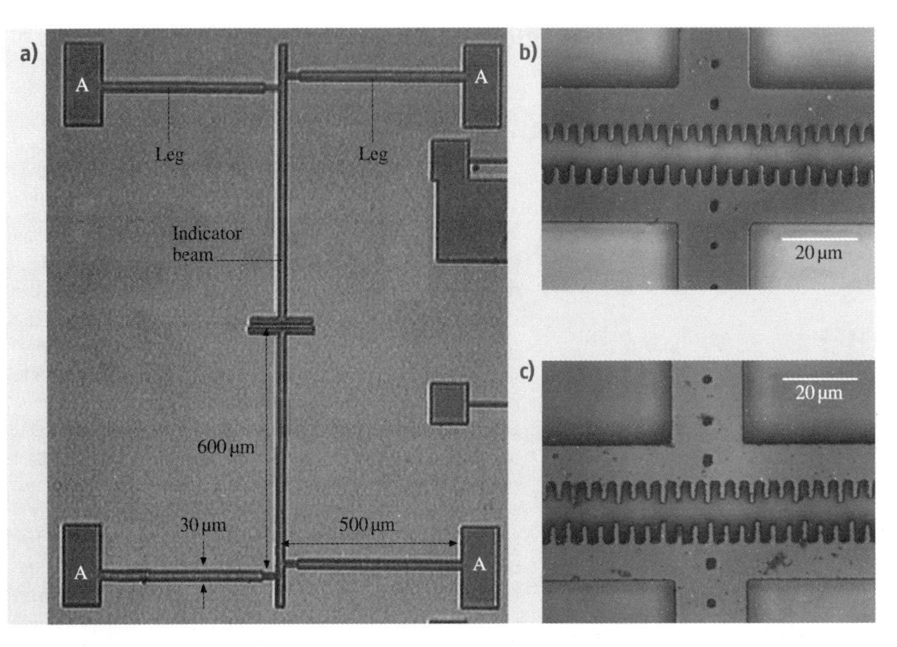

Fig. 50.2 (a) Microstrain gauge fabricated from polysilicon; panel (b) shows a close-up of the Vernier scale before release, and panel (c) shows the same area after release

Tensile stress in the outer ring will cause it to contract. This will lead to compressive stress in the central beam, even though the material was originally tensile before release. The amount of compression in the central beam can be determined analytically from the geometry of the device and the residual strain of the material. Again, by changing the length of the central beam it is possible to determine l_{cr}, and then the residual strain can be deduced.

Stress Gradient Measurements

For structures fabricated from thin deposited films, the stress gradient can be just as important as the stress itself. Figure 50.4 shows a portion of a silicon microactuator. The device is designed to be completely planar; however, stress gradients in the film cause the structures to bend. This figure illustrates the importance of characterizing and controlling stress gradients, and it also demonstrates that stress gradients are most easily

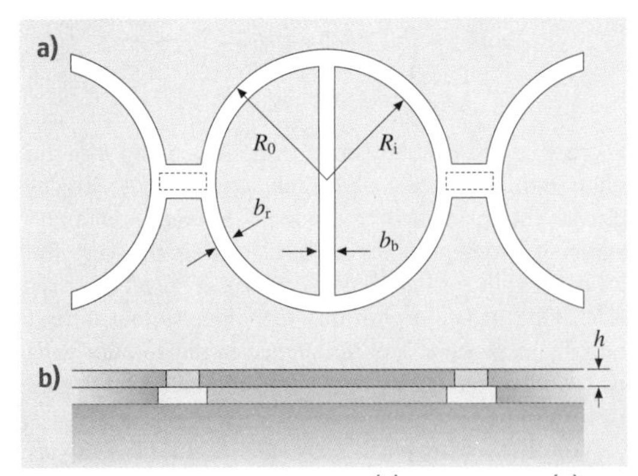

Fig. 50.3a,b Schematic showing (a) top view and (b) side view of Guckel ring structures (after [50.8]). The *dashed lines* in (a) indicate the anchors

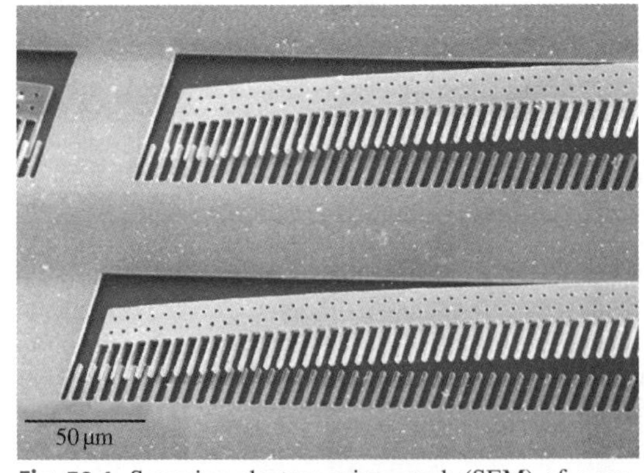

Fig. 50.4 Scanning electron micrograph (SEM) of a portion of a silicon microactuator. Residual stress gradients in the silicon cause the structure to bend

measured for a simple cantilever beam. By measuring the end deflection δ of a cantilever beam of length l and thickness t, the stress gradient $\mathrm{d}\sigma/\mathrm{d}t$ is determined by [50.9]

$$\frac{\mathrm{d}\sigma}{\mathrm{d}t}=\frac{2\delta}{l^2}\frac{E}{1-\nu}\,. \quad (50.4)$$

The magnitude of the end deflection can be measured by microscopy, optical interferometry, or any other technique.

Another useful structure for measuring stress gradients is a spiral. For this structure, the end of the spiral not anchored to the substrate will move out-of-plane. The diameter of the spiral will also contract, and the free end of the spiral will rotate when released [50.13].

Strength and Fracture Toughness Measurements

As mentioned above, if a doubly clamped beam contains a tensile stress, it will remain taut when released because it cannot relieve any of its stress by contracting. This tensile stress can be thought of as a tensile load being applied at the ends of the beam. If this tensile load exceeds the tensile strength of the material, the beam will break. Since the tensile stress can be measured, as discussed in the Sect. 50.2.1, this technique can be used to gather information on the strength of materials. Figure 50.5 shows two different beam designs that have been used to measure strength. The device shown in Fig. 50.5a was fabricated from a tensile polysilicon film [50.10]. Different beams were designed with varying lengths of the wider regions (marked l_1 in the figure). In this manner, the load applied to the narrow center beam was varied, even though the entire film contained a uniform residual tensile stress. For l_1 greater than a critical value, the narrow center beam fractured, giving a measurement for the tensile strength of polysilicon.

The design shown in Fig. 50.5b was fabricated from a tensile Si_xN_y film [50.11]. As seen in the figure, a stress concentration was included in the beam, to ensure the fracture strength would be exceeded. In this case, a notch was etched into one side of the beam. Since the stress concentration is not symmetric with regard to the beam axis, this results in a large bending moment at that position, and the test measures the bend strength of the material. Again, like the beams shown in Fig. 50.5a, the geometry of various beams fabricated from the same film were varied, to vary the maximum stress seen at the notch. By seeing which beams fracture at the stress concentration after release, the strength can be determined.

The fracture toughness of a material can be determined with a similar technique, but an atomically sharp pre-crack is used instead of a stress concentration. Sharp pre-cracks can be introduced into micromachined structures before release by adding a Vickers indent onto the substrate, near the device; the radial crack formed by the indent will propagate into the overlying structure [50.14]. Accordingly, the beam with a sharp pre-crack, shown in Fig. 50.6, was fabricated using polysilicon [50.12]. Due to the stochastic nature of indentation, the initial pre-crack length varies from

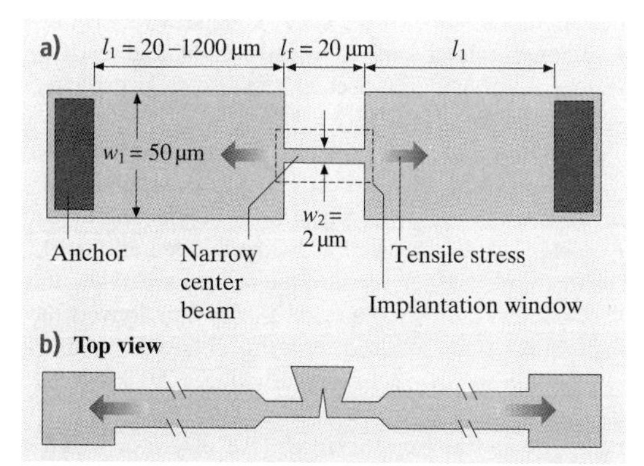

Fig. 50.5a,b Schematic designs of doubly clamped beams with stress concentrations used for measuring strength. (a) was fabricated from polysilicon (after [50.10]) and (b) was fabricated from Si_xN_y (after [50.11])

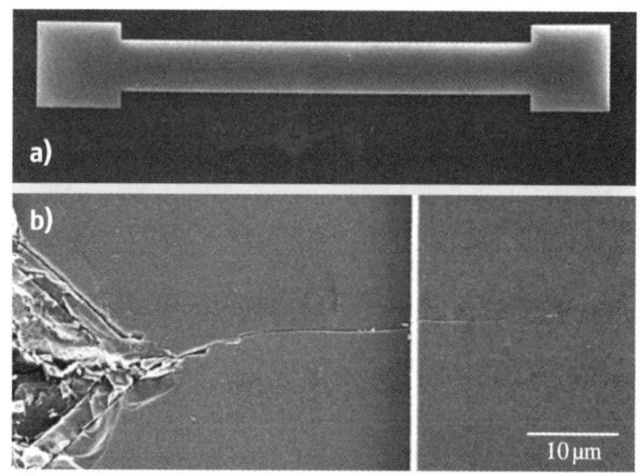

Fig. 50.6 (a) SEM of a 500 µm-long polysilicon beam with a Vickers indent placed near its center; (b) higher magnification SEM of the area near the indent showing the pre-crack traveling from the substrate into the beam (after [50.12])

beam to beam. Because of this, even though the geometry of the beam remains identical, the stress intensity K at the pre-crack tip will vary. Upon release, only those pre-cracks whose K exceeds the fracture toughness of the material K_{Ic} will propagate, and in this way upper and lower bounds for K_{Ic} can be determined for the material.

For materials that do not normally contain residual tensile stresses, composite beams can be fabricated. In this technique, a portion of the beam length is fabricated from a highly tensile material, and the rest of the beam from the material to be tested. The two materials must exhibit perfect adhesion. When released, the tensile material will tend to contract, putting the rest of the beam in tension. This has been demonstrated for evaporated Al films with highly tensile CVD SiN [50.16].

For all of the beams discussed in this section, finite element analysis is required to determine the stress concentrations and stress intensities. Even though approximate analytical solutions may exist for these designs, the actual fabricated structure will not have idealized geometries. For example, corners will never be perfectly sharp, and cracks will never be perfectly straight. This reinforces the idea that FEA is a powerful tool when determining mechanical properties of very small structures.

50.2.2 Active Structures

As discussed above, it is very convenient to design structures that act upon release to provide information on the mechanical properties of the structural materials. This is not always possible, however. For example, those passive devices just discussed rely on residual stresses to create the changes (rotation or fracture) that occur upon release, but many materials do not contain

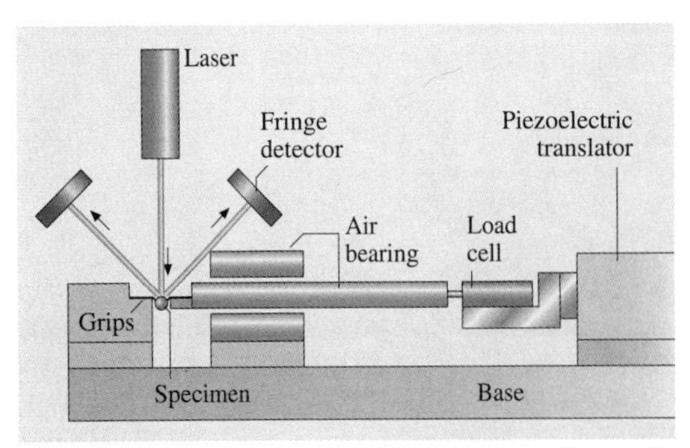

Fig. 50.7 Schematic of a measurement system for tensile loading of micromachined specimens (after [50.15])

high residual stresses as-deposited, or the processing scheme of the device precludes the generation of residual stresses. Also, some mechanical properties, such as fatigue resistance, require motion before they can be studied. Active devices are therefore used. These are acted upon by a force (the source of this force can be integrated into the device itself or can be external to the device) in order to create a change, and the mechanical properties are studied via the response to the force.

Young's Modulus Measurements

Young's modulus E is a material property critical to any structural device design. It describes the elastic response of a material and relates stress σ and strain ε by

$$\sigma = E\varepsilon . \tag{50.5}$$

In bulk samples, E is often measured by loading a specimen under tension and measuring displacement as a function of stress for a given length. While this is far more difficult for small structures, such as those fabricated from thin deposited films, it can be achieved with careful experimental techniques. Figure 50.7 shows a schematic of one such measurement system [50.15]. The fringe detectors in the figure detect the reflected laser signal from two gold lines deposited onto the polysilicon specimen, which act as gauge markers. This enables the strain in the specimen during loading to be monitored. Besides gold lines, Vickers indents placed in a nickel specimen can also serve as gauge markers [50.17], or a speckle interferometry technique [50.18] can be used to determine strain in the specimen. Once the stress-versus-strain behavior is measured, the slope of the curve is equal to E. By using a constant load, such as a dead weight, and resistive heating, high-temperature creep can also be investigated with this method [50.19].

In addition to the tensile test, Young's modulus can be determined by other measures of stress–strain behavior. As seen in Fig. 50.8, a cantilever beam can be bent by pushing on the free end with a nanoindenter [50.20]. The nanoindenter can monitor the force applied and the displacement, and simple beam theory can convert the displacement into strain in order to obtain E. A similar technique, shown schematically in Fig. 50.9 [50.21], involves pulling downward on a cantilever beam by means of an electrostatic force. An electrode is fabricated into the substrate beneath the cantilever beam, and a voltage is applied between the beam and the bottom electrode. The force acting on the beam is equal to the electrostatic force corrected to include the effects of

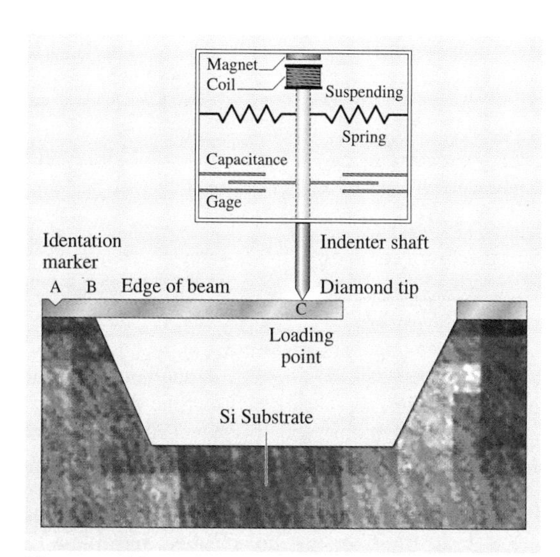

Fig. 50.8 Schematic of a nanoindenter loading mechanism pushing on the end of a cantilever beam (after [50.20])

fringing fields acting on the sides of the beam, namely

$$F(x)=\frac{\varepsilon_0}{2}\left(\frac{V}{g+z(x)}\right)^2\left(1+\frac{0.65[g+z(x)]}{w}\right), \tag{50.6}$$

where $F(x)$ is the electrostatic force at x, ε_0 is the dielectric constant of air, g is the gap between the beam and the bottom electrode, $z(x)$ is the out-of-plane deflection of the beam, w is the beam width, and V is the applied voltage [50.21]. In this work, the deflection of the beam as a function of position is measured using optical interferometry. These measurements combine to give stress–strain behavior for the cantilever beam. An extension of this technique uses doubly clamped beams instead of cantilever beams. In this case, the deflection of the beam at a given electrostatic force depends on the

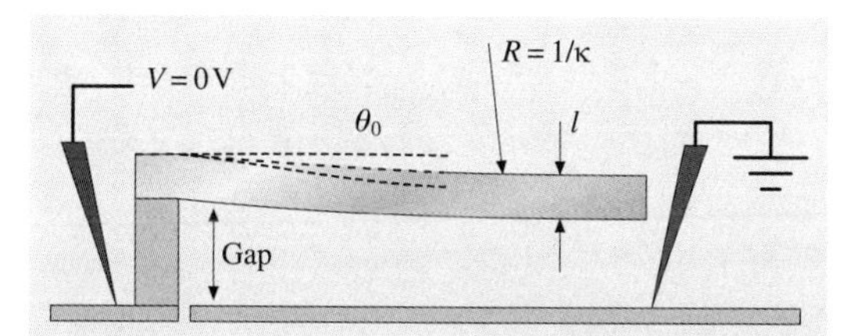

Fig. 50.9 Schematic of a cantilever beam bending test using an electrostatic voltage to pull the beam toward the substrate (after [50.21])

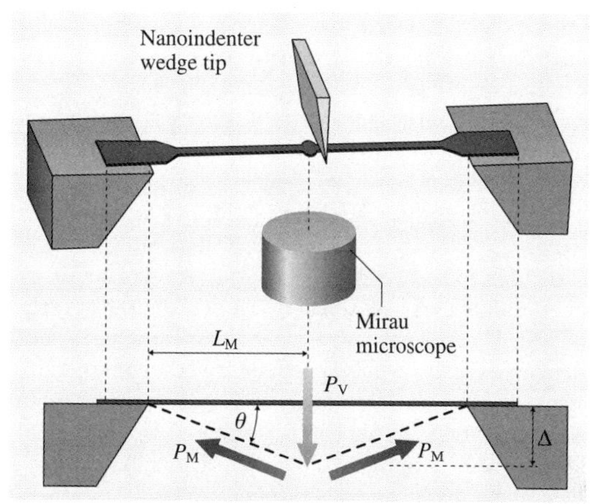

Fig. 50.10 Schematic of an externally loaded doubly-clamped beam (after [50.22])

residual stress in the material as well as Young's modulus. This method can therefore also be used to measure residual stresses in doubly clamped beams.

Similarly, an external load can be applied to the center of the doubly-clamped beam, instead of using an electrostatic force, as shown in Fig. 50.10 [50.22]. The beam deflection is monitored using optical interferometry. Yet another variation is illustrated in

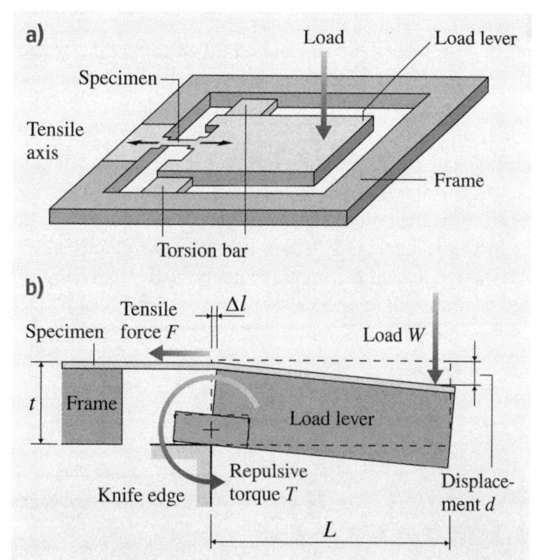

Fig. 50.11a,b Schematic of a device that transfers external vertical loads to in-plane tensile loads in deposited films (after [50.23])

Fig. 50.11 [50.23]. In this technique, the externally applied downward force is transferred to a film of any material deposited onto the device.

Another device that can be fabricated from a thin film and used to investigate stress–strain behavior is a suspended membrane, as shown in Fig. 50.12 [50.24]. As depicted in the schematic figure, the membrane is exposed to an elevated pressure on one side, causing it to bulge in the opposite direction. The deflection of the membrane is measured by optical or other techniques and related to the strain in the membrane. These membranes can be fabricated in any shape, typically square or circular. Both analytical solutions and finite element analyses have been performed to relate the deflection to the strain. Like the doubly clamped beams, both Young's modulus and residual stress play a role in the deflected shape. Both of these mechanical properties can therefore be determined by the pressure-versus-deflection performance of the membrane.

Another measurement besides stress–strain behavior that can reveal the Young's modulus of a material is the determination of the natural resonance frequency. For a cantilever, the resonance frequency f_r for free undamped vibration is given by

$$f_r = \frac{\lambda_i^2 t}{4\pi l^2}\left(\frac{E}{3\rho}\right)^{1/2} , \tag{50.7}$$

where ρ, l and t are the density, length and thickness of the cantilever; λ_i is the eigenvalue, where i is an integer that describes the resonance mode number; for the first mode $\lambda_1 = 1.875$ [50.26]. Given the geometry and density, measuring f_r allows E to be determined. The cantilever can be vibrated by a number of techniques, including a laser, loudspeaker or piezoelectric shaker. The frequency that produces the highest amplitude of vibration is the resonance frequency.

A micromachined device that uses an electrostatic comb drive and an AC signal to generate the vibration of the structure is known as a lateral resonator [50.27]. One example is shown in Fig. 50.13 [50.25]. When a voltage is applied across either set of the interdigitated comb fingers shown in Fig. 50.13, an electrostatic attraction is generated due to the increase in capacitance as the overlap between the comb fingers increases. The force F generated by the comb-drive is given by

$$F = \frac{1}{2}\frac{\partial C}{\partial x}V^2 = n\varepsilon\frac{h}{g}V^2 , \tag{50.8}$$

where C is capacitance, x is the distance traveled by one comb-drive toward the other, n is the number of pairs of comb fingers in one drive, ε is the permittivity of the fluid between the fingers, h is the height of the fingers, g is the gap spacing between the fingers, and V is the applied voltage [50.27]. When an AC voltage at the resonance frequency is applied across either of the two comb drives, the central portion of the device will vibrate. In fact, since force depends on the square of the voltage for electrostatic actuation, for a time t, a dependent drive voltage $v_D(t)$ (given by

$$v_D(t) = V_P + v_d \sin(\omega t) , \tag{50.9}$$

where V_P is the DC bias and v_d is the AC drive amplitude), the time-dependent portion of the force will scale with

$$2\omega V_P v_d \cos(\omega t) + \omega v_d^2 \sin(2\omega t) \tag{50.10}$$

[50.27]. Therefore, if an AC drive signal is used with no DC bias, at resonance, the frequency of the AC drive signal will be one half of the resonance frequency. For

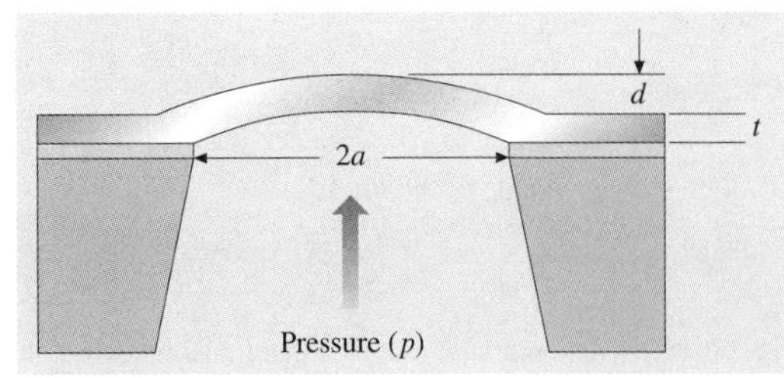

Fig. 50.12 Schematic cross section of a microfabricated membrane (after [50.24])

Fig. 50.13 SEM of a polysilicon lateral resonator (after [50.25])

this device, the resonance frequency f_r will be

$$f_r = \frac{1}{2\pi}\left(\frac{k_{sys}}{M}\right)^{1/2} , \quad (50.11)$$

where k_{sys} is the spring constant of the support beams and M is the mass of the portion of the device that vibrates. The spring constant is given by

$$k_{sys} = 24EI/L^3 , \quad (50.12)$$

$$I = \frac{hw^3}{12} , \quad (50.13)$$

where I is the moment of inertia of the beams, and L, h and w are the length, thickness, and width of each beam. Therefore, by combining these equations and measuring f_r, it is possible to determine E.

One distinct advantage of the lateral resonator technique and the electrostatically pulled cantilever technique for measuring Young's modulus is that they require no external loading sources. Portions of the devices are electrically contacted, and a voltage is applied. For the pure tension tests, as shown in Fig. 50.7, the specimen must be attached to a loading system, which can be extremely difficult for the very small specimens discussed here, and any misalignment or eccentricity in the test could lead to unreliable results. However, the advantage of the externally loaded technique is that there are no limitations on the type of materials that can be tested. Conductivity is not a requirement, nor is any compatibility with electrical actuation.

Strength and Fracture Toughness Measurements

As one might expect, any of the techniques discussed in the previous section that strain specimens in order to measure Young's modulus can also be used to measure fracture strength. The load is simply increased until the specimen breaks. As long as either the load or the strain is measured at fracture, and the geometry of the specimen is known, the maximum stress required for fracture σ_{crit} can be determined, either through analytical analysis or FEA. Depending on the geometry of the test, σ_{crit} will represent the tensile or bend strength of the material.

If the available force is limited, or if a localized fracture site is desired, stress concentrations can be added to the specimens. These are typically notches micromachined into the edges of specimens. Focused ion beams have also been used to carve stress concentrations into fracture specimens.

All of the external loading schemes, such as those shown in Figs. 50.7 and 50.8, have been used to measure fracture strength. Also, the electrostatically loaded doubly clamped beams can be pulled until they fracture. In this case, there is one complication. The electrostatic force is inversely proportional to the distance between the electrodes, and at a certain voltage, called the *pull-in voltage,* the attraction between the beam and the substrate will become so great that the beam will immediately be pulled into contact with the bottom electrode. As long as the fracture takes place before the pull-in voltage is reached, the experiment will give valid results.

Other loading techniques have been used to generate fracture of microspecimens. Figure 50.14 [50.28] shows one device designed to be pushed by the end of a micromanipulated needle. The long beams that extend from the sides of the central shuttle come into contact with anchored posts, and, at a critical degree of bending, the beams will break off. Since the applied force cannot be measured in this technique, the experiment is continuously optically monitored during the test, and the image of the beams just before fracture is analyzed to determine σ_{crit}.

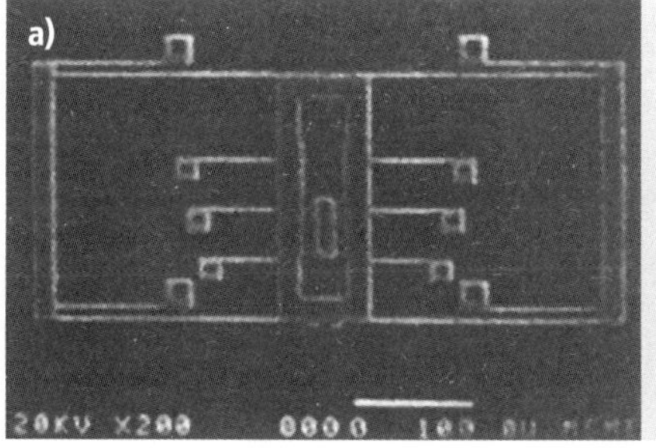

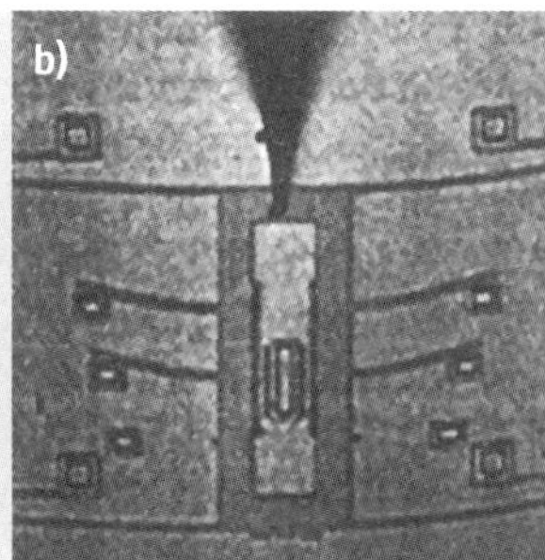

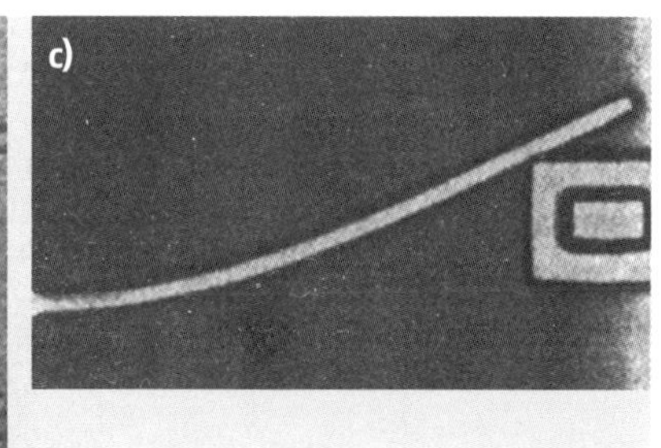

Fig. 50.14 **(a)** SEM of a device for measuring bend strength of polysilicon beams; **(b)** image of a test in process; **(c)** higher magnification view of one beam shortly before breaking (after [50.28])

Another loading scheme that has been demonstrated for micromachined specimens utilizes scratch drive actuators to load the specimens [50.29]. These types of actuators work like inchworms, traveling across a substrate in discrete advances as an electrostatic force is repeatedly applied between the actuator and the substrate. The stepping motion can be made on the nanometer scale, depending on the frequency of the applied voltage, and so it can be an acceptable approximation to continuous loading. One advantage of this scheme is that very large forces can be generated by relatively small devices. The exact forces generated cannot be measured, so (like the technique that used micromanipulated pushing) the test is continuously observed to determine the strain at fracture. Another advantage of this technique is that, like the lateral resonator and the electrically pulled cantilever, the loading takes place on-chip, and therefore the difficulties associated with attaching and aligning an external loading source are eliminated.

Another on-chip actuator used to load microspecimens is shown in Fig. 50.15, along with three different microspecimens [50.14]. Devices have been fabricated with each of the three microspecimens integrated with the same electrostatic comb-drive actuator. In all three cases, when a DC voltage is applied to the actuator, it moves downward, as oriented in Fig. 50.15. This pulls down on the left end of each of the three microspecimens, which are anchored on the right. The actuator contains 1486 pairs of comb fingers. The maximum voltage that can be applied is limited by the breakdown voltage of the medium in which the test takes place. In air, this limits the voltage to less than 200 V. As a result, given a finger height of 4 μm and a gap of 2 μm, and using (50.8), the maximum force generated by this actuator is limited to about 1 mN. Standard optical photolithography has a minimum feature size of about 2 μm. As a result, the electrostatic actuator cannot generate sufficient force to perform a standard tensile test on MEMS structural materials such as polysilicon. The microspecimens shown in Fig. 50.15 are therefore designed such that the stress is amplified.

The specimen shown in Fig. 50.15b is designed to measure bend strength. It contains a micromachined notch with a root radius of 1 μm. When the actuator pulls downward on the left end of this specimen, the notch serves as a stress concentration, and when the stress at the notch root exceeds σ_{crit}, the specimen fractures. The specimen in Fig. 50.15c is designed to test tensile strength. When the left end of this specimen is pulled downward, a tensile stress is generated in the upper thin horizontal beam near the right end of the specimen. As the actuator continues to move downward, the tensile stress in this beam will exceed the tensile strength, causing fracture. Finally, the specimen in Fig. 50.15d is similar to that in Fig. 50.15b, except

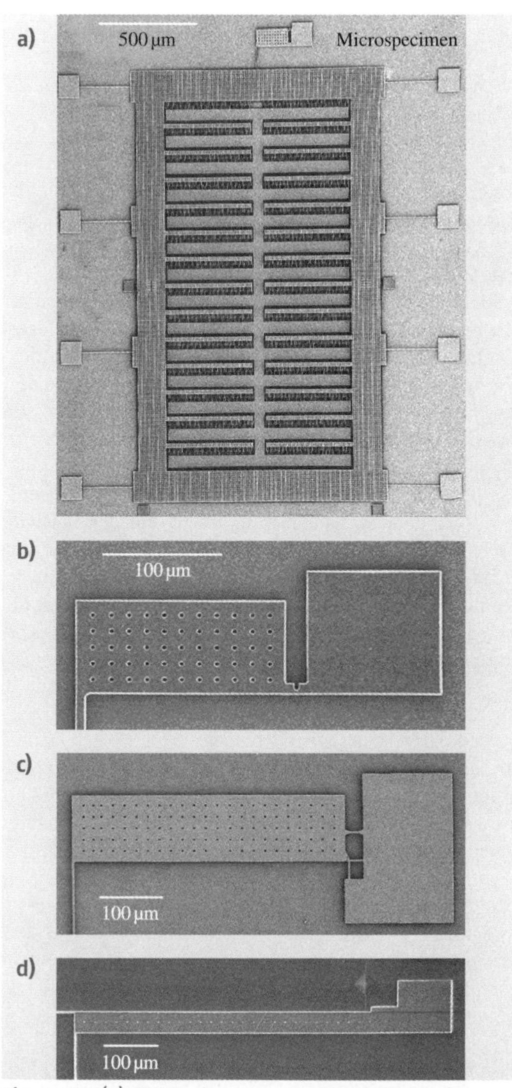

Fig. 50.15 (a) SEM of a micromachined device for conducting strength tests; the device consists of a large comb-drive electrostatic actuator integrated with a microspecimen; **(b–d)** SEMs of various microspecimens for testing bend strength, tensile strength and fracture toughness, respectively (after [50.14])

that the notch is replaced by a sharp pre-crack that was produced by the Vickers indent placed on the substrate near the specimen. When this specimen is loaded, a stress intensity K is generated at the crack tip. When the stress intensity exceeds a critical value K_{Ic}, the crack propagates. K_{Ic} is also referred to as the fracture toughness.

The force generated by the electrostatic actuator can be calculated using (50.8). However, (50.8) assumes a perfectly planar, two-dimensional device. In fact, when actuated, the electric fields extend out of the plane of the device, and so (50.8) is just an approximation. Instead, like many of the techniques discussed in this section, the test is continuously monitored, and the actuator displacement at the time of fracture is recorded. Then FEA is used to determine the magnitude of the stress or stress intensity seen by the specimen at the point of fracture.

In order to generate sufficient force to conduct tensile tests, a similar device to that shown in Fig. 50.15 has been designed which uses an array of parallel plate capacitors to provide the force, instead of comb-drives [50.31]. In this way the available force is increased but the maximum stroke is severely limited.

Fatigue Measurements

A benefit of the electrostatic actuator shown in Fig. 50.15 is that, besides monotonic loading, it can generate cyclic loading. This allows the fatigue resistance of materials to be studied. Simply by using an AC signal instead of a DC voltage, the device can be driven at its resonant frequency. The amplitude of the resonance depends on the magnitude of the AC signal. This amplitude can be increased until the specimen breaks; this will investigate the low-cycle fatigue resistance. Otherwise, the amplitude of resonance can be left constant at a level below that required for fast fracture, and the device will resonate indefinitely until the specimen breaks; this will investigate high-cycle fatigue. It should be noted that the resonance frequency of such a device is about 10 kHz. Therefore, it is possible to stress a specimen for over 10^9 cycles in less than a day. In addition to simple cyclic loading, a mean stress can be superimposed on the cyclic load if a DC bias is added to the AC signal. In this way, nonsymmetric cyclic loading (with a large tensile stress alternating with a small compressive stress, or vice versa) can be studied.

Another device that can be used to investigate fatigue resistance in MEMS materials is shown in Fig. 50.16 [50.30]. In this case, a large mass is attached to the end of a notched cantilever beam. The mass contains two comb drives on opposite ends. When an AC signal is applied to one comb drive, the device will resonate, cyclically loading the notch. The comb drive on the opposite side is used as a capacitive displacement sensor. This device contains many fewer comb fingers than the device shown in Fig. 50.15. As a result, it can apply cyclic loads by exploiting the resonance frequency of the device, but it cannot supply sufficient force to achieve monotonic loading.

Fatigue loading has also been studied using the same external loading techniques shown in Fig. 50.7. In this case, the frequency of the cyclic load is considerably lower, since the resonance frequency of the device is not being utilized. This leads to longer high-cycle testing times. Since the force is essentially unlimited, however, this technique allows a variety of frequencies to be studied to determine their effect on the fatigue behavior.

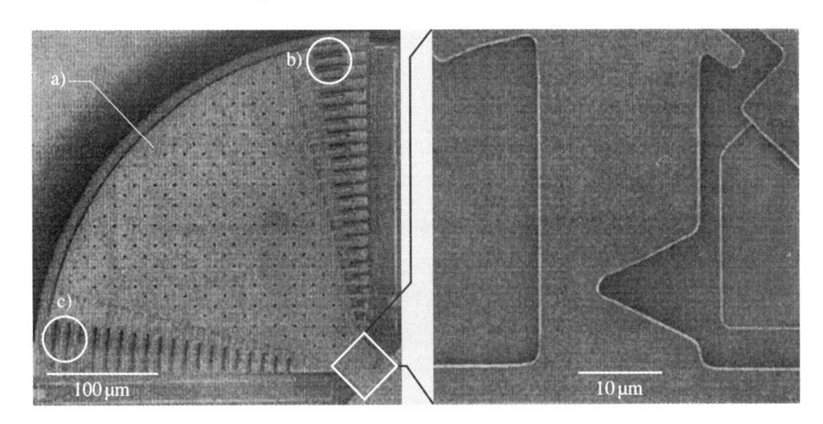

Fig. 50.16 SEM of a device used to investigate fatigue; the image *on the right* is a higher magnification view of the notch near the base of the moving part of the structure (after [50.30]): a) mass, b) comb-drive actuator, c) capacitive displacement sensor

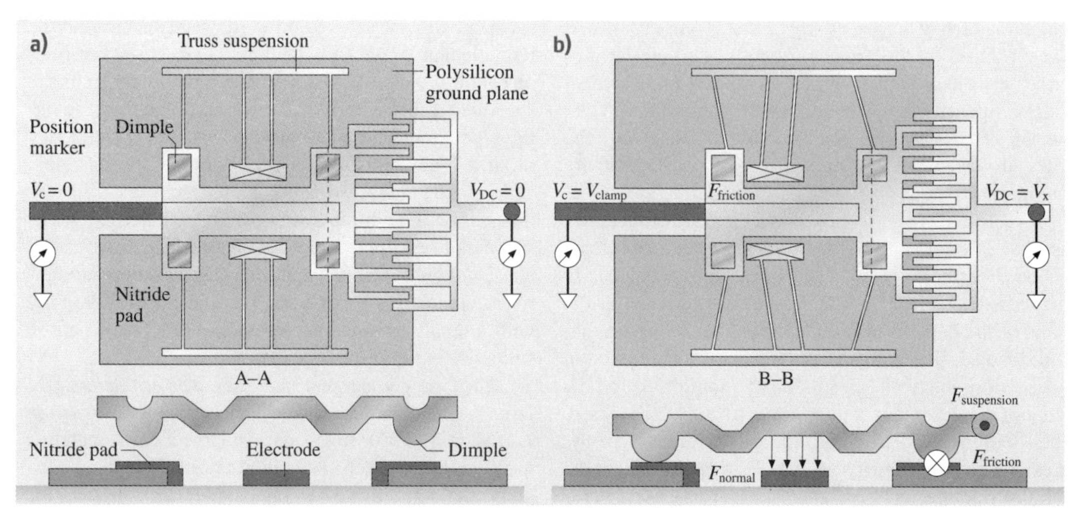

Fig. 50.17a,b Schematics depicting a device used to study friction. Panel (**a**) shows top and side views of the device in its original position, and panel (**b**) shows views of the device after it has been displaced using the comb-drive and clamped using the substrate electrode (after [50.32])

Friction and Wear Measurements

Friction is another property that has been studied in micromachined structures. To study friction, of course, two surfaces must be brought into contact with each other. This is usually avoided at all costs for these devices because of the risk of stiction. (Stiction is the term used when two surfaces that come into contact adhere so strongly that they cannot be separated.) Even so, a few devices have been designed that can investigate friction. One of these is shown schematically in Fig. 50.17 [50.32]. It consists of a movable structure with a comb-drive on one end and a cantilever beam on the other. Beneath the cantilever, on the substrate, is a planar electrode. The device is moved to one side using the comb-drive. Then a voltage is applied between the cantilever beam and the substrate electrode. The voltage on the comb-drive is then released. The device would normally return to its original position, to relax the deflection in the truss suspensions, but the friction between the cantilever and the substrate electrode holds it in place. The voltage to the substrate electrode is slowly decreased until the device starts to slide. Given the electrostatic force generated by the substrate electrode and the stiffness of the truss suspensions, it is possible to determine the static friction. For this device, bumps were fabricated on the bottom of each cantilever beam. This limited the surface area that came into contact with the substrate and so lowered the risk of stiction.

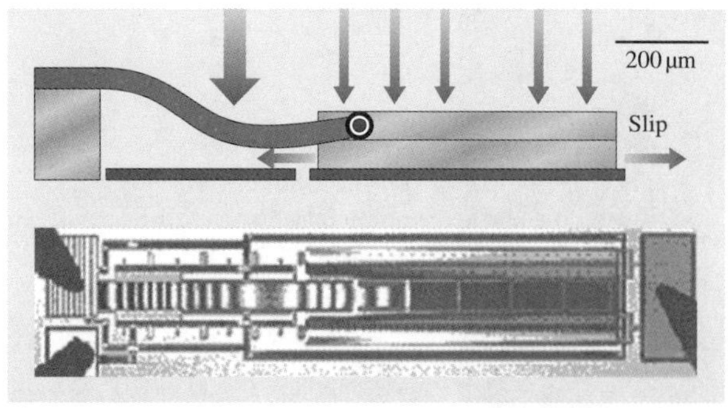

Fig. 50.18 Schematic cross-section and top-view optical micrograph of a hinged-cantilever test structure for measuring friction in micromachined devices (after [50.33])

Another device designed to study friction is shown in Fig. 50.18 [50.33]. This technique uses a hinged cantilever. The portion near the free end acts as the friction test structure, and the portion near the anchored end acts as the driver. The friction test structure is attracted to the substrate by means of electrostatic actuation, and when a second electrostatic actuator pulls down the driver, the friction test structure slips forward by a length proportional to the forces involved, including the frictional force. This distance, however, has a maximum of 30 nm, so all measurements must be exceedingly accurate in order to investigate a range of forces. This test struc-

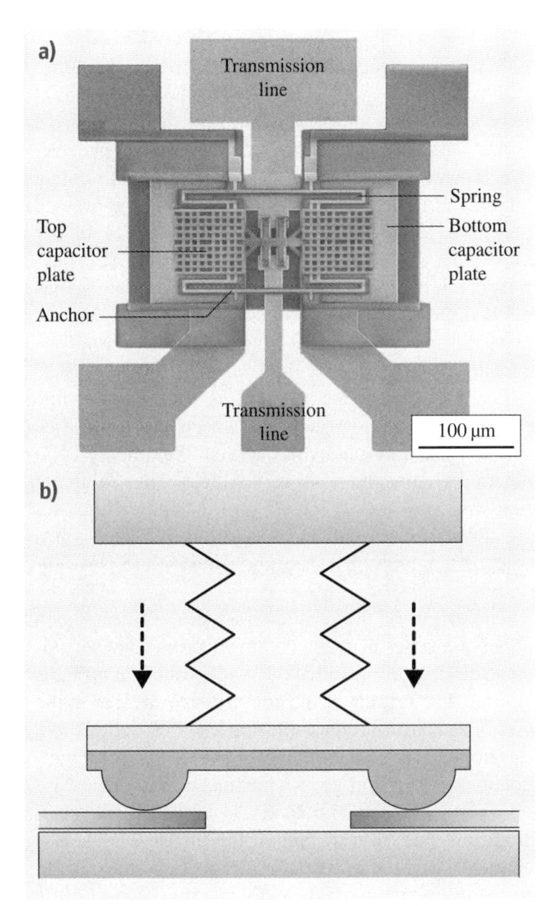

Fig. 50.19 (a) Micrograph of an RF MEMS switch designed by Rockwell Scientific, and (b) schematic drawing showing the movement of the metal bridge as it comes into contact with the underlying metal plate during switch operation (after [50.34]) ◄

ture can be used to determine the friction coefficients for surfaces with and without lubricating coatings.

Wear can occur whenever friction between two contacting components is nonzero. Wear mechanisms can entail plastic deformation or brittle fracture. Adhesive wear involves fusion between asperities of two different components, followed by fracture. Wear in polysilicon devices has been investigated using transmission electron microscopy [50.35].

Metal-to-Metal Contact

Recently, radiofrequency (RF) MEMS switches have been developed that are vast improvements over conventional technology. During operation, these devices endure unique mechanical metal-to-metal contacts that can be serious reliability concerns. An example of an RF MEMS switch is shown in Fig. 50.19 [50.34]. As illustrated in Fig. 50.19b, to close the switch the metal bumps are brought into contact with a metal plate, creating a short circuit. When this occurs, significant pressures are developed, depending on the roughness and asperity heights on the metal surfaces. Also, if currents are flowing, Joule heating and arcing can also take place. If the contact area becomes large, through surface roughness modification during operation, the adhesive forces between the two metal components can exceed the restoring force of the springs, and the switch can become stuck in the closed position. Alternatively, high currents can result in melting and short circuits [50.36]. To investigate this mechanical deformation during repeated impact, the RF switches themselves are usually examined before and after operation. A separate testing device has not been designed that could provide any accelerated testing conditions.

50.3 Measurements of Mechanical Properties

All of the techniques discussed in Sects. 50.1 and 50.2 have been used to measure the mechanical properties of MEMS and NEMS materials. As a general rule, the results from the various techniques have agreed well with each other, and the argument becomes which of the measurement techniques is easiest and most reliable to perform. It is crucial to bear in mind, however, that certain properties (such as strength) are process-dependent, and so the results taken at one laboratory will not necessarily match those taken from another. This will be discussed in more detail in Sect. 50.3.1.

50.3.1 Mechanical Properties of Polysilicon

In current MEMS technology, the most widely used structural material is polysilicon deposited by low-pressure chemical vapor deposition (LPCVD). One reason for the prevalence of polysilicon is the large body of processing knowledge for this material that has been developed by the integrated circuit community. Another reason, of course, is that polysilicon possesses a number of qualities that are beneficial to MEMS devices, in particular high strength and Young's modulus. Therefore, most of the mechanical properties investigations performed on MEMS materials have focused on polysilicon.

Residual Stresses in Polysilicon

The residual stresses of LPCVD polysilicon have been thoroughly characterized using both the wafer curvature technique, discussed in Sect. 50.1.1, and the microstrain gauges, discussed in Sect. 50.2.1. The results from both techniques give consistent values. Figure 50.20 summarizes the residual stress measurements as a function of deposition temperature taken from five different investigations at five different laboratories [50.37]. All five sets of data show the same trend. The stresses change from compressive at the lowest deposition temperatures to tensile at intermediate temperatures and back to compressive at the highest temperatures. The exact transition temperatures vary somewhat between the different investigations, probably due to differences in the deposition conditions: the silane or dichlorosilane pressure, the gas flow rate, the geometry of the deposition system, and the temperature uniformity. However, in each data set the transitions are easily discernible. The origin of these residual stress changes lies with the microstructure of the LPCVD polysilicon films.

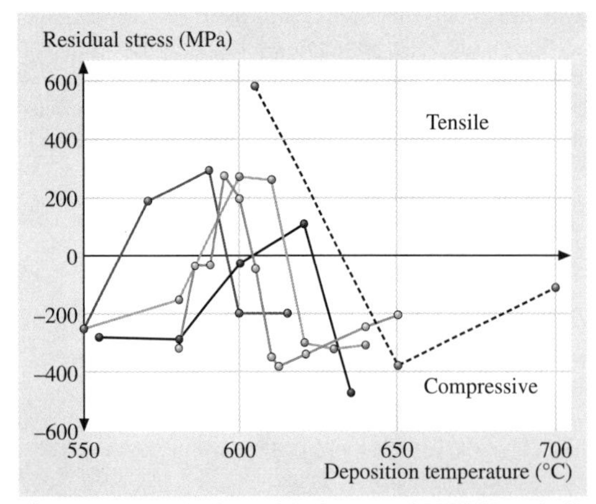

Fig. 50.20 Results for residual stress of LPCVD polysilicon films taken from five different investigations (after [50.37]). Data from the same investgation are connected by a *line*

As with all deposited films, the microstructure of the LPCVD polysilicon film is dependent on the deposition conditions. In general, the films are amorphous at the lowest growth temperatures (lower than $\sim 570\,°C$), display fine ($\sim 0.1\,\mu m$ diameter) grains at intermediate temperatures (~ 570 to $\sim 610\,°C$), and contain columnar (110)-textured grains with a thin fine-grained nucleation layer at the substrate interface at higher temperatures (~ 610 to $\sim 700\,°C$) [50.37]. The fine-grained microstructure results from the homogeneous nucleation and growth of silicon grains within an as-deposited amorphous silicon film. In this regime, the deposition rate is just slightly faster than the crystallization rate. The as-deposited films will be crystalline near the substrate interface and amorphous at the free surface. (The amorphous fraction can be quickly crystallized by annealing above 610 °C.) The columnar microstructure seen at higher growth temperatures results from the formation of crystalline silicon films as-deposited, with growth being fastest in the $\langle 110 \rangle$ directions.

The origin of the tensile stress in the fine-grained polysilicon arises from the volume decrease that accompanies the crystallization of the as-deposited amorphous material. The origins of the compressive stresses in the amorphous and columnar films are less well understood. One proposed explanation for compressive stress generation during thin film growth postulates that an increase in the surface chemical potential is caused by the deposition of atoms from the vapor; the increase in surface chemical potential induces atoms to flow into newly formed grain boundaries, creating a compressive stress in the film [50.38].

Stress gradients are also typical of LPCVD polysilicon films. The partially amorphous films contain large stress gradients since they are essentially bilayers of compressive amorphous silicon on top of tensile fine-grained polysilicon. The fully crystalline films also exhibit stress gradients. The columnar compressive films are most highly stressed at the film–substrate interface, with the compressive stresses decreasing as the film thickness increases; the fine-grained films are less tensile at the film–substrate interface, with the tensile stresses increasing as the film thickness increases [50.37]. Both stress gradients are associated with microstructural variations. For the columnar films, the initial nucleation layer corresponds to a very high compressive stress, which decreases as the columnar morphology develops. For the fine-grained films, the region near the film–substrate interface has a slightly smaller average grain size, due to heterogeneous nucleation at the interface. This region displays a slightly lower tensile stress than the rest of the film, since the increased grain boundary area reduces the local density.

Young's Modulus of Polysilicon

The Young's modulus of polysilicon films has been measured using all of the techniques discussed in Sect. 50.2.2. A good review of the experimental results taken from bulge testing, tensile testing, beam bending and lateral resonators are contained in [50.39]. All of the reported results are in reasonable agreement, varying from 130 to 175 GPa, though many values are reported with a relatively high experimental scatter. The main origin of the error in these results is the uncertainties involving the geometries of the small specimens used to make the measurements. For example, from (50.13), the Young's modulus determined by the lateral resonators depends on the cube of the tether beam width, typically about 2 μm. In general, the beam width and other dimensions can be measured via scanning electron microscopy to within about 0.1 μm; however, the width of the beam is not perfectly constant along the entire length or even throughout the thickness of the beam. These uncertainties in geometry lead to uncertainties in modulus.

In addition, the various experimental measurements lie close to the Voigt and Reuss bounds for Young's modulus calculated using the elastic stiffnesses and compliances for single-crystal silicon [50.39]. This strongly implies that Young's moduli of micro- and nanomachined polysilicon structures will be the same as for bulk samples made from polysilicon. This is not unexpected, since Young's modulus is a material property. It is related to interatomic interactions and should have no dependence on the geometry of the sample. It should be noted that polysilicon can display a preferred crystallographic orientation depending on the deposition conditions, and that this could affect the Young's modulus of the material, since the Young's modulus of silicon is not isotropic. However, the anisotropy is fairly small for cubic silicon.

A more recent investigation that utilized electrostatically actuated cantilevers and interferometric deflection detection yielded a Young's modulus of 164 GPa [50.21]. They found the grains in their polysilicon films to be randomly oriented, and calculated the Voigt and Reuss bounds to be 163.4–164.4 GPa. This appears to be a very reliable value for randomly oriented polysilicon.

Fracture Toughness and Strength of Polysilicon

Using the device shown in Fig. 50.15a and the specimen shown in Fig. 50.15d, the fracture toughness K_{Ic} of polysilicon has been shown to be $1.0 \pm 0.1\,\mathrm{MPa\,m^{1/2}}$ [50.40]. Several different polysilicon microstructures were tested, including fine-grained, columnar and multilayered. Amorphous silicon was also investigated. All of the microstructures displayed the same K_{Ic}. This indicates that, like Young's modulus, fracture toughness is a material property, independent of the material microstructure or the geometry of the sample.

A tensile test, such as that shown in Fig. 50.7 but using a sample with sharp indentation-induced pre-cracks, yields a K_{Ic} of $0.86\,\mathrm{MPa\,m^{1/2}}$ [50.41]. The passive, residual stress loaded beams with sharp pre-cracks shown in Fig. 50.6 gave a K_{Ic} of $0.81\,\mathrm{MPa\,m^{1/2}}$ [50.12].

Given that K_{Ic} is a material property for polysilicon, the measured fracture strength σ_{crit} is related to K_{Ic} by

$$K_{\mathrm{Ic}} = c\sigma_{\mathrm{crit}}(\pi a)^{1/2}\,, \qquad (50.14)$$

where a is the crack-initiating flaw size, and c is a constant of order unity. The value for c will depend on the exact size, shape and orientation of the flaw; for a semicircular flaw, c is equal to 0.71 [50.42]. Therefore, any differences in the reported fracture strength of polysilicon will be the result of changes in a.

A good review of the experimental results available in the literature for polysilicon strength is contained in [50.43]. The tensile strength data vary from about 0.5 to 5 GPa. Like many brittle materials, the measured strength of polysilicon is found to obey Weibull statistics [50.44, 45]. This implies that the polysilicon samples contain a random distribution of flaws of various sizes, and that the failure of any particular specimen will occur at the largest flaw that experiences the highest stress. One consequence of this behavior is that, since larger specimens have a greater probability of containing larger flaws, they will exhibit decreased strengths. More specifically, it was found that the most important geometrical parameter is the surface area of the sidewalls of a polysilicon specimen [50.43] and single crystal silicon [50.46] specimens. The sidewalls, as opposed to the top and bottom surfaces, are those surfaces created by etching the polysilicon film. This is not surprising since LPCVD polysilicon films contain essentially no flaws within the bulk, and the top and bottom surfaces are typically very smooth.

As a result, the etching techniques used to create the structures will have a strong impact on the fracture strength of the material. For single-crystal silicon specimens it was found that the choice of etchant could change the observed tensile strength by a factor of two [50.47] and by about 20% for polysilicon [50.48]. In addition, the bend strength of amorphous silicon was measured to be twice that of polysilicon for specimens processed identically [50.40]. It was found that the reactive ion etching used to fabricate the specimens produced much rougher sidewalls on the polysilicon than on the amorphous silicon.

The tensile strength of single crystal silicon has also been measured using a technique similar to that shown in Fig. 50.8. Sharp notches were introduced into the beams using a focused ion beam, and the apparent fracture toughness was measured for a variety of planes parallel to the notch front, along which the crack propagated. For the {110} notch plane, the fracture toughness was about $1\,\mathrm{MPa\,m^{1/2}}$, and for the {100} notch plane it was about $2\,\mathrm{MPa\,m^{1/2}}$ [50.49, 50] and for the {100} notch plane it was about $1\,\mathrm{MPa\,m^{1/2}}$ [50.50] or about $2\,\mathrm{MPa\,m^{1/2}}$ [50.49].

Fatigue of Polysilicon

Fatigue failure involves fracture after a number of load cycles, when each individual load is not sufficient by itself to generate catastrophic cracking in the material. For ductile materials, such as metals, fatigue occurs due to accumulated damage at the site of maximum stress and involves local plasticity. As a brittle material, polysilicon would not be expected to be susceptible to cyclic fatigue. However, fatigue has been observed for polysilicon tensile samples [50.41], polysilicon bend specimens with notches [50.30, 51], and polysilicon bend specimens with sharp cracks [50.52]. The exact origins of the fatigue behavior are still subject to debate, but some aspects of the experimental data are that the fatigue lifetime does not depend on the loading frequency [50.41], the fatigue behavior is affected by the ambient [50.12, 52], fatigue occurs faster in higher humidities [50.53], and the fatigue depends on the ratio of compressive to tensile stresses seen in the load cycle [50.12].

Fatigue in polysilicon has been found to either cause thickening of the native oxide in the highly stressed regions [50.54], or to cause no thickening of the native oxide in the highly stressed regions [50.55]. Even more strangely, fatigue cycling under certain conditions can lead to apparent strengthening of the polysilicon [50.56, 57]. While the exact mechanism of fatigue in polysilicon is not known, dislocations have been observed in tensile fracture of polysilicon at room temperature [50.58], which implies a role for dislocations in fatigue behavior.

Friction of Polysilicon

The friction of polysilicon structures has been measured using the techniques described in Sect. 50.2.2. The measured coefficient of friction was found to vary from 4.9 [50.32] to 7.8 [50.33].

The static and dynamic friction of polysilicon coated with monolayer lubricants has been measured with a device similar to that shown in Fig. 50.17, but using a scratch-drive actuator instead of a comb-drive [50.59]. The dynamic friction at $0.2\,\mathrm{m/s}$ was approximately 80% of the static friction value; the static friction at zero applied load was due to an adhesive force of $0.95\,\mathrm{nN/\mu m^2}$.

50.3.2 Mechanical Properties of Other Materials

As discussed above, of all the materials used for MEMS and NEMS, polysilicon has generated the most interest as well as the most research in mechanical properties characterization. However, measurements have been taken on other materials, and these are summarized in this section.

As discussed in Sect. 50.2.2, one advantage of the externally loaded tension test, as shown in Fig. 50.7, is that essentially any material can be tested using this technique. As such, tensile strengths have been measured to be 0.6–1.9 GPa for SiO_2 [50.60] and 0.7–1.1 GPa for titanium [50.61]. The yield strength for electrodeposited nickel was found to vary from 370 to 900 MPa, depending on the annealing temperature [50.17]. In addition, the yield strength was strongly affected by the current density during the electrodeposition process. Both the annealing and current density effects were correlated to changes in the microstructure of the material. The fatigue behavior of electrodeposited Ni was also investigated [50.62]. Microstructual changes due to thermal treatments also caused the yield strength of AlCuMgMn films to vary from 400 to 1000 MPa [50.63]. Young's moduli were determined to be 100 GPa for titanium [50.61] and 215 GPa for electrodeposited nickel [50.17] and 96 GPa for permalloy [50.23].

The tensile strength, Young's modulus, and Poisson's ratio of silicon nitride were measured to be 5.9 GPa, 0.23 and 257 GPa, respectively [50.64], and the same properties of amorphous diamondlike carbon were found to be 7.3 GPa, 0.17 and 759 GPa, respectively [50.65].

The technique of bending cantilever beams, shown in Fig. 50.8, can also be performed on a variety of materials. The yield strength and Young's modulus of gold were found to vary from 260 MPa and 57 GPa, respectively [50.20], to 300 MPa and 120 GPa, respectively [50.66], using this method. The same properties in Al were measured to be 150 MPa and 80 GPa, respectively [50.66], and in silicon nitride to be 6.9 and 260 GPa, respectively [50.67]. Using a technique similar to that shown in Fig. 50.8, except that a doubly-clamped beam was used instead of a cantilever beam, the fracture toughness of ultrananocrystalline diamond was measured to be 4.5 MPa $m^{1/2}$ [50.68]. Resonating cantilever beams revealed a Young's modulus for gold of 47 GPa [50.69], and Young's moduli of 1.8 and 14.4 GPa for silica and alumina aerogel thin films, respectively [50.70].

Another technique that can be used with a number of materials is the membrane deflection method, shown in Fig. 50.12. Silicon nitride measured with this technique revealed a Young's modulus of 258 GPa [50.64] to 325 GPa and a burst strength of 7.1 GPa [50.71]. A polyimide membrane gave a residual stress of 32 MPa, a Young's modulus of 3.0 GPa, and an ultimate strain of about four percent [50.24]. Membranes were also fabricated from polycrystalline SiC films with two different grain structures [50.72]. The film with (110)-texture columnar grains had a residual stress of 434 MPa and a Young's modulus of 349 GPa. The film with equiaxed (110)- and (111)-textured grains had a residual stress of 446 MPa and a Young's modulus of 456 GPa. Al and Ni membranes revealed Young's moduli of 70 and 200 GPa, respectively [50.73]. Polymeric SU-8 displayed a fracture strength of 73 MPa and a very low creep rate [50.74].

Other devices that are used to measure mechanical properties require more complicated micromachining, namely patterning, etching and release, in order to operate. These devices are more difficult to fabricate with materials that are not commonly used as MEMS structural materials. However, the following examples demonstrate work in this area. The structure shown in Fig. 50.5b was fabricated from Si_xN_y and revealed a apparent fracture toughness of 1.8 MPa $m^{1/2}$ [50.11]. The devices shown in Fig. 50.6 revealed a fracture toughness of SiC of 3.1 MPa $m^{1/2}$ [50.75]. Similar composite beams gave a yield strength of 625 MPa and a strain at crack initiation of 0.3 for Al [50.16]. Devices as shown in Fig. 50.10 showed a fracture toughness of diamond of 4.6 MPa $m^{1/2}$ [50.22], a Young's modulus of Cu of 115 GPa [50.76], and a Young's modulus and yield strength of Mo of 231 and 1.76 GPa, respectively [50.77].

Lateral resonators of the type shown in Fig. 50.13 were processed using single crystal and polycrystalline SiC, and the Young's modulus was determined to be between 360 GPa [50.78] and 426 GPa [50.79]. Resonating beams revealed the temperature coefficient of Young's modulus to be −53 ppm/K [50.80]. The device shown in Fig. 50.14 was fabricated from polycrystalline germanium, and used to measure a bend strength of 1.5 GPa for unannealed Ge and 2.2 GPa for annealed Ge [50.81]. The same device was also fabricated from SiC and revealed a bend strength of 23 GPa [50.82]. Devices similar to that shown in Fig. 50.15 revealed the tensile strength of silicon nitride to be 6.4 GPa [50.71], and the Young's modulus and yield strength of aluminium to be 74.6 GPa and 330 MPa, respectively [50.83].

The metal-to-metal contact RF switches, such as shown in Fig. 50.19 typically use Au as the contact metal, due to its low electrical resistance and high oxidation resistance. In these devices, adhesive forces were found to increase logarithmically with *cold* contact cycles (cycling with no current passing through the switch), consistent with creep as the underlying physical mechanism [50.34]. At high currents, no adhesion was found between the metal components, but melting led to extrusions and short circuits [50.36]. Both of these results are consistent with separate modeling investigations [50.84, 85].

References

50.1 W. Nix: Mechanical properties of thin films, Metall. Trans. A **20**, 2217–2245 (1989)

50.2 G.G. Stoney: The tension of metallic films deposited by electrolysis, Proc. R. Soc. A **82**, 172–175 (1909)

50.3 W. Brantley: Calculated elastic constants for stress problems associated with semiconductor devices, J. Appl. Phys. **44**, 534–535 (1973)

50.4 A. Ni, D. Sherman, R. Ballarini, H. Kahn, B. Mi, S.M. Phillips, A.H. Heuer: Optimal design of multi-

layered polysilicon films for prescribed curvature, J. Mater. Sci., **38**, 4169–4173 (2003)
50.5 J.A. Floro, E. Chason, S.R. Lee, R.D. Twesten, R.Q. Hwang, L.B. Freund: Real-time stress evolution during $Si_{1-x}Ge_x$ heteroepitaxy: Dislocations, islanding, and segregation, J. Electron. Mater. **26**, 969–979 (1997)
50.6 X. Li, B. Bhusan: Micro/nanomechanical characterization of ceramic films for microdevices, Thin Solid Films **340**, 210–217 (1999)
50.7 M.P. de Boer, H. Huang, J.C. Nelson, Z.P. Jiang, W.W. Gerberich: Fracture toughness of silicon and thin film micro-structures by wedge indentation, Mater. Res. Soc. Symp. Proc. **308**, 647–652 (1993)
50.8 H. Guckel, D. Burns, C. Rutigliano, E. Lovell, B. Choi: Diagnostic microstructures for the measurement of intrinsic strain in thin films, J. Micromech. Microeng. **2**, 86–95 (1992)
50.9 F. Ericson, S. Greek, J. Soderkvist, J.-A. Schweitz: High sensitivity surface micromachined structures for internal stress and stress gradient evaluation, J. Micromech. Microeng. **7**, 30–36 (1997)
50.10 M. Biebl, H. von Philipsborn: Fracture strength of doped and undoped polysilicon, Proc. Int. Conf. Solid-State Sens. Actuators, Stockholm 1995, ed. by S. Middelhoek, K. Cammann (Royal Swedish Academy of Engineering Sciences, Stockholm 1995) pp. 72–75
50.11 L.S. Fan, R.T. Howe, R.S. Muller: Fracture toughness characterization of brittle films, Sens. Actuators A **21-23**, 872–874 (1990)
50.12 H. Kahn, R. Ballarini, J.J. Bellante, A.H. Heuer: Fatigue failure in polysilicon is not due to simple stress corrosion cracking, Science **298**, 1215–1218 (2002)
50.13 L.S. Fan, R.S. Muller, W. Yun, R.T. Howe, J. Huang: Spiral microstructures for the measurement of average strain gradients in thin films, Proc. IEEE Micro Electro Mech. Syst. Workshop, Napa Valley 1990 (IEEE, New York 1990), 177–182
50.14 H. Kahn, N. Tayebi, R. Ballarini, R.L. Mullen, A.H. Heuer: Wafer-level strength and fracture toughness testing of surface-micromachined MEMS devices, Mater. Res. Soc. Symp. Proc. **605**, 25–30 (2000)
50.15 W.N. Sharpe Jr., B. Yuan, R.L. Edwards: A new technique for measuring the mechanical properties of thin films, J. Microelectromech. Syst. **6**, 193–199 (1997)
50.16 N. André, M. Coulombier, V. De Longueville, D. Fabrègue, T. Gets, S. Gravier, T. Pardoen, J.-P. Raskin: Microfabrication-based nanomechanical laboratory for testing the ductility of submicron aluminum films, Microelectron. Eng. **84**, 2714–2718 (2007)
50.17 H.S. Cho, W.G. Babcock, H. Last, K.J. Hemker: Annealing effects on the microstructure and mechanical properties of LIGA nickel for MEMS, Mater. Res. Soc. Symp. Proc. **657** (2001) EE5.23.1–EE5.23.6
50.18 W. Suwito, M.L. Dunn, S.J. Cunningham, D.T. Read: Elastic moduli, strength, and fracture initiation at sharp notches in etched single crystal silicon microstructures, J. Appl. Phys. **85**, 3519–3534 (1999)
50.19 C.-S. Oh, W.N. Sharpe: Techniques for measuring thermal expansion and creep of polysilicon, Sens. Actuators A **112**, 66–73 (2004)
50.20 T.P. Weihs, S. Hong, J.C. Bravman, W.D. Nix: Mechanical deflection of cantilever microbeams: A new technique for testing the mechanical properties of thin films, J. Mater. Res. **3**, 931–942 (1988)
50.21 B.D. Jensen, M.P. de Boer, N.D. Masters, F. Bitsie, D.A. La Van: Interferometry of actuated microcantilevers to determine material properties and test structure nonidealities in MEMS, J. Microelectromech. Syst. **10**, 336–346 (2001)
50.22 H.D. Espinosa, B. Peng: A new methodlolgy to investigate fracture toughness of freestanding MEMS and advanced materials in thin film form, J. Microelectromech. Syst. **14**, 153–159 (2005)
50.23 X. Li, G. Ding, T. Ando, M. Shikida, K. Sato: Micromechanical characterization of electroplated permalloy films for MEMS, Microsyst. Technol. **14**, 131–134 (2007)
50.24 M.G. Allen, M. Mehregany, R.T. Howe, S.D. Senturia: Microfabricated structures for the in situ measurement of residual stress, Young's modulus, and ultimate strain of thin films, Appl. Phys. Lett. **51**, 241–243 (1987)
50.25 H. Kahn, S. Stemmer, K. Nandakumar, A.H. Heuer, R.L. Mullen, R. Ballarini, M.A. Huff: Mechanical properties of thick, surface micromachined polysilicon films, Proc. IEEE Micro Electro Mech. Syst. Workshop, San Diego 1996, ed. by M.G. Allen, M.L. Redd (IEEE, New York 1996) pp. 343–348
50.26 L. Kiesewetter, J.-M. Zhang, D. Houdeau, A. Steckenborn: Determination of Young's moduli of micromechanical thin films using the resonance method, Sens. Actuators A **35**, 153–159 (1992)
50.27 W.C. Tang, T.-C.H. Nguyen, R.T. Howe: Laterally driven polysilicon resonant microstructures, Sens. Actuators A **20**, 25–32 (1989)
50.28 P.T. Jones, G.C. Johnson, R.T. Howe: Fracture strength of polycrystalline silicon, Mater. Res. Soc. Symp. Proc. **518**, 197–202 (1998)
50.29 P. Minotti, R. Le Moal, E. Joseph, G. Bourbon: Toward standard method for microelectromechanical systems material measurement through on-chip electrostatic probing of micrometer size polysilicon tensile specimens, Jpn. J. Appl. Phys. **40**, L120–L122 (2001)
50.30 C.L. Muhlstein, E.A. Stach, R.O. Ritchie: A reaction-layer mechanism for the delayed failure of micron-scale polycrystalline silicon structural films subjected to high-cycle fatigue loading, Acta Mater. **50**, 3579–3595 (2002)

50.31 A. Corigliano, B. De Masi, A. Frangi, C. Comi, A. Villa, M. Marchi: Mechanical characterization of polysilicon through on-chip tensile tests, J. Microelectromech. Syst. **13**, 200–219 (2004)

50.32 M.G. Lim, J.C. Chang, D.P. Schultz, R.T. Howe, R.M. White: Polysilicon microstructures to characterize static friction, Proc. IEEE Micro Electro Mech. Syst. Workshop, Napa Valley 1990 (IEEE, New York 1990) pp. 82–88

50.33 B.T. Crozier, M.P. de Boer, J.M. Redmond, D.F. Bahr, T.A. Michalske: Friction measurement in MEMS using a new test structure, Mater. Res. Soc. Symp. Proc. **605**, 129–134 (2000)

50.34 G. Gregori, D.R. Clarke: The interrelation between adhesion, contact creep, and roughness on the life of gold contacts in radio-frequency microswitches, J. Appl. Phys. **100**, 094904 (2006)

50.35 D.H. Alsem, E.A. Stach, M.T. Dugger, M. Enachescu, R.O. Ritchie: An electron microscopy study of wear in polysilicon microelectromechanical systems in ambient air, Thin Solid Films **515**, 3259–3266 (2007)

50.36 S.T. Patton, J.S. Zabinski: Fundamental studies of Au contacts in MEMS RF switches, Tribol. Lett. **18**, 215–230 (2005)

50.37 J. Yang, H. Kahn, A.Q. He, S.M. Phillips, A.H. Heuer: A new technique for producing large-area as-deposited zero-stress LPCVD polysilicon films: The MultiPoly process, J. Microelectromech. Syst. **9**, 485–494 (2000)

50.38 E. Chason, B.W. Sheldon, L.B. Freund, J.A. Floro, S.J. Hearne: Origin of compressive residual stress in polycrystalline thin films, Phys. Rev. Lett. **88** (2002) 156103-1–156103-4

50.39 S. Jayaraman, R.L. Edwards, K.J. Hemker: Relating mechanical testing and microstructural features of polysilicon thin films, J. Mater. Res. **14**, 688–697 (1999)

50.40 R. Ballarini, H. Kahn, N. Tayebi, A.H. Heuer: Effects of microstructure on the strength and fracture toughness of polysilicon: A wafer level testing approach, ASTM STP **1413**, 37–51 (2001)

50.41 J. Bagdahn, J. Schischka, M. Petzold, W.N. Sharpe Jr.: Fracture toughness and fatigue investigations of polycrystalline silicon, Proc. SPIE **4558**, 159–168 (2001)

50.42 I.S. Raju, J.C. Newman Jr.: Stress intensity factors for a wide range of semi-elliptical surface cracks in finite-thickness plates, Eng. Fract. Mech. **11**, 817–829 (1979)

50.43 J. Bagdahn, W.N. Sharpe Jr., O. Jadaan: Fracture strength of polysilicon at stress concentrations, J. Microelectromech. Syst., 302–312 (2003)

50.44 R. Boroch, J. Wiaranowski, R. Mueller-Fiedler, M. Ebert, J. Bagdahn: Characterization of strength properties of thin polycrystalline silicon films for MEMS applications, Fatigue Fract. Eng. Mater. Struct. **30**, 2–12 (2007)

50.45 B.L. Boyce, J.M. Grazier, T.E. Buchheit, M.J. Shaw: Strength distributions in polycrystalline silicon MEMS, J. Microelectromech. Syst. **16**, 179–190 (2007)

50.46 D.C. Miller, B.L. Boyce, M.T. Dugger, T.E. Buchheit, K. Gall: Characteristics of a commercially available silicon-on-insulator MEMS material, Sens. Actuators A **138**, 130–144 (2007)

50.47 T. Yi, L. Li, C.-J. Kim: Microscale material testing of single crystalline silicon: Process effects on surface morphology and tensile strength, Sens. Actuators A **83**, 172–178 (2000)

50.48 T. Alan, M.A. Hines, A.T. Zehnder: Effect of surface morphology on the fracture strength of silicon nanobeams, Appl. Phys. Lett. **89**, 091901 (2006)

50.49 X. Li, T. Kasai, S. Nakao, T. Ando, M. Shikida, K. Sato, H. Tanaka: Anisotropy in fracture of single crystal silicon film characterized under uniaxial tensile condition, Sens. Actuators A **117**, 143–150 (2005)

50.50 S. Koyama, K. Takashima, Y. Higo: Fracture toughness measurement of a micro-sized single crystal silicon, Key Eng. Mater. **297–300**, 292–298 (2005)

50.51 H. Kahn, R. Ballarini, R.L. Mullen, A.H. Heuer: Electrostatically actuated failure of microfabricated polysilicon fracture mechanics specimens, Proc. R. Soc. A **455**, 3807–3923 (1999)

50.52 W.W. Van Arsdell, S.B. Brown: Subcritical crack growth in silicon MEMS, J. Microelectromech. Syst. **8**, 319–327 (1999)

50.53 Y. Yamaji, K. Sugano, O. Tabata, T. Tsuchiya: Tensile-mode fatigue tests and fatigue life predictions of single crystal silicon in humidity controlled environments, Proc. MEMS 2007, Kobe (2007) pp. 267–270

50.54 D.H. Alsem, R. Timmerman, B.L. Boyce, E.A. Stach, J.T.M. De Hosson, R.O. Ritchie: Very high-cycle fatigue failure in micron-scale polycrystalline silicon films: effects of environment and surface oxide thickness, J. Appl. Phys. **101**, 013515 (2007)

50.55 H. Kahn, A. Avishai, R. Ballarini, A.H. Heuer: Surface oxide effects on failure of polysilicon MEMS after cyclic and monotonic loading, Scr. Mater. (2008), in press

50.56 H. Kahn, L. Chen, R. Ballarini, A.H. Heuer: Mechanical fatigue of polysilicon: effects of mean stress and stress amplitude, Acta Mater. **54**, 667–678 (2006)

50.57 R.E. Boroch, R. Müller-Fiedler, J. Bagdahn, P. Gumbsch: High cycle fatigue and strengthening in polycrystalline silicon, Scr. Mater. (2008) in press

50.58 S. Nakao, T. Ando, M. Shikida, K. Sato: Effects of environmental condition on the strength of submicron-thick single crystal silicon film, Proc. 14th Int. Conf. Solid-State Sens., Actuators Microsyst., Transducers Eurosens. '07, Lyon (2007) pp. 375–378

50.59 A. Corwin, M.P. de Boer: Effect of adhesion on dynamic and static friction in surface micromachining, Appl. Phys. Lett. **84**, 2451–2453 (2004)

50.60 T. Tsuchiya, A. Inoue, J. Sakata: Tensile testing of insulating thin films; humidity effect on tensile strength of SiO_2 films, Sens. Actuators A **82**, 286–290 (2000)
50.61 H. Ogawa, K. Suzuki, S. Kaneko, Y. Nakano, Y. Ishikawa, T. Kitahara: Measurements of mechanical properties of microfabricated thin films, Proc. IEEE Micro Electro Mech. Syst. Workshop (IEEE, New York 1997) pp. 430–435
50.62 Y. Yang, B.I. Imasogie, S.M. Allameh, B. Boyce, K. Lian, J. Lou, W.O. Soboyejo: Mechanisms of fatigue in LIGA Ni MEMS thin films, Mater. Sci. Eng. A **444**, 39–50 (2007)
50.63 R. Modlinski, R. Puers, I. De Wolf: AlCuMgMn micro-tensile samples mechanical characterization of MEMS materials at micro-scale, Sens. Actuators A **143**, 120–128 (2008)
50.64 R.L. Edwards, G. Coles, W.N. Sharpe: Comparison of tensile and bulge tests for thin-film silicon nitride, Exp. Mech. **44**, 49–54 (2004)
50.65 S. Cho, I. Chasiotis, T.A. Friedmann, J.P. Sullivan: Young's modulus, Poisson's ratio and failure properties of tetrahedral amorphous diamond-like carbon for MEMS devices, J. Micromech. Microeng. **15**, 728–735 (2005)
50.66 D. Son, J.-H. Jeong, D. Kwon: Film-thickness considerations in microcantilever-beam test in measuring mechanical properties of metal thin film, Thin Solid Films **437**, 182–187 (2003)
50.67 W.-H. Chuang, T. Luger, R.K. Fettig, R. Ghodssi: Mechanical property characterization of LPCVD silicon nitride thin films at cryogenic temperatures, J. Microelectromech. Syst. **13**, 870–879 (2004)
50.68 H.D. Espinosa, B. Peng: A new methodology to investigate fracture toughness of freestanding MEMS and advanced materials in thin film form, J. Microelectromech. Syst. **14**, 153–159 (2005)
50.69 C.-W. Baek, Y.-K. Kim, Y. Ahn, Y.-H. Kim: Measurement of the mechanical properties of electroplated gold thin films using micromachined beam structures, Sens. Actuators A **117**, 17–27 (2005)
50.70 R. Yokokawa, J.-A. Paik, B. Dunn, N. Kitazawa, H. Kotera, C.-J. Kim: Mechanical properties of aerogel-like thin films used for MEMS, J. Micromech. Microeng. **14**, 681–686 (2004)
50.71 A. Kaushik, H. Kahn, A.H. Heuer: Wafer-level mechanical characterization of silicon nitride MEMS, J. Microelectromech. Syst. **14**, 359–367 (2005)
50.72 S. Roy, C.A. Zorman, M. Mehregany: The mechanical properties of polycrystalline silicon carbide films determined using bulk micromachined diaphragms, Mater. Res. Soc. Symp. Proc. **657**, EE9.5.1–EE9.5.6 (2001)
50.73 A. Reddy, H. Kahn, A.H. Heuer: A MEMS-based evaluation of the mechanical properties of metallic thin films, J. Microelectromech. Syst. **16**, 650–658 (2007)
50.74 B. Schoeberle, M. Wendlandt, C. Hierold: Long-term creep behavior of SU-8 membranes: Application of the time-stress superposition principle to determine the master creep compliance curve, Sens. Actuators A **142**, 242–249 (2008)
50.75 J.J. Bellante, H. Kahn, R. Ballarini, C.A. Zorman, M. Mehregany, A.H. Heuer: Fracture toughness of polycrystalline silicon carbide thin films, Appl. Phys. Lett. **86**, 071920-1–071920-3 (2005)
50.76 Z.M. Zhou, Y. Zhou, C.S. Yang, J.A. Chen, W. Ding, G.F. Ding: The evaluation of Young's modulus and residual stress of copper films by microbridge testing, Sens. Actuators A **127**, 392–397 (2006)
50.77 J.-H. Kim, H.-J. Lee, S.-W. Han, J.-Y. Kim, J.-S. Kim, J.-Y. Kang, S.-H. Choa, C.-S. Lee: Mechanical behavior of freestanding Mo thin films for RF-MEMS devices, Int. J. Modern Phys. B **20**, 3757–3762 (2006)
50.78 W. Chang, C. Zorman: Determination of Young's moduli of 3C (110) single-crystal and (111) polycrystalline silicon carbide from operating frequencies, J. Mater. Sci. **43**, 4512–4517 (2008)
50.79 A.J. Fleischman, X. Wei, C.A. Zorman, M. Mehregany: Surface micromachining of polycrystalline SiC deposited on SiO_2 by APCVD, Mater. Sci. Forum **264–268**, 885–888 (1998)
50.80 M. Pozzi, M. Hassan, A.J. Harris, J.S. Burdess, L. Jiang, K.K. Lee, R. Cheung, G.J. Phelps, N.G. Wright, C.A. Zorman, M. Mehregany: Mechanical properties of a 3C-SiC film between room temperature and 600 °C, J. Phys. D: Appl. Phys. **40**, 3335–3342 (2007)
50.81 A.E. Franke, E. Bilic, D.T. Chang, P.T. Jones, T.-J. King, R.T. Howe, G.C. Johnson: Post-CMOS integration of germanium microstructures, Proc. Int. Conf. Solid-State Sens. Actuators (IEE Japan, Tokyo 1999) pp. 630–637
50.82 D. Gao, C. Carraro, V. Radmilovic, R.T. Howe, R. Maboudian: Fracture of polycrystalline 3C-SiC films in microelectromechanical systems, J. Microelectromech. Syst. **13**, 972–976 (2004)
50.83 M.A. Haque, M.T.A. Saif: A review of MEMS-based microscale and nanoscale tensile and bending testing, Exp. Mech. **43**, 248–255 (2003)
50.84 J. Song, D.J. Srolovitz: Adhesion effects in material transfer in mechanical contacts, Acta Mater. **54**, 5305–5312 (2006)
50.85 O. Rezvanian, M.A. Zikry, C. Brown, J. Krim: Surface roughness, asperity contact and gold RF MEMS switch behavior, J. Micromech. Microeng. **17**, 2006–2015 (2007)

51. High-Volume Manufacturing and Field Stability of MEMS Products

Jack Martin

Low-volume micro/nanoelectromechanical systems (MEMS/NEMS) production is practical when an attractive concept is implemented with business, manufacturing, packaging, and test support. Moving beyond this to high-volume production adds requirements on design, process control, quality, product stability, market size, market maturity, capital investment, and business systems. In a broad sense, this chapter uses a case study approach: It describes and compares the silicon-based MEMS accelerometers, pressure sensors, image projection systems, microphones, and gyroscopes that are in high-volume production. Although they serve several markets, these businesses have common characteristics; for example, the manufacturing lines use automated semiconductor equipment and standard material sets to make consistent products in large quantities. Standard well-controlled processes are sometimes modified for a MEMS product. However, novel processes that cannot run with standard equipment and material sets are avoided when possible. This reliance on semiconductor tools, as well as the organizational practices required to manufacture clean, particle-free products, partially explains why the MEMS market leaders are integrated circuit (IC) manufacturers. There are other factors. MEMS and NEMS are enabling technologies, so it can take several years for high-volume applications to develop. Indeed, market size is usually a strong function of price. This becomes a vicious circle, because low price requires low cost – a result that is normally achieved only after a product is in high-volume production. During the early years, IC companies reduce cost and financial risk by using existing facilities for low-volume MEMS production. As a result, product architectures are partially determined by capabilities developed for previous products. This chapter includes a discussion of MEMS product architecture with particular attention to the impact of electronic integration, packaging, and surfaces. Packaging and testing are critical, because they are significant factors in MEMS product cost. MEMS devices have extremely high surface-to-volume ratios, so performance and stability may depend on the control of surface characteristics *after packaging*. Looking into the future, the competitive advantage of IC suppliers is decreasing because MEMS foundries are growing and small companies are learning to integrate MEMS/NEMS devices with die from complementary metal-oxide-semiconductor (CMOS) foundries in one package. Packaging challenges still remain, because most MEMS/NEMS products must interact with the environment without degrading stability or reliability.

51.1 **Background** 1804
51.2 **Manufacturing Strategy** 1806
 51.2.1 Volume 1806
 51.2.2 Standardization 1807
 51.2.3 Production Facilities 1807
 51.2.4 Quality 1808
 51.2.5 Environmental Shield 1808
51.3 **Robust Manufacturing** 1808
 51.3.1 Design for Manufacturability 1808
 51.3.2 Process Flow and Its Interaction with Product Architecture 1809
 51.3.3 Microstructure Release 1818
 51.3.4 Wafer Bonding 1818
 51.3.5 Wafer Singulation 1820
 51.3.6 Particles 1820
 51.3.7 Electrostatic Discharge and Static Charges 1821
 51.3.8 Package and Test 1821
 51.3.9 Quality Systems 1824
51.4 **Stable Field Performance** 1825
 51.4.1 Surface Passivation 1825
 51.4.2 System Interface 1828
References 1828

51.1 Background

Solid-state pressure sensors were first reported in the 1960s and commercialized by a number of start-up companies in the 1970s. By 1983, a Scientific American cover story [51.1] described pressure sensors, accelerometers, inkjet printheads, and featured a gas chromatograph that combined an injection valve, a detector, and a chromatographic column on a silicon wafer. The future of the budding micromechanical device industry looked bright, but commercial reality seriously lagged the rosy market projections. The reason was simple: market forecasters failed to recognize the difference between laboratory prototypes and high-volume production of stable products. Many companies have demonstrated that small quantities can be produced for niche markets. However, routine high-volume production of reliable packaged devices is much more challenging. It requires technical skills to make prototypes plus the manufacturing disciplines that are critical to long-term stable production.

Successful companies understand and apply these lessons. Indeed, several hundred million MEMS pressure sensor, accelerometer, gyro, microphone, gas flow, and optical projection devices are shipped to customers annually. There is no universally accepted definition of MEMS. This chapter focuses on silicon-based products with movable micromechanical elements. It excludes disk-drive heads, inkjet printheads, hearing aids, microscale plastic, ceramic, quartz and metal components, and test strips for in vitro diagnostics that are sometimes included within the definition of MEMS. Many MEMS products do not integrate support electronics with the MEMS element on the same chip even though they are produced with IC equipment, materials, and processes. However, some companies utilize this capability to build leading-edge products. For example, it took over 9 years for Analog Devices to ship a hundred million surface-micromachined MEMS accelerometers with integrated electronics to customers. The second hundred million took less than 3 years, and by 2004, shipments were exceeding a million units a week. Texas Instruments experienced similar exponential growth for its integrated Digital Light Processing (DLP) products. Initial sales were low because it took 5 years (1996–2001) to ship one million units. However, three million had been shipped by early 2004 and five million by the end of 2004 [51.2].

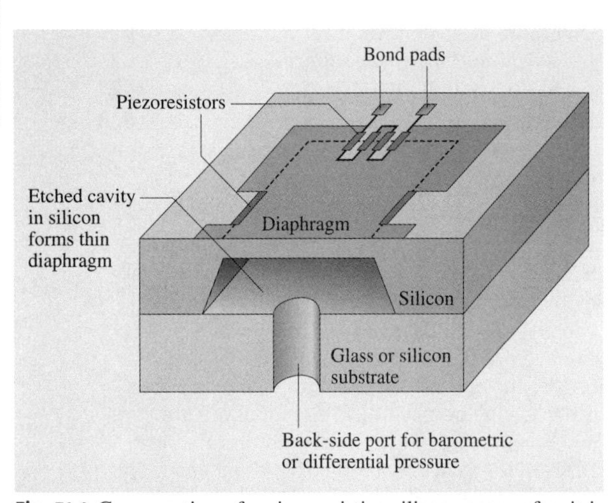

Fig. 51.1 Cross-section of a piezoresistive silicon sensor after it is bonded to a silicon or glass substrate. When pressure is applied, the thin silicon diaphragm deforms. This causes changes in four implanted resistors that form a Wheatstone bridge. The resistors are located and oriented to cause resistance to increase in two resistors and decrease in the other two

The literature related to high-volume production of MEMS products is quite uneven. Some suppliers have published detailed process and design descriptions, while others limit disclosures to market-focused publications. The two major suppliers of integrated MEMS products have discussed their designs and processes in detail. *Core* et al. [51.3], *Kuehnel* and *Sherman* [51.4], *Chau*, and *Sulouff* [51.5] and *Sulouff* [51.6] describe the Analog Devices *i*MEMS accelerometers and gyros. *Mignardi* et al. [51.7] provide an in-depth review of the Texas Instruments DLP products.

MEMS devices are components in larger systems so every MEMS product must define its system interface in a way that adds value from the perspective of the system designer. Some MEMS pressure-sensor products are simply Wheatstone bridge elements in a circuit (Fig. 51.1). The image projection systems that use the Texas Instruments Digital Light Processing technology could never be designed in this manner because up to two million mirrors are driven on each chip (Fig. 51.2). Such a system would be impractical unless electronics are integrated with the MEMS mirrors on the chip. Airbag sensors also illustrate how integration adds value. Non-MEMS ball-in-tube airbag sensors were often used in the early 1990s. Upon impact, a ball would be released and detected as it passed through the sensor tube. Unfortunately, these binary sensors could not always distinguish a frontal collision from the jolt

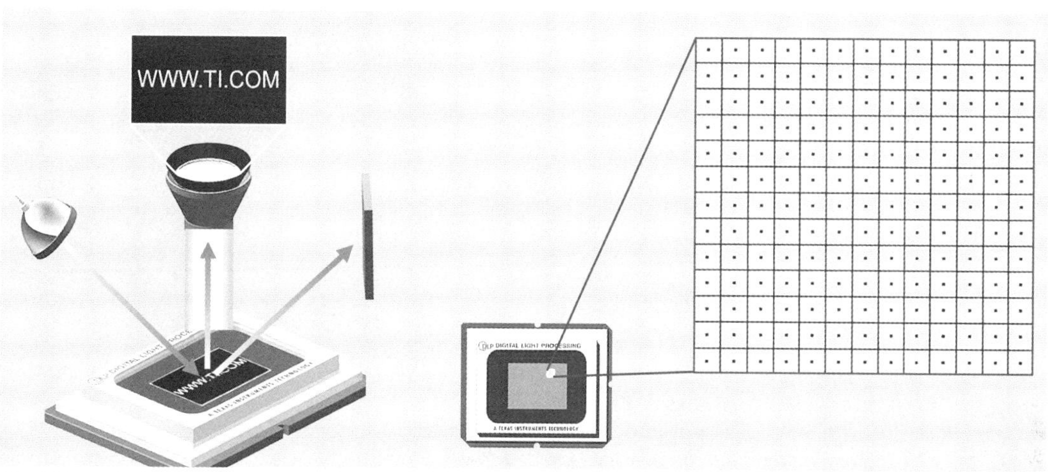

Fig. 51.2 Digital Light Processing technology is based on digital micromirror device (DMD) chips that address and drive an *array of mirrors*. The DMD is the key element in small computer-based projection systems. Each of the 14 or 17 μm mirrors represents one pixel. The *sketch on the left* represents light reflecting from a mirror to the system lens. The *illustration on the right* is an exploded view of a small region of mirrors in a packaged DMD (© Digital Light Processing, Texas Instruments, Inc.)

of a pothole or a side collision. Therefore, multiple ball-in-tube sensors were placed at different locations in a vehicle and wired together in order to reliably identify accidents that should result in airbag deployment. High cost limited the use of these distributed systems. Analog Devices changed the airbag market when it introduced the ADXL50 in 1993 (Figs. 51.3 and 51.4). The ADXL50 integrated electronics and an accelerometer on one chip to provide an analog output of deceleration versus time when an impact event occurred. This allowed automobile manufacturers to design airbag modules that compared the ADXL50 output with the known signature of a frontal collision. The result was greater reliability and lower cost, which led to installation of airbags in almost every new automobile.

This chapter will not recite detailed descriptions of particular commercial products. Instead, brief descriptions will be used to illustrate basic principles and concepts. Part of the chapter is organized according to

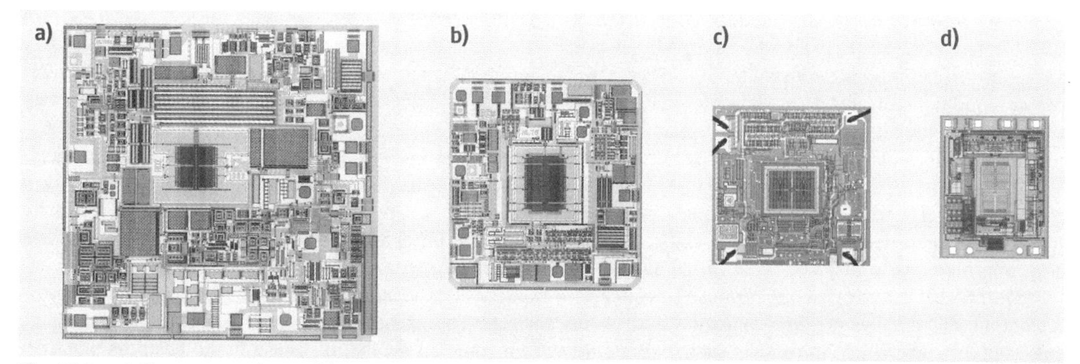

Fig. 51.3a–d Four generations of integrated accelerometers designed for single-axis airbag applications. Die size is reduced by a factor of two about every 4 years. A series of BiCMOS processes were used to manufacture the three products with polysilicon microstructures. The CMOS device (far right) is built on SOI wafers, so the microstructures are single-crystal silicon (© Micromachined Products Division, Analog Devices, Inc.)

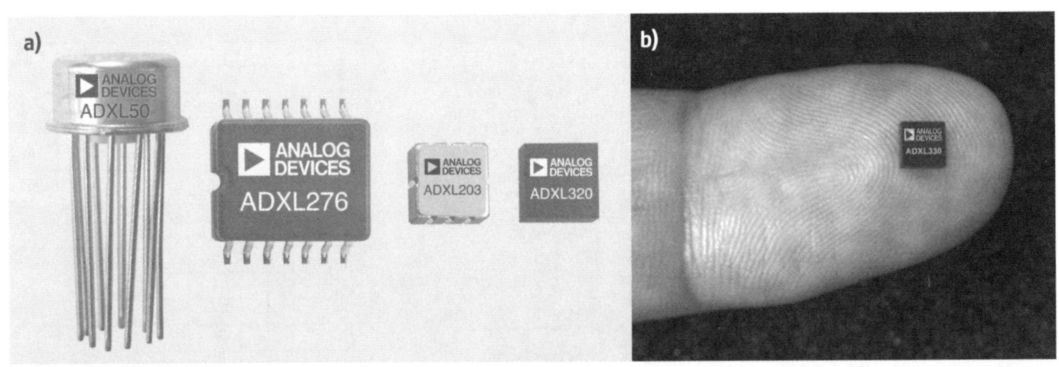

Fig. 51.4 (a) Package technology evolves to meet size, cost, and automated handling requirements. The ADXL50 was packaged in TO-100 seam-sealed metal headers. CERDIPs and CERPACs (ADXL276) were replaced by $5 \times 5\,mm^2$ LCC packages (ADXL203). The ADXL320 is molded in plastic. (b) The size of a three-axis accelerometer is illustrated for the ADXL330 (© Micromachined Products Division, Analog Devices, Inc.)

MEMS-specific issues such as wafer singulation. This is useful because some designs and unit processes are well suited for laboratory-scale or low-volume production, but impractical in a high-volume manufacturing environment. Complete reliance on such an organization would be artificial because partitioning omits the greatest challenge – combining these individual topics into a manufacturing flow that routinely produces products that meet cost, performance, and reliability expectations. The importance of process integration cannot be overemphasized because changes in one step almost always affect other steps. Solutions may involve design modifications so design for manufacturability is an important topic. Manufacturing and business skills are included to the extent that they apply to stable production of reliable MEMS products.

51.2 Manufacturing Strategy

MEMS manufacturing is based on existing IC technology. However, there are only a few examples of routine MEMS and IC production intermixed on the same equipment (or of MEMS devices with electronics integrated on the chip). Even when a MEMS process step is nominally the same as a standard IC process, the specific conditions or end point or some quality such as film stress will differ. Technologists from Analog Devices and Motorola (now Freescale) have discussed these thin-film issues in depth [51.8–11].

Manufacturers prefer to use dry etch processes when possible in order to achieve better control of features, minimize waste disposal, and utilize automated equipment developed for the IC industry. However, anisotropic wet etching provides unique value in the production of some MEMS products such as pressure, mass flow, and yaw-rate sensors [51.12].

Although MEMS production economics is closely tied to the semiconductor model, useful lessons can be drawn from many industries. Some of the key characteristics that affect success are as follows.

51.2.1 Volume

Semiconductor economics is rooted on the principle that expensive processes are cost-effective when many devices are made simultaneously. Obviously, a MEMS product that is sold in low quantities loses this critical advantage. MEMS components are *disruptive technologies* – they are the key element in higher-level products that were previously impractical. The Analog Devices and Texas Instruments data cited above show that it can take several years to develop those higher-level products and build a customer base. Therefore, it is prudent to include an incubation phase in the business plan for a novel MEMS device. Low sales volume during this phase causes high unit costs; batch-process cost benefits are achieved only after the market develops.

Startup companies often do not understand the implications of design for volume. It is a major challenge to transfer products made on laboratory tools to manufacturing equipment. The challenge is sufficient to justify the use of manufacturing people and equipment for product development.

Realistic price–volume estimates are also essential to a good business plan. The highest-volume MEMS products (accelerometers, pressure sensors, and microphones) all represent a small part of the total system cost, even though the functions they provide are absolutely critical to system performance.

Market-pull business models that respond to the needs of a specific large application in a particular industry are usually more successful than technology-driven models that seek applications for new devices. Given this reality, it is prudent to thoroughly understand the target industry and its quirks. Biotechnology, wireless communication, image projection, computer, and consumer markets offer tremendous growth potential. In the past, high-volume MEMS applications were driven by the automotive industry, with pressure sensors also benefiting from medical applications. *Sulouff* [51.13], *Weinberg* [51.14], *Marek* and *Illing* [51.15], *Eddy* and *Sparks* [51.16], and *Verma* et al. [51.17] discuss automotive applications and the challenges a successful supplier must master. The recent growth spurt extends beyond automotive because many new handheld consumer products contain multiaxis accelerometers or microphones.

51.2.2 Standardization

Mature industries develop standards in response to equipment vendor, material supplier, and customer interface requirements. The MEMS industry is fragmented, so even after 30 years it is not mature. Deviations from semiconductor industry standards are inevitable because MEMS products impose unique requirements on wafer processing, packaging, handling, and testing. For example, KOH wet etching raises mobile ion concerns. Gold used in some optical products is incompatible with CMOS. Any deviation from standard process and material sets is serious, because the price–volume relationship is a steep curve in high-volume applications.

51.2.3 Production Facilities

The semiconductor industry is divided into front-end (wafer fabrication) and back-end (product assembly and test) operations. IC design rules are based on prequalified fabrication, assembly, and test capabilities at specific sites. Consider how each of these functions matches the needs of MEMS production:

1. An IC product might be produced at an internal wafer fab, or designed to run on a standard foundry process. Until recently, the MEMS manufacturing model was quite different. Production challenges and proprietary *knowhow* caused MEMS suppliers to retain wafer fab production within their internal facilities. High-volume production in MEMS foundries was limited to bulk micromachined piezoresistive pressure sensors. To illustrate this point, Freescale (formerly Motorola) produced piezoresistive pressure sensors and accelerometers in an internal Arizona wafer fab from the early 1980s until 2003. Production of the mature pressure sensors was transferred to Dalsa in 2003 [51.18]. However, production of surface-micromachined accelerometers and tire pressure monitors was retained within Freescale; these products were transferred to Sendai, a wholly owned Freescale subsidiary in Japan. Knowles Acoustics launched their SiSonic surface-mount MEMS microphone in 2002. By the end of 2007, they had shipped over 400 million units – all manufactured at a foundry.
2. IC suppliers have shifted assembly operations to subcontractor facilities located in South East Asia. MEMS suppliers have followed this model. It is common to assemble low-volume products in North America, Europe or Japan. However, semiconductor assembly and package technologies are evolving rapidly. These changes are driven by subcontractors based in South East Asia. MEMS companies that do not utilize these resources pay a significant cost penalty. Obviously, the breakpoint at which offshore assembly is attractive differs from one product to the next. Products that have unique assembly requirements and support high market prices may never reach this point.
3. IC testing is often carried out at the assembly subcontractor site. MEMS package and testing comprise a major portion of the total product cost, so a high-volume manufacturing strategy should include South East Asia packaging and testing. This is complicated by the fact that MEMS testing requires unique stimuli that are not in the standard IC portfolio.

51.2.4 Quality

The goal in laboratory and low-volume production environments is to make functioning devices. As quantities increase, other factors become more important: lot-to-lot repeatability, yield, device performance that meets well-defined specifications, process stability to assure predictable on-time delivery, documentation and procedures that enable traceability and corrective actions when problems arise, etc. Methodologies that promote stable, high-quality manufacturing practices have been established in several industries. These standard procedures, such as ISO/TS16949 [51.19], ISO9000, QS9000, and cGMP, overlap to a large extent. QS9000:1998 [51.20] extended ISO9000:1994. It was developed to help automotive suppliers implement well-controlled processes, an essential requirement for long-term stable production of reliable products. Many of the QS9000:1998 features were incorporated into ISO9000:2000. ISO/TS16949 broadens this standard to more explicitly include business systems, management ownership, and customer needs. It is written for automotive suppliers and shifts the emphasis from procedures to processes. QS9000 and ISO9000:2000 are used as the model in this chapter, but ISO/TS16949 has elements that should be considered by any ISO9000:2000-compliant organization.

51.2.5 Environmental Shield

IC wafers are completely passivated before packaging in order to meet product stability and reliability requirements. Aside from microphones and inkjet printheads, the active regions of MEMS devices that are in high-volume production replicate this hermetic barrier in some manner. This passivation requirement is noteworthy because it has been a primary limitation on the growth of the MEMS industry. For example, it limits electrochemical sensors to benign or *throw-away* applications.

51.3 Robust Manufacturing

Stable production requires a manufacturing flow that is well controlled when measured against the product performance specifications. This has several implications.

51.3.1 Design for Manufacturability

Low-maintenance products and processes must be designed to run on standard equipment. The semiconductor industry has invested billions of dollars to develop equipment that is automated, maintainable, reliable, and capable of supporting well-controlled processes. Processes that are implemented on custom-designed equipment put this experience base aside and invariably have a long learning curve.

Equipment

Microstructure release illustrates this issue, and how it changes with time. Virtually all semiconductor processes operate in vacuum or at atmospheric pressure. In a brief, unpublished 1992 study related to ADXL-50 development, the author demonstrated the use of supercritical CO_2 mixtures for microstructure release and for particle cleaning. Unfortunately, supercritical processes were incompatible with the industry infrastructure because they require pressures in excess of 7.38 MPa (72.8 atm; 1070 psi). Implementation would have required process and equipment development, so a release process that ran on existing equipment was selected.

Supercritical equipment designed for MEMS release was introduced a few years later. However, maintenance and cycle-time concerns as well as single-wafer capabilities limited its adoption to university and low-volume manufacturing.

The intrinsic limitations of nonstandard equipment have caused high-volume MEMS suppliers to avoid supercritical processes. However supercritical equipment will continue to evolve and may become competitive in applications where design requirements make alternative processes unusable.

Materials

Process control can be no better than the materials used in the process. Self-assembled monolayer (SAM) coatings illustrate this limitation. These materials suppress stiction following aqueous release of microstructures by treating the wafers in a SAM solution before drying. Although straightforward in principle, high-volume manufacturers have not adopted SAM processes (Sect. 51.3.3 reviews the processes that are used commercially). There are several reasons for this reticence. Classical SAM materials are chlorosilanes that have at least one organic substituent. Organosilanes have been used to treat the surface of inorganic materials for many years. Applications include coupling agents on the surface of fillers and reinforcing agents in polymers as well as agglomeration control of particles. The chlorine sites hydrolyze when dissolved in solutions that contain a small amount of water. The resulting hydroxyl groups react with the microstructure surface oxide to produce a chemically bonded layer that reduces stiction due to the low surface energy and hydrophobic nature of the organic-rich surface. The concerns that cause manufacturers to use alternative stiction solutions include:

1. High variability due to the SAM chemical reactivity.
2. Particles are common by-products of the reactivity.
3. Many SAM process flows require organic baths. This raises health, safety and waste-disposal concerns.
4. The possibility of chloride corrosion on aluminum interconnects.

Maboudian and *Howe* [51.21], *Srinivasan* et al. [51.22], *Kim* et al. [51.23], and *Pamidighantam* et al. [51.24], describe the use of SAM solutions to suppress stiction. There is also ongoing research focused on solving the manufacturing concerns [51.25–29].

Note that stiction also arises in packaged MEMS products. This yield and reliability problem is discussed in *Mechanical – Stiction and Wear*.

51.3.2 Process Flow and Its Interaction with Product Architecture

MEMS products have structures and functions that do not exist in standard IC devices, so it is unrealistic to expect that every fab, assembly, and test step will reapply a previously qualified IC process. The challenge is to maximize utilization of available IC technology within the constraints of the product function and cost requirements. This leads to fundamental choices in product architecture, product design, and process flow, as discussed below.

Integration of MEMS and Signal Processing

Integration of MEMS and electronics on one chip has proven difficult. It adds little value in some MEMS products. However, as noted earlier, integration is essential to products based on the Texas Instruments (TI) Digital Light Processing (DLP) technology. The third (and largest) category consists of applications where integrated and nonintegrated products compete for market share. Pressure sensors and accelerometers are examples of this group.

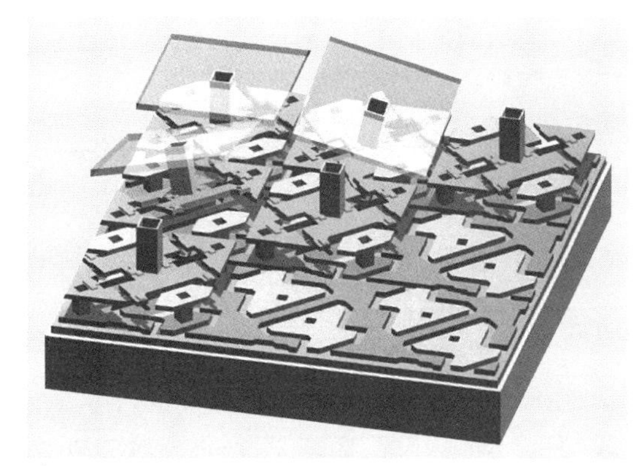

Fig. 51.5 DMD pixel array with tilted and nontilted mirrors (© Digital Light Processing, Texas Instruments, Inc.)

Image Projection. The first projectors based on the DLP technology were shipped in 1996. DLP chips are produced on CMOS wafers in a mature TI process. The circuitry addresses and drives aluminum mirror arrays that function as on/off pixels – two million mirrors in the larger arrays (Fig. 51.2). *Mignardi* et al. [51.7] describe the digital mirror device (DMD) fabrication process and illustrate some of the factors that must be considered when manufacturing flows are changed. The manufacturing flow patterns three aluminum depositions over static random-access memory (SRAM)

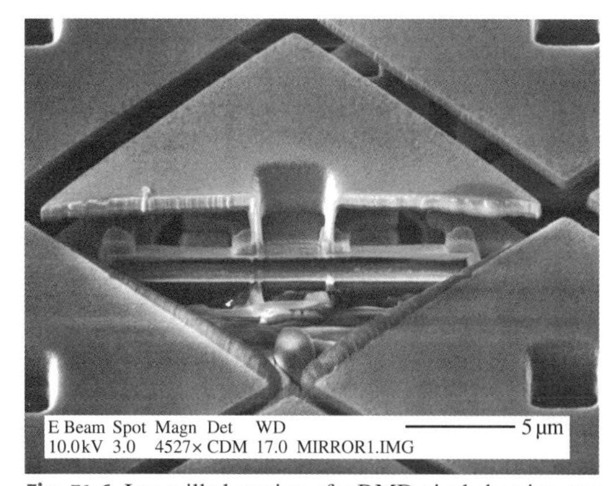

Fig. 51.6 Ion-milled section of a DMD pixel showing one mirror on its center support and its relation to the underlying layers. Note the close mirror spacing (© Digital Light Processing, Texas Instruments, Inc.)

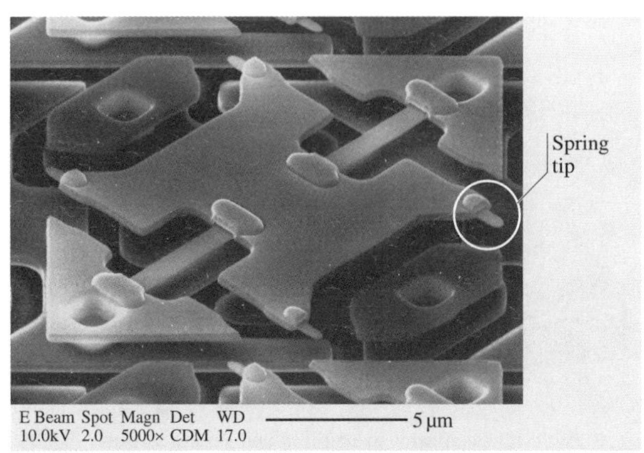

Fig. 51.7 DMD yoke and hinge layer. The spring tip touches landing pads when the drive electrode tilts the mirror. Elastic energy stored in the deformed spring is part of the restoring force that overcomes stiction when the drive electrode voltage is removed (© Digital Light Processing, Texas Instruments, Inc.)

cells that are positioned with 14 or 17 μm center-to-center spacing (Figs. 51.5 and 51.6). The first of these metal films form electrostatic drive electrodes and landing pads. The yoke and hinge layer has spring tips that land on the first metal layer when mirrors are rotated (Fig. 51.7). Mirrors are mechanically connected to the yokes through center pedestals. Sacrificial organic films separate the metal layers. These organics, along with a protective organic cover film, remain on the device until after the wafer is sawn and the chips are mounted in a ceramic package. They are removed by dry etching. Following a surface passivation, inspection, and testing, an optical glass subassembly is hermetically sealed to the ceramic package (Fig. 51.8). At this point, the device is ready for burn-in and final testing – a significant task because each mirror is actuated on chips with as many as two million mirrors.

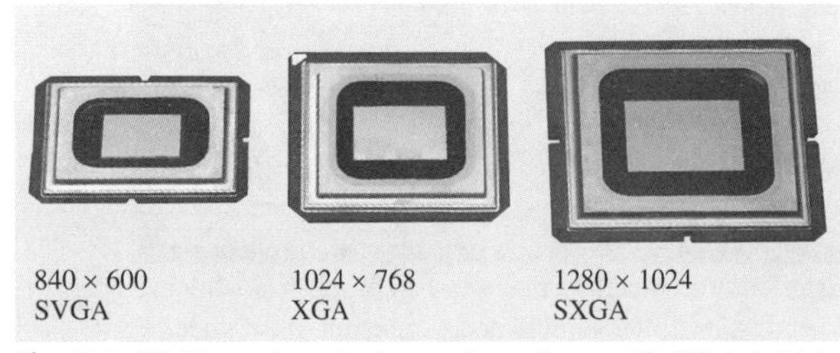

Fig. 51.8 DMD products in hermetic packages (© Digital Light Processing, Texas Instruments, Inc.)

In operation, a (nominal) 26 V electrode bias generates an electrostatic force that causes the mirror/yoke assembly to rotate either plus or minus 10° around the hinge.

Unlike the pressure sensors and accelerometers discussed below, TI had no choice but to integrate. It would be difficult, if not impossible, to devise a cost-effective multichip solution to control and drive a matrix of over a million mirrors. Successful integration of circuits and mirrors was only one step in the commercialization process. To remain competitive against lower-cost liquid-crystal display (LCD) and plasma products, TI continues to develop their package and test technologies, because package and test comprise a major part of the total product cost. This is discussed further in *Image Projection*.

Pressure Sensors. Early (1970s-era) bulk micromachined piezoresistive pressure sensors from Honeywell, ICT (later Foxboro-ICT), and Kulite were not integrated. However, in the 1980s, Motorola commercialized the first of its MPX5100 series piezoresistive pressure-sensor products. These products had bipolar signal conditioning electronics and temperature compensation on the sensor chip. Current MPX products are offered in various plastic packages (Figs. 51.9 and 51.10) with maximum pressure ranges between 10 and 300 kPa (1.45–44 psi). Medical and tire pressure products are also available. Note that Motorola's semiconductor products sector was spun off as a separate company (Freescale Semiconductor, Inc.) in 2004. This action split Motorola's MEMS business. Mo-

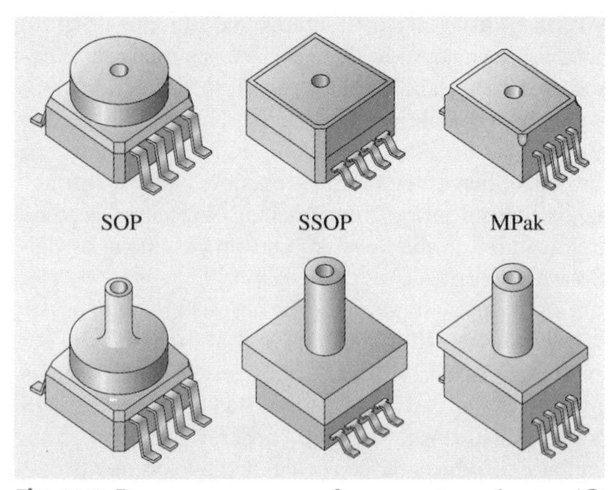

Fig. 51.9 Pressure sensor surface-mount packages (© Freescale Semiconductor, Inc.) (SOP – small outline package; SSOP – small shrink outline package)

torola retained the automotive module pressure-sensor business. Freescale has the device-level pressure sensors such as those sold for medical applications as well as tire pressure monitors. They also retained the accelerometer business. Over a 25 year period, Motorola/Freescale shipped more than 300 million pressure sensors [51.30].

Integrated Motorola and Freescale piezoresistive pressure-sensor products are cited here for illustrative purposes, but capacitive and piezoresistive silicon pressure sensors are available from many companies. Capacitive designs are less temperature sensitive and require less power than piezoresistive designs. However, die size tends to be larger. The fabrication process and interface electronics are also more involved. Piezoresistive products (sensor plus signal processing) are less expensive, smaller, easier to manufacture, and suitable for most applications. They rely on the fact that the resistance of silicon changes in response to strain. This effect is very temperature sensitive. However, configuring four piezoresistors as a Wheatstone bridge, judicious selection of doping level and junction depth, and use of a temperature-compensating resistor network minimize temperature error.

Piezoresistive pressure sensors (Fig. 51.1) require double-sided polished wafers and front-to-back alignment. The piezoresistors are implanted into the front side of (100) wafers and precisely located with respect to the edges of thin diaphragms. The diaphragms are produced by anisotropically wet etching of cavities into the back surface of the wafers. KOH etch solutions are usually used to avoid the safety, toxicity, and waste-disposal concerns that arise when large quantities of organic etchants such as ethylenediamine/pyrocatechol/water (EDP) are used. Front-to-back-side wafer alignment is critical in order to achieve precise resistor location with respect to the cavity edges. Alignment to the wafer crystal axis is also important, because the slow-etching (111) planes determine the cavity wall, and thus the location of the diaphragm edges. The piezoresistors are commonly placed near these edges to maximize the signal (the effect of pressure on diaphragm stress is greatest at the edges). An etch stop may be used at additional cost because diaphragm thickness variations also have a large effect on sensitivity.

Control of mount stress is critical to achieving predictable performance. After wafer fab, some suppliers hermetically bond pressure-sensor wafers to a backup silicon wafer using glass frit. Others anodically bond them to a borosilicate glass wafer that has a thermal expansion coefficient close to that of silicon (Pyrex 7740 and Schott 8330 are two glasses used for this purpose). Absolute pressure sensors are produced when this bonding step is done in vacuum. As shown in Figs. 51.1 and 51.10, differential pressure and gage pressure products incorporate through holes in the backup wafers. These ports allow fluid pressure to be applied through the cavities to the back side of the diaphragms.

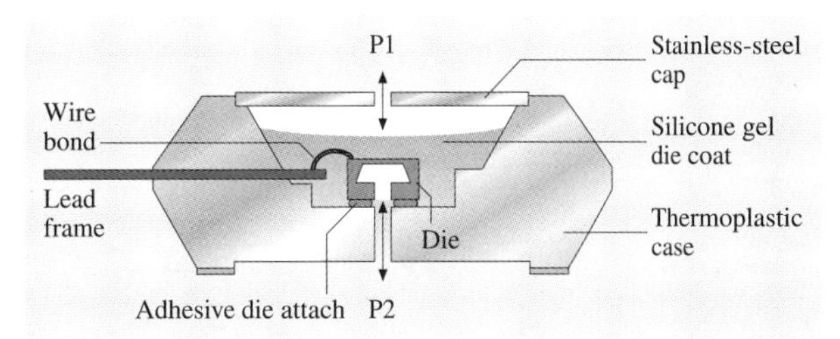

Fig. 51.10 Cross-section of a typical gage or differential pressure sensor. The silicon pressure sensor chip is bonded to a silicon substrate. This bonded unit is mounted in a premolded plastic package and protected from the ambient environment by silicone gel (© Freescale Semiconductor, Inc.)

After wafer singulation, the chips are mounted and sealed in cavity packages. Soft die-attach materials, often silicone based, are used to decouple the sensor from package and substrate stresses. In some products, the sensor chip is directly mounted to a substrate with a soft die attach without the intermediate backup wafer. Many package variations of MEMS pressure sensors are available, as summarized in *Pressure Sensors*.

The MEMS pressure sensor market is believed to exceed a hundred million pressure sensors annually [51.31]. Automotive applications such as manifold absolute pressure (MAP) sensors, barometric absolute pressure (BAP) sensors, and a wide range of gas and liquid pressure sensors form the largest market segment. Motorola, Bosch, and Denso are the largest automotive market suppliers [51.32]. Automotive applications are the primary market for integrated pressure sensors. Health care and medical uses such as disposable blood pressure transducers and sensors to measure pressure in angioplasty catheters, infusion pumps, and intrauterine products are the second largest market. Industrial products such as process control pressure and differential pressure transmitters, household appliance, and aeronautical products are also significant. *Maudie* and *Wertz* [51.33] review the performance and reliability issues related to appliance applications such as household washing machines.

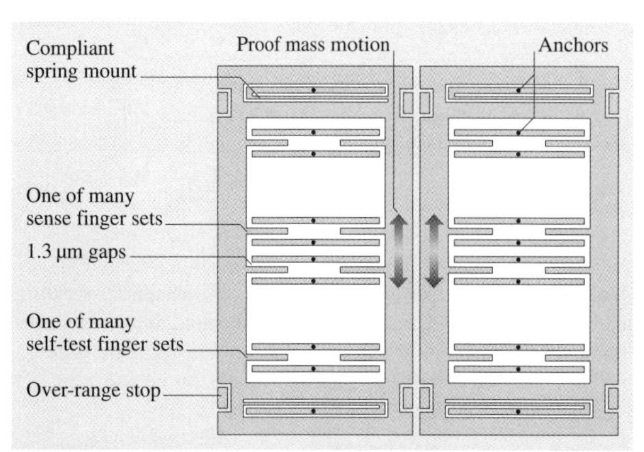

Fig. 51.11 Block diagram of the MEMS element in an ADXL78 single-axis accelerometer. The detailed design applies design-for-manufacturability principles to ensure close control of critical dimensions and residue-free removal of the sacrificial layer (© Micromachined Products Division, Analog Devices, Inc.)

Accelerometers. Most MEMS accelerometer designs apply some variant of Newton's law, $F = ma$, to sense the response of a proof mass. Many sensing principles, including piezoresistive, capacitive, piezoelectric, and resonant, have been examined [51.34]. Figure 51.11 shows a conceptual schematic of the capacitive design used in Analog Devices ADXL accelerometers.

SensoNor was an early pioneer in silicon accelerometer manufacturing. Their product had a mass bonded to the end of a silicon piezoresistive element in an oil-filled package. However, the first MEMS accelerometer product to achieve large-scale market acceptance was the integrated ADXL50 airbag sensor from Analog Devices. When it was fully qualified in 1993, the ADXL50 was the first integrated surface micromachined MEMS device of any type in production. The MEMS element in the ADXL50 was part of a closed-loop differential capacitance circuit and was designed to allow routine self-testing after the airbag module was installed in vehicles. This self-test capability eased concerns relative to adoption of a new technology in a critical safety application. The analog output feature also allowed the industry to use single-point sensors – a significant system cost saving. *Core* et al. [51.3] and *Sulouff* [51.6] describe the ADXL wafer fabrication process. In essence, circuits were fabricated with a well-established bipolar CMOS (BiCMOS) process. The sensor polysilicon was deposited after these high-temperature steps, followed by the lower-temperature metal and passivation processes. The in-situ doped polysilicon was connected to the circuit through n^+-doped runners. Following wafer fabrication, thin film resistors were laser-trimmed in an automated trim system to meet the specific end user's requirements. After wafer singulation (Sect. 51.3.5), the chips were assembled in hermetic cavity packages and screened in an automated test system to ensure compliance to the performance specification (see *Accelerometers and Gyros*).

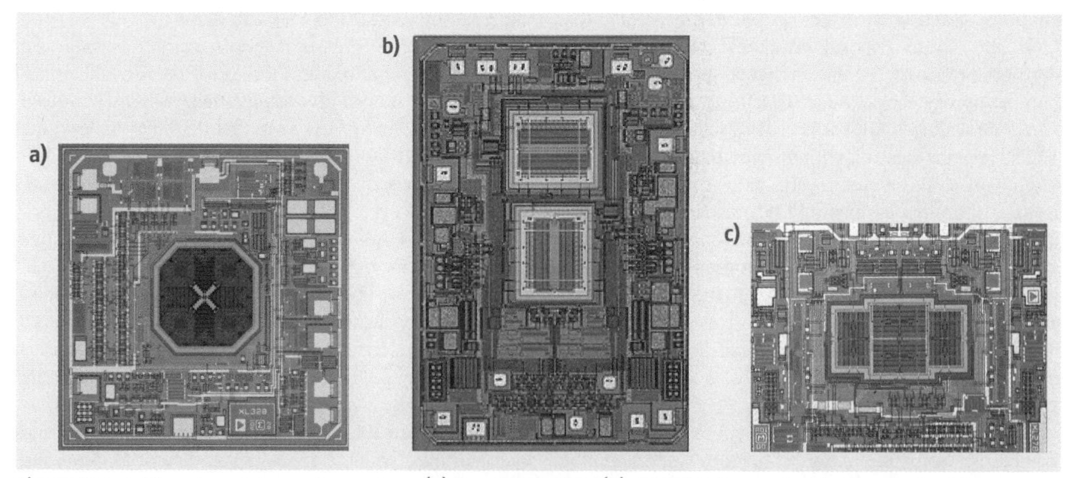

Fig. 51.12a–c Three two-axis accelerometers. **(a)** 2-*g* ADXL203. **(b)** ADXL276 was the first two-axis airbag sensor. The ADXL278 **(c)** family replaced it. The ADXL203 and ADXL278 are packaged in the LCC shown in Fig. 51.4 (© Micromachined Products Division, Analog Devices, Inc.)

Front airbag sensors have an output of 2 V at a full-scale acceleration that is specified by the user (typically 35–50 *g*). Satellite airbag sensors are located in door pillars and near the front of the automobile and typically have higher ranges (250 *g*). Regardless of range, all airbag sensors must meet stringent cross-axis specifications, i. e., a front airbag sensor must not react to a side collision or to potholes in the road. This cross-axis requirement also applies to products that are designed to sense in two axes (Fig. 51.12).

Progressive reductions in chip and package sizes (Figs. 51.3 and 51.4) led to cost reductions and penetration into new markets. One example started in 1998 with the ADXL202 and evolved to the ADXL203 and ADXL320 6 years later (Figs. 51.4 and 51.12). These two-axis, low-*g* accelerometers had a full scale of 2*g* in each axis. Production volume was relatively small in the first few years as consumer, industrial, and automotive users learned how to design products based on low-*g* multiaxis accelerometers. It is worth noting that these products were developed using an organizational philosophy that emphasized design for manufacturability; for example, development was conducted on production equipment and directly involved production personnel. This may appear to be expensive and unwieldy. However, it ensured that product designs remained within the bounds of practical manufacturing. It also minimized the difficulties that are normally encountered when new products are transferred to production.

This moderate growth initial phase changed by the end of 2006 as suppliers of video games, cellphones, and handheld devices learned how to design features based on low-*g* multiaxis accelerometers into their products. STMicroelectronics and Kionix made significant shipments of three-axis low-*g* accelerometers in 2005. By 2007, 15 suppliers were offering or sampling two- and three-axis accelerometers [51.35]. This report showed Analog Devices and STMicroelectronics dominating the market, with each having about 30% market share.

Three-axis accelerometers are routinely used in many products, but Nintendo's Wii raised MEMS visibility to a new level. Analog Devices and STMicroelectronics three-axis accelerometers are the motion sensors in the Wii controller and Nunchuck, respectively. What gave the Wii its tremendous market appeal? Microsoft and Logitech pioneered the use of Analog Devices motion sensors in 1998, while Nintendo had been using them since 2001. None of those products generated the consumer excitement produced by the Wii. The answer appears to be related to system architecture. Wii controllers are treated as an extension of the player's arm rather than part of the game console, so they give players a sense of reality that was not present in earlier-generation games.

Prior to introduction of the ADXL50, most MEMS airbag sensors were nonintegrated, bulk-micromachined piezoresistive sensors [51.4]. However, Freescale and Bosch soon followed with an integrate-in-the-package strategy. Denso, Delphi, and VTI Hamlin were also leaders in some market sectors. The Freescale and Bosch products are nonintegrated capacitive MEMS products that are hermetically sealed at the wafer level (see *Glass Frit Bonding*). By interconnecting a separate circuit chip and a sealed accelerometer chip in one package, they are able to use low-cost CMOS for signal processing and near-standard plastic packaging (see *Accelerometers and Gyros*).

In contrast to the ADXL and Bosch sensors, the early Freescale MMA airbag products measured acceleration normal to the plane of the chip (*z*-axis). These differential capacitance sensors are formed from three layers of polysilicon. The middle layer is a *proof mass* that is suspended with ligaments and is free to move in response to an accelerating force. Ranges and outputs are similar to the ADXL and Bosch products. Freescale later expanded their product line, adding *x*-

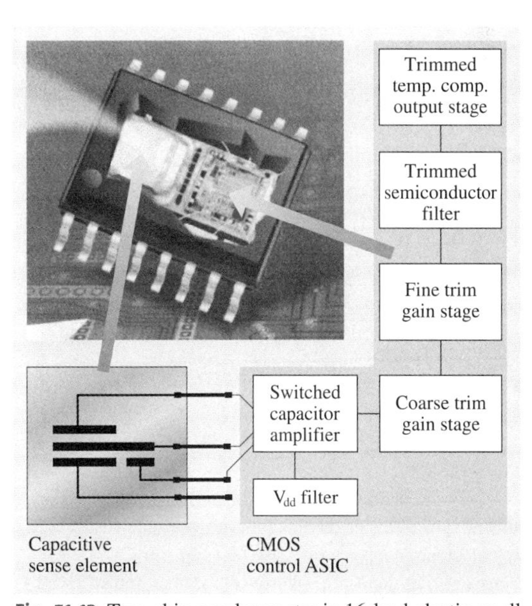

Fig. 51.13 Two-chip accelerometer in 16-lead plastic small outline integrated circuit (SOIC) package (© Freescale Semiconductor, Inc.)

axis airbag products and a broad line of low-g products (Fig. 51.13).

The Bosch x-axis sensor has interdigitated fingers that are patterned in a thick (11 μm) polysilicon film that is deposited in an epi reactor. Tight control of stress and stress gradients in these films is essential in order to meet sensor performance requirements [51.36, 37]. After patterning, the polysilicon is dry etched using fluorine-based chemistry [51.38] to form the differential capacitance sensor. A HF vapor process is used to remove the sacrificial oxide and release the microstructures. These thick structures have working capacitances near 1 pF [51.39]. *Laermer* et al. [51.40] explain how Bosch applies deep silicon etch processing to several production and experimental inertial sensors. Better known as the *Bosch etch*, this deep reactive-ion etching technique has pioneered the field of high-aspect-ratio silicon-based MEMS.

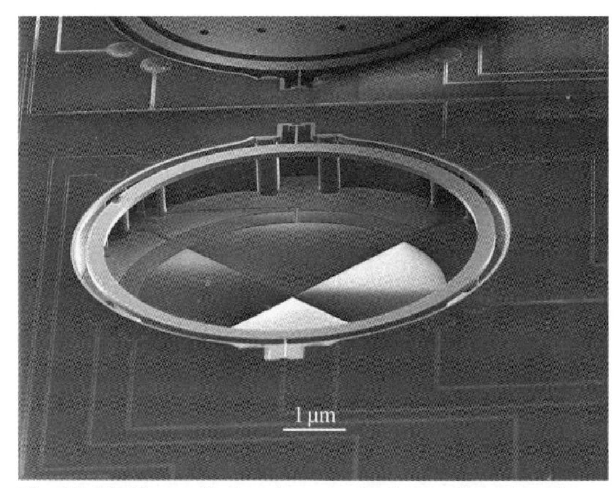

Fig. 51.15 View of the deep etched trenches, interlayer connections, and electrodes under a mirror of the SOI MEMS device illustrated in Fig. 51.14 (© Micromachined Products Division, Analog Devices, Inc.)

The Future of Integration. MEMS sensors deliver value when the quantity being sensed is converted into a useful analog or digital output. The ability to produce both MEMS sensors and the related signal processing was limited to large semiconductor companies that have in-house signal-processing expertise and the willingness to invest substantial resources over a number of years. Each company identified a large-volume market that could support a profitable integrated product. This *large semiconductor company* barrier to entry is diminishing as companies learn to design two-chip products using a foundry application-specific integrated circuit (ASIC) and a MEMS chip. The general approach to integration will continue to be two-chip solutions with MEMS and circuit die assembled in one package but the key to success is selecting a strategy that develops cost-effective manufacturing and quality expertise (Sect. 51.3.9).

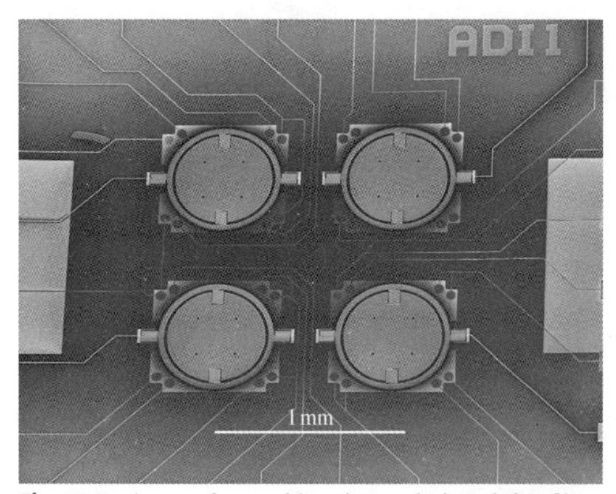

Fig. 51.14 Array of steerable mirrors designed for fiber-optic network switching applications. The gimbal-mounted mirrors in this experimental product are electrostatically driven. Closed-loop control based on capacitive sense electrodes ensures that the proper angle is maintained. High- and low-voltage circuits are integrated with mirrors on one chip (© Micromachined Products Division, Analog Devices, Inc.)

In the late 1990s, optical communication companies funded MEMS companies to develop the components required to build new systems. One result was the development of silicon-on-insulator (SOI) technology that integrates steerable mirror arrays, high-voltage drive electronics, and low-voltage control electronics on one chip (Figs. 51.14 and 51.15). The communications market did not develop as projected. However, the use of SOI wafers for other MEMS applications is occurring, as illustrated by the new generation of accelerometers from Analog Devices and Freescale. The ADXL180 and ADXL950 are integrated CMOS devices that use trench isolation to form electrically isolated regions within suspended mechanical microstructures. See [51.41] for a more detailed discussion. The Freescale SOI accelerometers are initially being targeted at automotive applications such as vehicle stability control (VSC) and electrical parking brake (EPB) [51.42]. These applications require high sensitivity and low zero-g error. The Freescale integration strategy uses MEMS-only SOI die that are capped and stacked on a 0.25 μm ASIC chip with an interposer die between them. Die stacking allows the product to be molded in $6 \times 6\,\mathrm{mm}^2$ quad flat no-lead (QFN) packages (Fig. 51.16).

Poly-SiGe is a competing integration strategy. Interuniversity MicroElectronics Center (IMEC) scientists have demonstrated a plasma-enhanced chemical vapor deposition (PECVD) process that deposits poly-SiGe films at temperatures near 400 °C [51.43–45]. University of California technologists developed a similar low-pressure chemical vapor deposition (LPCVD) process for poly-SiGe [51.46, 47]. Since poly-SiGe process temperatures are within the thermal budget of most CMOS products, a MEMS-only company can postprocess CMOS foundry wafers by depositing and patterning poly-SiGe microstructures on the wafers. This capability is not possible with polysilicon due to higher process temperatures. Ge and poly-SiGe with high Ge levels are convenient sacrificial materials because they can be removed in peroxide etchants [51.48].

Risk

High risk is normal in research programs. However, the tremendous financial investments required to manufacture new products make risk of failure a key consideration. This section discusses how the perception of manufacturing *risk* changes with time and is influenced by available knowledge and expertise.

Every development program makes fundamental product architecture decisions. These decisions can lead competing companies down different paths. For example, Analog Devices pursued chip-level integration while Bosch and Freescale chose package-level integration. Why did these companies choose different paths? The answer is related to available resources, cost, time to market, and technical maturity. Analog Devices chose to integrate because of its expertise in process innovation and signal processing. This path did have risks. When the ADXL50 was designed, the fatigue life of polysilicon MEMS devices was unknown. Therefore, a closed-loop circuit architecture was chosen in order to keep the proof mass motion < 100 Å. Polysilicon actually has excellent fatigue life. Once this was established, Analog Devices changed their ADXL products to an open-loop architecture that can detect capacitance changes as low as 20 zF (10^{-21} F). Most companies do not have the expertise to work in this regime, but the decision to integrate on one chip allowed Analog Devices to make full use of its signal-processing knowledge.

The two-chip Freescale and Bosch designs allow the circuit chip to be fabricated on any IC process. This product concept also removed the concern that integrating MEMS and circuits on one chip might reduce overall product yield. However, it does require higher output signals from the MEMS chip in order to overcome parasitics that arise in the bond wires between the circuit and sensor chips. This requirement led Freescale towards large-area z-axis designs, while Bosch utilized its deep-etch process capability. To minimize cost and ensure reliability, both companies seal the sensors with glass at the wafer level. Sensor and circuit chips are then molded together in one plastic package. Each company built on expertise that it already had in order to minimize risk and cost. Freescale had in-house manufacturing expertise in wafer-level glass frit sealing of pressure sensors that are subsequently molded in plastic packages. Bosch had deep-etch process capability and the world's largest hybrid manufacturing plant, so screen-printing and glass sealing were well established. These considerations plus the concern that MEMS yield loss would undermine product yield contributed to their two-chip decision.

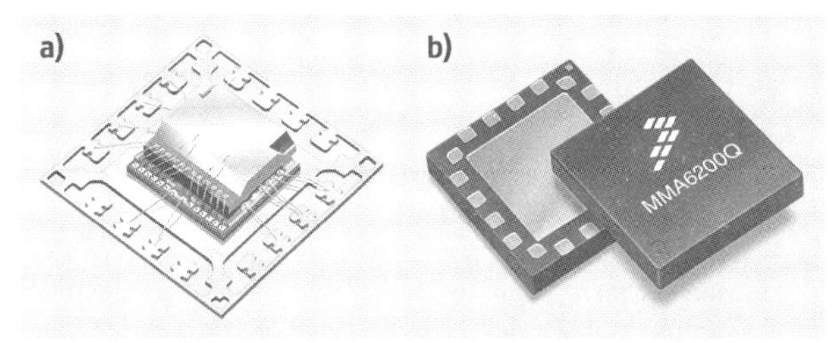

Fig. 51.16a,b The Freescale QFN has four layers of silicon. (**a**) The ASIC die is bonded to the leadframe; an interposer die is bonded onto the ASIC. A capped accelerometer die is then bonded onto the interposer die prior to plastic overmolding. The final product is shown in (**b**) (© Freescale Semiconductor, Inc.)

Semiconductor companies continually drive down cost and serve evolving markets by shrinking product size. Analog Devices chose to deal with one-chip risks rather than package uncertainties. ADXL50 chips were capped and molded in plastic long before the first qualified products were shipped in 1993 [51.49]. However, production-worthy capping equipment did not exist at that time. By introducing the initial product in TO-100 hermetic metal packages, the risks associated with custom equipment and plastic package stress were removed. To reduce cost, subsequent products were packaged in ceramic or molded in plastic. Analog Devices continues to integrate electronics and MEMS on one chip for most inertial products. However, some of their three-axis accelerometers are two-chip products. Product size is minimized by using one microstructure to sense acceleration in all three axes (other three-axis accelerometer suppliers have a MEMS chip with two microstructures).

Reusable Engineering and Facilities to Achieve Economies of Scale

The IC industry produces many products on each process flow. Foundries and packaging subcontractors routinely apply this *economy-of-scale* principle. Only basic items such as mask sets and bond pad diagrams are changed. The MEMS industry has been less successful in this area, although techniques such as wafer-level laser trim and blowing poly fuses allow accelerometer suppliers to offer *specialty* products without losing the advantages of high-volume production.

If only a few devices are produced in a capital-intensive facility, unit costs are high. This issue is particularly critical during the first years of product life when sales volume is normally low. One solution is to build the market by manufacturing the new product in an existing facility. In 1998, the two-axis low-g ADXL202 was built using a new mask set with an existing airbag sensor process and package. The low-g market niche developed slowly because companies had to learn how to design this new capability into their products. Once the initial phase passed, sales grew rapidly (see *Integration of MEMS and Signal Processing*).

TI is following an incremental improvement path. As they gain manufacturing experience, they introduce digital imaging products with progressively better resolution or lower cost. This strategy has resulted in strong growth in digital projector applications and a new high-definition television (HDTV) market. Experience gained in these products allows TI to build high-end products with more than two million mirrors for the digital cinema market.

Market Dynamics Drive New Product Opportunities

In the 1980s, low cost and high reliability allowed MEMS pressure sensors to penetrate automotive applications such as the measurement of manifold vacuum (MAP sensors). Once a technology is accepted, new opportunities usually develop. For example, tire pressure monitoring systems (TPMS) with radiofrequency (RF) communication capability are now available in most car models. The wireless communication drives solutions based on integration of a sensor, a communications chip, and a battery into the tire stem. This challenging application became a legal mandate in the USA as a result of legislation passed in 2001. Adoption is spreading rapidly, with global demand exceeding 120 million units per year by 2008 [51.50]. The pressure sensor in each tire must survive exposure to road salts, water, noxious organic vapors, temperature extremes, and high levels of shock and vibration for 10 years. Small size and long-term stability are also critical. Low power is even more important because the TPMS unit (sensor plus ASIC) inside a tire is expected to function on battery power for 10 years. *Osajda* [51.51] describes the approach that Freescale is using to meet these challenges. Many observers expect that TPMS products will migrate to non-battery-powered systems when these systems are developed to the point where they meet automotive cost and reliability expectations.

The TPMS market is effectively limited to tire stem suppliers such as Schraeder, Siemans, Lear, and Beru. Early market leaders for the pressure sensor element appear to be SensoNor (a subsidiary of Infineon Technologies) and GE NovaSensor. SensoNor uses a piezoresistive pressure sensor and a piezoresistive accelerometer sandwiched between anodically bonded glass wafers (see *Anodic Bonding*). One of the glass layers forms a vacuum-sealed cap over the implanted piezoresistors in the accelerometer and the pressure sensor. Their pressure port is on the back side of the bulk-micromachined pressure sensor in order to avoid contact between the piezoresistors and tire vapors. The relatively high power required by piezoresistive technology is partially mitigated by having the accelerometer *wake up* the system when the automobile moves. SensoNor TPMS products are assembled with a programmable ASIC to allow calibration. In the future this type of monitor will be a safety feature expected by all consumers. Indeed, wireless communication capabilities will also be a common MEMS product requirement.

This example illustrates the fact that MEMS products are merely system components. They must support seamless integration into the larger system. MEMS opportunities in automobile safety systems will be affected by both communication protocols and by interconnect technology (copper, fiber optic, and wireless). The 1990s growth spurt started with front-airbag crash sensors located in the center module of automobiles. It evolved to include satellite crash sensors behind bumpers, side-impact sensors in the B pillar (the vertical post between doors), impact sensors inside doors, rollover sensors, and vehicle dynamic control sensors to improve ride and handling. Each of these applications has unique characteristics that can favor one technology over another. For example, side-impact sensing systems must respond faster than front-impact systems. This can favor the use of pressure sensors that monitor air pressure inside door cavities. An alternative is to place an accelerometer in the B pillar, thin structural members that have very little space.

High-technology markets are seldom static. For example, conventional airbag systems use information below 400 Hz to detect inertial motion of the body frame. The satellite sensors described above provide early detection of a collision. Early detection allows the system to quickly activate passive restraints such as belt pretensioners and airbags. It is essential to distinguish minor events that do not warrant airbag deployment from serious collisions, so the airbag is triggered after a second sensor confirms that a serious collision is occurring. Smart airbag systems modulate the airbag inflation power by using these algorithms to interpret the nature of the collision. A new system design reduces cost by replacing satellite sensors with a wideband sensor located in the center module [51.52]. The wideband sensor primes the system by measuring acoustic energy that propagates through the vehicle structure at frequencies up to 20 kHz. It also senses the slower inertial motion. This type of system may shift demand away from 400 Hz sensors to those that are designed to operate up to 20 kHz.

AT&T Bell Laboratories patents issued in 1985 describe the advantages of using IC technology to make low-cost integrated microphones [51.53–55]. These MEMS microphones were not commercialized because lower-cost electret microphones also gave acceptable performance.

Electret and MEMS microphones both use thin diaphragms that move in response to sound pressure. Both also have a back plate near the diaphragm, allowing these elements to function as a dynamic capacitor. The charge in electret microphones (several hundred volts) is permanently implanted in the polymer surface. In contrast, the charge in MEMS microphones (typically 12 V) is maintained by the CMOS circuit when the unit is powered. Electret microphones have assembly and size limitations. Their 85 °C temperature limit requires that they be hand-inserted into sockets after the module is assembled. Exposure to the solder temperatures used in low-cost automated assembly processes would remove the implanted charge, thus destroying device performance. Over time, technology advances have reduced MEMS microphone size and manufacturing costs. Applications that value the functionality enabled by ASIC integration have also developed.

Knowles was the first company to recognize this change and has responded with substantial shipments of MEMS microphones. Their development efforts began in 1988 but volume production did not start until 2003 [51.56]. This two-chip product has a MEMS chip and an ASIC mounted on a small circuit board that is subsequently capped. Akustica, Sonion, and Analog Devices [51.57] have also introduced MEMS microphone products. Akustica uses Carnegie Mellon University technology for their one-chip integrated microphones [51.58]. The diaphragms in these devices utilize existing CMOS metal layers that are subsequently etched and treated.

Gyros – New Products Produced with Preexisting Manufacturing Capabilities

Gyroscopes are transducers for rotational motion. They have a wide range of uses in platform stabilization and robotics, as well as automotive applications [dead reckoning back-up for global positioning system (GPS) navigation systems when satellite communication links are lost, skid control systems, and roll-over detection]. Most non-MEMS and resonant quartz gyros used in military and aerospace applications are too expensive for many high-volume applications, but BEI Systron Donner tailored their high-performance quartz gyro to meet automotive needs and has shipped several million gyros. Other lower-cost gyro and angular rate products such as the Murata vibrating ceramic bimorph products are used in cameras. However, as a group, these low-cost products do not meet the performance required for many applications.

Analog Devices and Bosch reapplied their accelerometer capabilities to develop and manufacture low-cost gyroscopes. *Geen* et al. [51.59] describe the Analog Devices product. *Funk* et al. [51.60] and *Lutz* et al. [51.61] describe the Bosch product. These three publications reference previous work and discuss the challenges of producing gyros that meet performance, reliability, and cost targets.

The second-generation Analog Devices ADXRS610 gyro (Fig. 51.17) illustrates the technology. It has a resonant microstructure that is integrated with two electronic systems on one chip. One of the electronic systems drives the MEMS resonator. When the ADXRS610 is rotated, Coriolis acceleration is generated in a direction perpendicular to the vibration axis of the resonator. This motion is detected by the second accelerometer system. Integration of electronics

Fig. 51.17 ADXRS610 gyro with integrated electronics in $7 \times 7\,mm^2$ BGA package (© Micromachined Products Division, Analog Devices, Inc.)

on the chip, microstructure modeling, and predictable manufacturing were essential to success because the full-scale Coriolis motion is only ≈ 1 Å. The combination of differential capacitance detection, area averaging, and correlation techniques allow this motion to be resolved from thermal noise down to about 0.00016 Å [51.59].

As noted above, the high cost of non-MEMS and resonant quartz gyros restricted them to specialty applications. Most require expensive vacuum packaging to achieve high mechanical Q values at resonance. Unfortunately, vacuum packaging makes them susceptible to mechanical damage. Viscous damping in gas-filled devices such as the ADXRS610 suppresses the destructive impact events that occur in vacuum-packaged MEMS products during shipping, handling, and in-use conditions. Equally important are design features that allow the measurement to reject mechanical shock, vibration, and other environmental noise sources. As a result, under-the-hood location is practical in automotive applications, with significantly reduced system cost.

51.3.3 Microstructure Release

Release stiction occurs when microstructures stick together after the sacrificial material is removed in a wet etch. It can cause considerable yield loss and must be considered in devices such as capacitive pressure sensors and accelerometers that have closely spaced, mechanically compliant microstructures. Piezoresistive pressure sensors are not susceptible to release stiction, because these sensing diaphragms are not near other surfaces.

Release stiction is most frequently observed in silicon MEMS devices that use a sacrificial oxide. The oxide is removed by etching in wet HF, rinsed in deionized water, and then dried. Water promotes growth of a hydrophilic surface oxide. It also has a high surface tension. To minimize energy, the high surface tension in the shrinking water droplets causes the microstructures to be pulled together. When they touch, clean dry oxide surfaces stick, thus destroying device functionality. *Maboudian* and *Howe* [51.21] review release stiction mechanisms, as well as the solutions that have been reported in the literature.

Some manufacturers do not publicly discuss their fab process. Those that do avoid release stiction by using gas-phase processes to remove the sacrificial material. Texas Instruments uses photoresist rather than silicon dioxide as the sacrificial material in their DMD manufacturing process. Two photoresist layers are under the aluminum mirrors. The wafer is also covered with a third organic layer to provide mechanical damage protection and allow for particle cleaning after the wafer is singulated into individual devices. By removing these organics in a dry etch process, the stiction problems that characterize wet etch processes are avoided.

Analog Devices avoids release stiction by dividing its accelerometer process flow into several steps, each of which is well controlled with standard manufacturing equipment. After the microstructure is formed, a few small channels are etched in the sacrificial oxide and filled with photoresist. These photoresist pedestals hold the microstructures in place when the remaining sacrificial oxide is etched, rinsed, and dried. The pedestals are then removed in a dry etch process. *Core* and *Howe* [51.62] and *Sulouff* [51.6] describe this technique in more detail.

Bosch avoids release stiction by holding accelerometer wafers above an HF solution. HF vapors from the solution react with the sacrificial oxide, converting it to gaseous SiF_4 and water. The wafers are heated to prevent water from condensing. This process is noted in *Offenberg* et al. [51.63]. *Anguita* and *Briones* [51.64] describe a similar process.

51.3.4 Wafer Bonding

The growth of MEMS wafer bonding has been driven by three factors:

- Sealing microstructures by bonding wafers together addresses two sources of yield loss in MEMS manufacturing: wafer singulation and particles
- Wafer-level mounting can be a cost-effective way to control stress in products that are sensitive to variations in mount stress
- Fabrication of three-dimensional (3-D) microstructures from elements that are formed on multiple wafers.

About 80% of commercial MEMS wafer bonding in 2004 used either glass frit or anodic bonding [51.65]. Silicon direct bonding, polymer bonding, and several metal bonding techniques were used in a few applications. This technology mix is evolving as new products impose limits on process temperatures and contamination levels.

Piezoresistive pressure sensor wafers have been anodically bonded to borosilicate glass wafers for about 25 years. The earliest products were manufactured using little more than a hot plate and a high-voltage power supply. The era of custom-built manual equipment has passed, because both EVG and SUSS MicroTec offer automated production-worthy tools for anodic, glass frit, organic, metallic, and silicon direct wafer bonding. *Schmidt* [51.66], *Mirza* [51.67, 68], *Matthias* et al. [51.69], *Dragoi* [51.70], and *Farrens* [51.71, 72] review wafer–wafer bonding processes, applications, and equipment.

Anodic Bonding

Anodic bonding is commonly used to seal glass wafers to the back side of bulk-micromachined pressure sensor wafers. The process requires that a silicon wafer be placed in intimate contact with a glass wafer that contains a mobile ion at elevated temperature. An applied electric field causes mobile ions (usually sodium, Na^+) in the glass to move away from the silicon interface towards the cathode at the other side of the glass. Bound negative charges remain in the glass near the silicon interface. These charges produce an electric field that pulls the wafers together and drives oxygen across the interface to anodically oxidize the silicon surface. Hermetic seals are routinely produced between flat wafers when particles are rigorously excluded. Process conditions depend on the glass composition and thickness, but 500–1000 V at 400 °C with 10 min cycle times is typical. Borosilicate glasses with thermal expansion coefficients close to silicon are used to minimize stress. Anodic bonding promotes surface conformation so hermetic glass–silicon seals can be formed with surface grooves as deep as 50 nm [51.73]. This process has a long history and has been demonstrated to work with many material combinations [51.74].

Glass Frit Bonding

Glass frit bonding reapplies techniques and materials developed for hybrid processes and CERDIP packages. It is often used to seal pressure sensor wafers to silicon substrate wafers. A second common application is to glass-frit-seal bulk-micromachined silicon cap wafers over the microstructures in accelerometer wafers. In this process, fine glass powder is dispersed in an organic binder to form a paste. This paste is screen-printed or stenciled in the desired pattern on a wafer. The wafer is then passed through a furnace (typically at 430–500 °C). This causes the organics to burn off and the individual glass particles to coalesce into the desired pattern.

The wafer bonding process starts with a wafer prepared as described in the preceding paragraph. This wafer is aligned to a second wafer in a fixture. The fixture with its aligned wafer pair is then placed in the bonder, where moderate pressure is applied and temperature is raised beyond the glass-softening point. Liquid glass readily wets the surface of the adjacent wafer when at least trace amounts of oxygen are present. The liquid-wetting mechanism produces hermetic seals even when the wafer surfaces are relatively rough. However, seal widths less than 100 μm are difficult to achieve reliably due to the resolution limits of screen-printing. Low-melting lead oxide and boron oxide glasses are commonly used in processes that typically run at 450 °C. Glass wet-out occurs quickly but process cycle times are limited by the cooling rate (rapid cooling can cause stress gradients and cracking in the glass if proper design practices are not followed). Glass frit bonding to semiconductor wafers requires a good understanding of glass composition effects and process conditions [51.75].

Silicon Direct Wafer Bonding

Silicon direct bonding has been demonstrated in many IC and MEMS applications over the last 30 years. These processes bring highly polished wafers into intimate contact and are often promoted by a thin hydrophilic oxide that is left on the surfaces by the aqueous cleaning process. Clean, well-polished wafers with dielectric surfaces bond spontaneously when brought into contact at room temperature. The mechanism is believed to involve hydrogen bonding through adsorbed water molecules. Subsequent annealing (800–1200 °C) decomposes these bonds and forms Si–O–Si covalent bonds that increase the wafer bond strength by an order of magnitude. When electrically conductive bonds are required, wafers are treated in aqueous HF to form hydrophobic surfaces with Si–H terminal groups. The Si–H bonds are broken during the anneal step, forming direct Si–Si bonds.

EVG broadened the application range for direct bonding when it introduced equipment that plasma-treats wafers prior to bonding [51.77]. The low power plasma activates the surfaces. As a result, the required anneal temperature is reduced to 200–400 °C [51.69, 70, 78, 79]. SUSS MicroTec has a similar system [51.77]. Plasma activation has also been applied to bonding of various polymers and glass surfaces [51.71, 80].

A fundamental difference between MEMS silicon direct bonding and either anodic or glass frit bonding is that it is used to produce complex microstructures that cannot be made on one wafer. In contrast, most anodic and glass frit bonding applications address assembly and packaging issues. A primary limitation is the need for smooth flat surfaces, so wafer bow can be an issue when direct bonding is applied to circuit wafers. Chemical–mechanical polishing (CMP) is used to meet the surface finish requirement in some applications.

51.3.5 Wafer Singulation

The cooling water and the particles generated in a standard IC diamond saw will destroy the microstructures on MEMS wafers unless some form of protection is used. Many manufacturers solve this problem by glass-frit-bonding cap wafers to surface-micromachined wafers. The caps protect the microstructures from water, particles, and mechanical damage so the wafers can be sawn with standard equipment.

The optical function of DMD chips required Texas Instruments to find a different solution for DMD wafers. Originally, they used a partial saw process on wafers that were protected with an organic film. This allowed wafer-level testing, but caused particle contamination and die loss when wafers were broken along partial saw cuts. They now use a standard saw process, but do it before the mirrors are released. The wafers have the protective organic film over the mirrors so normal cleaning processes are used. The singulated chips are mounted in ceramic packages before the organic layers are dry etched. This flow eliminates a primary source of particle contamination and problems associated with handling partially sawn wafers [51.7].

Analog Devices singulates uncapped MEMS wafers using an upside-down saw process and standard equipment and fixtures that are slightly modified (Fig. 51.18) (*Roberts* et al. [51.76]). A major attraction is that it allows wafer-level test and trim, so only chips that meet the product specification are assembled. It also avoids the added cost and yield loss associated with wafer capping. Although originally developed for accelerometers, this technology was extended to gyro wafers and optical mirror arrays. They also singulate some MEMS wafers after wafer capping [51.81].

51.3.6 Particles

Particles cause yield loss in every product that has closely spaced movable microstructures. They also cause large-area defects in anodic and silicon direct-wafer-bonded processes. Commercial MEMS suppliers use several solutions:

1. Cap the MEMS wafers before they leave the wafer fab clean area. This approach is commonly used for accelerometer products (Figs. 51.13 and 51.16). It solves the wafer saw dilemma faced by most MEMS suppliers and allows the use of plastic packaging. Disadvantages include yield loss due to cap misalignment and stress, the cost of the added capping steps, and nonstandard hermeticity testing.
   Capping protects MEMS from particle contamina-

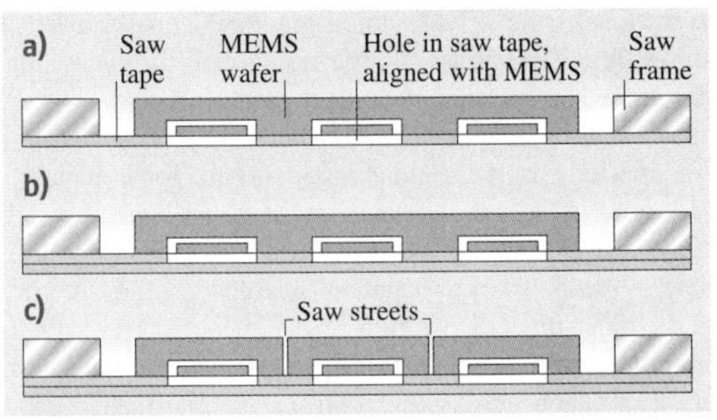

Fig. 51.18a–c Analog Devices *upside-down* saw process flow. **(a)** Saw tape is mounted on a saw frame. Holes are punched in the tape to match the microstructures on the wafer. The wafer is mounted upside down on the saw tape and aligned so that the microstructures are in the holes. **(b)** A second layer of tape is placed over the first layer to form watertight pockets. **(c)** Wafer is aligned and sawn using standard equipment. (After [51.76])

tion during assembly, but does not prevent wafer fab contamination. A greater concern is that standard glass-frit seal processes and equipment do not meet fab cleanliness standards. Once capped, particle inspection is essentially impossible.
2. Analog Devices and Texas Instruments assemble their MEMS products in cleanrooms. This option increases capital costs. TI invested in environmental and process control, rather than develop optical-quality capping for the DMD. Analog Devices had solved the wafer saw dilemma, so environmental and process control was the low-risk path for particle control. Some of the recent Analog Devices products are capped and packaged in plastic (Fig. 51.4).

Particle control is central to high yield and reliability. The first control level is to minimize particle contamination through proper handling procedures, equipment maintenance, environment controls, and process design. The DMD protective organic film illustrates how process design can be used to control particles. This film allows the wafer to be cleaned after wafer sawing and die attach.

Even with the best controls, some particle contamination occurs in every cleanroom. Electrical tests in the automated wafer- and package-level test programs are designed to detect particles in Analog Devices products. Texas Instruments also finds particles in DMD products by driving each mirror in an array to ensure proper operation. Such screens are good, but imperfect. Attempts have been made to replace visual inspection with automated particle inspection based on pattern-recognition software. To date, however, machine-vision systems have not been able to match the human eye and mind aided by a high-quality microscope.

51.3.7 Electrostatic Discharge and Static Charges

Susceptibility to electrostatic discharges (ESD) led semiconductor manufacturers to make ESD avoidance a central criterion in equipment and fixture design as well as handling procedures. Control of ESD events on the manufacturing floor is only the first step. Analog Devices and Texas Instruments incorporate ESD protective circuitry into their MEMS products as a standard practice to avoid failures in their customer assembly lines, as well as in the final system.

Most MEMS products use electrostatic force to actuate or control suspended microstructures. This design characteristic introduces performance parasitics and failure mechanisms that are not encountered in normal IC devices. For example, charges that build up on a dielectric surface will deflect nearby microstructures, even when they are several millimeters apart. Such effects are insidious because closed-loop control systems do not always correct for electrostatic-induced errors.

Section 51.4.1 discusses surface treatments designed to suppress stiction caused by surface forces. These treatments often produce a dielectric surface. If the dielectric properties of the treatment and the device design cause the surface to hold a charge, that antistiction coating may actually promote electrostatic stiction. Thus, any *solution* must be critically evaluated in order to identify and remove undesirable side-effects.

51.3.8 Package and Test

IC packages are environmental (mechanical, chemical, electromagnetic, optical) barriers that protect chips from surrounding media. Only power and electrical signals and inertial forces pass unimpeded, so most MEMS sensors and actuators require that this barrier function be selectively penetrated. Packaging and test functions are unique to each company's design and MEMS product type. Standard semiconductor packaging equipment and processes are modified when possible. However, even the line of demarcation that separates wafer fab from packaging is unique to each MEMS product. *Hsu* [51.82] edited a book that gathers information on assembly, packaging, and testing on many types of MEMS devices. Similarly, *Tabata* and *Tsuchiya* [51.83] include several chapters that focus on evaluation and testing of the thin-film materials which are commonly used in MEMS.

High-volume MEMS testing is even more challenging than MEMS packaging. In essence, the task is to measure and trim the response of products mounted in nonstandard packages to calibrated stimuli over a range of temperatures. Automated IC test equipment does not have the calibrated pressure, acceleration or optical test functionality required for this task. *Maudie* et al. [51.84] detail the test system elements that are required to support volume production of MEMS products.

Pressure Sensors

Pressure sensors are extremely susceptible to stress, so die-attach stress variations broaden performance distributions. The uniformity produced by wafer-level bonding is a primary reason why these processes are used to make the first-level package in many pressure sensor products. Soft silicone and fluorosilicone rubber products are often used to attach the sensor chip to the package to maximize isolation from mechanical stresses transmitted through the mount.

Most semiconductor products are assembled in plastic packages. The automated plastic presses transfer viscous epoxy compounds into multicavity molds at 175 °C and pressures near 6900 kPa (1000 psi). The hot liquid plastic creates high shear stress as it flows around devices in the mold cavities. Further stresses develop as the plastic hardens and cools. Even if this technology were adapted to create pressure ports in packages, the mold stresses would substantially affect pressure sensor performance.

Premolded plastic packages avoid the stress problem and incorporate pressure ports (Fig. 51.10). Polyphenylene sulfide, liquid-crystal polymer (LCP), and medical-grade polysulfone are high-temperature thermoplastics commonly used to make premolded pressure sensor packages.

The liquid or gas that is being measured can cause corrosion of bond wires and chip metallurgy, as well as parasitic leakage paths in the sensor. For this reason, a barrier such as silicone or fluorosilicone gel is often applied over the sensor. The soft gel transmits pressure with high fidelity over a wide temperature range and passivates the chip against many types of chemical attack. Parylene is also used for this purpose. This vapor-deposited organic coating is much stiffer than the gels, so thickness is typically controlled to near 1 μm.

Petrovic et al. [51.85] tested the ability of fluorosilicone gels and several parylene C thicknesses to protect powered pressure sensors against automotive and white-goods benchmark liquids. One purpose of their study was to propose and demonstrate a formal media compatibility test protocol for pressure sensors similar to the IC industry standard tests. They found that the lifetime of sensors coated with both gel and parylene was considerably longer than sensors coated with either gel or parylene alone. *Petrovic* [51.86] later summarized the advantages and limitations of gels, parylene, and other techniques.

Figure 51.9 shows examples of commercial plastic-packaged pressure sensors. Pressure calibration equipment and software must be custom designed and built for these products. However, most pressure products are dimensioned to match IC packages in order to maximize compatibility with standard IC test handlers.

The harsh environment of some under-the-hood applications has been used to justify the high cost of metal cavity packages. For example, Bosch introduced a piezoresistive MAP sensor with signal conditioning that was hermetically sealed in an evacuated T08 header [51.87]. In-package trim of planar thyristors was used to bring the sensor output within specification limits. The design avoided chip metallurgy corrosion and surface electrical parasitics by using the back surface of the silicon diaphragm as the fluid interface.

Much more expensive packaging is used in the pressure and differential pressure transmitters that are designed for industrial process control. These products seal the sensor in silicone oil behind thin metal diaphragms. Such products can measure pressure differences of 20 kPa superimposed on a pressure of 20 000 kPa (3000 psi) over wide temperature ranges with high accuracy and stability. Soft die-attach materials decouple the sensor from thermal stresses that arise when they are mounted in metal housings. Note, however, that silicone die-attach materials cannot be used if the transmitter is filled with silicone oil. Fluorosilicones are one alternative. Piezoresistive sensors have been used in pressure and differential pressure transmitters since the 1980s. *Fung* et al. [51.88] describe a relatively new version. This product uses a piezoresistive polysilicon sensor to measure both absolute and differential pressure on the same chip. See *Chau* et al. [51.89] for early work on a multirange version of this technology.

Image Projection

The Texas Instruments DMD poses unusual challenges because the package lid must be an optical-quality glass with antireflection coatings to improve optical performance and reduce heat load. The glass lid has opaque borders around the image area to minimize stray-light effects and create a sharp edge on the projected image. It is fused to Kovar frames to produce lid subassemblies. The package base is a multilayer alumina substrate with cofired tungsten to provide the electrical interconnects. The package sidewalls are formed by brazing a Kovar seal ring to the substrate. Heat dissipation through this substrate is a significant consideration because the service life of DMD mirror hinges (Fig. 51.7) is largely determined by operating temperature [51.90]. Excessive high-temperature creep results in *hinge memory* and would cause a gradual drift in mirror orientation.

The DMD chips are die-attached in the package cavity before the mirrors are released. A dry etch process is used to remove the organic sacrificial layers. The die are then passivated and tested. Getter strips are attached on the inside surface of the glass at the sides to control vapor composition in the package cavity before the lid is aligned and seam-sealed to the package seal ring. Image quality requirements place stringent requirements on handling, alignment, and spacing tolerances. Thermal stresses also arise when dissimilar materials are joined in high-temperature processes to produce the final hermetic package. The DMD packages (Fig. 51.8) and the related processes are discussed further by *O'Connor* [51.91], *Mignardi* et al. [51.7], *Bang* et al. [51.92], and *Poradish* and *McKinley* [51.93].

DMD mirror release does not occur until after die attach. This has advantages with respect to wafer sawing and particle suppression, but it sacrifices the economic attraction of wafer-level testing. There is no commercial test system that is capable of combining CMOS electrical testing with 100% testing of optical mirrors. Therefore, a custom electrooptic test system was built around an x–y–θ translation stage with a charge-coupled device (CCD) camera, light source, control hardware and software, and test programs [51.7]. A characteristic of papers published by the DMD group is their effective use of this test system to examine problems and statistically validate the solutions.

Assembly of the DMD on the projector electronics board requires alignment with the system optics, in addition to electrical connections. Initially, the DMD was held with a plastic clamp and electrically connected through elastomer pads that had alternating layers of conductive and nonconductive material. This allowed easy replacement if a DMD was damaged during assembly. However, the impedance of this connector system was too high for new, higher-speed products. Electrical intermittents were also observed. Therefore, it was replaced [51.94] by a grid of c-shaped springs (cLGA, Intercon Systems, Inc., Harrisburg, PA).

The DMD has about 40% of the business and entertainment image projection market [51.95]. Products based on liquid-crystal technology serve the balance of the market. Market share is largely driven by price, so considerable effort has been placed on reducing the cost of DMD package and board-level assembly. *Migl* [51.94] reports on the assembly benefits achieved by replacing the epoxied heat sink with a mechanically attached heat sink. *Jacobs* et al. [51.95] describes the effort to replace the seam-sealed window mount with a lower-cost epoxy-bonded design. In theory, such a bond is not hermetic. However, the team realized that proper material selection, design, analysis, and use of moisture getters would produce a low-humidity package through the product life. This change was not released to production. If it is implemented in the future, success will require that adhesive bond integrity also be maintained, because the adhesive joint sustains thermal expansion and mechanical clamp stress cycling each time the projector is used.

Accelerometers and Gyros

Inertial sensor capping is a first-level package that is applied at the end of wafer fab. After capping, the wafers can be sawn using standard equipment and do not require cleanroom assembly conditions. The capped sensors are compatible with standard plastic packaging when the cap and/or the MEMS wafer is backlapped. Molding stresses were a serious problem in early airbag accelerometers that were molded in dual inline packages (DIP) and single inline packages (SIP) for assembly on through-hole circuit boards. The shift to smaller surface-mount packages reduces the quantity of plastic in the package, so the stresses applied to the sensor chip of current products are reduced.

Analog Devices uses a near-standard saw process (Sect. 51.3.5 and Fig. 51.18) and hermetic cavity packages for its uncapped MEMS products. Such packages are standard, but more expensive than molded plastic. A major attraction of cavity packages is that they eliminate plastic package stress (see *Accelerometers and Gyros*) because the only mechanical connection between the chip and the package is through the die attach and bond wires. As a result, ADXL accelerometer and ADXRS gyro products are fully tested and trimmed on automated systems before the wafers are sawn. A lot-tracking system transfers this data to automated die-attach systems. These systems are programmed to pick only chips that meet performance specifications from the wafer-saw film frames. Thus, reject die are not assembled into packages. After packaging, devices are tested to ensure conformance to specification. However, package-level trim is not required because the assembly process and cavity packages do not appreciably shift device parametrics.

The evolution of ADXL cavity packages is illustrated in Fig. 51.4. Initial products were packaged in TO-100 metal packages. This seam-sealed package is useful for development purposes. However, it is expensive and incompatible with the automated equipment used to assemble electronic circuit boards. Therefore, the early airbag sensors were soon switched to CERDIPs. CERDIPs are made from low-cost molded ceramic bases that have glass seal surfaces. The high process temperature requires use of a silver–glass die attach product. The full assembly flow includes two or three furnace passes near 450 °C to produce hermetic cavity packages.

IC products have been packaged in CERDIPs for decades, but through-hole circuit boards are seldom used today. A simple change in lead forming allowed CERDIPs to be fully compatible with standard surface-mount boards (*CERPACs*), so CERDIPs and cerpacs were assembled on the same equipment. Most of the ADXL products shipped in the late 1990s were packaged in cerpacs. However, small leadless chip carriers (LCCs) became the package of choice for one- and two-axis accelerometers. The LCC uses solder-sealed ceramic bases and metal lids. Organic die-attach materials are practical, because the furnace gas is nitrogen, rather than air, and temperatures are about a hundred degrees lower than cerpac furnace temperatures. More recent ADXL products are wafer-capped before molding in plastic while the ADXRS gyros are assembled in a solder-sealed $7 \times 7\,mm^2$ ceramic ball grid array (BGA) package.

The fundamental message of this section is that IC package technology – both plastic and hermetic – is evolving very rapidly. Cost-competitive suppliers must remain cognizant of these trends in order to use them to best advantage. MEMS products that are packaged in a way that is not compatible with standard IC equipment and interfacing standards bear a significant cost premium. High package cost seriously limits market size because the price–volume curve is steep for most products. The package-size trend is also critical to new market penetration. In summary, customer interface and new application requirements will continue to drive down both the cost and size of most MEMS products.

A similar message applies to developing the test strategy, although the challenge of procuring custom-automated MEMS test equipment is diminishing. For example, Multitest GmbH now makes handlers with integrated shakers that are specifically designed for accelerometer testing. The leading assembly subcontractors are also gaining expertise in testing conventional MEMS products such as pressure sensors and accelerometers.

51.3.9 Quality Systems

Quality systems are an intrinsic part of stable MEMS production. Management must drive a systemic approach to quality and set continuous improvement as a high-priority goal.

Several quality systems define continuous improvement methodologies. In general, they formalize the process used to minimize defects and variations in products and in the processes used to manufacture them. Automotive supply companies must implement quality systems that meet QS-9000:1998 [51.20] before product volume ramps up, and to continue them through the product lifecycle. Four elements that are worthy of note are:

1. *Failure-mode effect analysis* (*FMEA*). Early in the development phase, a team with representatives from several disciplines reviews the design (or process) in order to identify possible causes of failure. Each potential failure mode is given three numerical rankings. One ranking represents the likelihood of occurrence, while the others rank the severity of that result and detectivity. These numerical scores are combined for each potential failure mode in order to identify the issues that merit the most attention before they become problems. FMEA spreadsheets are periodically updated as the product or process moves into production. More targeted versions are used in products or processes that are in stable production.
2. *Process control.* Every significant process must have a *short loop* monitor to ensure stability. This measurement is tracked and statistically analyzed with respect to the control limits to identify changes before the process strays beyond the control limits.
3. *Review boards.* Every change, unusual occurrence, and proposed solution is assessed by boards that meet on a regular basis to ensure that it does not put product quality at risk. Affected production material is put aside until the appropriate board approves its release. This may appear to be expensive and bureaucratic. However, the cost of a scraped wafer lot is insignificant when compared with the cost of a field replacement program.
4. *Processes, procedures, and specifications.* Each step must be fully specified and identified in the process flow, along with the appropriate metrics for each lot. The compilation of this data in a retrievable form is an essential part of every continuous improvement program. By combining these product and process databases with lot-tracking software, Pareto charts that link yield loss and test failures to process and material variations are generated. Experience has shown that unexpected second-order effects are present in every production line. However, data-driven decisions on these indirect effects

cannot be made on small test populations. They only become evident when production-volume data is analyzed.

Implementation of a continuous improvement program requires well-informed failure analysis teams, methodologies, and programs. Such analyses start with gathering facts and making relatively simple tests. Often this is sufficient. However, MEMS products are susceptible to uncommon failure modes, so *Walraven* et al. [51.96] gathered examples of more powerful analytical techniques and show how they are applied to MEMS devices.

Each supplier uses the information and insights gathered from product performance, as well as control and yield data at different points in the process flow, to refine their operations. *Douglass* [51.97], for example, outlined the yield loss and failure mechanisms observed in early Texas Instruments DMD products. Concerns such as hinge memory, hinge fatigue, particles, stiction, and environmental robustness were each addressed and mitigated by focused teams. For example, *Mignardi* et al. [51.7] describe the partial wafer saw process used in the original DMD manufacturing flow process. Breaking these delicate wafers into individual product chips generated particles that caused yield loss and were potential sources of field failures. A new flow based on a standard full saw process eliminates this particle source. The result was an increase in both yield and reliability. Hinge fatigue characterization showed that bulk metal fatigue models do not properly describe thin-film behavior. The hinge memory effort required an understanding of how thin-film metal creep is affected by alloy composition and environment [51.90]. Stiction control involves surface passivation, spring design, mirror dynamics, and moisture level [51.97, 98]. These publications give an insight into the quantities of data and the time required to bring robust products to market. They also illustrate how data-based evaluations can uncover unexpected effects such as the acceleration of hinge creep by adsorbed moisture [51.98].

The second continuous-improvement example applies to MEMS integration. High yield loss is perceived to be a serious risk when electronics and MEMS are integrated on one chip. Indeed, the initial yields on the Analog Devices integrated MEMS accelerometers were not impressive. However, it increased each year due to the work of many mission-oriented teams. Yield has reached defect-density-limited levels because the teams eliminated all significant failure mechanisms.

Coincident with the annual yield increases were reductions in customer failure rates. Failure rates in Analog's automotive accelerometer products are in the low single-digit ppm range. Quality does pay.

51.4 Stable Field Performance

Some topics discussed in Sect. 51.3 such as particles and ESD are equally relevant to long-term stability. Most microstructure products have elements that are in close proximity, so stability is also affected by surface characteristics and mechanical shock.

51.4.1 Surface Passivation

Early, 1970s-era piezoresistive pressure sensors were not passivated. Like early ICs, these wafers only had oxide over the piezoresistors. Performance was inconsistent and drifted with time, because surface interactions with moisture and other atmospheric gases created a variety of shunts, parasitics, and charging issues.

Electrical Surface Passivation

The introduction of silicon nitride passivation and conductive field plates led to stable products that hitherto were unobtainable. The ability to achieve measurement stability was absolutely critical to growth of the MEMS pressure sensor industry.

Aside from pressure sensors, few MEMS products are passivated. Indeed, standard plasma nitride processes are usually impractical, because they produce dielectric coatings that support static charges. Such charges can cause electrical drift, or be the source of electrostatic forces that cause stiction.

There have been published reports attributing instability to the lack of passivation on microstructures. For example, Analog Devices implemented a special process in ADXL50 accelerometers before the product was released in order to suppress high-temperature electrical drift [51.99, 100]. This was not a moisture effect because it occurred in dry nitrogen packages. Later ADXL designs eliminated the root cause.

Scientists at Lucent Technologies observed anodic oxidation in polysilicon electrodes used to electrostatically drive mirrors in optical cross-connect products [51.101]. If allowed to occur, this corrosion mechanism would cause the mirror position to drift and the product to eventually fail. The study concluded that, within the limits of the test, maintaining low moisture in these optical packages eliminates anodic oxidation.

The preceding paragraphs suggest that lack of MEMS passivation can affect long-term product stability. This issue is very design related and can be driven by factors other than humidity. Even when low package humidity is determined to be adequate, design reviews should consider how normal manufacturing variations, outgassing, and diffusion over product life affect the moisture level in the gas adjacent to the microstructures.

Mechanical – Stiction and Wear

In-use stiction is difficult to predict and may not become an issue until manufacturing volumes increase, or when end users handle the part in ways that are not anticipated. The resulting product liability and field replacement programs carry substantial financial costs and have caused MEMS suppliers to withdraw from the market. End users have little tolerance for failure. Since in-use stiction is a failure mechanism in MEMS devices, a company that solves this problem creates an effective market barrier against competitors. Although product-level data are not available, published work on the fundamental causes of stiction indicates that there is a strong link between attractive surface forces and the topography of contacting surfaces [51.102–105].

The significance of stiction varies between product areas. For example, the Texas Instruments DMD has moving mirrors that are designed to touch down on a substrate and later release. This product must address both stiction and wear. In contrast, the accelerometer products from Analog Devices, Bosch, and Freescale have proof masses that are suspended on compliant springs. By design, an automobile collision only displaces the proof mass by a few percent of the gap that separates it from adjacent surfaces. It is difficult to generate a shock wave in an airbag module that is sufficient to cause these proof masses to contact adjacent surfaces (acceleration levels of several thousand g are required). However, handling of the discrete packaged parts during test, shipment, and module assembly frequently cause shock events that bring MEMS surface into contacts. For example, *Li* and *Shemansky* [51.106] used both math modeling and tests to show that dropping a packaged MEMS accelerometer from the height of a tabletop generates several tens of thousands of g in the device when it lands on the floor. Thus, noncontacting MEMS products must be designed to withstand at least transient stiction.

Design. When shock, vibration or functional operation causes MEMS elements to touch, the mounting springs are designed to pull them apart (Fig. 51.11). High-stiffness springs are desirable to suppress stiction, but low-stiffness springs increase measurement sensitivity and signal-to-noise ratio. Thus stiction is a fundamental MEMS design constraint that often limits product performance. Each manufacturer has proprietary design practices to address this performance–reliability tradeoff. In general, designers suppress stiction by minimizing the sources and applying supplementary techniques to ensure recovery when contact occurs:

1. Thorough analysis and testing to move harmful resonances beyond the range where they might be excited. Note that MEMS component-level analysis is insufficient. This is a system issue based on the packaged part as it is handled through manufacturing, system assembly, and end use.
2. Surface modification (discussed in the next section).
3. Minimizing contact area by integrating bumps and *stoppers* into the design [51.107].
4. Elimination of dielectric surfaces that may accumulate surface charges and result in electrostatic attraction.
5. Release of DMD mirrors from the touchdown position is assisted by pulsing the reset voltage to excite a mirror resonant frequency [51.97, 108].
6. Use of gas-filled packages to reduce contact velocity, as discussed later in this chapter.
7. ESD-protected designs that prevent voltage transients that may cause electrostatic attraction.

The only product in significant production that contains MEMS elements that are designed to touch is the Texas Instruments DMD. The proprietary aluminum alloy springs used in the mirror contact points touch and release millions of times during the product life. Considerable development was required to overcome metal creep that occurred in early designs.

Surface Modification. In addition to the touchdown springs, early DMD die were treated with perfluorodecanoic acid vapor before the glass lid was sealed onto the package. This treatment [51.109] created a low-energy monolayer coating that suppressed initial stiction. However, improvements were required to achieve longer wear life, higher production rates, and lower contamination levels. A new process [51.110] places capsules inside DMD packages. After cleaning, the packages are immediately sealed. An oven bake releases antistiction vapor from the capsules to create a monolayer organic coating on surfaces inside these sealed packages. Excess antistiction material in the capsules maintains a low vapor pressure in the package cavities throughout product life. Therefore, wear damage at the mirror contact springs is continuously repaired to prevent stiction failures before they occur.

Maintaining a low level of organic vapor inside the DMD reapplies a concept used with other organics to prevent corrosion of electronic metallurgy and contact fretting of separable connectors [51.111, 112]. These treatments have been shown to add years to the useful life of electronic systems that are in enclosures, but also exposed to aggressive environments.

Stiction can be minimized by design and process adjustments – stiff springs, control of contact area, etch residues, etc. Some suppliers design within these limits while others modify the MEMS surfaces. The requirements of any surface-modification technology are substantial. In addition to normal manufacturing requirements (stable, scalable process and materials), the treatment process must be conformal and must uniformly modify all areas, including microstructure bottom and side surfaces. If a conductive material is used, it must somehow be patterned in order to avoid electrical shorting. If a dielectric surface is formed, it must be extremely thin because thick dielectrics support surface charges. Such charges cause performance shifts and induce electrostatic stiction.

Since devices with movable microstructures are particularly susceptible to stiction, most suppliers who develop production-worthy surface-modification processes consider this information as trade secret. Analog Devices chose to patent portions of its technology. The following paragraphs give an overview of the published Analog Devices surface-treatment technologies. *Martin* [51.99] describes the evolution of these technologies from 1992 to 1999. Further information is contained in *Martin* and *Zhao* [51.113] and *Martin* [51.100, 114].

The earliest of the published Analog Devices surface-treatment technologies was implemented after stiction was observed during automated testing and module assembly of airbag sensors that were packaged in CERDIPs. CERDIP assembly typically includes at least two furnace processes in air at 440–500 °C, so they seldom contain organic materials. The high-temperature furnace air thermo-oxidizes adsorbed materials and removes them from the MEMS surface. The resulting surfaces are extremely clean. Unfortunately, clean inorganics have high surface energies. Therefore, adjacent microstructures readily stick together if shock and vibration during handling cause them to touch. CERDIPs (and the surface-mount cerpacs) have very low moisture levels. Mil specs allow up to 5000 ppm moisture, but < 100 ppm moisture is common in a well-controlled manufacturing line. Deliberately raising the moisture level was found to be a potential solution because, within limits, moisture adsorption reduces surface energy. However, the solution adopted by Analog Devices in the mid-1990s involved dispensing a controlled amount of a pure phenylmethyl siloxane liquid into each package immediately before sealing. Selection of this liquid was based on thermo-oxidative stability, low volatility, purity, and the requirement that it be liquid at the dispensing temperature. As packages are heated in the furnace, material adsorbed on the microstructures is removed, leaving chemically active, high-energy surfaces. Further heating volatilizes the phenylmethyl siloxane. Before escaping into the furnace, the siloxane vapor surrounds the reactive MEMS surfaces and chemically bonds to them. The result is phenyl group infusion into the native oxide surface. Organic-rich surfaces are unreactive and have low surface energy. Thus the treatment is self-limiting after the first vapor molecules react. Obviously, this volatilization process must be complete before the glass softens and seals the package. Some of the phenylmethyl siloxane degrades in this process. However, silicone thermo-oxidation simply adds a few angstroms of silicon oxide to the native oxide, so the process, by design, is noncontaminating.

Although variations and refinements of this process were developed, this antistiction technology required that each individual package be treated. Automated dispensing equipment made this requirement tolerable when product volumes were moderate. However, as production volumes grew beyond several million per month, it became evident that a wafer-level antistiction process was required.

Assessment of candidate processes and materials led to a vapor treatment based on a custom-synthesized solid polymeric siloxane. This polymer is volatilized in a standard chemical vapor deposition (CVD) furnace that treats up to a hundred wafers at a time. Repeatability is extraordinary because thickness variation within a furnace run, and from run to run, is only ≈ 1 Å.

One of the side-benefits of this treatment is that it suppresses adsorption of atmospheric gases on MEMS surfaces. This leads to tighter distributions in the electrical parameters of packaged products.

Package Environment Effects. The viscosity and density of the gas that contacts MEMS resonators limits the device Q value, so resonant devices are often designed to operate in vacuum. Unfortunately, this makes them susceptible to handling damage. Viscous fill gases reduce destructive excursions of high-Q microstructures when they are mechanically shocked. They also provide squeeze film damping to cushion the contact event as surfaces approach each other. Susceptibility to damage, as well as the high cost of vacuum packaging, are the reasons why few high-volume MEMS products are evacuated.

Many investigators believe that suppression of in-use stiction requires low humidity in the package. This perception is not surprising because most MEMS research groups have observed stiction caused by aqueous surface tension during microstructure release. However, in-use stiction and wear test results published by several investigators lead to a different conclusion [51.113, 115–120]. In general, they find high stiction in both dry and moist environments. However, stiction is substantially suppressed at intermediate humidity levels ($\approx$ 15–40% relative humidity at room temperature). This is equivalent to a moisture level of 3000–10 000 ppm. Note that the limits cited by different authors vary considerably due to the use of different materials and test conditions. Stiction at low humidity is probably caused by high surface energy on inorganic oxides that do not have adsorbed surface films. The low stiction and wear rates observed at intermediate humidity are attributed to the passivating and lubricating effects of adsorbed water. Capillary forces caused by the high surface tension of water become dominant at high humidity.

51.4.2 System Interface

MEMS products are mounted in systems that have much greater mass than the device itself. For this reason, it is difficult to transmit a mechanical shock wave to a MEMS sensor that is sufficient to cause failure after the device is installed. However, failures are possible when devices are handled during module assembly. Improper functional response is also possible if the module mount system has a mechanical resonance that amplifies or reduces mechanical transmission under certain conditions. This is a concern in automotive safety applications because every automobile platform has a different mechanical signature. Proper module mount design is required to ensure that the impact signal from an automobile crash is properly transmitted to airbag modules.

Suppliers of large IC die closely examine board-level stresses in order to minimize solder joint fatigue. Such stresses are more serious in MEMS products because the device, as well as the package, is susceptible to mounting conditions. Furthermore, customer processes impose these stresses, so a supplier may have little information with respect to their origin or existence. One solution is to define and qualify a product-specific mount system as Texas Instruments has done with the DMD. The ideal, however, is to devise products that are compatible with standard board footprint and assembly processes. This challenge became more difficult with the shift from through-hole to surface-mount technology because the mechanical isolation provided by the package pins is no longer present. Small package size, in-package and within-chip isolation techniques, as well as thorough package-board stress analysis are used to address this concern.

References

51.1 J.B. Angell, S.C. Terry, P.W. Barth: Silicon micromechanical devices, Sci. Am. **248**(4), 44–50 (1983)
51.2 Texas Instruments, Inc., Texas Instruments DLP™ Products Announces 5 Million Units Shipped, Press Release, Dec. 13, 2004, www.dlp.com/about_dlp/about_dlp_press_release.asp?id=1222&bhcp=1
51.3 T.A. Core, W.K. Tsang, S.J. Sherman: Fabrication technology for an integrated surface-micromachined sensor, Solid State Technol. **36**(10), 39–47 (1993)
51.4 W. Kuehnel, S. Sherman: A surface micromachined silicon accelerometer with on-chip detection circuitry, Sens. Actuators A **45**, 7–16 (1994)
51.5 K.H.-L. Chau, R.E. Sulouff Jr.: Technology for the high-volume manufacturing of integrated surface-micromachined accelerometer products, Microelectron. J. **29**, 579–586 (1998)
51.6 B. Sulouff: Integrated surface micromachined technology. In: *Sensors for Automotive Technology, Vol. 4 of Sensors Applications*, ed. by J. Marek,

H.-P. Trah, Y. Suzuki, I. Yokomori (Wiley-VCH, Weinheim 2003), Chap. 5.2
51.7 M.A. Mignardi, R.O. Gale, D.J. Dawson, J.C. Smith: The digital micromirror device – A micro-optical electromechanical device of display applications. In: *MEMS and MOEMS Technology and Applications*, ed. by P. Rai-Choudhury (SPIE, Bellingham 2000), Chap. 4
51.8 K. Nunan, G. Ready, J. Sledziewski: LPCVD & PECVD operations designed for iMEMS sensor devices, Vac. Technol. Coat. **2**(1), 26–37 (2001)
51.9 M. Williams, J. Smith, J. Mark, G. Matamis, B. Gogoi: Development of low stress, silicon-rich nitride film for micromachined sensor applications, micromachining and microfabrication process technology VI, Proc. SPIE, Vol. 4174, ed. by J. Karam, J. Yasaitis (SPIE, Bellingham 2000) pp. 436–442
51.10 Z. Zhang, K. Eskes: Elimination of wafer edge die yield loss for accelerometers, micromachining and microfabrication process technology VI, Proc. SPIE, Vol. 4174, ed. by J. Karam, J. Yasaitis (SPIE, Bellingham 2000) pp. 477–484
51.11 G. Bitco, A.C. McNeil, D.J. Monk: Effect of inorganic thin film material processing and properties on stress in silicon piezoresistive pressure sensors, MRS Fall Meet., Proc. Mater. Res. Soc. Symp., Vol. 444, ed. by M. Reed, M. Elwenspoek, S. Johansson, E. Obermeier, H. Fujita, Y. Uenishi (Materials Research Society, Warrendale 1997) pp. 221–226
51.12 A. Hein, S. Finkbeiner, J. Marek, E. Obermeier: Material related effects on wet chemical micromachining of smart MEMS devices, micromachined devices and components V, Proc. SPIE, Vol. 3876, ed. by P. French, E. Peeters (SPIE, Bellingham 1999) pp. 29–36
51.13 B. Sulouff: Commercialization of MEMS automotive accelerometers, 7th Int. Conf. Commer. Micro Nano Syst. (COMS) (Micro and Nano Technology Commercialization Education Foundation, Albuquerque 2002) pp. 267–270
51.14 H. Weinberg: MEMS sensors are driving the automotive industry, Sensors **19**(2), 36–41 (2002)
51.15 J. Marek, M. Illing: Microsystems for the automotive industry, Electron Devices Meet., IEDM Technical Digest International, San Francisco 2000 (IEEE, New York 2000) pp. 3–8
51.16 D.S. Eddy, D.R. Sparks: Application of MEMS technology in automotive sensors and actuators, Proc. IEEE **86**(8), 1747–1755 (1998)
51.17 R. Verma, I. Baskett, B. Loggins: Micromachined electromechanical sensors for automotive applications, SAE Special Pub. **1312**, 55–59 (1998)
51.18 P. Adrian: Sensor companies can use foundries to efficiently boost their ability to serve high-volume markets, Sensor Business Digest (Oct. 2002), www.sensorsmag.com/resources/businessdigest/sbd1002.shtml
51.19 International Organization for Standardization: ISO16949:2002, Quality Management Systems – Particular Requirements for the Application of ISO 9001:2000 for Automotive Production and Relevant Service Part Organizations (2002) (International Organization for Standardization, Geneva)
51.20 *Quality Systems Requirements QS-9000*, 3rd edn. (Automotive Industry Action Group of the American Society for Quality, Milwaukee 1998)
51.21 R. Maboudian, R.T. Howe: Critical review: Adhesion in surface micromechanical structures, J. Vac. Sci. Technol. B **15**(1), 1–20 (1997)
51.22 U. Srinivasan, M.R. Houston, R.T. Howe, R. Maboudian: Alkyltrichlorosilane-based self-assembled monolayer films for stiction reduction in silicon micromachines, J. Microelectromech. Syst. **7**(2), 252–260 (1998)
51.23 B.H. Kim, T.D. Chung, C.H. Oh, K. Chun: A new organic modifier for anti-stiction, J. Microelectromech. Syst. **10**(1), 33–40 (2001)
51.24 S. Pamidighantam, W. Laureyn, A. Salah, A. Verbist, H. Tilmans: A novel process for fabricating slender and compliant suspended poly-Si micromechanical structures with sub-micron gap spacing, 15th IEEE 2002 Micro Electro Mech. Syst. (MEMS) Conf. (IEEE, New York 2002) pp. 661–664
51.25 B.C. Bunker, R.W. Carpick, R.A. Assink, M.L. Thomas, M.G. Hankins, J.A. Voigt, D. Sipola, M.P. de Boer, G.L. Gulley: Impact of solution agglomeration on the deposition of self-assembled monolayers, Langmuir **16**, 7742–7751 (2000)
51.26 Y. Jun, V. Boiadjiev, R. Major, X.-Y. Zhu: Novel chemistry for surface engineering in MEMS, Materials and devices characterization in micromachining III, Proc. SPIE, Vol. 4175, ed. by Y. Vladimirsky, P. Coane (SPIE, Bellingham 2000) pp. 113–120
51.27 R. Maboudian, W.R. Ashurst, C. Carraro: Self-assembled monolayers as anti-stiction coatings for MEMS: Characteristics and recent developments, Sens. Actuators A **82**(1), 219–223 (2000)
51.28 W.R. Ashurst, C. Yau, C. Carraro, R. Maboudian, M.T. Dugger: Dichlorodimethylsilane as an anti-stiction monolayer for MEMS: A comparison to the octadecyltrichlorosilane self-assembled monolayer, J. Microelectromech. Syst. **10**(1), 41–49 (2001)
51.29 W.R. Ashurst, C. Yau, C. Carraro, C. Lee, G.J. Kluth, R.T. Howe, R. Maboudian: Alkene based monolayer films as anti-stiction coatings for polysilicon MEMS, Sens. Actuators A **91**(3), 239–248 (2001)
51.30 Freescale Semiconductor: Sensor Solutions (2005) www.freescale.com/files/sensors/doc/fact_sheet/SNSRSOLUTNTMFS.pdf
51.31 Nexus Task Force Report: Market Analysis for Microsystems 1996–2002 (NEXUS, Grenoble 1998) www.nexus-mems.com
51.32 G. Dahlmann, G. Holzer, S. Hering, U. Schwarz: A modular CMOS foundry process for integrated piezoresistive pressure sensors. In: *Advanced Mi-*

crosystems for Automotive Applications 2005, ed. by J. Valldorf, W. Gessner (Springer, Berlin 2005) pp. 413–424

51.33 T. Maudie, J. Wertz: Pressure sensor performance and reliability, IEEE Ind. Appl. Mag. **3**(3), 37–43 (1997)

51.34 L. Ristic, R. Gutteridge, B. Dunn, D. Mietus, P. Bennett: Surface micromachined polysilicon accelerometer, IEEE 1992 Solid State Sens. Actuator Workshop, IEEE 5th Technical Digest (IEEE, New York 1992) pp. 118–121

51.35 M. Potin: Motion sensing applications, new opportunities for development of 3-axis accelerometers?, Micronews **68** (Yole Développement SARL, Lyon April 2008) pp. 6/7

51.36 M. Furtsch, M. Offenberg, H. Münzel, J.R. Morante: Influence of anneals in oxygen ambient on stress of thick polysilicon layers, Sens. Actuators A **76**, 335–342 (1999)

51.37 P. Lange, M. Kirsten, W. Riethmüller, B. Wenk, G. Zwicker, J.R. Morante, F. Ericson, J.Å. Schweitz: Thick polycristalline silicon for surface micromechanical applications: Deposition, structuring and mechanical characterization, Proc. 8th Int. Conf. Solid State Sens. Actuators Eurosens. IX, Transducers '95 (IEEE, New York 1995) pp. 202–205

51.38 F. Laermer, A. Schilp: Method of Anisotropically Etching Silicon, Patent 5501893 (1996)

51.39 M. Offenberg, H. Münzel, D. Schubert, O. Schatz, F. Laermer, E. Müller, B. Maihöfer, J. Marek: Acceleration sensor in surface micromachining for airbag applications with high signal/noise ratio, SAE Special Pub. **1133**, 35–41 (1996)

51.40 F. Laermer, A. Schilp, K. Funk, M. Offenberg: Bosch deep silicon etching: Improving uniformity and etch rate for advanced MEMS applications, Proc. 12th IEEE Int. Conf. Micro Electro Mech. Syst. MEMS (IEEE, New York 1999) pp. 211–216

51.41 G.K. Fedder, J. Chae, K. Najafi, T. Denison, J. Kuang, S. Lewis: Monolithically integrated inertial sensors. In: *CMOS-MEMS 2*, ed. by O. Brand, G.K. Fedder (Wiley-VCH, Weinheim 2005), Chap. 3

51.42 M. Reze, J. Hammond: Low g inertial sensor based on high aspect ratio MEMS. In: *Advanced Microsystems for Automotive Applications*, ed. by J. Valldorf, W. Gessner (Springer, Berlin Heidelberg 2005) pp. 459–471

51.43 A. Witvrouw, M. Gromova, A. Mehta, S. Sedky, P. De Moor, K. Baert, C. Van Hoof: Poly-SiGe, a superb material for MEMS, Proc. Mater. Res. Soc. Symp. Micro Nanosyst. **782**, 25–36 (2003)

51.44 A. Mehta, M. Gromova, C. Rusu, R. Olivier, K. Baert, C. Van Hoof, A. Witvrouw: Novel high growth rate processes for depositing poly-SiGe structural layers at CMOS compatible temperatures, Proc. 17th IEEE Int. Conf. Micro Electro Mech. Syst. MEMS (IEEE, New York 2004) pp. 721–724

51.45 T. Van Der Donck, J. Proost, C. Rusu, K. Baert, C. Van Hoof, J.-P. Celis, A. Witvrouw: Effect of deposition parameters on the stress gradient of CVD and PECVD poly-SiGe for MEMS applications, Proc. SPIE, **5342**, 8–18 (2004)

51.46 A.E. Franke, J.M. Heck, T.-J. King, R.T. Howe: Polycrystalline silicon-germanium films for integrated microsystems, Journal MEMS **12**(2), 160–171 (2003)

51.47 B.C.-Y. Lin, T.-J. King, R.T. Howe: Optimization of poly-SiGe deposition processes for modular MEMS integration, Proc. Mater. Res. Soc. Symp. Micro Nanosyst. **782**, 43–48 (2003)

51.48 B.L. Bircumshaw, M.L. Wasilik, E.B. Kim, Y.R. Su, H. Takeuchi, C.W. Low, G. Liu, A.H. Pisano, T.-J. King, R.T. Howe: Hydrogen peroxide etching and stability of p-type poly-SiGe films, Proc. 17th IEEE Int. Conf. Micro Electro Mech. Syst. MEMS (IEEE, New York 2004) pp. 514–519

51.49 J.R. Martin, C.M. Roberts Jr.: Package for sealing an integrated circuit die, Patent 6323550 (2001), Patent 6621158 (2003), Patent 6911727 (2005)

51.50 D. Marsh: Safety check: Wireless sensors eye tyre pressure, EDN Europe (2004) pp. 31–37

51.51 M. Osajda: Highly integrated tire pressure monitoring solutions. In: *Advanced Microsystems for Automotive Applications* (Springer, Berlin Heidelberg 2004) pp. 23–37

51.52 M. Feser, C. Wieand, C. Schmidt, T. Brandmeier: Crash impact sound sensing (CISS) – Higher crash discrimination performance at lower cost, 7th Int. Symp. Sophistic. Car Occup. Saf. Syst. (Karlsruhe 2004), Paper V17

51.53 W.S. Lindenberger, T.L. Poteat, J.E. West: Integrated electroacoustic transducer with built-in bias, Patent 4524247 (2003)

51.54 J.C. Baumhauer, H.J. Hershey, T.L. Poteat: Integrated electroacoustic transducer, Patent 4533795 (2003)

51.55 I.J. Busch-Vishniac, W.S. Lindenberger, W.T. Lynch, T.L. Poteat: Integrated capacitive transducer, Patent 4558184 (2004)

51.56 Anonymous: *The Prismark Wireless Report (March 2005)* (Prismark Partners LLC, Cold Spring Harbor 2005) pp. 34–44

51.57 J.W. Weigold, T.J. Brosnihan, J. Bergeron, X. Zhang: A MEMS condenser microphone for consumer applications, 19th IEEE Int. Conf. Micro Electro Mech. Syst., MEMS 2006 (IEEE, New York 2006) pp. 86–89

51.58 J.J. Neumann, K.J. Gabriel: CMOS-MEMS acoustic devices. In: *CMOS-MEMS 2*, ed. by O. Brand, G.K. Fedder (Wiley-VCH, Weinheim 2005), Chap. 4

51.59 J.A. Geen, S.J. Sherman, J.F. Chang, S.R. Lewis: Single-chip surface micromachined integrated gyroscope with 50°/h Allan deviation, IEEE J. Solid-State Circuits **37**(12), 1860–1866 (2002)

51.60 K. Funk, H. Emmerich, A. Schilp, M. Offenberg, R. Neul, F. Larmer: A surface micromachined silicon gyroscope using a thick polysilicon layer, Proc.

12th Int. Conf. Micro Electro Mech. Syst. MEMS (IEEE, New York 1999) pp. 57–60

51.61 M. Lutz, W. Golderer, J. Gerstenmeier, J. Marek, B. Maihöfer, S. Mahler, H. Münzel, U. Bischof: A precision yaw rate sensor in silicon micromachining, 11th Int. Conf. Solid State Sens. Actuators, Transducers '97 (IEEE, New York 1997) pp. 847–850

51.62 T.A. Core, R.T. Howe: Method for fabricating microstructures, Patent 5314572 (1994)

51.63 M. Offenberg, F. Lärmer, B. Elsner, H. Münzel, W. Riethmüller: Novel process for a monolithic integrated accelerometer, Proc. 8th Int. Conf. Solid State Sens. Actuators Eurosens. IX, Transducers '95 (IEEE, New York 1995) pp. 589–592

51.64 J. Anguita, F. Briones: HF/H_2O vapor etching of SiO_2 sacrificial layer for large-area surface-micromachined membranes, Sens. Actuators A **64**(3), 247–251 (1998)

51.65 P. Lindner, V. Dragoi, S. Farrens, T. Glinsner, P. Hangweier: Advanced techniques for 3d devices in wafer-bonding processes, Solid State Technol. **47**(6), 55–58 (2004)

51.66 M.A. Schmidt: Wafer-to-wafer bonding for microstructure formation, Proc. IEEE **86**(8), 1575–1585 (1998)

51.67 A.R. Mirza: Wafer-level bonding technology for MEMS, Proc. 7th Int. Conf. Therm. Thermomech. Phenom. Electron. Syst. 2000, ITHERM 2000, Vol. 1 (IEEE, New York 2000) pp. 113–119

51.68 A.R. Mirza: One micron precision, wafer-level aligned bonding for interconnect, MEMS and packaging applications, Proc. 50th Electron. Compon. Technol. Conf. (IEEE, New York 2000) pp. 676–680

51.69 T. Matthias, V. Dragoi, S. Farrens, P. Lindner: Aligned fusion wafer bonding for wafer-level packaging and 3-D integration, Proc. Int. Symp. Microelectron. – IMAPS 2005 (2005) pp. 715–725

51.70 V. Dragoi: From magic to technology: Materials integration by wafer bonding, Integrated optics: Devices, materials, and technologies, Proc. SPIE, Vol. 6123, ed. by Y. Sidorin, C.A. Waechter (SPIE, Bellingham 2006) pp. 330–343

51.71 S. Farrens: Packaging methods and techniques for MOEMS and MEMS, Reliability, Packaging, Testing and Characterization of MEMS/MOEMS IV, Proc. SPIE, Vol. 5716, ed. by D.M. Tanner, R. Ramesham (SPIE, Bellingham 2005) pp. 9–18

51.72 S. Farrens: Wafer bonding technologies and strategies for 3-D ICs. In: *Wafer Level 3-D ICs Process Technology*, ed. by C.S. Tan, R.J. Gutmann, L.R. Reif (Springer, Berlin Heidelberg 2008)

51.73 S. Mack, H. Baumann, U. Gösele: Gas tightness of cavities sealed by silicon wafer bonding, Proc. 10th Int. Conf. Micro Electro Mech. Syst., MEMS; IEEE Micro Electro Mech. Syst. (MEMS) (IEEE, New York 1997) pp. 488–493

51.74 G. Wallis: Field assisted glass sealing, SAE Automot. Eng. Congr. (Society of Automotive Engineers, New York 1971), Paper 71023

51.75 S.A. Audet, K.M. Edenfeld: Integrated sensor wafer-level packaging, Proc. Int. Conf. Solid State Sens. Actuators, Transducers '97, Vol. 1 (IEEE, New York 1997) pp. 287–289

51.76 C.M. Roberts Jr., L.H. Long, P.A. Ruggerio: Method for separating circuit dies from a wafer, Patent 5362681 (1994)

51.77 M. Gabriel, M. Eichler, M. Reiche: On the effect of plasma treatments on low-temperature wafer bonding, Electrochem. Soc. 203rd Meet., 7th Int. Symp. Semicond. Wafer Bond. Sci., Technol. Appl. (2003), Paper 1024

51.78 V. Dragoi, P. Lindner: Plasma activated wafer bonding of silicon: In situ and ex situ processes, ECS Transactions **3**(6), 147–154 (2006)

51.79 V. Dragoi, G. Mittendorfer, C. Thanner, P. Lindner: Wafer-level plasma activated bonding: new technology for MEMS fabrication, Microsyst. Technol. **14**(4/5), 509–515 (2008)

51.80 S. Pargfrieder, P. Kettner, V. Dragoi, S. Farrens: New low temperature bonding technologies for the MEMS industry, 6th Korean MEMS Conf. (2004)

51.81 L.E. Felton, P.W. Farrell, J. Luo, D.J. Collins, J.R. Martin, W.A. Webster: MEMS capping method and apparatus, Patent 6893574 (2005)

51.82 T.-R. Hsu (Ed.): *MEMS Packaging* (INSPEC, London 2004)

51.83 O. Tabata, T. Tsuchiya (Eds.): *Reliability of MEMS Adv. Micro Nanosyst*, Vol. 6 (Wiley-VCH, Weinheim 2008)

51.84 T. Maudie, T. Miller, R. Nielsen, D. Wallace, T. Ruehs, D. Zehrbach: Challenges of MEMS device characterization in engineering development and final manufacturing, Proc. 1998 IEEE AUTOTESTCON (IEEE, New York 1998) pp. 164–170

51.85 S. Petrovic, A. Ramirez, T. Maudie, D. Stanerson, J. Wertz, G. Bitko, J. Matkin, D.J. Monk: Reliability test methods for media-compatible pressure sensors, IEEE Trans. Ind. Electron. **45**(6), 877–885 (1998)

51.86 S. Petrovic: Progress in media compatible pressure sensors, Proc. InterPACK'01, Pac. Rim/Int. Intersoc. Electron. Packag. Tech./Bus. Conf. Exhib. (ASME, New York 2001), IPACK2001-15517

51.87 H.-J. Kress, J. Marek, M. Mast, O. Schatz, J. Muchow: Integrated pressure sensors with electronic trimming, Automot. Eng. **103**(4), 65–68 (1995)

51.88 C. Fung, R. Harris, T. Zhu: Multifunction polysilicon pressure sensors for process control, Sensors **16**(10), 75–79, 83 (1999)

51.89 H.L. Chau, C.D. Fung, P.R. Harris, J.G. Panagou: High-stress and overrange behavior of sealed-cavity polysilicon pressure sensors, IEEE 4th Technical Digest, Solid State Sensor and Actuator Workshop (IEEE, New York 1990) pp. 181–183

51.90 A.B. Sontheimer: Digital micromirror device (DMD) hinge memory lifetime reliability modeling, Proc. 40th Ann. IEEE Int. Reliab. Phys. Symp. (IEEE, New York 2002) pp. 118–121

51.91 J.P. O'Connor: Packaging design considerations and guidelines for the digital micromirror device™, Proc. InterPACK'01, Pac. Rim/Int. Intersoc. Electron. Packag. Tech./Bus. Conf. Exhib. (ASME, New York 2001), IPACK2001-15526

51.92 C. Bang, V. Bright, M.A. Mignardi, D.J. Monk: Assembly and test for MEMS and optical MEMS. In: *MEMS and MOEMS Technology and Applications*, ed. by P. Rai-Choudhury (SPIE, Bellingham 2000), Chap. 7

51.93 F. Poradish, J.T. McKinley: Package for a semiconductor device, Patent 5293511 (1994)

51.94 T.W. Migl: Interfacing to the digital micromirror device for home entertainment applications, Proc. InterPACK'01, Pac. Rim/Int. Intersoc. Electron. Packag. Tech./Bus. Conf. Exhib. (ASME, New York 2001), IPACK2001-15712

51.95 S.J. Jacobs, J.J. Malone, S.A. Miller, A. Gonzalez, R. Robbins, V.C. Lopes, D. Doane: Challenges in DMD™ assembly and test, Proc. Mater. Res. Soc. **657**, EE6.1.1–EE6.1.12 (2001)

51.96 J.A. Walraven, B.A. Waterson, I. De Wolf: Failure analysis of micromechanical systems (MEMS). In: *Microelectronic Failure Analysis, Desk Reference*, 4th edn. (ASM, Materials Park 2002), 2002 Supplement

51.97 M.R. Douglass: Lifetime estimates and unique failure mechanisms of the digital micromirror device (DMD), Proc. 36th Ann. IEEE Int. Reliab. Phys. Symp. (IEEE, New York 1998) pp. 9–16

51.98 S.J. Jacobs, S.A. Miller, J.J. Malone, W.C. McDonald, V.C. Lopes, L.K. Magel: Hermeticity and stiction in MEMS packaging, 40th Ann. IEEE Int. Reliab. Phys. Symp. (IEEE, New York 2002) pp. 136–139

51.99 J.R. Martin: Surface characteristics of integrated MEMS in high volume production. In: *Nanotribology: Critical Assessment and Research Needs*, ed. by S.M. Hsu, Z.C. Ying (Kluwer, Dordrecht 2002), Chap. 14

51.100 J.R. Martin: Stiction suppression in high volume MEMS products, Proc. 2003 STLE/ASME Jt. Int. Tribol. Conf. (ASME, New York 2003), 2003TRIB-266

51.101 H.R. Shea, A. Gasparyan, C.D. White, R.B. Comizzoli, D. Abusch-Magder, S. Arney: Anodic oxidation and reliability of MEMS poly-silicon electrodes at high relative humidity and high voltages, MEMS reliability for critical applications, Proc. SPIE, Vol. 4180 (SPIE, Bellingham 2000) pp. 117–122

51.102 F.W. Delrio, M.P. De Boer, J.A. Knapp, E.D. Reedy Jr., P.J. Clews, M.L. Dunn: The role of van der Waals forces in adhesion of micromachined surfaces, Nat. Mater. **4**, 629–634 (2005)

51.103 A. Hariri, J.W. Zu, R.B. Mrad: Modeling of dry stiction in micro-electro-mechanical systems (MEMS), J. Micromech. Microeng. **16**, 1195–1206 (2006)

51.104 E.J. Thoreson, J. Martin, N.A. Burnham: The role of few-asperity contacts in adhesion, J. Colloid Interface Sci. **298**(1), 94–101 (2006)

51.105 D.-L. Liu, J. Martin, N.A. Burnham: Optimal roughness for minimal adhesion, Appl. Phys. Lett. **91**, 043107 (2007)

51.106 G.X. Li, F.A. Shemansky Jr.: Drop test and analysis on micro-machined structures, Sens. Actuators A **85**, 280–286 (2000)

51.107 R.T. Howe, H.J. Barber, M. Judy: Apparatus to minimize stiction in micromachined structures, Patent 5542295 (1996)

51.108 L.J. Hornbeck, W.E. Nelson: Spatial light modulator and method, Patent 5096279 (1992)

51.109 L.J. Hornbeck: Low reset voltage process for DMD, Patent 5331454 (1994)

51.110 E.C. Fisher, R. Jascott, R.O. Gale: Method of passivating a micromechanical device within a hermetic package, Patent 5936758 (1999)

51.111 B.A. Miksic: Use of vapor phase inhibitors for corrosion protection of metal products, Proc. 1983 NACE Ann. Conf., Corrosion 83 (National Assoc. of Corrosion Engineers, Houston 1983), Paper 308

51.112 D. Vanderpool, S. Akin, P. Hassett: Corrosion inhibitors in the electronics industry: Organic copper corrosion inhibitors, Proc. 1986 NACE Ann. Conf., Corrosion 86 (National Assoc. of Corrosion Engineers, Houston 1986), Paper 1

51.113 J.R. Martin, Y. Zhao: Micromachined device packaged to reduce stiction, Patent 5694740 (1997)

51.114 J.R. Martin: Process for wafer level treatment to reduce stiction and passivate micromachined surfaces and compounds used therefor, Patent 6674140 (2004), Patent 7220614 (2007), Patent 7364942 (2008)

51.115 M.P. de Boer, P.J. Clews, B.K. Smith, T.A. Michalske: Adhesion of polysilicon microbeams in controlled humidity ambients, Proc. Mater. Res. Soc. Symp., Vol. 518 (Materials Research Society, Warrendale 1998) pp. 131–136

51.116 D.M. Tanner, J.A. Walraven, L.W. Irwin, M.T. Dugger, N.F. Smith, W.P. Eaton, W.M. Miller, S.L. Miller: The effect of humidity on the reliability of a surface micromachined microengine, Proc. 1999 37th Ann. IEEE Int. Reliab. Phys. Symp. (IEEE, New York 1999) pp. 189–197

51.117 S.T. Patton, W.D. Cowan, J.S. Zabinski: Performance and reliability of a new MEMS electrostatic lateral output motor, Proc. 1999 37th Ann. IEEE Int. Reliab. Phys. Symp. (IEEE, New York 1999) pp. 179–188

51.118 S.T. Patton, W.D. Cowan, K.C. Eapen, J.S. Zabinski: Effect of surface chemistry on the tribological performance of a MEMS electrostatic lateral output motor, Tribol. Lett. **9**, 199–209 (2000)

51.119 S.T. Patton, K.C. Eapen, J.S. Zabinski: Effects of adsorbed water and sample aging in air on the μN level adhesion force between Si(100) and silicon nitride, Tribol. Int. **34**(7), 481–491 (2001)

51.120 S.T. Patton, J.S. Zabinski: Failure mechanisms of a MEMS actuator in very high vacuum, Tribol. Int. **35**(6), 373–379 (2002)

52. Packaging and Reliability Issues in Micro-/Nanosystems

Yu-Chuan Su, Jongbaeg Kim, Yu-Ting Cheng, Mu Chiao, Liwei Lin

The potential of MEMS technologies has been viewed as a revolution comparable or even bigger than that of microelectronics. These scientific and engineering advancements in micro-/nano-electromechanical systems (MEMS/NEMS) could bring previously unthinkable applications to reality, from space systems, environmental instruments, to daily-life appliances. As presented in previous chapters, the development of core MEMS processes has already demonstrated a lot of commercial applications as well as future potentials with elaborate functionalities. However, creating a low-cost reliable package for the protection of these MEMS products is still a very difficult task. Without addressing these packaging and reliability issues, no commercial products can be sold on the market. Packaging design and modeling, packaging material selection, packaging process integration, and packaging cost are the main issues to be considered. In this chapter, we will present the fundamentals of MEMS packaging technology, including packaging processes, hermetic and vacuum encapsulations, polymer-MEMS assembly and encapsulation, thermal issues, packaging reliability, and future packaging trends. The future development of MEMS packaging will rely on the success of the implementation of several unique techniques, such as packaging design kits for system and circuit designer, low-cost wafer-level and chip-scale packaging techniques, effective testing techniques, and reliable fabrication of an interposer [52.1] with vertical through-interconnects for device integrations.

52.1 **Introduction MEMS Packaging** ... 1835
52.1.1 MEMS Packaging Fundamentals ... 1836
52.1.2 Contemporary MEMS Packaging Approaches ... 1838
52.1.3 Bonding Processes for MEMS Packaging Applications ... 1838

52.2 **Hermetic and Vacuum Packaging and Applications** ... 1841
52.2.1 Integrated Micromachining Processes ... 1842
52.2.2 Postpackaging Processes ... 1843
52.2.3 Localized Heating and Bonding Processes ... 1846
52.2.4 Localized Heating and Bonding for Polymer-MEMS Packaging ... 1849

52.3 **Thermal Issues and Packaging Reliability** ... 1851
52.3.1 Thermal Issues in Packaging ... 1851
52.3.2 Packaging Reliability ... 1853
52.3.3 Long-Term and Accelerated MEMS Packaging Tests ... 1854

52.4 **Future Trends and Summary** ... 1858

References ... 1859

52.1 Introduction MEMS Packaging

MEMS are miniaturized systems that may have mechanical, chemical, or biomedical features with or without integrated circuits (IC) for sensing or actuation applications [52.2]. For example, pressure [52.3], temperature, flow [52.4], accelerometers [52.5], gyroscopes [52.6] and chemical sensors [52.7] can be fabricated by modern technologies for sensing applications. Fluidic valves [52.8], pumps [52.9], and inkjet printer heads are examples of actuation devices for medical, environmental, office and industrial applications. Silicon is typically used as the primary substrate material for MEMS fabrication because it can provide unique electrical, thermal, and mechanical properties but can also be easily micromachined in a form of batch processing and be incorporated with microelectronic circuit by using most of the conventional semiconductor manufacturing processes and tools. As a result, smaller size, lighter weight, lower power consumption and cheaper fabrication cost are advantages of MEMS devices. For example, with the advances of microfabrication technologies in the past decades, the MEMS market at the component level is currently in excess of 5 billion and is driving end-product markets larger than US$ 100 billion [52.10].

Nevertheless, the road to the commercialization of MEMS does not look as promising as expected. Many industrial companies took advantage of MEMS technologies for product integrations and cost efficiency is the key toward commercialization. Several MEMS devices have been developed for and applied in the automotive industry and information technology and they dominate the market due to their high production volumes. On the other hand, most custom-designed MEMS products are still very diverse, aiming for different applications with small- to medium-scale production. Based on past experience of the IC industry, the cost of packaging processes is about 30%, and sometimes can be more than 70%, of the total production expenses. MEMS packaging processes are expected to be even more costly because of the stringent packaging requirements for MEMS components, in addition to the microelectronic circuitry, in a typical MEMS product [52.11].

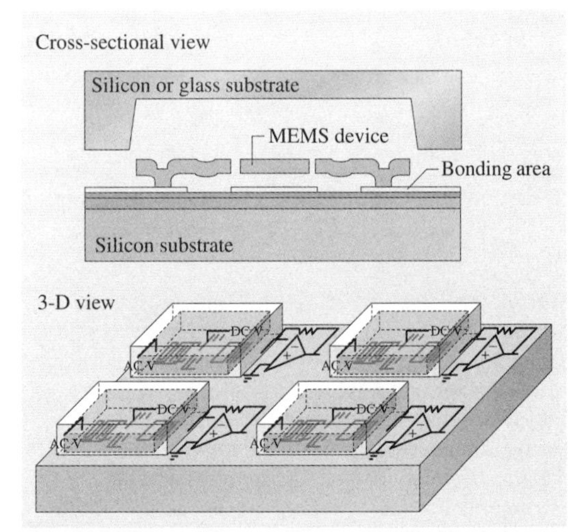

Fig. 52.1 A typical MEMS packaging illustration where a MEMS structure is encapsulated and protected by the packaging cap

52.1.1 MEMS Packaging Fundamentals

Sealing or encapsulation is an important step in either IC or MEMS packaging process to protect working devices. In traditional IC packaging procedures, the overall packaging steps often involve: (1) wafer dicing, (2) pick-and-place; (3) electrical connections, such as wire bonding, and (4) plastic molding or housing for sealing [52.12, 13]. With the increasing needs of high-performance multifunctional consumer electronic products, IC packaging processes have incorporated more-complex designs and advanced fabrication technologies, such as Cu interconnects [52.14], flip-chip bonding [52.15], ball grid arrays [52.16], wafer-level chip-scale packaging [52.17], and three-dimensional (3-D) packaging [52.18] to satisfy the necessity for high input/output (I/O) density, large die area, and high clock frequencies. The functions of conventional IC packaging are to protect, power, and cool the microelectronic chips or components and provide electrical and mechanical connection between microelectronic parts and the outside world. Unlike regular ICs, the diversity of MEMS products complicates the sealing issues. MEMS packaging processes cannot directly follow the procedures set by the IC packaging industry due to possible free-standing physical micro/nanostructures or chemical substances which cannot survive the dicing or pick-and-place steps. Moreover, MEMS components may need to interface with the outside environment (for example, fluidic interconnectors [52.19]), some other components may need to be hermetically sealed in vacuum (for example, inertial sensors [52.5]) in addition to the needs of electrical interconnects. Therefore, MEMS packaging processes have to provide more functionalities, including better mechanical protection, thermal management, hermetic sealing and complex electrical and signal distribution.

It has been suggested that MEMS packaging should be incorporated in the device fabrication stage as part of the micromachining process. Although this approach solves the packaging need for some specific devices, it does not solve the packaging need for general microsystems. In particular, many MEMS devices are now fabricated by various foundry services [52.20, 21] and there is a tremendous need for a uniform packaging process. Figure 52.1 shows a typical MEMS device being encapsulated by a packaging cap. The most fragile part on this device is the suspended mechanical sensor, which is a freestanding mechanical, mass-spring microstructure. It is desirable to protect this mechanical structure during the packaging and handling processes. Moreover, vacuum encapsulation may be required for these microstructures in applications such as resonant accelerometers [52.5] or gyroscopes [52.6, 22]. A *packaging cap* with a properly designed microcavity is to be fabricated to encapsulate and protect the fragile MEMS structure as the first-level postpackaging process. The wafer can be diced afterwards and the well-established packaging technology in IC industry can be followed to finish the packaging steps. A hermetic seal may be required to assure that no moisture or contamination can enter the package and affect the functionality of the micro/nanostructures. This increases the difficulty of common IC packaging processes tremendously. Although some MEMS chips can employ typical IC packaging techniques, such as die-attached processes and wiring interconnects for packaging [52.12], advanced packaging techniques, especially wafer-level packaging, are required for easy integration for multifunctional applications. For example, if chemical or biomedical substances are present [52.23], the sealing process must be carried out at a low processing temperature. If there are optical devices [52.24], the sealing process should provide good optical paths. If there are mechanical resonators [52.25], vacuum sealing might be required for better device performance and the desirable vacuum level depends on the specification of the device.

Before the state-of-the-art MEMS packaging processes are discussed, several primary microfabrication processes for packaging applications are briefly summarized, including the flip-chip (FC) technique, ball grid arrays (BGA), through-wafer etching, and plating. Other silicon-based processes such as thin-film deposition, wet and dry chemical etching, lithography, lift-off, and wire bonding can be found in the textbooks [52.26].

Flip-Chip Technique (FC)

This technique is commonly used in the assembly process between a chip with microelectronics and a package substrate [52.15]. The microelectronic chip is *flip-joined* with the packaging substrate, and metal solder bumps are used as both the bonding agents and electrical paths between the microelectronic chip and the package substrate. Because the vertical bonding space can be very small, as controlled by the heights of the solder bumps, and the lateral distributions of bond pads can be on the whole chip instead of being only on the edge, this technique can provide a high density of input/output (I/O) connections. In the FC technique, solder bumps are generally fabricated by means of electroplating. Before the bumping process, multiple metal layers, such as TiW/Cu, Cr/Cu, Cr/Ni, TaN/Ta/Ni, have to be deposited as a seed layer for electroplating and as a diffusion barrier to prevent the diffusion of solder into the electrical interconnects underneath.

Ball Grid Arrays (BGA)

This technology is very similar to the FC technique. An area array of solder balls on a single- or multi-chip module are used in the packaging process as electrical, thermal and mechanical connects to join the module with the next-level package, usually a printed circuit board [52.16]. The major difference between typical BGA and FC chips is the size of the solder bump. In BGA chips, the bumps are on the order of 750 μm in diameter, that is 10 times larger than those commonly used in FC chips.

Through-Wafer Etching

This is a chemical etching process to make through-wafer channels on a silicon substrate for the fabrication of vertical through-wafer interconnects. The chemical etching process can be either a wet or dry process. Anisotropic or isotropic etching solutions can be used in the wet etching process. The dry etch process is based on plasma and ion-assisted chemical reactions, which can cause either isotropic or anisotropic etching. In order to create high-density high-aspect-ratio through-wafer vias, deep reactive-ion etching (DRIE) is typically used. Two popular DRIE approaches, Bosch, and Cyro, are well described in the literature [52.27].

Electroplating

Electroplating is another common microfabrication process. It can be conducted for the deposition of an adherent metallic layer onto a conductive or nonconductive substrate. The process on a conductive substrate is called electrolytic plating that utilizes a seed layer as the anode to transfer metal ions onto the cathode surface when a direct current (DC) is passed through the plating solution. Plating processes that occur without applying an electrical current are called electroless plating, and can be used for both conducive and nonconductive surfaces. Electroless plating processes required a layer of a noble metal such as Pd, Pt, or Ru on the substrate to act as the catalyst to trigger the self-decomposition reaction in the plating solution. These electroplating processes are very important for electrical interconnect and solder-bump fabrication for packaging applications because of the low process temperature and cost.

These processes have generally been developed to provide electrical and thermal paths for various IC/MEMS packaging approaches.

52.1.2 Contemporary MEMS Packaging Approaches

Several MEMS packaging issues and approaches before 1985 were discussed in the book *Micromachining and Micropackaging of Transducers* [52.28] and researchers have been working on MEMS packaging approaches continuously. For example, *Senturia* and *Smith* [52.29] discussed packaging and partitioning issues for microsystems. *Smith* and *Collins* [52.23] used epoxy to bond glass and silicon for chemical sensors. Several multi-chip module (MCM) methods have been proposed. *Butler* et al. [52.30] proposed adapting multi-chip module foundries using the chip-on-flex (COF) process. *Schünemann* et al. [52.31] introduced a 3-D stackable packaging concept for top–bottom ball grid arrays (TB-BGA) that includes electric, fluidic, optic, and communication interfaces. *Lee* et al. [52.32] and *Ok* et al. [52.10] presented a direct-chip-attach MEMS packaging method using through-wafer electrical interconnects. *Laskar* and *Blythe* [52.33] developed an MCM-type packaging process using epoxy. *Reichl* [52.34] discussed different materials for bonding and interconnection. *Grisel* et al. [52.35] designed a special process to package microchemical sensors. Special processes have also been developed for MEMS packaging, such as packaging for microelectrodes [52.36], packaging for biomedical systems [52.37] and packaging for space systems [52.38]. These specially designed, device-oriented packaging methods are aimed at individual systems. There is no reliable method yet that would qualify as a versatile postpackaging process for MEMS with the rigorous process requirements of low temperature, hermetic sealing and long-term stability.

Previously, an integrated process using surface-micromachined microshells has been developed [52.39]. This process applies the concepts of sacrificial layer and low-pressure chemical vapor deposition (LPCVD) sealing to achieve wafer-level postpackaging. Similar processes have been demonstrated. For example, *Guckel* et al. [52.40] and *Sniegowski* et al. [52.41] developed a reactive sealing method to seal vibratory micromachined beams. *Ikeda* et al. [52.42] adopted epitaxial silicon to seal microstructures. *Mastrangelo* and *Muller* [52.43] used silicon nitride to seal mechanical beams as light sources. *Smith* et al. [52.44] accomplished a new fabrication technology by embedding microstructures and complementary metal oxide semiconductor (CMOS) circuitry. All of these methods have integrated the MEMS process with the postpackaging process such that no extra bonding process is required. However, these schemes are highly process-dependent and not suitable for prefabricated circuitry.

New efforts for MEMS postpackaging processes have been reported. *Butler* et al. [52.30] demonstrated an advanced MCM packaging scheme. It adopts the high-density interconnect (HDI) process consisting of embedding bare die into premilled substrates. Because the MEMS structures have to be released after the packaging process, this approach is undesirable for general microsystems. *Van der Groen* et al. [52.45] reported a transfer technique for CMOS circuits based on epoxy bonding. This process overcomes the surface roughness problem but epoxy is not a good material for hermetic sealing. In 1996, Cohn et al. demonstrated a wafer-to-wafer vacuum packaging process by using silicon/gold eutectic bonding with a 2 μm thick polysilicon microcap. However, experimental results showed substantial leakage after a period of 50 days. *Cheng* et al. [52.46] developed a vacuum packaging technology using localized aluminum/silicon-to-glass bonding. In 2002, *Chiao* and *Lin* [52.47] demonstrated vacuum packaging of microresonators by rapid thermal processing. These recent and ongoing research efforts indicate the strong need for a versatile MEMS postpackaging process.

52.1.3 Bonding Processes for MEMS Packaging Applications

Previously, silicon-bonding technologies have been used in many fabrication and packaging applications, where two types of bonding processes are commonly used: (1) direct bonding processes such as anodic bonding and fusion bonding, and (2) bonding processes with intermediate layers such as epoxy bonding, eutectic bonding and solder bonding. Direct wafer-bonding processes are procedures that facilitate permanent attachments between two wafers without any intermediate layer. A permanent bond between two wafers can also be accomplished by using intermediate layers. Joining processes using intermediate layers have been used extensively by the ceramic industry to form metal-to-metal and metal-to-ceramic joints [52.48, 49] and can be characterized as [52.50]: (1) fusion or melting of two materials to form a stable intermediate compound which facilitates the bond; (2) diffusion, in which pressurized joint parts are heated to 70% of the material's melting temperature and a stable intermediate compound is formed at the interface; and (3) brazing, in which a filler material is fitted into the two parts to be joined and, upon heating, a stable intermediate compound is formed. These processes are commonly used when lower bonding temperatures or a stronger bonding interface is required but cannot be achieved by the direct bonding process. Furthermore, the intermediate layers may reflow during the bonding process and fill the gaps between two bonding surfaces to overcome the surface roughness problem commonly encountered during the direct wafer-bonding processes. As such, the requirement for fine surface roughness for the direct wafer-bonding processes is greatly relieved for wafer-bonding processes with intermediate layers.

There have been many MEMS applications of both direct bonding and bonding processes with the assistance of intermediate layers. For example, devices such as pressure sensors, micropumps, biomedical sensors or chemical sensors require mechanical interconnectors to be bonded on the substrate (see for example [52.7, 19, 51]). Glass has commonly been used as the bonding material, using anodic bonding at a temperature of about 300–450 °C (see for example [52.52, 53]). *Klaassen* et al. [52.54] and *Hsu* and *Schmidt* [52.55] have demonstrated different types of silicon fusion bonding and Si/SiO_2 bonding processes at very high temperatures of over 1000 °C. *Ko* et al. [52.28], *Tiensuu* et al. [52.56], *Lee* et al. [52.57] and *Cohn* et al. [52.58] have used eutectic bonding for different applications. All of these bonding techniques have different mechanisms that determine the individual bonding characteristics and process parameters. This section discusses the details of these processes.

Fusion Bonding

Silicon fusion bonding is an important fabrication technique for silicon on insulator (SOI) devices. The bonding is based on Si–O, Si–N, or Si–Si strong covalent bonds. However, very high bonding temperature (higher than 1000 °C), flat bonding surfaces (less than 6 nm) and intimate contact are the three basic requirements for strong, uniform, and hermetic bonding. The common silicon-to-silicon fusion bonding process starts with wafer hydration (soaking the wafers in a H_2O_2/H_2SO_4 mixture, diluted H_2SO_4, boiling nitric acid or oxygen plasma) to create a hydrophilic top layer consisting of O–H bonds [52.59]. Prebonding is accomplished when the two wafers are brought into intimate contact and Van der Waals forces creates bonds between the two wafers. An annealing step at elevated temperature is required to strengthen the bond. Although hydrophilic surface treatment can lower the bonding temperature, an annealing step higher than 800 °C is still needed to remove possible bubble formation at the bonding interface. *Bower* et al. [52.60] proposed that low-temperature Si_3N_4 fusion bonding could be achieved at less than 300 °C. *Takagi* et al. [52.61] proposed that silicon fusion bonding could be carrier out at room temperature by using Ar^+-beam treatment on the wafer surface with a bond strength comparable to that of conventional fusion bonding.

Anodic Bonding

The invention of anodic bonding dates back to 1969 when *Wallis* and *Pomerantz* [52.62] found that glass and metal could be bonded together at about 200–400 °C below the melting point of glass with the aid of a high electrical field. This technology has been widely used for protecting onboard electronics in biosensors (see for example [52.63–65]) and sealing cavities in pressure sensors (see for example [52.66]). Many reports have also discussed the possibility of lowering the bonding temperature by different mechanisms [52.67, 68]. Anodic bonding forms Si–O or Si–Si covalent bonds and is one of the strongest chemical bonds available for silicon-based systems. The bonding process can be accomplished on a hot plate with temperature of 180–500 °C in atmosphere or a vacuum environment. When a static electrical field is built up within the Pyrex glass and silicon, the sodium ions in the glass migrate away from the silicon–glass interface, creating a locally high electrical field and a bond is formed by electrochemical effects [52.62]. In order to create high electrical fields, a flat bonding surface with less than 50 nm roughness is required. In addition, the electrical field required for bonding is larger than 3×10^6 V/cm [52.28]. Such a high electrical field is generated by a power supply of 200–1000 V. Figure 52.2

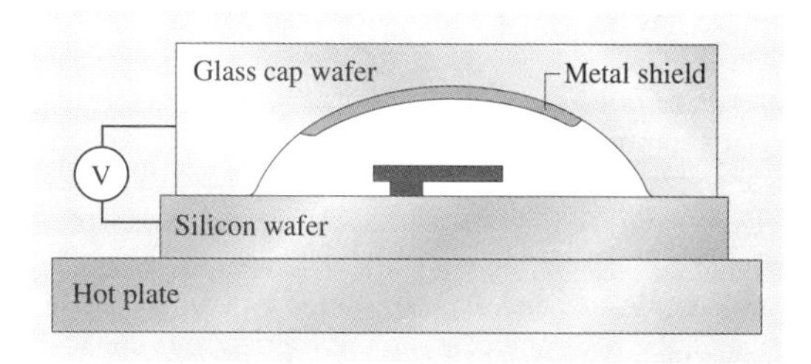

Fig. 52.2 Schematic diagram of the set up for the silicon-to-glass anodic bonding process

shows the setup for anodic bonding where two bonding wafers are brought together and heated to an elevated temperature to supply the bonding energy. If there are freestanding, conductive micromechanical structures on any bonding wafer, care must be taken as the high voltage tends to pull the micromechanical structure and damage may occur. A thin-film metal pattern on the glass cap can be formed to provide shielding to solve this problem, as shown in Fig. 52.2. Furthermore, Corning 7740 Pyrex is commonly used in the silicon-to-glass bonding system because it has a thermal expansion coefficient close to that of single-crystalline silicon in the range 200–300 °C. The induced residual-stress problem can be minimized in that temperature range. *Hanneborg* et al. [52.69] have successfully bonded silicon to other thin solid films, such as silicon dioxide, nitride, and polysilicon together with an intermediate glass layer using anodic bonding technique. *Chavan* and *Wise* [52.70] have reported on absolute pressure sensors fabricated by using the anodic bonding technique. In the process, a silicon cap with a thin heavily doped boron layer and a recess cavity was bonded in a vacuum environment to a glass substrate with prefabricated interconnection lines. The problem of oxygen outgassing due to the high electrical field in the anodic bonding process presents a challenge for the vacuum sealing process [52.71]. A thin Ti/Pt layer predeposited onto the glass surface has been shown to provide a good diffusion barrier, and the resulting pressure in the cavity can reach 200 mTorr [52.70]. In another example, microgyroscopes have been fabricated by *Hara* et al. [52.72] using the anodic bonding technique.

In practice, electrostatic bonding has become widely accepted in MEMS fabrication and packaging applications as described above. Unfortunately, possible contamination due to excess alkali metal in the glass; possible damage to microelectronics due to the high electrical field; and the requirement for flat surfaces for bonding limit the application of anodic bonding to MEMS postpackaging applications [52.73].

Epoxy Bonding (Adhesive Bonding)

Epoxy comprises four major components: epoxy resin, a filler such as silver slake, solvent or reactive epoxy diluent, and additives such as hardeners and catalysts [52.74, 75]. The bonding mechanism for epoxy is very complicated and depends on the type of epoxy. In general, the main source of bonding strength is the Van der Waals force. Because epoxy is a soft polymer material and its curing temperature for bonding is only around 150 °C, low residual stress and process temperature are the major advantages of epoxy bonding. However, the properties of epoxy can easily change with environmental humidity and temperature so the bonding strength decays over time. In addition, epoxy bonding has low moisture resistance and is a dirty process due to its additives. These disadvantages have made epoxy unfavorable for the special hermetic or vacuum sealing MEMS packaging requirements.

Eutectic Bonding

In many binary systems, there is a eutectic point corresponding to the alloy composition with the lowest melting temperature. If the environmental temperature is kept higher than the eutectic point, two contacted surfaces containing two elements with the eutectic composition can form a liquid-phase alloy. The solidification of the eutectic alloy forms *eutectic bonding* at a temperature lower than the melting temperature of either element in the alloy. Eutectic bonding can be a strong metal bonding. For example, in the case of the Au/Si alloy system, the eutectic temperature is only 363 °C when the composition is at the atomic ratio of 81.4% Au to 18.6% Si and bonding strength is higher than 5.5 GPa [52.76]. Because other alloy systems may have lower eutectic temperatures than the Al/Si system, they present great potential for MEMS packaging applications. In addition to the Au/Si system, the Al/Ge/Si, Au/SnSi, and Au/Ge/Si systems have been applied for MEMS packaging.

Solder Bonding

Solder bonding has been widely applied in microelectronic packaging [52.77]. Its low bonding temperature and high bonding strength are good characteristics for packaging. Furthermore, there are a variety of choices of solder material for specific applications. *Singh* et al. [52.78] have successfully applied solder bump bonding in the integration of electronic components and mechanical devices for MEMS fabrication [52.79]. In this case, indium metal was used to bond two separated silicon surfaces together by applying 350 MPa pressure; the bonding strength was as strong as 10 MPa. Glass frits can also be treated as a solder material and have been extensively used for vacuum encapsulation in MEMS industry. Glass frits are ceramic materials that can provide strong bonding strength with silicon with good hermeticity. Its bonding temperature is lower than 400 °C and is suitable for electronic components. However, a bonding area more than 200 μm wide, is required to achieve good bonding results and this may

Table 52.1 Summary of bonding mechanisms

Bonding methods	Temperature	Roughness	Hermeticity	Post-packaging	Reliability
Fusion bonding	Very high	Highly sensitive	Yes	Yes by LH	Good
Anodic bonding	Medium	Highly sensitive	Yes	Difficult	Good
Epoxy bonding	Low	Low	No	Yes	???
Integrated process	High	Medium	Yes	No	Good
Low-temperature bonding	Low	Highly sensitive	???	No	???
Eutectic bonding	Medium	Low	Yes	Yes by LH	???
Brazing	Very high	Low	Yes	Yes by LH	Good

???: no conlusive data
LH: localized heating

become a drawback because area is the measure of manufacturing cost in IC industry. Nevertheless, glass frit is the most popular bonding process used in current products.

Localized Heating and Bonding

Low bonding temperature and short process time are desirable process parameters in MEMS packaging fabrication to provide a lower thermal budget and high throughput. However, most chemical bonding reactions require a minimum and sufficient thermal energy to overcome the reaction energy barrier, also called the activation energy, to start the reaction and to form a strong bond. As a result, a high bonding temperature generally results in a shorter processing time to reach the same bonding quality at a lower bonding temperature [52.80]. The common limitations for the above bonding techniques are their individual bonding characteristics and temperature requirements. In general, MEMS packaging requires good bonding for hermetic sealing while the processing temperature must be kept low at the wafer level to reduce thermal effects on the existing devices. For example, a MEMS device may have prefabricated circuitry, biomaterial or other temperature-sensitive materials such as organic polymers, magnetic metal alloys, or piezoceramics. Since the packaging step comes after the device fabrication processes, the bonding temperature should be kept low to avoid the effects of high temperature on the system. Possible temperature effects include residual stress due to the mismatch of thermal expansion coefficient of bonding materials and substrates, electrical contact failure due to atomic interdiffusion at the interface, and contamination due to the outgassing or evaporation of materials. In addition to the control of bonding temperature, the magnitude of the applied force required to create intimate contact for bonding and atmospheric environment control are other factors that should be considered. Based on heat-transfer simulation studies [52.81], it is possible to confine the high-temperature area to a small region by localized heating without heating the whole substrate. Therefore assembly steps can always be processed after device fabrication without having detrimental effects. As such, localized heating and bonding techniques have been introduced for post-processing approaches [52.80, 82].

Table 52.1 summarizes these MEMS packaging technologies and their limitations, including the localized heating and bonding approach. The localized heating approach introduces several new opportunities. First, better and faster temperature control can be achieved. Second, higher temperature can be applied to improve the bonding quality. Third, new bonding mechanisms that require high temperature such as brazing [52.83] may now be explored. As such, it has potential applications for a wide range of MEMS devices and is expected to advance the field of MEMS packaging.

52.2 Hermetic and Vacuum Packaging and Applications

Hermetic packaging is important because it provides a moisture-free environment to avoid charge separation in capacitive devices, corrosion in metallization, or electrolytic conduction in order to prolong the lifetime of electronic circuitry. Especially for MEMS packaging, hermeticity of the packaging is desirable in most cases since one of the main failure mechanisms is humidity. Furthermore, the surface tension of water could cause stiction of micromechanical structures, leading to malfunction. In several device applications, vacuum encapsulation is necessary but can be costly. Many resonant devices, such as comb-shaped μ-resonators and ring-type μ-gyroscopes that have very large surface-to-volume ratios and vibrate in a very tight space [52.22, 42], need a vacuum to improve their performance. Two important approaches to hermetic and vacuum packaging of MEMS devices have been demonstrated: (1) the integrated encapsulation approach and (2) the postprocess packaging approach, as discussed in this section. Moreover, vacuum encapsulation by means of localized heating and bonding is discussed separately as a packaging example.

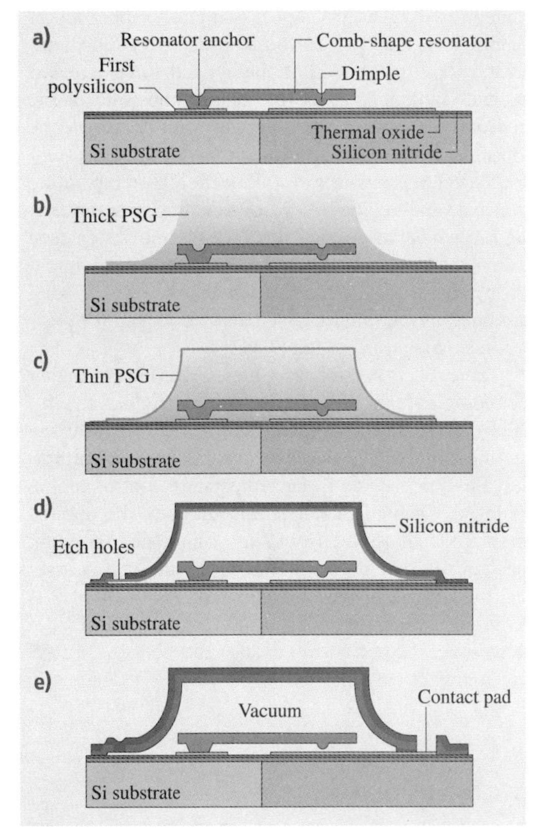

Fig. 52.3a–e An integrated vacuum encapsulation process using LPCVD nitride sealing to package micromechanical resonators [52.81]

52.2.1 Integrated Micromachining Processes

Several MEMS hermetic and vacuum packaging processes have been demonstrated based on integrated micromachining processes where the construction of sealing or protection caps is integrated with device manufacturing. The integrated approach has the advantage of sealing mechanical components in situ prior to the chip dicing and handling steps to avoid contamination. An integrated vacuum sealing process by LPCVD is presented here as an illustration. This integrated process can encapsulate comb-shaped microresonators [52.84] in vacuum at the wafer level. Figure 52.3 illustrates the cross-sectional view of the manufacturing process. First, a standard surface-micromachining process [52.85] is conducted by using four masks to define the first polysilicon layer, the anchors to the substrate, the dimples and the second polysilicon layer, as shown in Fig. 52.3a. The process so far is similar to the multiuser MEMS processes (MUMPs) [52.21] and comb-shaped microstructures are fabricated at the end of these steps. In the standard surface-micromachining process, the sacrificial layer (oxide) is etched away to release the microstructures. In the postpackaging process, a thick (7 μm) phosphorus-doped glass (PSG) is deposited to cover the microstructure and patterned by using 5 : 1 buffered HF (BHF) to define the microshell area, as shown in Fig. 52.3b. A thin (1 μm) PSG layer is then deposited and defined to form etch channels, as illustrated in Fig. 52.3c. The microshell material, low-

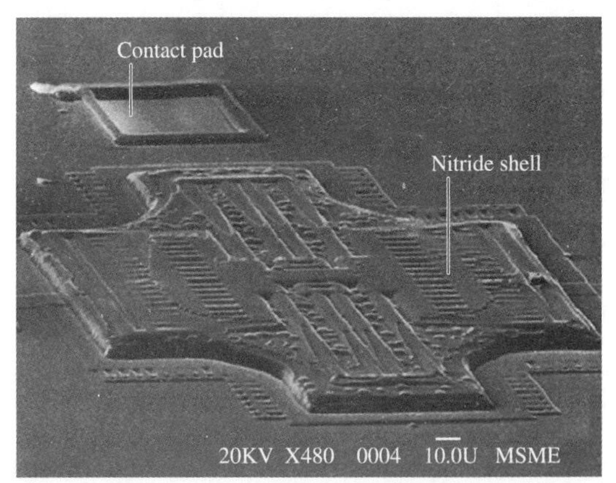

Fig. 52.4 SEM micrograph showing a vacuum packaged MEMS mechanical comb-shaped resonator packaged using an the integrated LPCVD sealing process as depicted in Fig. 52.3 [52.81]

stress silicon nitride, is now deposited with a thickness of 1 μm. Etch holes are defined and opened on the silicon nitride layer by using a plasma etcher. Silicon dioxide inside the packaging shell is now etched away by concentrated HF and the wafer is dried by using the supercritical CO_2 drying process [52.86]. After these steps, Fig. 52.3d applies. A 2 μm thick LPCVD low-stress nitride is now deposited at a deposition pressure of 300 mTorr to seal the shell in the vacuum condition. Finally, the contact pads are opened, as shown in Fig. 52.3e. Figure 52.4 is the scanning electron microscope (SEM) micrograph of a finished device with the protected microshell on top. The total packaging area (microshell) is about $400 \times 400\,\mu m^2$. A contact pad is shown with the covering nitride layer removed. The shape of the microresonator, with beams 150 μm long and 2 μm wide is reflected on the surface of the microshell due to the integrated packaging process. The total height of the nitride shell is 12 μm, as seen standing above the substrate. Spectral measurement of the comb resonator inside the packaging reveals that a vacuum level of about 200 mTorr has been accomplished [52.87].

Similarly, *Aigner* et al. [52.88] reported a Bi-CMOS-compatible integrated vacuum sealing process to package a polysilicon microaccelerometer. The protecting shell was a polysilicon layer with supporting pillars anchored on the structural polysilicon. The release process was done in a HF gas-phase etching process to remove the sacrificial oxide layer and the release holes were sealed in a vacuum environment. The device was then injection-molded into a plastic package at a pressure of 10 MPa; the supporting pillars were strong enough to hold the polysilicon shell under the high-pressure molding process.

An integrated sealing process using evaporation of aluminum has also been reported [52.89], as shown in Figs. 52.3 and 52.4. A silicon substrate was deposited with an n-type, 4 μm thick epitaxial silicon layer. A controlled plasma etch and oxidation process formed a sharp tip and a layer of boro-phosphosilicate glass (BPSG) was used to fill the trench as the sacrificial layer. The nitride sealing cap was deposited and patterned and a 290 nm thick PSG sacrificial release via was deposited and patterned, followed by the deposition and patterning of the polysilicon anode. After the release etching process, aluminum evaporation was done in a 2×10^{-6} Torr vacuum chamber and an 800 nm thick aluminum layer was deposited to seal the release via. The resulting pressure was estimated as 1 mPa by measuring the vacuum diode characteristics.

Other similar processes have been demonstrated based on the integrated encapsulation concept. For example, *Sniegowski* et al. [52.41] developed a reactive sealing method to seal vibratory micromachined beams; *Ikeda* et al. [52.42] used epitaxial silicon to seal microstructures; *Mastranglo* et al. [52.90] used silicon nitride to seal mechanical beams as light sources; *Smith* et al. [52.44] used the technology of embedding microstructures and CMOS circuitry. All of these approaches integrated the encapsulation process within the MEMS fabrication processes. The typical advantage of this approach is that these devices could be ready for standard IC packaging processes such as dicing, pick-and-place etc. once the wafer-level integrated sealing processes are established.

Although this vacuum sealing processes successfully achieves hermetic and vacuum packaging, it has drawbacks. For example, these postpackaging processes are highly process-dependent and are not suitable for generic MEMS postpackaging processes. Companies or researchers have to adapt these postpackaging processes for their own device manufacturing process. Currently, standard foundry services do not support any of these integrated processes. Also, integrated encapsulation does not allow the cavity pressure to be controlled, although it can achieve low pressure by wafer-level fabrication and provide lower manufacturing cost.

52.2.2 Postpackaging Processes

The second approach is defined as postpackaging process. The packaging process starts when the device fabrication processes are completed, so this approach has great flexibility for various microsystems. For example, Fig. 52.5 shows a common industrial posthermetic packaging called dual-in-line packaging (DIP) [52.91, 92]. A die is placed inside a ceramic holder cov-

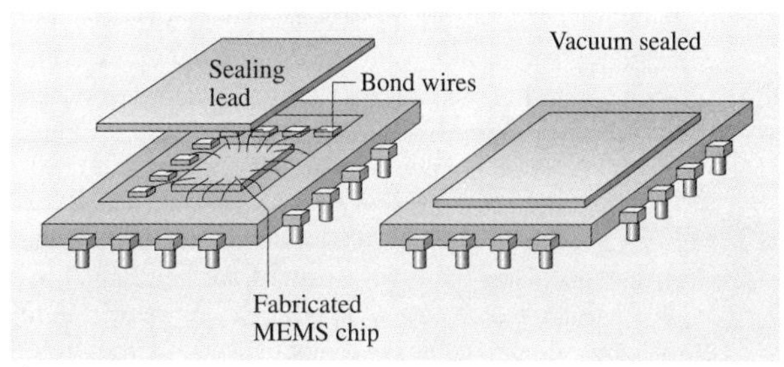

Fig. 52.5 A schematic diagram of industrial postpackaging (DIPS) using a ceramic holder to be covered by a sealing lid

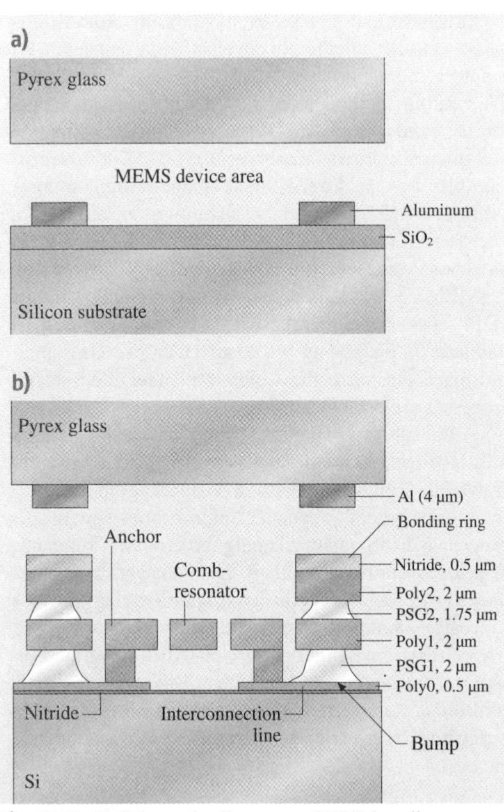

Fig. 52.6a,b Schematic diagrams of RTP bonding experiment. (**a**) The concept of aluminum-to-glass bonding. (**b**) Aluminum-to-nitride bonding with comb resonators. Drawn not to scale

ered by a sealing lid. Solder or ceramic joining is generally used to assemble the lid and holder under a pressure-controlled environment. Its high cost is the major drawback of this method because of the expensive ceramic holder and the low fabrication throughput. Another example of the postpackaging method is based on wafer-bonding techniques combined with microshell encapsulation. Devices are sealed by stacking another micromachined silicon or glass substrate, as illustrated in Fig. 52.1. Integrated microsystems and protection shells are fabricated on different wafers, made of either silicon or glass, at the same time. After the two substrates are assembled together using silicon fusion, anodic, or low-temperature solder bonding to achieve the final encapsulation, these microshells will provide mechanical support, thermal path, or electrical contact for the MEMS devices. Low packaging cost can be expected due to wafer-level processing.

A special heating method using rapid thermal processing (RTP) for wafer bonding applications is explained here to illustrate the roles of various control parameters such as temperature, time and intermediate bonding materials. *Chiao* and *Lin* [52.93] reported a wafer-bonding process based on the melting of an intermediate filler material to facilitate sealing of micromechanical structures. Figure 52.6a shows the concept of the bonding and sealing scheme using the aluminum-to-glass bonding system while Fig. 52.6b shows the aluminum-to-nitride bonding experimental setup with integrated comb resonators inside. Aluminum with a thickness of 3–4 μm was used and patterned to form sealing rings that surround the micromechanical structures on the device wafer. The width of a typical aluminum sealing ring was ≈ 100–200 μm and the sealing area was $600 \times 600\,\mu m^2$. A glass (Pyrex, Corning 7740) wafer was used as the cap to cover the MEMS devices. The heating and bonding energy was provided by RTP and the typical heating history is shown in Fig. 52.7 where the overall heating process can be completed in 1 min, during which the temperature rose from room temperature to 990 °C and cooled down to 350 °C. The bonding and joining process of aluminum to glass was accomplished by heating at 990 °C for 2 s in the RTP chamber. It was demonstrated that aluminum could extract oxygen to form aluminum oxide to assist the bonding process [52.94]. Figure 52.8 shows

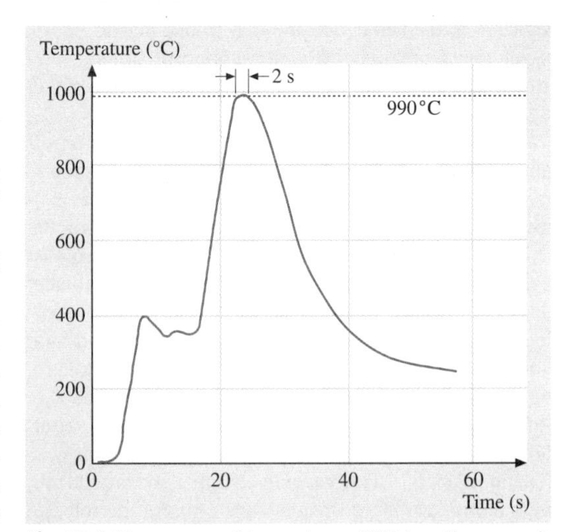

Fig. 52.7 Temperature history in an RTP bonding experiment

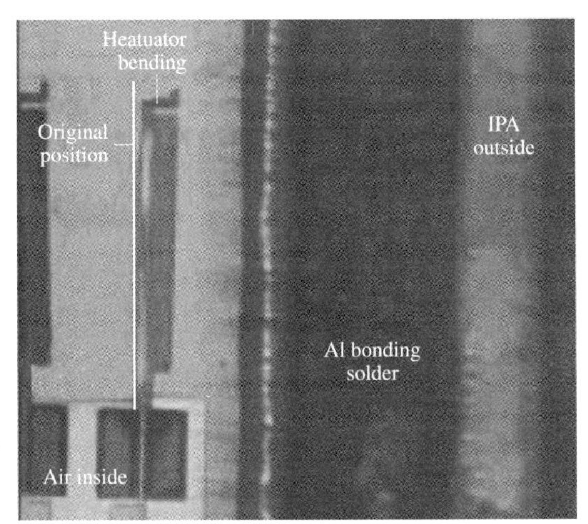

Fig. 52.8 A microheatuator that was hermetically packaged and was operational when the package was immersed in liquid

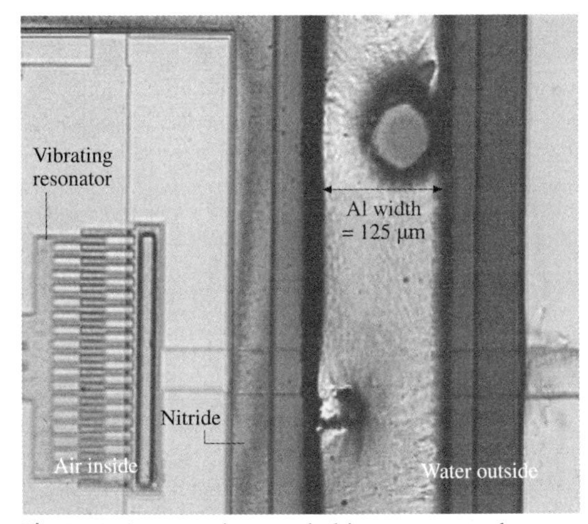

Fig. 52.9 A resonating comb-drive resonator that was sealed in the package chip and was operational after the package is immersed in water

a microheatuator that has been successfully packaged using this bonding process and the surrounded liquid [in this case, isopropanol alcohol (IPA)] was sealed from penetrating inside the package.

Other material systems have also been bonded by the RTP bonding process such as the aluminum-to-silicon nitride joining shown in Fig. 52.6b [52.95]. In this case, a 5000 Å thick LPCVD silicon nitride layer was deposited and patterned on top of sealing ring structures that encompassed surface-micromachined comb-shaped resonators [52.85]. Using a process based on 10 s at the peak temperature of 750 °C achieved by RTP, a stable bond was formed at the aluminum–nitride interface. Figure 52.9 shows the packaged comb resonator that was resonating at 19.6 kHz when immersed in deionized (DI) water as seen under an optical microscope. The aluminum-to-nitride seal successfully blocked water from entering the package. In order to examine the bond strength, the package was forcefully broken, as shown in Fig. 52.10. The glass debris was attached to the sealing ring surrounding the comb-drive resonator on the silicon substrate. This shows that the bonding strength in the aluminum-to-nitride system is greater than the glass fracture strength, which is estimated to be around 270 MPa [52.96].

A vacuum sealing process by means of RTP bonding is discussed here in detail to address the technical issues in vacuum sealing processes. *Chiao* and *Lin* [52.97] have reported a vacuum sealing process by RTP aluminum-to-nitride bonding. The RTP bonding process was conducted in a vacuum quartz tube, as shown in Fig. 52.11. Both device and cap wafers must be baked in vacuum at 300 °C for at least 4 h to drive out water and gas species that may adhere to the wafer surface [52.98]. This prebaking process in vacuum was necessary to minimize the outgassing effect during the bonding process in order to achieve a high-quality vacuum. Afterwards, the device and cap wafers were flip-chip assembled immediately, loaded

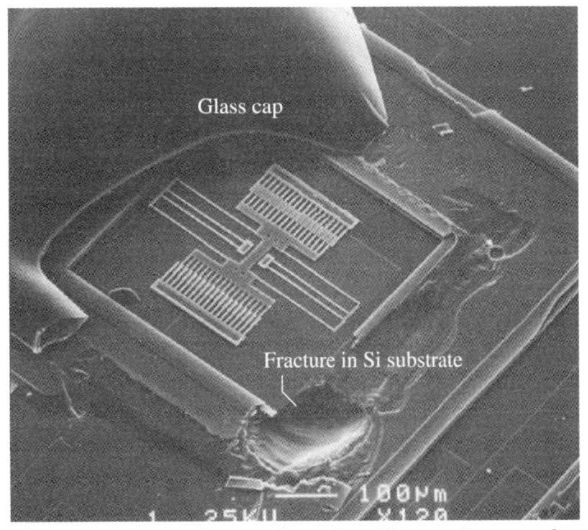

Fig. 52.10 SEM micrograph of the silicon substrate after forcefully breaking the aluminum-to-nitride bond. Glass debris is found attached to the silicon substrate

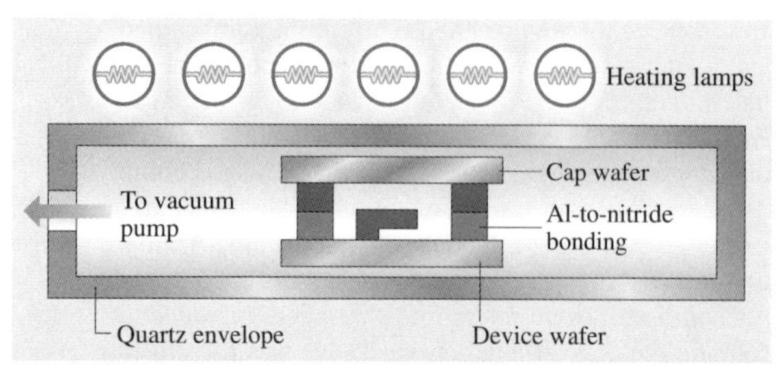

Fig. 52.11 Vacuum packaging apparatus using RTP aluminum-to-nitride bonding. Drawn not to scale

into a sample holder and put inside a quartz chamber, as shown in Fig. 52.11. The system was then placed inside the RTP equipment and the base pressure was pumped down to about 1 mTorr by using a turbo pump. The vacuum was held steady for 4 h to drive out trapped gas inside the package cavity [52.99]. The bonding and vacuum sealing process was done by RTP heating for 10 s at 750 °C to complete the bonding process.

Figure 52.12 shows the measured spectrum of a vacuum-packaged, double-folded beam comb-drive resonator by using a microstroboscope. The central resonant frequency is at about 18 625 Hz and the quality factor is extracted as 1800 ± 200, corresponding to a pressure level of about 200 mTorr inside the package [52.99]. This type of postpackaging process at the wafer level has become the favorite approach to fabricate a hermetic encapsulation because it can provide lower cost and more process flexibility. However, this packaging process relies on good bonding techniques. A strong and reliable bonding between two substrates should be provided and this bonding procedure should be compatible with the other microsystem fabrication processes.

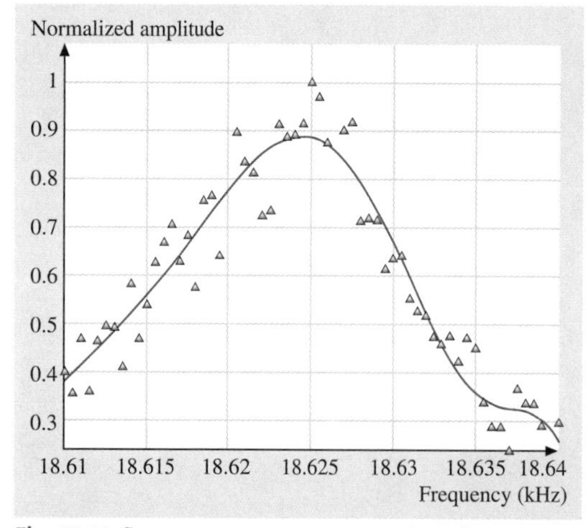

Fig. 52.12 Spectrum measurement results of a vacuum encapsulated comb-shaped resonator by using the RTP aluminum-to-nitride bonding method [52.47]

52.2.3 Localized Heating and Bonding Processes

The approach to MEMS postpackaging based on localized heating and bonding has been proposed to address the problems of global heating effects. In this section, resistive microheaters are used as an example to provide localized heating, although several other means of localized heating have been demonstrated, including laser welding [52.100], inductive heating [52.101] and ultrasonic bonding [52.102]. The principle of localized heating is to achieve high temperatures for bonding while maintaining low temperatures globally at the wafer level. Resistive heating by using microheaters on top of the device substrate is applied to form a strong bond with silicon or the glass cap. According to the results of a two-dimensional (2-D) heat-conduction finite-element analysis, as shown in Fig. 52.13, the steady-state heating region of a 5 μm-wide polysilicon microheater capped with a Pyrex glass substrate can be confined locally as long as the bottom of the silicon substrate is constrained to the ambient temperature. The physics of localized heating behind this design can be understood by solving the govern-

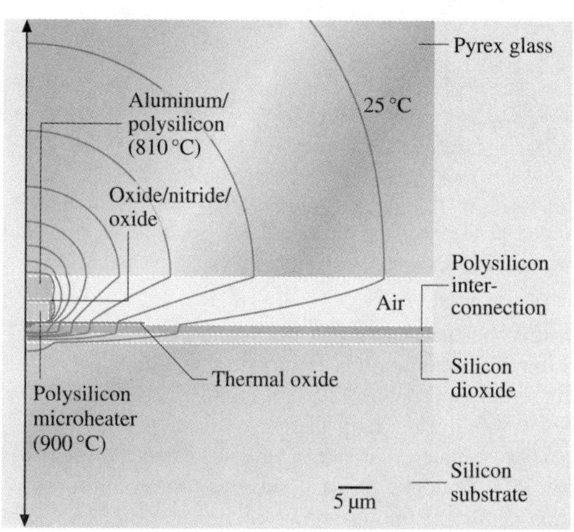

Fig. 52.13 A schematic diagram of the 2-D heat-transfer model, geometry and boundary conditions

ing heat-conduction equations of the device structure without a cap [52.99]. As long as the width of the microheater and the thickness of the silicon substrate are much smaller than the die size and a good heat sink is placed underneath the silicon substrate, the heating can be confined locally. The temperature of the silicon substrate can be kept low or close to room temperature. Several localized resistive heating and bonding techniques have been successfully developed for packaging applications, including localized silicon-to-glass fusion bonding, gold-to-silicon eutectic bonding and localized solder bonding. Several solder materials have been successfully tested, including PSG, indium, and aluminum alloy [52.103].

The vacuum packaging example presented here is based on the localized aluminum/silicon-to-glass solder bonding technique. Built-in folded-beam comb-drive μ-resonators are used to monitor the pressure of the package. Figure 52.14 shows the fabrication process of the package and resonators. Thermal oxide (2 μm) and LPCVD Si_3N_4 (3000 Å) are first deposited on a silicon substrate for electrical insulation followed by the deposition of 3000 Å of LPCVD polysilicon. This polysilicon is used as both the ground plane and the electrical interconnect to the μ-resonators, as shown in Fig. 52.14a. Figure 52.14b shows a 2 μm LPCVD SiO_2 layer that is deposited and patterned as a sacrificial layer for the fabrication of polysilicon μ-resonators using a standard surface-micromachining process. A 2 μm thick phosphorus-doped polysilicon is used for both the structural layer of the microresonators and the on-chip microheaters. This layer is formed over the sacrificial oxide in two steps to achieve a uniform doping profile. Lower input power and better process compatibility are two major advantages to using the on-chip microheater in the glass package. The resonators are separated from the heater by a short distance (30 μm) to effectively prevent their exposure to the high heater temperature, as shown in Fig. 52.14c. This concludes the fabrication of the μ-resonators.

In order to prevent the current supplied to the microheater from leaking into the aluminum solder during bonding, an LPCVD Si_3N_4 (750 Å)/SiO_2 (1000 Å)/Si_3N_4 (750 Å) sandwich layer is grown and patterned on top of the microheater, as shown in Fig. 52.14d. Figure 52.14e,f shows that aluminum (2.5 μm) and polysilicon (5000 Å) bonding materials are deposited and patterned. The sacrificial release is the final step to form freestanding μ-resonators. Figure 52.14f shows that a thick AZ 9245 photoresist is applied to cover the aluminum/silicon-to-glass bonding system to ensure that the system withstands the attack

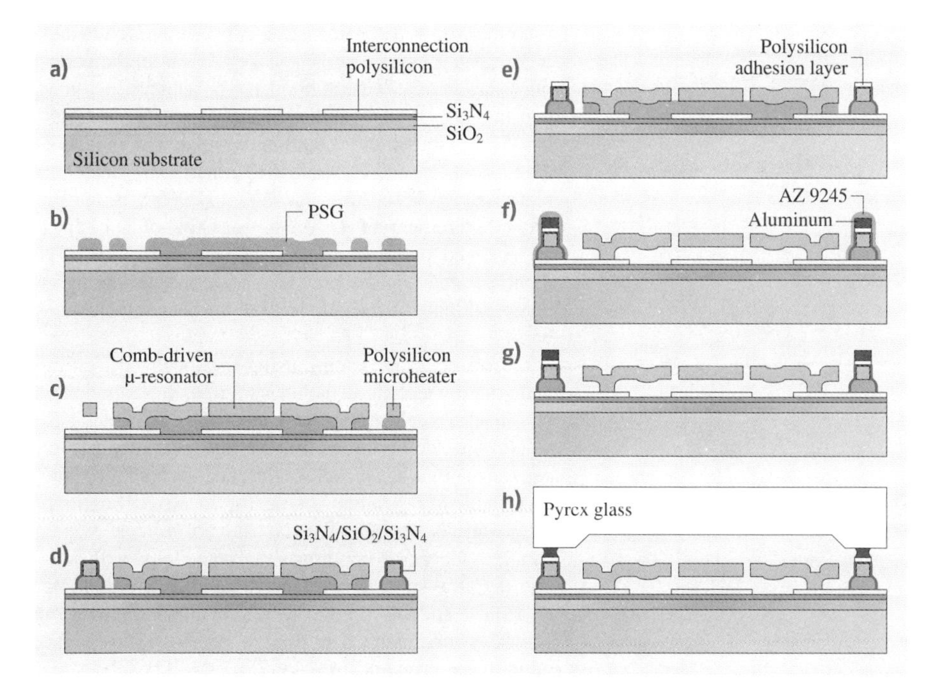

Fig. 52.14a–h The schematic process flow of vacuum encapsulation using localized aluminum/silicon-to-glass bonding

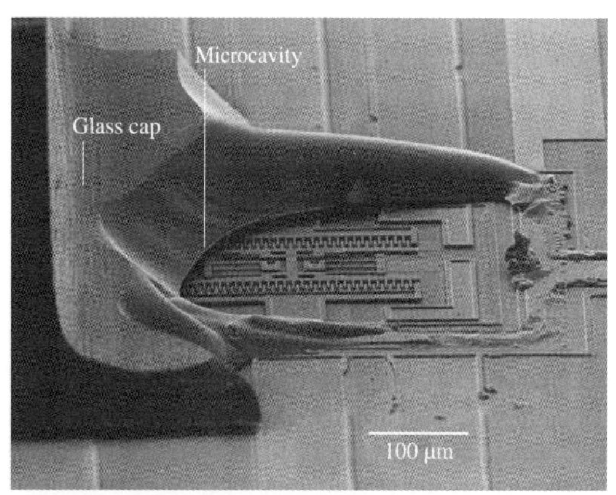

Fig. 52.15 The SEM micrograph of encapsulated microresonators after the glass cap is forcefully broken

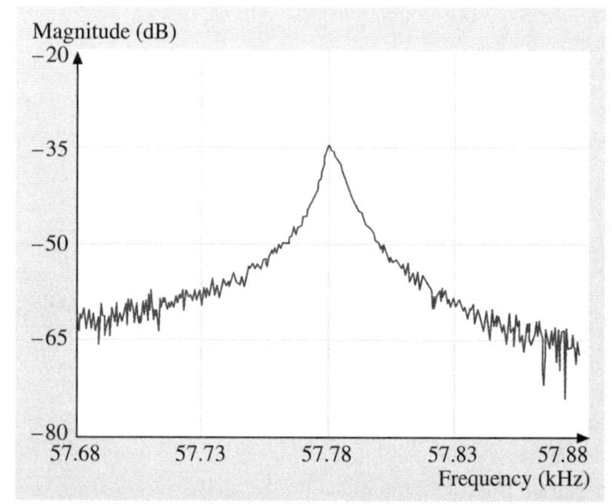

Fig. 52.16 The transmission spectrum of a glass-encapsulated μ-resonator after 120 min pump-down time in vacuum environment ($Q = 9600$)

from concentrated hydrofluoric acid. After 20 min of sacrificial release in concentrated HF, the system as shown in Fig. 52.14g is ready for vacuum packaging. A Pyrex glass cap with a 10 μm deep recess is then placed on top with an applied pressure of ≈ 0.2 MPa under a 25 mTorr vacuum, and the heater is heated using 3.4 W input power (the exact amount depends on the design of the microheaters) for 10 min to complete the vacuum packaging process, as shown in Fig. 52.14h.

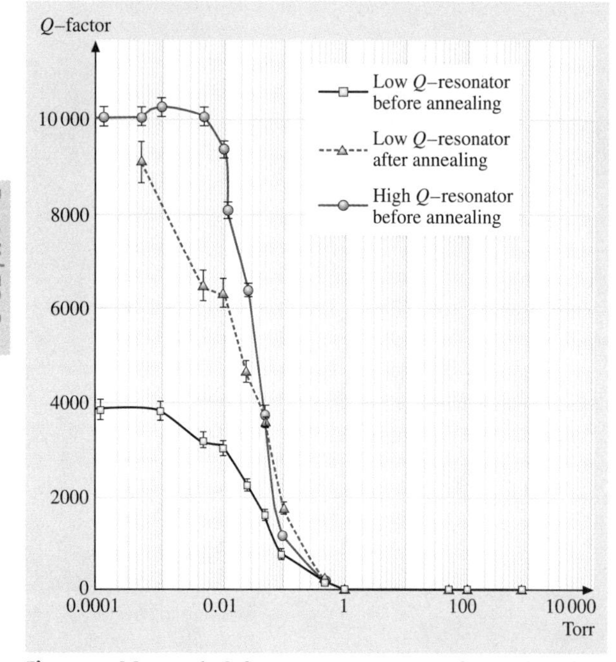

Fig. 52.17 Measured Q-factor versus pressure of unpackaged μ-resonators

To evaluate the integrity of the resonators packaged using localized aluminum/silicon-to-glass solder bonding, the glass cap is forcefully broken and removed from the substrate. It is observed that no damage is found on the μ-resonator and a part of the microheater is stripped away, as shown in Fig. 52.15, demonstrating that a strong uniform bond can be achieved without detrimental effects on the encapsulated device. Figure 52.16 shows a vacuum-encapsulated unannealed μ-resonator (≈ 57 kHz) after 120 min of wait time. The measured Q-factor after packaging is 9600. Based on the measurement of Q versus pressure of a high-Q unpackaged μ-resonator as shown in Fig. 52.17, it is demonstrated that the pressure inside the packaging is comparable to the vacuum level of th epackaging chamber.

Postprocess packaging using the localized heating and bonding technique includes four basic components: (1) an electrical and thermal insulation layer such as silicon dioxide or silicon nitride should be used for localized heating, (2) resistive microheaters are fabricated to provide the heating source for localized bonding, (3) materials, including metal and polysilicon, which can provide good bonding and hermeticity with silicon or glass substrates are considered as the bonding materials, and (4) a good heat sink under the device substrate for localized heating is provided during the bonding experiments. MEMS devices will be fabricated

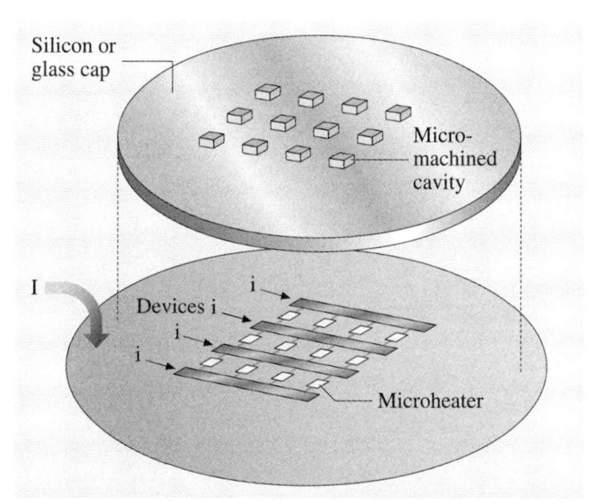

Fig. 52.18 Illustration of wafer-level vacuum packaging at the wafer level using the localized heating and bonding technique

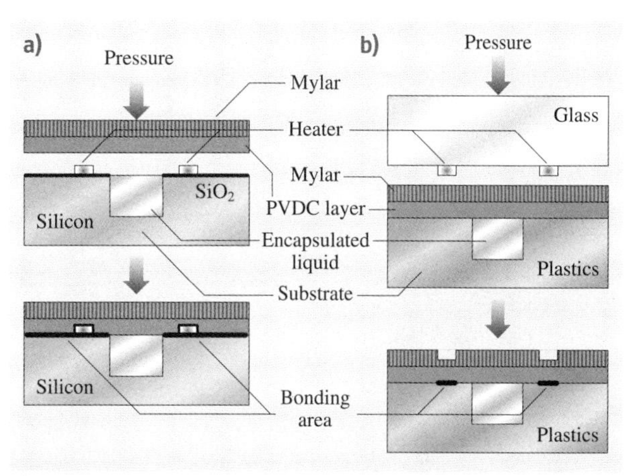

Fig. 52.19a,b Schematic illustration of polymer bonding processes. (a) Built-in heater configuration. (b) Reusable heater configuration

on the device chip and hermetically sealed in the cavity formed by the device chip, resistive microheaters, and protection cap. The process can be either die level or wafer level. The schematic design of the wafer-level packaging process is shown in Fig. 52.18. The resistive microheaters are parallel to each other and connected together in order to ensure that identical current density is applied to individual packages at the same time. These heaters can either be fabricated on the chip or protection cap and can be built into a larger wafer for current inputs. The interconnections for these packaging cavities can be built into the dicing area such that no extra space is required for the packaging process.

52.2.4 Localized Heating and Bonding for Polymer-MEMS Packaging

Using microheaters to provide heat for the bonding process locally, the global temperature can be significantly reduced to satisfy the strict low-temperature processing requirement for a variety of biomaterials and polymers. Polymeric materials are extensively utilized in microfluidic systems for applications in life sciences [52.104–106]. The advantage of polymer-MEMS includes broad-range choices of material properties, low raw material cost, and availability for mass production. To facilitate the fabrication and application of polymer-MEMS, effective assembly and packaging processes are highly desired. In this chapter, two localized bonding and assembly schemes are introduced, which employ soft thermoplastic materials with low glass transition temperatures as intermediate layers for packaging demonstrations in various systems, including polymer-to-silicon, polymer-to-glass, and polymer-to-polymer. The two bonding schemes are illustrated in Fig. 52.19a bonding by built-in-type heaters, and Fig. 52.19b bonding by external, reusable heaters. A Mylar film (DuPont Teijin Films) coated with a thermoplastic polyvinylidene chloride (PVDC) copolymer layer on one surface has been bonded to either silicon, glass, or another polymeric substrate. The scheme with built-in microheaters is suitable for bonding thick polymeric materials with various types of substrates including silicon, glass, and polymer because microheaters can easily provide the activation energy locally for bonding. Meanwhile, the scheme with external, reusable heaters is best suitable for the bonding of thin polymer films to various substrates. Heat is transferred through the thin polymer film and a heated zone is generated locally in the bonding interface where the bond is formed. In this case, it is preferable that the thin polymer films have two layers. The top layer should have high melting temperature to prevent bonding with the external reusable heaters such that heaters can be easily removed after the completion of the bonding process and be used repeatedly. Experimentally, aluminum wires of 30–70 μm in width and 3 μm in thickness have been fabricated by the lift-off process. They are employed as the heating elements to form bonding loops of 1 to 2 mm in diameter on either silicon or glass substrates. Because of the small bonding area (less than $1\times10^{-6}\,\mathrm{m}^2$), about 0.4 N of force is enough to provide a bonding pressure of 0.4 MPa. These aluminum heaters

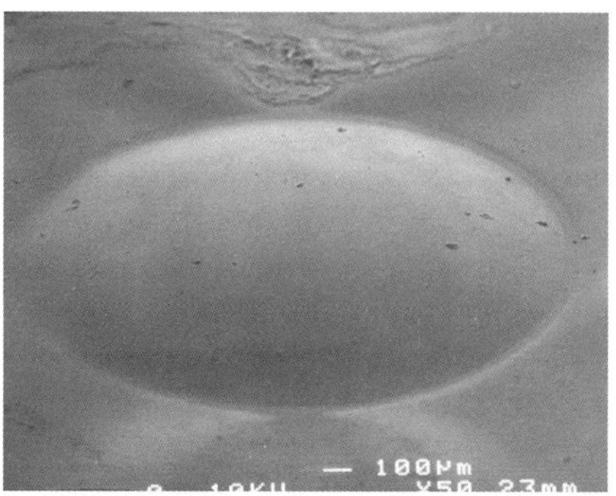

Fig. 52.20 Dome shape is formed under SEM micrograph of the Mylar membrane to PMMA demonstrating good seal [52.106]

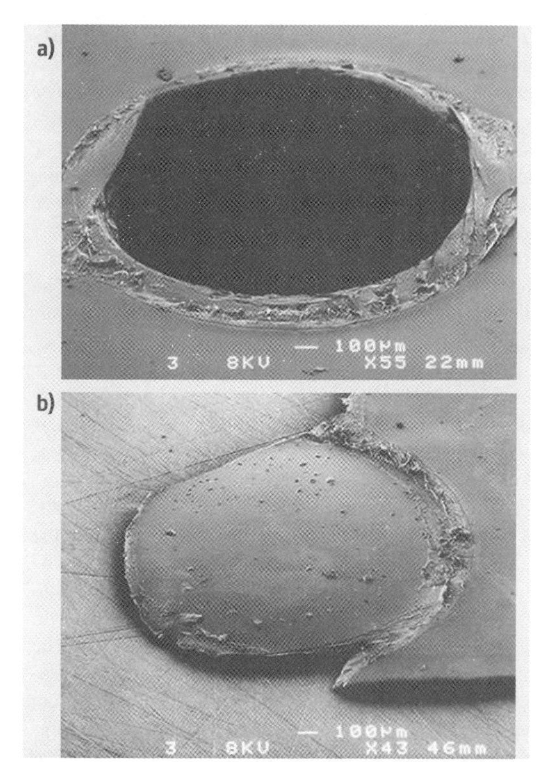

Fig. 52.21a,b SEM micrographs of bonding interface. (**a**) PMMA substrate. (**b**) Mylar layer [52.106]

have resistances of $\approx 0.8\,\Omega$ and when voltages of 3.5 V are applied, about 15 W of power is generated locally

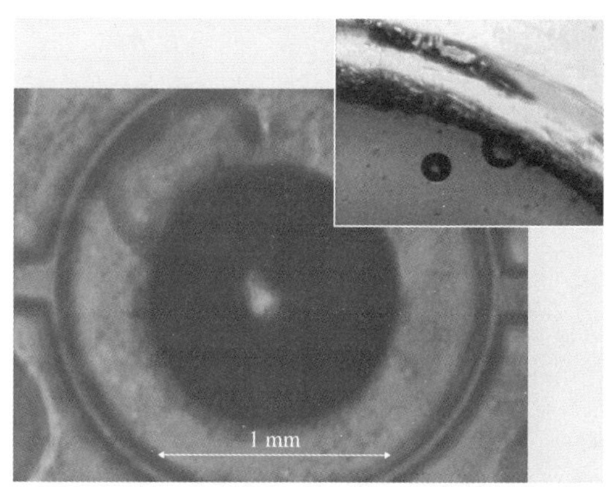

Fig. 52.22 Photograph of water encapsulation result [52.106]

to increase the temperature for bonding within about 0.25 s.

In order to test the quality of the bonding processes, several experiments have been designed. In the first experiment, bonded systems are put into a vacuum chamber and observed under an optical microscope. It is found that the top of the encapsulated chamber expands to form a dome shape due to the 1 atm pressure difference across the membrane. The diameter of the bonding ring is 1.4 mm in this case which corresponds to an effective area of $1.6 \times 10^{-6}\,\mathrm{m}^2$ and the air permeability of Mylar is $8\,\mathrm{cc/m^2\,d\,atm}$. Based on simplified approximation, it takes about 18 days for the air inside the cavity to diffuse out, and the membrane should return to flat. Figure 52.20 is a SEM photograph showing the result of Mylar bonded on PMMA. The dome shape can be observed as the proof that a good seal is achieved. Afterwards, the polymer-to-polymer bond (Mylar-to-PMMA) is forcefully broken to examine the bonding interface under SEM as shown in Fig. 52.21. It is observed that the bonding result is uniform, and part of the Mylar film is attached to the bonding substrate. Another experiment shows that direct encapsulation of water using localized heating can be achieved as the proof-of-concept demonstration of low-temperature processing. In this case, Mylar-to-PMMA bonding with reusable heaters is performed to encapsulate 0.18 l of water in the cavity as shown in Fig. 52.22. The close-view photograph on top shows two small bubbles in the water-filled and encapsulated chamber. These bubbles provide an easy way to verify the existence of water in the chamber. No leakage path can be identified and water inside the package escape mainly through evaporation and diffusion through the top Mylar membrane as well as the plastic substrate.

In summary, among the tested bonding systems using built-in aluminum heaters, it is found that plastics-to-plastics bonding results have the highest bonding strength. In the bonding systems using silicon or glass as the bonding substrates, the bonding interface happens between plastics and aluminum heaters and the bonding strength is lower than that of plastics-to-plastics bond but higher than that of plastics-to-silicon or plastics-to-glass bonds. Meanwhile, plastics-to-plastics bonding process using reusable heaters has the best bonding results as compared to plastics-to-silicon and plastics-to-glass systems. In the cases of using external, reusable heaters for plastic-to-silicon and plastics-to-glass bonding systems, it is suggested that thin plastic films with (1) high adhesion chemistry with silicon and glass and (2) low melting temperature should be employed as the intermediate bonding layers to achieve high bonding quality.

52.3 Thermal Issues and Packaging Reliability

52.3.1 Thermal Issues in Packaging

The two key thermal issues related to MEMS packaging are: (1) heat dissipation from actuators and integrated circuitry components, and (2) thermal stress generated during the packaging process. These two topics are discussed separately.

Heat Dissipation Issues

In microelectronic chips, heat dissipation is a serious problem as the size of transistors continues to shrink and the density of transistors on a chip continues to increase with advances in IC fabrication technology. The trend of power packing into smaller packages has created increasing thermal management challenges [52.107]. Since the electrical characteristics of transistors change with working temperature, inefficient heat dissipation could raise the working temperature and affect device performance. Present MEMS devices do not need high-power high-performance microprocessors, so power dissipation is not a problem. Nevertheless, some functional components in packaged MEMS are very sensitive to temperature variation, such as biomaterials or laser diodes. Several chemical sensors and other applications such as micro-polymerase chain reaction (PCR) chambers for deoxyribonucleic acid (DNA) replication actually require elevated temperature for operation and microthermal platforms are built for these devices. Thermal management to maintain the working temperature on these chips for stable operation is still an essential issue for packaging considerations. The geometrical complexity of MEMS resulting from packing various functional components in a tight space increases the difficulty of thermal management. As packaging integration process becomes more complex, the fabrication constraint in the packaging process will have a great impact on the heterogeneous integration process in the front-end MEMS and IC processes. For example, the requirement for low temperatures in the packaging process generally limits the possible choices of materials in the back-end process. In general, conventional IC packaging employs a heat sink attached to the chip to remove heat. The heat sink is generally made of a copper or stainless-steel bar with an array of fin structures on one side for better natural or forced heat convection to dissipate heat into the environment. In addition to heat sinks, thermal vias, heat pipes, immersion cooling, and thermoelectric cooling can also be used for effective heat removal. Because most MEMS packages still follow the typical IC packaging architecture, one promising thermal management method, the heat pipe, is discussed for possible MEMS packaging applications.

A heat pipe is a sealed slender tube containing a wick structure and a working fluid, typically water in electronics cooling. It is composed of three sections, the evaporator section at one end, the condenser section at the other end and the adiabatic section in the middle. In the evaporator section, heat is absorbed by the working fluid via a phase transformation from liquid to vapor. In the condenser section, heat dissipates into the outside environment. Thus, the fluid goes back to the liquid phase. The vapor phase is in a high-pressure high-temperature state that forces the vapor to flow into the condenser section at a lower temperature. Once the vapor condenses and gives up its latent heat, the condensed fluid is then pumped back to the evaporator section by the capillary force developed in the wick structure. Therefore, the middle adiabatic section contains two phases, the vapor phase in the core region and the liquid phase in the wick, flowing in opposite directions to each other and with no significant heat transfer between the fluid and the surrounding medium. Silicon has good thermal conductivity (1.41 W/(cm K)) and is easily micromachined to fabricate the heat pipe. Therefore, there is a great potential for the implementation of the silicon micro-heat pipe in IC and MEMS packaging and several approaches have been proposed on this topic [52.108–110].

Packaging-Induced Thermal Stresses

Thermal-based bonding processes have been used in MEMS packaging applications for many years, as discussed in this chapter. Thermal management is extremely important during the bonding process to avoid fracture in the substrate or MEMS device itself. Extremely high temperatures or rapid cooling conditions are may cause damage and should be carefully evaluated both analytically and experimentally. There are many ways to provide heating energy including electrical resistive heating, oven heating or induction heating [52.101]. These bonding processes may be put into two categories: (1) localized bonding, where the heat is directly applied only to the adhesive material used to bond the package to the device, and (2) global heating, where the entire system (MEMS device, adhesive, and packaging material) is heated to produce bonding of the materials, which is the most common approach. Therefore, this section focuses on the thermal stress effects during the heating and cooling procedures in MEMS packaging. The RTP aluminum-to-glass bonding process is used as a specific example for the discussion of thermal stresses [52.93]. The bonding process heats up the packaging system to 750 °C for 10 s, and then cools it back to room temperature. To simulate this process, an ANSYS program [52.111] was established to examine the shear stress due to coefficient of thermal expansion (CTE) variations in the bonding system as a result of temperature changes. The shear stress was recorded from the ANSYS analysis on the aluminum–Pyrex glass interface and the aluminum–silicon interface.

Two different models were analyzed. The first was the quartz–aluminum–silicon bonding system and the second was the Pyrex glass–aluminum–silicon bonding system [52.112]. The results of the ANSYS calculation were then analyzed and compared with experimental observations. Figure 52.19 shows the ANSYS results for a Pyrex glass bonding system, where the width of the aluminum solder is 100 μm and the maximum residual stress is 60 MPa in glass, which is slightly lower than the fracture strength of Pyrex glass at 70 MPa. It was discovered that increasing the aluminum width can lower these residual stresses. For example, the maximum residual stress analyzed from ANSYS in the Pyrex glass bonding systems is 74.5, 58, and 60 GPa for aluminum widths of 30, 50, and 100 μm, respectively [52.112]. Pyrex glass has a documented strength of around 69 GPa [52.113]. Fracture should always occur with an aluminum width of 30 μm or less, according to the ANSYS analysis. Fracture may occur sporadically at widths of 50 or 100 μm, depending on the amount and magnitude of the flaws in the Pyrex glass. Experiments were done on the Pyrex glass bonding systems with a width of 100 μm. The samples were heated up to 750 °C and then cooled down by taking them out of the oven. In all four experimental cases, small cracks were observed in the Pyrex glass, as shown in Fig. 52.20. These cracks may have occurred consistently for several reasons. Firstly, they may be a result of handling of the Pyrex glass before bonding. The Pyrex glass samples were kept in containers with each other, which may have resulted in abrasive contact. This may have caused flaws in the materials, which could have reduced the fracture strength. Secondly, it was observed that the cracks are small, only occurring tens of microns away from the aluminum and not propagating completely through the Pyrex glass. These cracks could be caused by the high stress applied, but the crack has not reached a critical size, and therefore has not propagated completely through the Pyrex glass. Therefore, the strength remains at the theoretical strength of 69 GPa, and the Pyrex glass is only partially cracked. Experimental analysis done by *Chaio* and *Lin* [52.93] shows that fracture was not observed when using aluminum widths greater than 150 μm. This is consistent with the results of the ANSYS analysis that showed that, as the width of the aluminum is increased, the residual stress decreases.

The ANSYS stress predictions for the quartz bonding system yielded a maximum stress of 207, 117, and 100 GPa, for aluminum widths of 30, 50, and 100 μm, respectively. All three of these stresses are much larger than the theoretical strength of quartz at 48 GPa, and therefore fracture should always occur. Quartz has a much larger coefficient of thermal expansion (CTE) than silicon, which is why this was predicted. Experimentally, a quartz substrate was tested for the silicon–aluminum–quartz bonding test and the result is shown in Fig. 52.21. It is observed that cracks occur all over and cause serious damage on the quartz wafer. These cracks could be the failure mechanism of the hermetic package. Therefore, Pyrex glass is identified as a better bonding substrate than quartz.

The thermal stresses generated in the packaging process in quartz are much larger than the stresses in Pyrex glass, because of the difference between the CTE in the two systems. Quartz has a low CTE ($0.54\times10^{-6}\,K^{-1}$) compared to the aluminum CTE ($23\times10^{-6}\,K^{-1}$) and silicon CTE ($3.5\times10^{-6}\,K^{-1}$). On the other hand Pyrex glass has a much closer CTE ($3.2\times10^{-6}\,K^{-1}$) to silicon and aluminum, resulting in smaller stresses. The practical implications of the ANSYS results and the preceding information are that materials must be chosen carefully when carrying out the bonding process. Materials with a much higher or lower CTE than silicon should not be used to ensure fracture will not occur.

52.3.2 Packaging Reliability

Packaging is one of the key issues to be addressed for the evaluation of the reliability of MEMS products. Any defects created during the sealing and packaging process may cause immediate device failure or may degrade the device performance over time. For example, microaccelerometers that are used to deploy air bags in automobile safety applications require excellent reliability. If any leakage path is created during the sealing process between the two bonding interfaces, moisture may enter the sealed microcavity and cause device failure over time. Furthermore, thermal stress induced by the CTE mismatch is one of the main factors that affect the packaging reliability. In fact, the formation of the stress can happen not only during packaging process but also during the operation of devices. In particular, during device operation, the package will go through various temperature cycles because of environmental change. Such temperature variation causes the expansion of packaging materials when they are constrained by the packaged assembly. As a result of such thermal mismatch, significant stresses are induced in the package and may cause the device to fail. In addition to thermal mismatch, corrosion, creep, fracture, fatigue crack initiation and propagation, and delamination of thin films are all possible factors that may cause the failure of packaged devices [52.114]. These failure mechanisms could be prevented or deferred by using proper packaging designs. For instance, the thermally induced strain inside the packaging material is generally below the tolerance of the material and cannot cause immediate catastrophic damage. However, cyclic loading can generate and accumulate stresses and eventually cause failure. Several common designs in IC packaging have been used to prolong the lifetime of devices. For example, the strain in solder interconnects of BGA or flip-chip packaging can be effectively reduced by introducing a polymer underfill material between the chip and the substrate to distribute well the thermal stress induced by CTE mismatch [52.115]. The strain can be further reduced if excellent thermal paths are built around interconnects to alleviate thermal stress originating from the temperature gradient between the ambient and operation temperatures. On the other hand, delamination phenomena occur in the interface of adjacent material layers such as components made of dissimilar materials that are subsequently bonded together. Delamination can result in electrical or mechanical failures of devices such as mechanically cracking through the electrical via wall to make an electrical open because of the propagation of the delamination of metal line from the dielectric layer or overheating of the die because of delamination of the underside of the die, causing openings in the heat dissipation path. Because of the stress and thermal loading, the geometry, and the material properties are complex in MEMS, the development of the packaging designs to increase the reliability is very important and requires extensive investigations.

Reliability testing is required before a new device can be delivered into the market. The test results can provide information for the improvement of packaging design and fabrication. Hence, how to analyze the failure data, which is called the reliability metrology, is very important in the packaging industry. The analysis method is to use the mathematical tools of probability and statistical distributions to evaluate data to understand the patterns of failure and to identify the sources of failure. For example, a failure density function is defined as the time derivative of the cumulative failure function

$$f(t)=\frac{\mathrm{d}F(t)}{\mathrm{d}t}\,, \tag{52.1}$$

$$F(t)=\int_0^t f(s)\,\mathrm{d}s\,. \tag{52.2}$$

The cumulative failure function $F(t)$ is the fraction of a group of original devices that has failed at time t. The Weibull distribution function is one of the analytic mathematic models commonly used in the packaging reliability evaluation to represent the failure density function [52.12].

$$f(t)=\frac{\beta}{\lambda}\left(\frac{t}{\lambda}\right)^{\beta-1}\exp\left[-\left(\frac{t}{\lambda}\right)^{\beta}\right]\,, \tag{52.3}$$

where β and λ are the Weibull parameters. The parameter β is called the shape factor and measures how the failure frequency is distributed around the average lifetime. The parameter λ is called the lifetime parameter and indicates the time at which 63.2% of the devices failure. By integrating both sides of (52.3), $F(t)$ becomes

$$F(t) = 1 - \exp\left[-\left(\frac{t}{\lambda}\right)^{\beta}\right] . \tag{52.4}$$

Using the Weibull distribution function with the two parameters extrapolated by experimental data, one can estimate the number of failures at any time during the test. Moreover, knowing the meaning and values of the parameters, one can compare two sets of test data. For example, greater value of λ indicate that samples have a longer lifetime. Because all of the mathematical models are statistical approximations based on real experimental data, more testing samples provide better estimation accuracy.

52.3.3 Long-Term and Accelerated MEMS Packaging Tests

The capability to estimate the reliability or lifetime of a device provides valuable information for the manufacturer to maximize the profit margin by balancing the cost and the quality of the product. Moreover, a warranty period given by the manufacturer has to be built on the reliability information of the product. The reliability of MEMS packages is best characterized using long-term tests with statistical data analyses. However, it is very difficult to measure the reliability or lifetime of a device in real time because testing over a prolonged time period may be required to prompt many devices to fail. In order to evaluate the reliability of a device in a timely fashion, accelerated testing is normally conducted in order to speed up the device aging process and thus shorten the required testing time. Accelerated testing, from the packaging and sealing point of view, is a testing method that emphasizes failure of the seal when foreign elements that may affect device performance leak into the microcavity. For a hermetic package, the lifetime of a MEMS device is essentially an estimation of the time required for water to penetrate into the package. For vacuum encapsulated MEMS devices, in addition to water penetration through the seal, gas penetration or outgassing from within the packaging materials such as the substrate, cap and sealing materials over time can degrade the vacuum level and thus the device performance. Therefore, the lifetime of a vacuum encapsulated MEMS package can be evaluated by the time that gas evolves into the package from either the seal or within the device materials, whichever comes first. Unfortunately, there are not many research publications that deal with long-term and accelerated test issues for MEMS packaging reliability. In the conventional IC packaging industry, reliability estimation is carried out by accelerated tests and statistical predictions [52.12]. Accelerated tests often utilize high temperatures and humidities, such as autoclave tests [52.116] to speed up corrosion against the sealing boundary to accelerate the failure of the packages. The industry could use very similar accelerated tests to estimate the lifetime of a package because the basic assumptions of the failure mode and humidity issues are similar to those for conventional IC packages. Several research groups have reported reliability studies for MEMS packages formed by different bonding methods and materials [52.94, 95, 117]. In this section, two packaging examples that aim to address long-term and accelerated tests are discussed.

Figure 52.22 shows long-term measurements of the Q-factor of a vacuum packaged μ-resonators using localized aluminum/silicon-to-glass bonding [52.94]. The vacuum encapsulation process is described in detail in Fig. 52.8. It was found that vacuum packaging by means of localized heating and bonding provides stable vacuum environments for the μ-resonator, and a quality factor of 9600 has been achieved with no degradation for at least one year. Since the performance of high-Q μ-resonators is very sensitive to environmental pressure, as shown in Fig. 52.17, any leakage can be easily detected. The fact that this high Q value can be maintained for one year indicates that the packaging process has been well performed and that both aluminum and Pyrex glass are suitable materials for vacuum packaging applications. According to a previous study of hermeticity in different materials, metal has a lower permeability to moisture than other materials such as glass, epoxy, and silicon. With a width of 1 μm, metal can effectively block moisture for more than 10 years [52.12]. In this vacuum package system, the bonding width is 30 μm such that it can sufficiently block the diffusion process of moisture. On the other hand, the diffusion effects of air molecules into these tiny cavities have not been studied extensively and the design guidelines for vacuum encapsulations are not clearly defined. Further investigations will be needed in this area and the example presented here serves as a good starting point.

On the other hand, accelerated testing puts a large number of samples in harsh environment, such as elevated temperature, elevated pressure and 100% humidity, to accelerate the corrosion process. The statistical failure data are gathered and analyzed to predict the lifetime of packages under a normal usage environment. As a result, the long-term reliability of the package can be predicted without going through true long-term tests. Unfortunately, accelerated tests have been an area that has not been addressed in MEMS research papers. Although the industry must have done some extensive reliability tests, they do not publish their results, probably due to liability concerns. Among the very limited publications, this section uses a specific MEMS packaging system that has gone through accelerated tests as an illustrative example [52.47].

The package is accomplished by means of rapid thermal processing (RTP) bonding as described previously in this chapter. The goal of the accelerated test is to examine the failure rate at the bonding interface. The accelerated tests start by putting the packaged samples into an autoclave chamber filled with high-temperature pressurized steam at 130 °C, 2.7 atm and 100% relative humidity for accelerated tests. The pressurized steam can penetrate small crevasses if there is a defect at the bonding interface [52.65]. Elevated temperature and humid environment speed up the corrosion process. A package is considered as a failure if water condenses or diffuses into the package. The statistical data gathered from accelerated tests in this case has been categorized as *right-censored* data [52.118]. Statistical failure data are gathered every 24 h under optical examination for a period of 864 h when new failure is seldom observed (therefore, right-censored on the time axis). In practice, this method was easier and more economical to implement than other methods. Owing to the robustness of the sample, it is difficult to conduct the tests to the point where all packages fail. The cumulative failure function $F(t)$ is defined as

$$F(t) = \frac{\text{Number of cumulative failures}}{\text{Number of samples } N}, \qquad (52.5)$$

where N is the sample size at the beginning of the test. A package was considered as a failure if water condensed inside or diffused into the package. For example, water was found to diffuse into the cavity after 240 h into the test, as shown in Fig. 52.23. However, no leakage path could be identified under an optical microscope in this case. Figure 52.24 shows that the function $F(t)$ (in %) is plotted versus the logarithm of time. In general, most of the failures occurred in the first 96 h [$\ln(t) \approx 4.56$] and such a high number of early failures reflects the yielding issue of the sealing process. Moreover, packages with a smaller bonding width and larger bonding areas showed a higher percentage

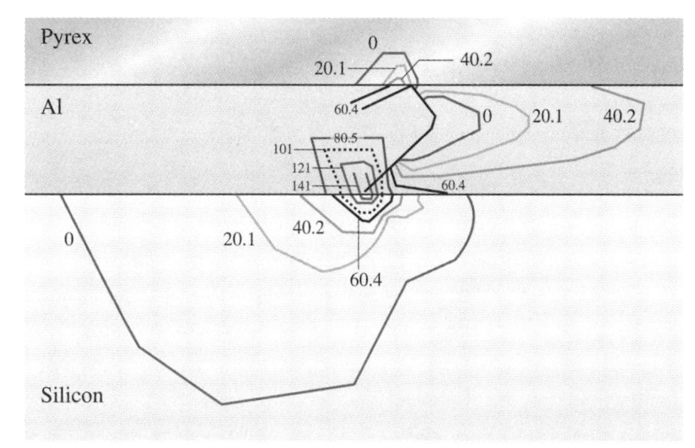

Fig. 52.23 Residual stress (GPa) for an aluminum solder width of 100 μm in the RTP silicon–aluminum–glass bonding system

of failure. Both Weibull and log-normal statistical models [52.118] were used and compared to analyze the collected data to predict the lifetime of the packages, and the least-square fit method was used to determine the best fitting model. It was found that R^2, the *coefficient of determination* [52.118], values were generally in the range of 0.8 using the log-normal model, compared to values of 0.5 using the Weibull model. Therefore, the log-normal model was used to predict the life time of packages.

Figure 52.25 shows the inverse standard normal distribution function versus $\ln(t)$ and the maximum likelihood estimator (MLE) is then used to predict the mean, standard deviation and the mean time to failure (MTTF). Table 52.2 shows the MLE calculation results for the MTTF. The wide confidence level interval comes from the fact that only a small number of samples failed

Fig. 52.24
Micrograph of the experimental result on the Pyrex glass–aluminum–silicon system. Small cracks can be observed

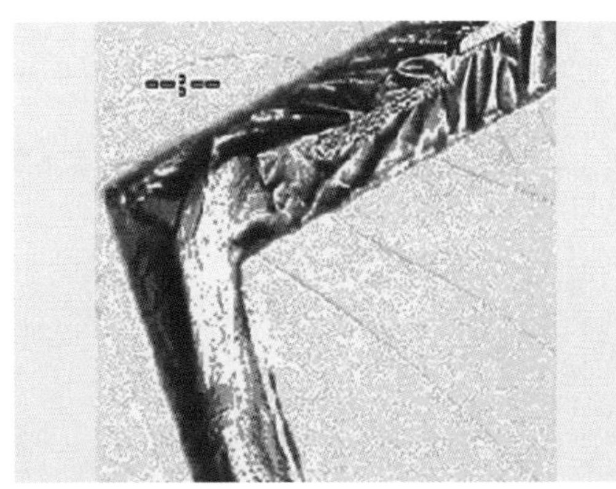

Fig. 52.25 Bonding result of the quartz–aluminum–silicon system where fracture can be observed

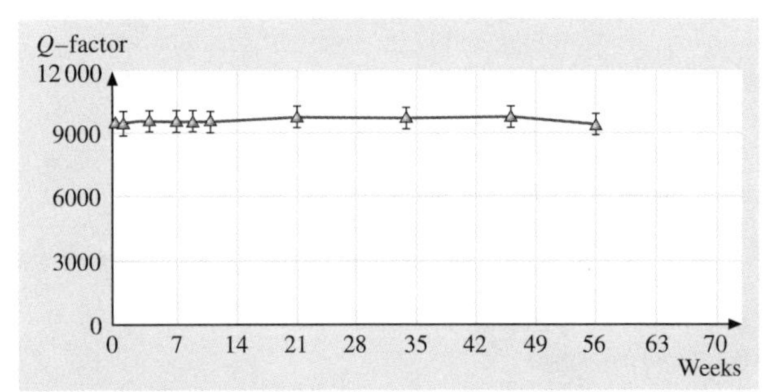

Fig. 52.26 Long-term measurement of encapsulated μ-resonators. No degradation of Q-factors is found after 56 weeks

at the end of the test. It is also observed that packages with larger bonding widths and smaller bonding areas have larger MTTF values. The lower bound on the MTTF provides the worst-case scenario. For example, only 4 out of 31 samples failed by the end of the test in the case of a bonding ring width of 200 μm and a sealing area of 450×450 μm^2. The MTTF predicts, in the worst-case scenario, that there is a 90% chance that a package will fail in 0.57 years in the autoclave environment.

It is widely accepted that the acceleration factor (AF) for autoclave tests follows the Arrhenius equation [52.12] and can be modeled as

$$\mathrm{AF} = \frac{(\mathrm{RH}^{-n}\,\mathrm{e}^{\Delta E_\mathrm{a}/k_\mathrm{B}T})_\mathrm{normal}}{(\mathrm{RH}^{-n}\,\mathrm{e}^{\Delta E_\mathrm{a}/k_\mathrm{B}T})_\mathrm{accelerated}}\,, \tag{52.6}$$

where RH is the relative humidity (85%, RH = 85), k_B is the Boltzmann constant and T is the absolute temperature. The recommended value for n, an empirical constant, is 3.0 [52.119], and ΔE_a, the activation energy, is 0.9 eV for a plastic dip package and 0.997 eV for an anodically bonded glass-to-silicon package [52.65]. If $\Delta E_\mathrm{a} = 0.9$ is used to estimate the AF for the accelerated testing conditions compared with the jungle conditions (35 °C, 1 atm and 95% RH), the AF is about 3000 and the worst-case lifetime values in the jungle conditions are also listed in Table 52.2. The high values of estimated MTTF in the jungle condition could be a result of overestimation of the AF because plastic dip package may have smaller AFs than those of glass

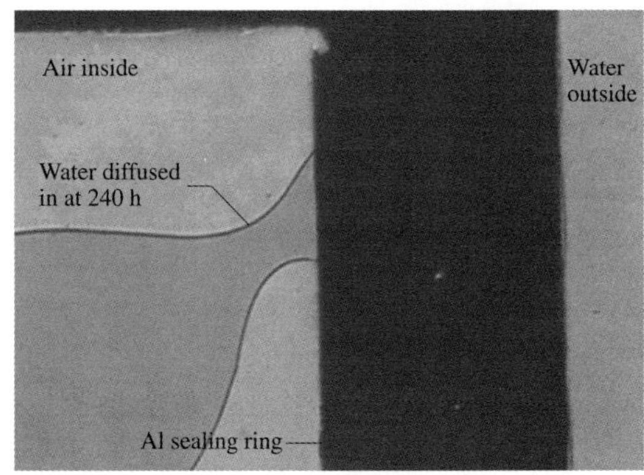

Fig. 52.27 A particular device failed at 240 h of testing time

Table 52.2 The maximum likelihood estimation for mean time to failure (MTTF)

MLE calculation resuls of MMTF[a]				
Bonding width	Area	MTTF		Worst cases
W (μm)	A (μm^2)	UB (years)	LB (years)	in jungle condition (years)
200	450×450	1.8×10^7	0.57	1700
100	450×450	5.3	0.10	300
200	1000×1000	6.5×10^3	0.09	270
150	1000×1000	0.50	0.017	50

[a] UB is the upper bound, and LB is the lower bound of the 90% confidence interval, respectively. The MTTF LB times AF is the worst-case MTTF used in the jungle condition

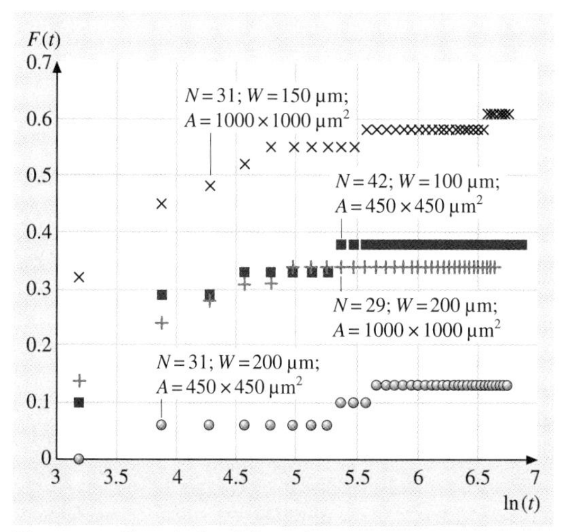

Fig. 52.28 Cumulative failure data

packages. Nevertheless, these data and analyses provide important guidelines in the area of accelerated tests for MEMS packages.

For vacuum packaged MEMS devices, the lifetime can be evaluated by monitoring the quality factor of microresonators inside their sealed cavities. Again, vacuum packaged MEMS resonators by using RTP aluminum-to-nitride bonding are discussed in detail here for better illustration of the various factors relevant to reliability. It was found that, under normal condition at room-temperature storage, the quality factor of resonators remained constant after 37 weeks, as shown in Fig. 52.26. Furthermore, the vacuum quality under a harsh environment was characterized by putting a vacuum-encapsulated comb-resonator into an autoclave chamber (130 °C, 2.7 atm and 100% RH) for accelerated testing. The result is shown in Fig. 52.27 and the quality factor stayed at 200 after 24 h in the autoclave testing chamber. Since the slight differences between the two spectra are within the normal experimental errors, a conclusion may be drawn that this harsh environment test did not affect the vacuum seal.

In order to characterize the vacuum lifetime of the packages, two vacuum packaged comb-resonators were put in the harsh environment for continuous testing for up to 1008 h and the results are summarized in Table 52.3. The quality factor of each package has been

Table 52.3 Summary of the accelerated testing results of two vacuum packaged resonators

	Package 1	Package 2
Q-value	400	200
Hours under testing (h)	> 1008	576
MTTF, lower bond (h)	769	149

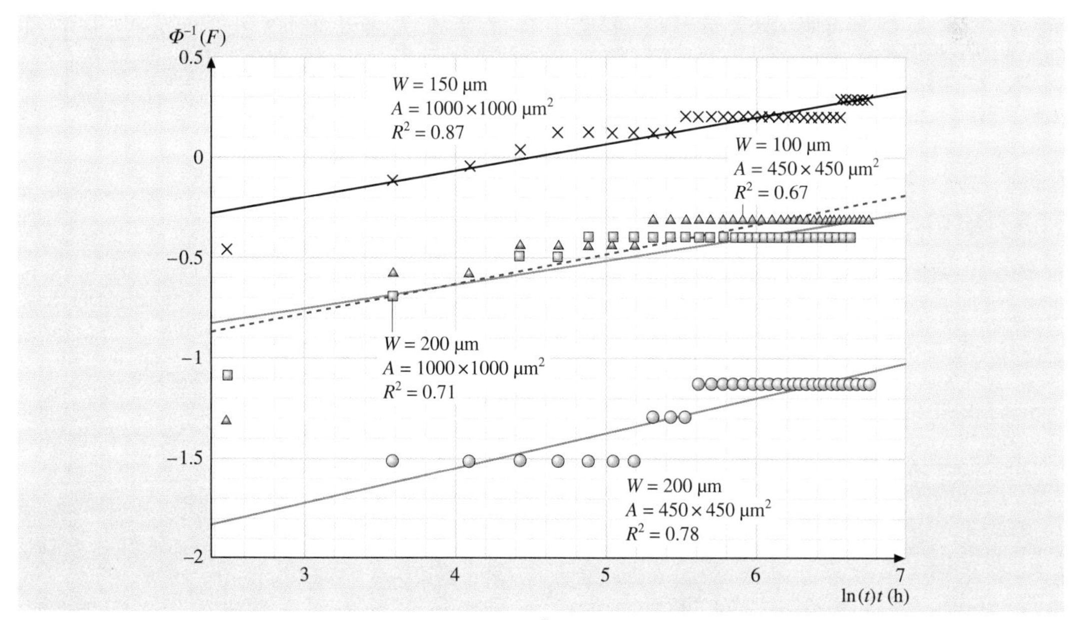

Fig. 52.29 Life data fitted by the log-normal distribution. R^2 is the coefficient of determination

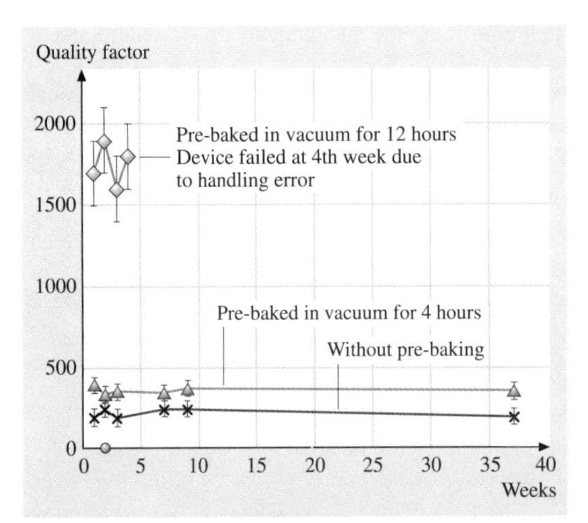

Fig. 52.30 Long-term stability tests to 37 weeks. The Q-value increases with the prebaking time

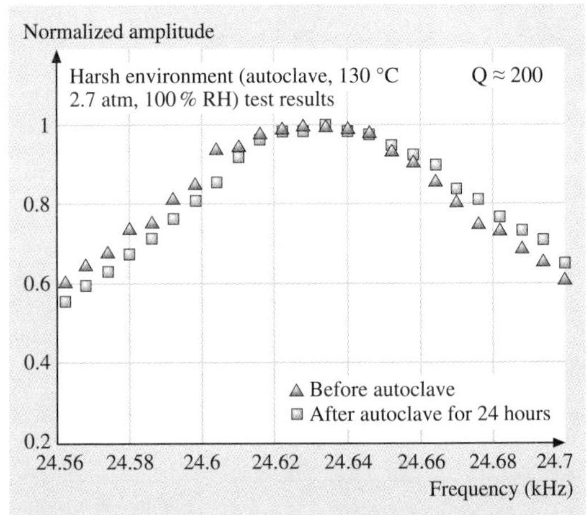

Fig. 52.31 Spectrum measured before and after accelerated testing for 24 h

measured in every 24 h interval and they were found to be maintained at 400 and 200, respectively, before failure. The first packaged resonator with a Q-value of 200 had an aluminum sealing ring with a width of 75 μm and a sealing area of $650 \times 650\,\mu m^2$. The second packaged microresonator had a Q-value of 400 with an aluminum sealing ring width of 200 μm and a sealing area of $550 \times 550\,\mu m^2$. If the accelerated testing results on water penetration as discussed previously [52.95] were applied here for gas penetration as a measure of the vacuum sealing characteristics, the accelerated lifetime of the first and second packaged resonators would be 0.017–0.1 years (149–876 h) and 0.09–0.57 years (769–4993 h), respectively. Experimentally, the first packaged resonator failed at 576 h into the test in the autoclave chamber. This corresponds well to the lifetime prediction from the vapor-penetration work [52.95]. The second packaged resonator survived in the autoclave chamber for more than 1008 h (the device did not fail) and this result also verified the prediction made from the vapor-penetration work. However, it is noted that these results are preliminary data and more packaged devices and tests should be conducted to establish meaningful statistical analyses.

52.4 Future Trends and Summary

In the past, the development of MEMS packaging mainly originated from IC packaging advancements because existing packaging techniques can significantly reduce the development cost of MEMS. However, it is expected that the situation will change very soon so that MEMS packaging approaches will assist IC packaging development. Recent progress in IC packaging is aimed at providing high I/O density and more chip integration capability for the needs of high-speed high-data-rate communications. In order to satisfy those requirements, several packaging concepts and techniques have been developed, such as 3-D packaging, wafer-level packaging, BGA, and the flip-chip technique. Although all of these concepts and methods can provide a package with more I/O density, flexibility in chip integration, and lower manufacturing cost for IC fabrication, they are still insufficient to provide solutions for future applications because of the increasing complexity and requirements of MEMS packaging. On the contrary, with the progress of MEMS fabrication technologies, several key processes such as deep reactive-ion etching (DRIE), wafer bonding, and thick-photoresist processes [52.120] have been utilized for IC packaging fabrication. Therefore, technologies developed in MEMS fabrication can also assist the development of new IC packaging approaches.

In order to address the future needs of process integration, adaptive multi-chip module (MCM) [52.31] or 3-D packaging combined with vertical through-substrate interconnects [52.10, 121] are promising approaches for the development of future MEMS packaging processes. Based on low-temperature flip-chip solder bonding techniques, these packaging methods can provide more flexibility in device fabrication and packaging. Devices can be fabricated before they are integrated together to form the microsystems to dramatically reduce the packaging cost. Vertical through-substrate interconnects can have higher I/O density, smaller resistance and parasitic capacitance, and mutual inductance. Although this approach provides many possible advantages, technical challenges exist. For instance, metal is commonly used as the fill material inside the vertical vias to form electrical interconnects. This may introduce a large thermal mismatch with respect to the silicon substrate and generate huge thermal stress that could cause packaging reliability problems. Moreover, it will be an interesting engineering challenge to fill materials into those high-aspect-ratio vias.

The future development of MEMS packaging depends on the successful implementation of unique techniques as described in the following:

1. Development of mechanical, thermal, and electrical models for packaging designs and fabrication processes
2. Wafer-level chip-scale packaging with low packaging cost and high yields
3. Effective testing techniques at the wafer level to reduce testing costs
4. Device integration by vertical through-interconnects as an interposer [52.1] to avoid thermal mismatch problems.

In addition to these approaches and challenges, there are many other possibilities that have not been listed but also require dedicated investigation. For example, several key nanotechnologies have been introduced in the previous chapters but packaging solutions for these NEMS devices have not been addressed. Because it is feasible to use MEMS as a platform for NEMS fabrication, all the packaging issues discussed in the chapter can be directly applied to NEMS devices. On the other hand, nanotechnology may introduce new opportunities for MEMS packaging applications by providing superior electrical, mechanical and thermal properties [52.122–125]. For example, carbon nanotubes have very high thermal conductivity [52.124] and may enable improved cooling effects for IC/MEMS packaging applications.

In summary, MEMS packaging issues have been introduced in the areas of fabrication, application, reliability and future development. Packaging design and modeling, packaging material selection, packaging process integration, and packaging cost are the main issues to be considered when developing a new MEMS packaging process.

References

52.1 M. Matsuo, N. Hayasaka, K. Okumura, E. Hosomi, C. Takubo: Silicon interposer technology for high-density package, IEEE ECTC (2000) pp. 1455–1459

52.2 K.E. Peterson: Silicon as a mechanical material, Proc. IEEE **70**, 420–457 (1982)

52.3 L. Lin, H.-C. Chu, Y.-W. Lu: A simulation program for the sensitivity and linearity of piezoresistive pressure sensors, IEEE/ASME J. Microelectromech. Syst. **8**, 514–522 (1999)

52.4 Y.C. Tai, R.S. Muller: Lightly-doped polysilicon bridge as a flow meter, Sens. Actuators **15**, 63–75 (1988)

52.5 D. Hicks, S.-C. Chang, M.W. Putty, D.S. Eddy: Piezoelectrically activated resonant bridge microaccelerormenter, Solid-State Sens. Actuator Workshop (Transducers Research Foundation, Hilton Head Island 1994) pp. 225–228

52.6 J. Berstein, S. Cho, A. King, A. Kourepenis, P. Maciel, M. Weinberg: A micromachined comb-drive tuning fork rate gyroscope, 6th IEEE Int. Conf. MEMS (IEEE, Fort Lauderdale 1993) pp. 143–148

52.7 M. Madou: Compatibility and incompatibility of chemical sensors and analytical equipment with micromachining, Solid-State Sens. Actuator Workshop (Transducers Research Foundation, Hilton Head Island 1994) pp. 164–171

52.8 M.J. Zdeblick, J.B. Angell: A microminiature electric-to-fluidic valve, Proc. Transducers'87, 4th Int. Conf. Solid-State Transducers Actuators (IEEE, Tokyo 1987) pp. 827–829

52.9 J.H. Tsai, L. Lin: A thermal bubble actuated micro nozzle-diffuser pump, IEEE/ASME J. Microelectromech. Syst. **11**, 665–671 (2002)

52.10 S.J. Ok, D. Baldwin: High density, aspect ratio through-wafer electrical interconnect vias for low cost, generic modular MEMS packaging, 8th IEEE Int. Symp. Adv. Packag. Mater. (IEEE, New York 2002) pp. 8–11

52.11 H. Reichl, V. Grosser: Overview and development trends in the field of MEMS packaging, 14th IEEE Int. Conf. MEMS (IEEE, Interlaken 2001) pp. 1–5

52.12 R.R. Tummala, E.J. Rymaszewski, A.G. Klopfenstein: *Microelectronics Packaging Handbook, Semiconductor Packaging* (Chapman Hall, New York 1997)

52.13 C.A. Harper (Ed.): *Electronic Packaging and Interconnection Handbook* (McGraw-Hill, New York 1991)

52.14 P. Kapur, P.M. McVittie, K. Saraswat: Technology and reliability constrained future copper interconnects, part II: performance implication, IEEE Trans. Electron. Dev. **49**, 598–604 (2000)

52.15 J. Lau: *Flip Chip Technologies* (McGraw-Hill, New York 1996)

52.16 R. Prasad: *Surface Mount Technology: Principles and Practice*, 2nd edn. (Chapman Hall, New York 1989)

52.17 M. Töpper, J. Auersperg, V. Glaw, K. Kaskoun, E. Prack, B. Keser, P. Coskina, D. Jäger, D. Petter, O. Ehrmann, K. Samulewicz, C. Meinherz, S. Fehlberg, C. Karduck, H. Reichl: Fab integrated packaging (FIP) a new concept for high reliability wafer-level chip size packaging, IEEE ECTC (Las Vegas 2000) pp. 74–81

52.18 S. Savastiouk, O. Siniaguine, E. Korczynski: 3D wafer level packaging, Int. Conf. High-Density Interconnect Syst. Packag., Vol. 9 (2000) pp. 26–31

52.19 J.H. Tsai, L. Lin: Micro-to-macro fluidic interconnectors with an integrated polymer sealant, J. Micromech. Microeng. **11**, 577–581 (2001)

52.20 T.A. Core, W.K. Tsang, S. Sherman: Fabrication technology for an integrated surface-micromachined sensor, Solid State Technol. **36**, 39–47 (1993)

52.21 D. Koester, R. Majedevan, A. Shishkoff, K. Marcus: Multi-user MEMS processes (MUMPS) introduction and design rules, rev. 4, MCNC MEMS Technology Applications Center, Research Triangle Park, NC 27709, July (1996)

52.22 R. Lengtenberg, H.A.C. Tilmans: Electrically driven vacuum-encapsulated polysilicon resonantor, Part I: design and fabrication, Sens. Actuators A **45**, 57–66 (1994)

52.23 R.L. Smith, S.D. Collins: Micromachined packaging for chemical microsensors, IEEE Trans. Electron. Dev. **35**, 787–792 (1988)

52.24 M.C. Wu: Micromachining for optical and optoelectronic systems, Proc. IEEE **85**(11), 1833–1856 (1997)

52.25 M. Putty, K. Najafi: A micromachined gyroscope, Solid-State Sens. Actuator Workshop (Transducers Research Foundation, Hilton Head Island 1994) pp. 212–220

52.26 S. Wolf: *Silicon Processing for the VLSI Era, Vol. I: Process Technology* (Lattice, Sunset Beach 1995)

52.27 J.K. Bhardway, H. Ashraf: Advanced silicon etching using high density plasmas, SPIE Micromach. Fab. Technol. **2639**, 224–233 (1995)

52.28 W.H. Ko, J.T. Suminto, G.J. Yeh: *Bonding Techniques for Microsensors, Micromachining and Micropackaging of Transducers* (Elsevier Science, New York 1985)

52.29 S.D. Senturia, R.L. Smith: Microsensor packaging and system partitioning, Sens. Actuators **15**, 221–234 (1988)

52.30 J.T. Butler, V.M. Bright, P.B. Chu, R.J. Saia: Adapting multichip module foundries for MEMS packaging, IEEE Int. Conf. Multichip Modules High-Density Packag. (IEEE, Denver 1998) pp. 106–111

52.31 M. Schünemann, A.J. Kourosh, V. Grosser, R. Leutenbauer, G. Bauer, W. Schäfer, H. Reichl: MEM modular packaging and interfaces, IEEE ECTC (2000) pp. 681–688

52.32 D.W. Lee, T. Ono, T. Abe, M. Esashi: Fabrication of microprobe array with sub-100 nm nano-heater for nanometric thermal imaging and data storage, Proc. IEEE Micro Electro Mech. Syst. Conf. (IEEE, Interlaken 2001) pp. 204–207

52.33 A.S. Laskar, B. Blythe: Epoxy multichip modules, a solution to the problem of packaging and interconnection of sensors and signal-processing chips, Sens. Actuators A **36**, 1–27 (1993)

52.34 R. Reichl: Packaging and interconnection of sensors, Sens. Actuators A **25-27**, 63–71 (1991)

52.35 A. Grisel, C. Francis, E. Verney, G. Mondin: Packaging technologies for integrated electrochemical sensors, Sens. Actuators **17**, 285–295 (1989)

52.36 J.L. Lund, K.D. Wise: Chip-level encapsulation of implantable CMOS microelectrode arrays, Solid-State Sens. Actuator Workshop (Transducers Research Foundation, Hilton Head Island 1994) pp. 29–32

52.37 T. Akin, B. Siaie, K. Najafi: Modular micromachined high-density connector for implantable biomedical systems, Micro Electro Mech. Syst. Workshop (IEEE, San Diego 1996) pp. 497–502

52.38 L. Muller, M.H. Hecht: Packaging qualification for MEMS-based space systems, Micro Electro Mech. Syst. Workshop (IEEE, San Diego 1996) pp. 503–508

52.39 L. Lin, K. McNair, R.T. Howe, A.P. Pisano: Vacuum encapsulated lateral microresoantors, 7th Int. Conf. Solid State Sens. Actuators (IEEE, Yokohama 1993) pp. 270–273

52.40 H. Guckel: Surface micromachined pressure transducers, Sens. Actuators A **28**, 133–146 (1991)

52.41 J.J. Sniegowski, H. Guckel, R.T. Christenson: Performance characteristics of second generation polysilicon resonating beam force transducers, IEEE Solid-State Sens. Actuator Workshop (Transducers Research Foundation, Hilton Head Island 1990) pp. 9–12

52.42 K. Ikeda, H. Kuwayama, T. Kobayashi, T. Watanabe, T. Nishikawa, T. Oshida, K. Harada: Three dimensional micromachining of silicon pressure sensor integrating resonant strain gauge on diaphragm, Sens. Actuators A **21-23**, 1001–1010 (1990)

52.43 C.H. Mastrangelo, R.S. Muller: Vacuum-sealed silicon micromachined incandescent light source, IEEE IEDM (1989) pp. 503–506

52.44 J. Smith, S. Montague, J. Sniegowski, R. Manginell, P. McWhorter, R. Huber: Characterization of the embedded micromechanical device approach to the monolithic integration of MEMS with CMOS, Proc. SPIE **2879**, 306–314 (1996)

52.45 S. Van der Groen, M. Rosmeulen, P. Jansen, K. Baert, L. Deferm: CMOS compatible wafer scale adhesive bonding for circuit transfer, Int. Conf. Solid-State Sens. Actuators Transducers'97 (IEEE, Chicago 1997) pp. 629–632

52.46 Y.T. Cheng, Y.T. Hsu, L. Lin, C.T. Nguyen, K. Najafi: Vacuum packaging using localized aluminium/silicon-to-glass bonding using localized aluminium/silicon-to-glass bonding, IEEE Int. Conf. MEMS (IEEE, Interlaken 2001) pp. 18–21

52.47 M. Chiao, L. Lin: Vacuum packaging of microresonators by rapid thermal processing, Proc. SPIE Smart Electron. MEMS Nanotechnol. (SPIE, San Diego 2002) pp. 17–21

52.48 J.H. Partridge: *Glass-to-Metal Seals* (The Society of Glass Technology, Sheffield 1949)

52.49 P. Kumar, V.A. Greenhut: *Metal-to-Ceramic Joining* (The Minerals Metals and Materials Society, Warrendale 1991)

52.50 M.G. Nicholas, D.A. Mortimer: Ceramic/metal joining for structural applications, Mater. Sci. Technol. **1**, 657–665 (1985)

52.51 M. Esashi, S. Shoji, A. Nakano: Normally closed microvalve and micropump fabricated on a silicon wafer, Sens. Actuators **20**, 163–169 (1989)

52.52 M.E. Poplawski, R.W. Hower, R.B. Brown: A simple packaging process for chemical sensors, Solid-State Sens. Actuator Workshop (Transducers Research Foundation, Hilton Head Island 1994) pp. 25–28

52.53 S.F. Trautweiler, O. Paul, J. Stahl, H. Baltes: Anodically bonded silicon membranes for sealed and flush mounted microsensors, Micro Electro Mech. Syst. Workshop (IEEE, San Diego 1996) pp. 61–66

52.54 E.H. Klaassen, K. Petersen, J.M. Noworolski, J. Logan, N.I. Malfu, J. Brown, C. Storment, W. McCulley, G.T.A. Kovac: Silicon fusion bonding and deep reactive ion etching: A new technology for microstructures, Sens. Actuators A **52**, 132–139 (1996)

52.55 C.H. Hsu, M.A. Schmidt: Micromachined structures fabricated using a wafer-bonded sealed cavity process, Solid State Sens. Actuator Workshop (Transducers Research Foundation, Hilton Head Island 1994) pp. 151–155

52.56 A.L. Tiensuu, J.A. Schweitz, S. Johansson: In situ investigation of precise high strength micro assembly using Au-Si eutectic bonding, Int. Conf. Solid-State Sens. Actuators Eurosens. IX (IEEE, Stockholm 1995) pp. 236–239

52.57 A.P. Lee, D.R. Ciarlo, P.A. Krulevitch, S. Lehew, J. Trevino, M.A. Northrup: Practical microgripper by fine alignment, eutectic bonding and SMA actuation, Int. Conf. Solid-State Sens. Actuators Eurosens. IX (IEEE, Stockholm 1995) pp. 368–371

52.58 M.B. Cohn, Y. Liang, R. Howe, A.P. Pisano: Wafer to wafer transfer of microstructures for vacuum package, Solid-State Sens. Actuator Workshop (Transducers Research Foundation, Hilton Head Island 1996) pp. 32–35

52.59 Q.-Y. Tong, U. Gosele: *Semiconductor Wafer Bonding, Science and Technology* (Wiley, New York 1999)

52.60 R.W. Bower, M.S. Ismail, B.E. Roberds: Low temperature Si_3N_4 direct bonding, Appl. Phys. Lett. **62**, 3485–3487 (1993)

52.61 H. Takagi, R. Maeda, T.R. Chung, T. Suga: Low temperature direct bonding of silicon and silicon dioxide by the surface activation method, Int. Conf. Solid-State Sens. Actuators Transducer 97 (IEEE, Chicago 1997) pp. 657–660

52.62 G. Wallis, D. Pomerantz: Filed assisted glass-metal sealing, J. Appl. Phys. **40**, 3946–3949 (1969)

52.63 L. Bowman, J. Meindl: The packaging of implantable integrated sensors, IEEE Trans. Biomed. Eng. **BME-33**, 248–255 (1986)

52.64 M. Esashi: Encapsulated micro mechanical sensors, Microsyst. Technol. **1**, 2–9 (1994)

52.65 B. Ziaie, J. Von Arx, M. Dokmeci, K. Najafi: A hermetic glass-silicon micropackages with high-density on-chip feedthroughs for sensors and actuators, J. Microelectromech. Syst. **5**, 166–179 (1996)

52.66 Y. Lee, K. Wise: A batch-fabricated silicon capacitive pressure transducer with low temperature sensitivity, IEEE Trans. Electron. Dev. **ED-29**, 42–48 (1982)

52.67 S. Shoji, H. Kicuchi, H. Torigoe: Anodic bonding below 180 °C for packaging and assembling of MEMS using lithium aluminosilicate-beta-quartz glass-ceramic, Proc. 1997 10th Annu. Int. Workshop Micro Electro Mech. Syst. (IEEE, Nagoay 1997) pp. 482–487

52.68 M. Esashi, N. Akira, S. Shoji, H. Hebiguchi: Low-temperature silicon-to-silicon anodic bonding with intermediate low melting point glass, Sens. Actuators A **23**, 931–934 (1990)

52.69 A. Hanneborg, M. Nese, H. Jakobsen, R. Holm: Silicon-to-thin film anodic bonding, J. Micromech. Microeng. **2**, 117–121 (1992)

52.70 A.V. Chavan, K.D. Wise: Batch-processed vacuum-sealed capacitive pressure sensors, ASME/IEEE J. Microelectromech. Syst. **10**(4), 580–588 (2001)

52.71 H. Henmi, S. Shoji, Y. Shoji, K. Yoshimi, M. Esashi: Vacuum packaging for microsensors by glass-silicon anodic bonding, Sens. Actuators A Phys. **43**(1-3), 243–248 (1994)

52.72 T. Hara, S. Kobayashi, K. Ohwada: A new fabrication method for low-pressure package with glass-silicon-glass structure and its stability, 10th

Int. Conf. Solid-State Sens. Actuators Transducers'99, Digest of Technical Papers, Vol. 2 (IEEE, Sendai 1999) pp. 1316–1319
52.73 S.A. Audet, K.M. Edenfeld: Integrated sensor wafer-level packaging, Int. Conf. Solid-State Sens. Actuators Transducers'97 (IEEE, Chicago 1997) pp. 287–289
52.74 R.C. Benson, N. deHaas, P. Goodwin, T.E. Phillips: Epoxy adhesives in microelectronic hybrid applications, Johns Hopkins APL Tech. Dig. **13**, 400–406 (1992)
52.75 M. Shimbo, J. Yoshikawa: New silicon bonding method, J. Electrochem. Soc. **143**, 2371–2377 (1996)
52.76 P.M. Zavracky, B. Vu: Patterned eutectic bonding with Al/Ge thin film for MEMS, Proc. SPIE **2639**, 46–52 (1995)
52.77 G. Humpston, D.M. Jacobson: Principles of soldering and brazing, ASM Int. (1993) pp. 241–244
52.78 A. Singh, D. Horsely, M.B. Cohn, R. Howe: Batch transfer of microstructures using flip-chip solder bump bonding, Int. Conf. Solid State Sens. Actuators Transducer 97, Vol. 1 (IEEE, Chicago 1997) pp. 265–268
52.79 M.M. Maharbiz, M.B. Cohn, R.T. Howe, R. Horowitz, A.P. Pisano: Batch micropackaging by compression-bonded wafer-wafer transfer, 12th Int. Conf. MEMS (IEEE, Orlando 1999) pp. 482–489
52.80 Y.T. Cheng, L. Lin, K. Najafi: Localized silicon fusion and eutectic bonding for MEMS fabrication and packaging, IEEE/ASME J. Microelectromech. Syst. **9**, 3–8 (2000)
52.81 L. Lin: Selective encapsulations of MEMS: micro channels, needles, resonators, and electromechanical filters. Ph.D. Thesis (University of California, Berkeley 1993)
52.82 Y.C. Su, L. Lin: Localized plastic bonding for micro assembly, packaging and liquid encapsulation, Proc. IEEE Micro Electro Mech. Syst. Conf. (IEEE, Interlaken 2001) pp. 50–53
52.83 M. Schwartz: *Brazing* (Chapman Hall, London 1995)
52.84 L. Lin, R.T. Howe, A.P. Pisano: Microelectromechanical filters for signal processing, IEEE/ASME J. Microelectromech. Syst. **7**, 286–294 (1998)
52.85 W.C. Tang, C.T.-C. Nguyen, R.T. Howe: Laterally driven polysilicon resonant microstructures, Sens. Actuators A **20**, 25–32 (1989)
52.86 G.T. Mulhern, D.S. Soane, R.T. Howe: Supercritical carbon dioxide drying of microstructures, 7th Int. Conf. Solid State Sens. Actuators (IEEE, Yokohama 1993) pp. 296–299
52.87 M. Judy: Micromechanisms using sidewall beams. Ph.D. Thesis (University of California, Berkeley 1994) p. 162
52.88 R. Aigner, K.-G. Oppermann, H. Kapels, S. Kolb: Cavity-micromachining technology: zero-package solution for inertial sensors, Transducers '01, Eurosensors XV, 11th Int. Conf. Solid-State Sens. Actuators, Digest of Technical Papers, Vol. 1 (IEEE, Munich 2001) pp. 186–189
52.89 M. Bartek, J.A. Förster, R.F. Wolfenbuttel: Vacuum sealing of microcavities using metal evaporation, Sens. Actuators A **61**, 364–368 (1997)
52.90 C.H. Mastrangelo, R.S. Muller, S. Kumar: Microfabricated incandescent lamps, Appl. Opt. **30**, 868–873 (1993)
52.91 A.M. Leung, J. Jones, E. Czyzewska, J. Chen, B. Woods: Micromachined accelerometer based on convection heat transfer, 11th Int. Conf. MEMS (IEEE, Heidelberg 1998) pp. 627–630
52.92 D.R. Spark, L. Jordan, J.H. Frazee: Flexible vacuum-packaging method for resonating micromachines, Sens. Actuators A **55**, 179–183 (1996)
52.93 M. Chiao, L. Lin: Hermetic wafer bonding based on rapid thermal processing, Sens. Actuators A **91**, 398–402 (2001)
52.94 Y.T. Cheng, L. Lin, K. Najafi: Fabrication and hermeticity testing of a glass-silicon packaging formed using localized aluminium/silicon-to-glass bonding, IEEE/ASME J. Microelectromech. Syst. **10**, 392–399 (2001)
52.95 M. Chiao, L. Lin: Accelerated hermeticity testing of a glass-silicon package formed by rapid thermal processing aluminum-to-silicon nitride bonding, Sens. Actuators A Phys. **97/98**, 405–409 (2002)
52.96 M.K. Keshavan, G.A. Sargent, H. Conrad: Statistical analysis of the Hertzian fracture of pyrex glass using the Weibull distribution function, J. Mater. Sci. **15**, 839–844 (1980)
52.97 M. Chiao, L. Lin: A wafer-level vacuum packaging process by RTP aluminum-to-nitride bonding, Solid-State Sens. Actuator Microsyst. Workshop, Technical Digest (Transducers Research Foundation, Hilton Head Island 2002) pp. 81–85
52.98 F. Rosebury: *Handbook of Electron Tube and Vacuum Techniques* (Addison-Wesley, New York 1965)
52.99 Y.T. Cheng, W.T. Hsu, K. Najafi, C.T. Nguyen, L. Lin: Vacuum packaging technology using localized aluminium/silicon-to-glass bonding, J. Microelectromech. Syst. **11**, 556–565 (2002)
52.100 C. Luo, L. Lin: The application of nanosecond-pulsed laser welding technology in MEMS packaging with a shadow mask, Sens. Actuators A **97-98**, 398–404 (2002)
52.101 A. Cao, M. Chiao, L. Lin: Selective and localized wafer bonding using induction heating, Solid-State Sens. Actuators Workshop, Technical Digest (Transducers Research Foundation, Hilton Head Island 2002) pp. 153–156
52.102 J.B. Kim, M. Chiao, L. Lin: Ultrasonic bonding of In/Au and Al/Al for hermetic sealing of MEMS packaging, Proc. IEEE Micro Electro Mech. Syst. Conf. (IEEE, Las Vegas 2002) pp. 415–418
52.103 Y.T. Cheng, L. Lin, K. Najafi: Localized bonding with PSG or indium solder as intermediate layer, 12th Int. Conf. MEMS (1999) pp. 285–289

52.104 J. Voldman, M.L. Gray, M.A. Schmidt: Microfabrication in biology and medicine, Annu. Rev. Biomed. Eng. **1**, 401425 (1999)
52.105 H. Becker, L.E. Locascio: Polymer microfluidic devices, Talanta **56**, 267287 (2002)
52.106 Y.-C. Su, L. Lin: Localized bonding processes for assembly and packaging of polymeric MEMS, IEEE Trans. Adv. Packag. **28**, 635–642 (2005)
52.107 G. Thyrum, E. Cruse: Heat pipe simulation, a simplified technique for modeling heat pipe assisted heat sinks, Adv. Packag. **115**, 23–27 (2001)
52.108 G.P. Peterson, A.B. Duncan, M.H. Weichold: Experimental investigation of micro heat pipes fabricated in silicon wafers, J. Heat Transfer **115**, 751–756 (1993)
52.109 L. Jiang, M. Wong, Y. Zohar: Forced convection boiling in a microchannel heat sink, IEEE/ASME J. Microelectromech. Syst. **10**, 80–87 (2001)
52.110 F. Arias, S.R.J. Oliver, B. Xu, E. Holmlin, G.M. Whitesides: Fabrication of metallic exchangers using sacrificial polymer mandrils, IEEE/ASME J. Microelectromech. Syst. **10**, 107–112 (2001)
52.111 SAS IP: *ANSYS Modeling and Meshing Guide*, 3rd edn. (SAS IP, Canousburg 2002)
52.112 D. Bystrom, L. Lin: Residual stress analysis of silicon-aluminum-glass bonding processes, ASME Int. Mech. Eng. Congr. Expo. MEMS Symp, New Orleans (ASME, New Orleans 2002)
52.113 H. Scholze: *Glass: Nature, Structure, and Properties*, 1st edn. (Springer, Berlin Heidelberg 1991)
52.114 R.R. Tummala: *Fundamentals of Microsystems Packaging* (McGraw-Hill, New York 2001)
52.115 S.J. Adamson: BGA, CSP, and flip chip, Adv. Packag. **11**, 21–24 (2002)
52.116 JESD22-A102C: *Accelerated Moisture Resistance-Unbiased Autoclave* (JEDEC Solid State Technology Association, Arlington 2000)
52.117 M. Dokmeci, K. Najafi: A high-sensitivity polyimide capacitive relative humidity sensor for monitoring anodically bonded hermetic micropackages, IEEE/ASME J. Microelectromech. Syst. **10**(2), 197–204 (2001)
52.118 E.E. Lewis: *Introduction to Reliability Engineering*, 2nd edn. (Wiley, New York 1996)
52.119 W.D. Brown: *Advanced Electronic Packaging* (IEEE, New York 1999)
52.120 F. Niklaus, P. Znoksson, E. Käluesten, G. Stemme: Void free full wafer adhesion bonding, Proc. IEEE Micro Electro Mech. Syst. Conf. (IEEE, Miyazaki 2000) pp. 241–252
52.121 C.H. Cheng, A.S. Ergun, B.T. Khuri-Yakub: Electrical through-wafer interconnects with sub-picofarad parasitic capacitance, IEEE ECTC (2002) pp. 18–21
52.122 D. Routkevitch, A.A. Tager, J. Haruyama, D. Almawlawi, M. Moskovits, J.M. Xu: Nonlithographic nano-array arrays: fabrication, physics, and device applications, IEEE Trans. Electron. Dev. **43**, 1646–1658 (1996)
52.123 J. Gou, M. Lundstrom, S. Datta: Performance projections for ballistic carbon nanotube fieldeffect transistors, Appl. Phys. Lett. **80**, 3192–3194 (2002)
52.124 S.U.S. Choi, Z.G. Zhang, W. Yu, F.E. Lockwood, E.A. Grulke: Anomalous thermal conductivity enhancement in nanotube suspensions, Appl. Phys. Lett. **79**, 2252–2254 (2001)
52.125 K. Velikov, A. Moroz, A. Blaaderen: Photonic crystals of core-shell colloidal particles, Appl. Phys. Lett. **80**, 49–51 (2002)

Part I

Technolo

Part I Technological Convergence and Governing Nanotechnology

53 Governing Nanotechnology: Social, Ethical and Human Issues
William Sims Bainbridge, Arlington, USA

53. Governing Nanotechnology: Social, Ethical and Human Issues

William Sims Bainbridge

This chapter is a human-centered survey of nanotechnology's broader implications, reporting on the early phase of work by social scientists, philosophers, and other scholars. It begins with the social science agenda developed by governments, and the heritage of research on technology and organizations that social science brings to this mission. It then outlines current thinking about nanotechnology's economic impacts, health or environmental impacts, and social contributions. It discusses how technology can be regulated by a combination of informal ethics and formal law, then concludes by considering the shape of popular nanotechnology culture, as reflected in science fiction, public perceptions, and education.

53.1 **Social Science Background** 1867
53.1.1 The Scope of Societal Implications Research 1867
53.1.2 Technological Determinism Theory 1869
53.1.3 Organization Theory 1870
53.2 **Human Impacts of Nanotechnology** 1871
53.2.1 Economic Impacts 1872
53.2.2 Health and Environment 1872
53.2.3 Social Scenarios 1873
53.3 **Regulating Nanotechnology** 1874
53.3.1 Ethics 1874
53.3.2 Law and Governance 1875
53.4 **The Cultural Context for Nanotechnology** 1876
53.4.1 Science Fiction 1876
53.4.2 Public Perceptions 1878
53.4.3 Education 1879
53.5 **Conclusions** 1879
References 1880

53.1 Social Science Background

The role that social sciences will play in the future management of nanotechnology will depend both upon the tasks that society wishes to assign to it and upon the histories of social science itself, which have shaped their current scopes, theories, and research methodologies. Very recent governmental and professional publications suggest the tasks that social science will be asked to undertake.

53.1.1 The Scope of Societal Implications Research

In the United States, the President's budget request for the fiscal year 2006 called for significant efforts to understand the *practical implications and cultural context of nanotechnology research and development* [53.1, p. 28], listing three subtopics:

1. Research directed at environmental, health, and safety (EHS) impacts of nanotechnology development, and risk assessment of such impacts
2. Education-related activities such as development of materials for schools, undergraduate programs, technical training, and public outreach
3. Research directed at identifying and quantifying the broad implications of nanotechnology for society, including social, economic, workforce, educational, ethical, and legal implications.

Social and behavioral science research is potentially relevant to all three of these topics, but primarily for the third. Much of the research concerning EHS impacts will, of necessity, involve chemical and biological studies of the toxicity of nanostructured materials, and research on how such materials may travel in the natural environment, for example in groundwater. However, health and environmental protection are also socioeconomic issues, involving studies of how human behavior affects the impact that such materials may have over the lifecycle of manufactured products, from production to ultimate disposal.

The nanoscience curricula developed for schools and colleges must, of course, be based on the findings and theories of physics, chemistry, materials science, and nanoscale biology. However, education will be more effective to the extent that it is also designed on the basis of a correct understanding of human cognition and the learning process. Phenomena distinctive to the nanoscale are remote from people's daily experience, and thus it will be important to develop effective means to help them to visualize these processes in a way that is not only accurate but also intelligible to students at a range of levels of sophistication.

The list in the third subtopic – social, economic, workforce, educational, ethical, legal – is a set of research areas, but it also implies a roster of scientific and scholarly approaches. The social topics naturally belong to the traditional social sciences: sociology, political science, cultural anthropology, social psychology, and linguistics. Economic issues naturally belong to economics, with some input from other sciences, notably the psychology of decision-making and the sociology of organizations. The workforce topic is shared primarily by sociology and economics. Studies in education draw upon all of the fields, notably cognitive science to understand the learning process itself, and the social sciences to understand the functioning of schools.

There is some controversy over which fields of science and scholarship have the most to contribute to an understanding of ethics. Within philosophy, generally considered one of the humanities rather than one of the sciences, ethics is an established discipline, and some philosophers – called *ethicists* – specialize in analyzing real-world problems. The processes by which norms, customs, and ethical principles become established in society, and their variations across societies, are studied by sociologists, social psychologists, cultural anthropologists, and historians. Throughout human existence, religion has been the primary institution of society establishing ethical standards, but its role in modern society is hotly debated.

Some would say that modern society has replaced the informality of traditional ethics with the formality of law and government regulations. Legal implications of nanotechnology are the domain of political science, sociolegal studies, jurisprudence, and criminology. However, government does not monopolize decisions about right and wrong, so the field of ethics remains largely autonomous, rather than being subsumed by law. The professional ethics of nanoscientists and nanotechnologists is important in its own right, as are ethical standards developed by corporations and business groups interacting with the general public.

Another starting point is the categorization of fields suggested by the organization of research programs at the National Science Foundation (NSF), which has taken a leading role in the US National Nanotechnology Initiative. The chief home for social research is NSF's Directorate for Social, Behavioral and Economic Research, which contains two divisions whose missions are to support university-based scientific research: the Division of Behavioral and Cognitive Sciences, and the Division of Social and Economic Sciences. The Directorate for Engineering, which has played a leading role in the NNI, supports some socioeconomic research on industrial management. The Directorate for Computer and Information Science and Engineering supports research on social implications of information technology, which will increasingly be affected by nanotechnology in such areas as sensors, mobile computing, and high-density data storage. The Directorate for Education and Human Resources has a responsibility for research as well as curriculum development and specific educational programs across science, engineering, and mathematics. One difference between the NSF and its counterparts in other countries is that the humanities are not included, except for a little support for the history and philosophy of science.

A European report, *The Social and Economic Challenges of Nanotechnology*, issued by the Economic and Social Research Council in the UK, suggests a rather different three-part research agenda for the social sciences [53.2]:

1. The governance of technological change
2. Social learning and the evaluation of risk and opportunity under uncertainty
3. The role of new technology in ameliorating or accentuating inequity and economic divides.

The first item suggests a more aggressive, centralized role of government in setting research priorities, imposing regulations (concerning health, safety and intellectual property), and supporting education than generally envisioned in the US. The second item suggests the challenges faced by corporations and individuals as they learn from experience about the costs and benefits of various technical possibilities. The third suggests the larger political and economic context that mediates between the inventions and their ultimate human impact.

The following pages will consider such issues from the perspective of social science, especially informed by a pair of major conferences on the societal implications of nanoscience and nanotechnology that the author helped organize and edit. The first major gathering to examine the societal implications of nanoscience and nanotechnology was held at NSF, from the 28th to the 29th of September 2000. At that time, the social sciences had not yet been mobilized to address nano issues, so the emphasis was on a somewhat pragmatic level: how nanotechnology could contribute specific innovations to a range of applications meeting human needs. However, the resulting book-length report – which we can call *Societal Implications I* – was able to identify a large number of social science hypotheses and the beginnings of a rich research agenda.

Over the following five years, a large number of smaller conferences, publications, and individual efforts have helped build a new field dedicated to the study of the societal implications of nano, but the most significant single event was a second NSF conference held from the 3rd to the 5th of December 2003. For the sake of concision, we can call its report *Societal Implications II*. By the time it was prepared, some real social science research had been performed on the topic, a variety of new theory-based hypotheses had been framed, and data collection activities had been launched. To provide a context for understanding recent developments, it is necessary to survey at least a small part of the historical background that formed the consciousness of contemporary social scientists.

53.1.2 Technological Determinism Theory

Social science has a very long and fruitful history of research into the societal implications of technology, and it has also addressed the social wellsprings of ethics and legal institutions. This background will be useful when addressing new issues concerning nanotechnology, but it is essential to recognize that a diversity of viewpoints have survived a century of debate, so a consensus should not be expected soon, if ever. Nonetheless, many insights, theories, research methodologies, and historical analogies have been developed that can be bought to bear on nanotechnology, as we attempt to understand and deal with its implications for human society.

The classic sociological theory about the role of technology in human history is *technological determinism*, the view that technological innovation is essentially self-causing and largely determines social trends. Although this perspective is certainly not fully correct, it still has much to teach us and is a benchmark against which other theories are measured. The best statement of technological determinism is probably still the one that *Ogburn* expressed the better part of a century ago in his 1922 textbook *Social Change* [53.3]. For Ogburn, there were four primary steps in the process of development:

1. Invention
2. Accumulation
3. Diffusion
4. Adjustment.

Invention is the process by which new technologies are created. Many people have the romantic notion that inventions are the products of unusually creative individuals, but Ogburn said instead that it was a largely impersonal process. He noted that a very large fraction of all worthwhile inventions are developed almost simultaneously by two or more different people, and suggested that inventions were the natural result of the particular level of science and technology at that moment. If all of the pieces exist, somebody is bound to put them together. For example, the telegraph is a combination of existing elements: electricity, coils, batteries, signaling, and alphabet codes. Once all of these were in place, several individuals simultaneously invented variants of the telegraph, and it joined the existing cultural base that would make still further developments possible, like the telephone, the Internet, and the World Wide Web. There is no single inventor for any major branch of nanotechnology.

Accumulation is the growth of technical culture, as new inventions are conceived at a faster rate than old ones are forgotten. Today, humans invest great effort into preserving the technical expertise required to develop and apply inventions, especially through a myriad of educational programs in such fields as engineering, computer science, mathematics, and the natural sciences. In ancient times, this was accomplished through apprenticeship relationships, which were one of the most significant societal institutions of the Renaissance. Today, information about nanotechnology accumulates in specialist journals, and is also communicated at conferences and within major research centers.

Diffusion is the transmission of an invention from one context to another. This can occur from one field of activity to another, from one geographic location to another, or even across cultural boundaries. Another, related term is *technology transfer*, which sometimes has the connotation of applying a scientific discovery to a practical problem or commercializing it, or it may mean applying an engineering approach from one industry to another. Much diffusion takes place when people move, for example when a student trained in some branch of nanoscience takes a job in industry.

Adjustment is the process by which the nontechnical aspects of a culture respond to invention. For Ogburn, this was the one step where social factors really mattered, and he felt that this adjustment was often carried out by social movements. Today, government-sponsored activities like the *Societal Implications* conferences anticipate nanotechnology's impacts, in order to prepare a swift and hopefully painless adjustment.

Like many other American sociologists of the middle of the twentieth century, but hardly any 21st century sociologists, Ogburn was a *functionalist*. He believed that societies naturally developed harmonious cultures, in which the major institutions (including beliefs and values as well as formal organizations) were well-adapted to each other and to surrounding conditions such as the economy or the physical environment. Occasionally, a new invention would be so powerful that it would upset this equilibrium, and there might be an uncomfortable period of *cultural lag* before the balance was reestablished. Foresight minimizes cultural lag in the case of nanotechnology.

Ogburn's four-step model is remarkably clear and compelling, but a few moments of thought and the research of eight decades suggest modifications. First of all, there would logically be feedback loops between some of the four steps. The accumulation of inventions makes more inventions possible, because many new ideas are fresh combinations of old ones. Likewise, diffusion promotes more innovation, because it brings together many ideas that may join to form new ideas. Furthermore, some inventions promote accumulation and diffusion – such as the development of writing in ancient times and the creation of the Internet today – which then, in turn, stimulate more invention. Taken together, these ideas suggest that under favorable circumstances, the innovation process can be self-reinforcing, leading to exponential rates of change.

However, Ogburn's notion that invention is self-generating is questionable. The sociologist *Merton* argued that different areas of endeavor, including fields of science, receive greater or lesser attention as society's values change, thus increasing or decreasing the rate of invention in a particular field [53.4]. Several other sociologists argued that the variable structure of social relationships in society retards or advances the diffusion that can enable further invention [53.5–7]. Not surprisingly, economists like *Schmookler* argued that invention is an economic activity which will vary depending upon how much the market invests in it [53.8]. Using the specific example of the space program, *Bainbridge* suggested that some major technological developments are the results of social movements that push their goals despite the indifference of most scientists or investors [53.9], while *Freiberger* and *Swaine* described the birth of the personal computer in similar terms [53.10]. Perhaps nanotechnology can be understood as a social movement, too [53.11].

At the present time, nanotechnology seems to be riding the crest of several enthusiasm waves at once. It integrates innovations from several traditional fields (chemistry, physics, materials science), bringing together ideas and methods that can be combined in various ways to produce many new inventions. Research is going on worldwide, with rapid diffusion of innovations across national boundaries and technology transfer from one area to another. Nanotechnology concepts have started to affect the wider culture, and heavy investment indicates that the field has significant value.

It is clear that Ogburn's model must be modified to take account of the roles that markets, government agencies, social movements, and the wider culture will play in accelerating or decelerating progress in particular areas, by investing in research or discouraging innovations of particular kinds through regulation. This observation raises questions regarding both policy and research, specifically directed at the twenty-first century's premier technological revolution, nanotechnology.

53.1.3 Organization Theory

Among all the many well-established social science approaches relevant to societal implications of technology, one deserves to be mentioned here: the study of formal organizations. Nanoscience and nanotechnology are being developed within organizations, notably universities and corporations. Today, we understand that an organization is a complex system of social roles (notably leadership and labor), commitments (including formal rules), positions (statuses, jobs), and information (not only technical knowledge but also economic data and information about customers and competitors). In an extremely influential 1981 textbook, *Scott* [53.12] described three different *system* perspectives on organizations that had guided both scholars and managers alike, that emerged in the following order over the preceding decades: *rational*, *natural*, and *open*.

The *rational system* perspective assumes that corporations and other formal organizations are dedicated to achieving explicit goals and are formally structured in a manner that should help the organization achieve its goals efficiently and effectively. Among the most important principles of rational management are the following:

1. All of the organization's personnel should be linked into a single hierarchy of authority in the form of a pyramid.
2. Each person should receive orders from only one superior.
3. No superior should have more subordinates than he or she can effectively control.
4. All routine business should be handled by subordinates according to a set of well-established rules, and superiors should concentrate on dealing with exceptions that fall outside the routine.

Another way of putting this is to say that top management makes decisions, perhaps based on information passed up from below, and it issues commands about the work to be done. Many managers find this a very comfortable way of thinking about their organization, and it presents a relatively simple context for setting and following policies concerning nanotechnology. Management decides which nanotechnologies to invest in, which lines of inquiry their nanoscientists are to explore, and which safeguards should be in place to ensure an adequate level of safety, health, and environmental protection. By the middle of the twentieth century, however, social scientists had discovered that this *idealized* picture of organizations did not fit reality well.

The *natural system* perspective argued that organizations are collections of people who are affected by formal goals and rules but not entirely dominated by them. These people have their own goals and needs, which must be met reasonably well if the organization is to function. Indeed, the fundamental goal of the organization and its parts is simply survival, and specific goals set by management are only secondary. To understand an organization, one should study the informal relationships among the individual people, and the formal organizational chart may be quite misleading. Attempting to impose greater rationalization may be dysfunctional, and a better way to manage is to build leadership at all levels, motivate workers, and strengthen human relationships. This suggests that an understanding of social and ethical issues concerning nanotechnology must be widely shared by people throughout the organization, rather than merely possessed by top management.

The *open system* perspective challenged both of the earlier perspectives for wrongly imagining that organizations are stable structures with consistent long-term goals and programs. Instead, it says, organizations are shifting coalitions of interest groups, strongly influenced by external factors that also shift unpredictably. To use the terminology of the early twenty-first century, organizations are *complex systems* that may or may not be *adaptive* and that are susceptible to *chaotic behavior*. A open system organization does not chart a fixed course toward a solidly established goal, but employs feedback to adjust course constantly toward moving goals. In such a variegated and unstable context, there is no single best organizational form or management style, and different methods work better in different environments.

To the extent that the open system model correctly describes organizations in the modern world, it may be impossible to establish firm policies about how to exploit nanotechnologies for maximum human benefit, but it will be necessary to constantly reexamine situations and respond flexibly. However, all three perspectives probably describe modern organizations to some extent. Thus the best overall strategy will be for management to establish firm goals and policies to the extent that this is possible, invite all members of the organization to participate in the social and ethical debates, and to follow flexible strategies guided by constant feedback from both inside and outside the organization.

53.2 Human Impacts of Nanotechnology

Researchers in a number of fields have begun to monitor the economic, environmental, and social effects of nanotechnology. Fundamental to this effort are data regarding how extensively and in what areas nanoscience is being pursued and nanotechnology is being developed. For example, a team at the University of California, Los Angeles, combining expertise in both economics and sociology, is creating

NanoBank [53.13]. This will be a massive database for use both by specialists and by members of the public, beginning with information about nano research publications, patents, industries, and eventually social impacts. Meanwhile, a number of participants in a growing community of scientists have begun to theorize and collect data about the economic, health, and social impacts.

53.2.1 Economic Impacts

Most participants in the *Societal Implications* conferences were confident that nanotechnology could enable a wide range of economically profitable applications progressively over the coming decades as techniques are devised and investments are made in the infrastructure required for research, development, and production [53.14–18]. At first, nanotechnology would enable improvements in existing industries. Radical innovations that launch entirely new industries would be likely further in the future.

Among the contributors to *Societal Implications I* was economist Irwin Feller, who was at the time a professor at Pennsylvania State University and chairman of the advisory committee for the NSF Directorate for Social, Behavioral and Economic Sciences. He argued that nanotechnology, in all its many forms, is likely to enter the marketplace first in applications where performance is especially important and people are willing to pay somewhat higher prices for incremental improvements [53.19]. An interesting example is the announced use of carbon nanotubes in the heads of golf clubs manufactured by the Wilson Golf company; here they are used to achieve the most effective balance by separating strength from weight, supposedly allowing golfers to hit balls farther. Golf and other sports greatly reward small improvements in performance. Other examples of fields where there is a premium on performance include medicine, aviation and space technology, as well as military technology.

Freeman, of Harvard University, has offered a set of three predictions about the economic impact of nanotechnology, based on an analysis of its nature as well as on economic theory [53.20]. First, the chief beneficiary will be labor, in the form of increased wages for workers. Second, nano will gradually increase labor demand and thus wages across the entire economy, rather than only in a few high-tech sectors. Third, the effect on the economy will be smooth, rather than causing sudden changes in production and employment. In great measure, these conclusions rest on the observation that the forms of nanotechnology we can envision over the next decade or two include a very large number and variety of developments that can reinvigorate many different industries to some degree, rather than a single revolutionary innovation. We should note, however, that all analyses and predictions at this early point are uncertain, and nothing can substitute for careful, empirical research.

Among the questions that need to be studied concerning nanotechnology's economic impact, the following stand out:

1. Can the technologically most advanced nations capture the profits from their own innovations, or will nanotechnology chiefly enrich the front rank of developing nations (notably China) that may be able to manufacture quality products at lower cost?
2. Can nanotechnology reassert the importance of manufacturing industries (over service and information industries), or will the declining relative significance of manufacturing reduce nano's potential economic impact?
3. Will the economic benefits of nanotechnology be shared widely, or will they primarily accrue to owners and investors of large corporations?
4. Will the great cost of retooling prevent the introduction of potentially important nanotechnology innovations, such as the replacement of current *silicon chip* microelectronics by molecular electronics based (for example) on carbon nanotubes?

53.2.2 Health and Environment

Beginning in 2003, evidence began to be published that some nanomaterials could be toxic or otherwise harmful to health under some circumstances [53.21–24]. This would not be particularly surprising, of course, because many materials can be harmful, depending upon the dosage and the organism's ability to clean small quantities out of its system. Risk, in nanotechnology as in other fields, is a factor of both the harm an event might cause and the probability that the event would actually occur [53.25]. A toxic substance that is safely locked away in a durable container, such as perhaps carbon nanotubes inside a composite material, is not harmful. Also crucial to an understanding of these effects is research into how the substance is transported in the natural environment, for example in the atmosphere [53.26]. Thus, knowledge about the potential danger to human health of a nanostructured material is a good first step toward taking the precautions necessary to ensure that it has no opportunity to cause that harm [53.27]. Organizations with a responsibility in this area, such as the US Environmental Protection Agency, have begun exploring such issues [53.28, 29].

Already in *Societal Implications I*, Lester Lave noted that it was important to consider the costs and benefits of a particular kind of nanotechnology in terms of its entire lifecycle – from manufacture, through performance in use, to disposal – and in terms of the entire technical system of which it is a part. For example, the use of carbon nanotubes in parts of vehicles may entail some health hazards in the workplace before they are safely embedded in composite materials, but it could reduce health hazards in use by reducing the weight of the vehicle, thereby also reducing fuel use and the associated air pollution. The savings and other benefits in use could pay for the extra costs of manufacture, and some costs involved in safe disposal or recycling once the vehicle was worn out. Even with all the costs figured in, *Lave* argued that the net benefit of nanotechnology could be enormous [53.30]:

> *It promises to reduce by orders of magnitudes the inputs of energy and materials and associated environmental discharges required to produce a device that can perform a particular task. The result could be perhaps an order of magnitude increase in real income for the current world population without requiring more energy, materials, or resulting in additional discharges. Thus, nanotechnology offers the prospect of giving poor nations much higher standards of living and making the world economy sustainable.*

Nanotechnology can reduce the use of energy and other resources, thus simultaneously promoting conservation of these resources while reducing the pollution that is caused when they are used. The economic gains achieved by nano can not only be used partly to mitigate or prevent any harmful health or environmental consequences of the technology itself, but also can be invested directly into improving physical quality of life. The other side of this argument is that economic growth is likely to be very costly to the environment if industry is unable to employ nanotechnology. Naturally, such claims are open to challenge, and the only way to be certain is to carry out extensive, rigorous research, not only into the possible toxicity of some nanomaterials, but also on the sociotechnical systems that may employ them, to identify dangerous points in the lifecycle where precautions would be advisable.

53.2.3 Social Scenarios

Participants in the *Societal Implications II* task force on the *quality of life* predicted that nanotechnology could contribute significantly with respect to food, water, energy, and the environment [53.31]:

- *Food*: improving inventory storage; growing a diversity of high-yield crops
- *Water*: low-energy purification and desalination; reducing water waste in manufacturing and farming
- *Energy*: reduced dependence on fossil fuels; solar photovoltaic energy production; renewable energy systems; energy distribution with hydrogen
- *Environment*: remediating waste and pollution; permitting systems and materials that use resources most efficiently; recycling pollution into raw materials; ensuring safety and sustainability of new materials.

Another task force, with the responsibility to frame alternative social scenarios, pointed out that we do not really know how adequate our current institutions are to manage rapid technological change in a graceful and maximally beneficial manner. Indeed, the open systems model of organizations, discussed above, suggests that institutions will need to change in order to adapt to the new technological possibilities. Given our lack of knowledge about how well our institutions will perform over the short run, we cannot be sure about which of two very different scenarios will actually happen:

1. The transition from current technologies to much more capable nano-enabled technologies will be smooth and benign, bringing benefits quickly to the majority of people, with relatively few unwanted consequences
2. The transition will be rough, with many bad decisions that either fail to take advantage of good technologies or apply hazardous technologies unthinkingly, causing many conflicts among societal institutions and triggering an anti-nano backlash.

With this second, chilling scenario in mind, the group recommended research into how our current institutions function when dealing with questions about new technologies, with the aim to use the results of such studies to redesign faulty institutions. One starting point would be an inventory of existing institutions, evaluating how they cope with uncertainty, change, and conflict.

Looking further into the future, *Hanson* [53.32] has argued that nanoscience may ultimately transform the structure of manufacturing industries in a series of five steps:

1. Methods are successfully developed for efficient manufacturing atom-by-atom.
2. General-purpose factories are established, using methods analogous to three-dimensional printing (stereolithography) to make a great variety of products from common raw and recycled materials, following software instructions.
3. A large number of small factories of this kind are built everywhere, near their customers, dominating manufacturing to the exclusion of today's large corporations and huge centralized factories.
4. These local, generalized factories are usually idle, because they are cheap and efficient.
5. These small factories can copy themselves.

Note that this is not the infamous *gray goo* scenario, because the factories copy themselves only when human labor is invested to run off and assemble the necessary parts for a new factory. The net result would be not only increased prosperity but also a further reduction of the economic significance of manufacturing, and the same might be the case for the transportation industry and many kinds of raw material extraction and production as well. This would further increase the relative importance of service and information industries, which themselves may be local (in the case of services) or highly distributed (in the case of information). If all this occurred, as *Tonn* [53.33] has pointed out, local communities would become much more important socially, and large governmental units such as nations could weaken.

53.3 Regulating Nanotechnology

Given the alternative models of formal organizations developed by social scientists, deciding how to *govern* nanotechnology could present problems. In the rational system model, we need only decide which policies to follow, and organizations will carry them out. However, that model is faulty, and both the natural system and open system models predict that control will never be perfect. Thus, it will be important to instill a proper sense of ethics into engineers and decision-makers. The challenges for law and government are only beginning, but will require much work in the coming decades.

53.3.1 Ethics

A long-standing debate in philosophy concerns the extent to which ethical rules can be objectively defined, versus being judgments made by human beings from their particular standpoints and interests [53.34–36]. A modern view is that morality is negotiated between people, in order to enhance cooperation and to serve both their mutual interests and their enlightened self-interests [53.37, 38]. At a joint US–European workshop on the societal implications of nanotechnology, held in 2002, *Mosterin* noted that ethical standards are constantly changing as society and our technical capabilities change [53.39]. Thus, we cannot expect to settle nanorelated ethical issues today, once and for all. Instead, each generation will need to engage these issues afresh.

Such observations cast doubt on the role of the philosophy of ethics itself, as at least three participants in *Societal Implications II* noted. Writing on *The Ethics of Ethics*, *Von Ehr* argued that professional ethicists who comment on nanotechnology may merely be expressing their personal, political opinions as if they were the result of objective philosophical analysis [53.40]. Similarly, *McGinn* doubted that ethicists were sufficiently knowledgeable about nanotechnology to be able to apply philosophical analysis to it, and *Trumpbour* speculated that whatever one wants to do with nano, it would be possible to find some ethicist or other who would endorse it [53.41, 42].

To see our way around this impasse, we need to recall the origins of philosophy in ancient Greece. Socrates did not pretend to know the exact truth himself, and with respect to ethics he ultimately deferred to his society when he drank that cup of hemlock that ended his own life. The role of the philosopher is not to proclaim truth but to help other people seek it, perhaps through that method of exploratory discussion that even today is called *Socratic dialog*. This is what philosopher *Weil* did when she contributed to both *Societal Implications* conferences [53.43, 44]. In a similar spirit, in *Societal Implications II*, philosopher *Berne* explored the meanings of the concept of quality of life in order to better understand how nanotechnology might affect it [53.45].

53.3.2 Law and Governance

Relatively little research has been carried out up to this point regarding legal issues that might distinctively concern nanotechnology, but one area of fairly active investigation is patents. The US Patent and Trademark office has established patent class 977 for nanotechnology [53.46]:

> *related to research and technology development at the atomic, molecular or macromolecular levels, in the length of scale of approximately* 1–100 nm *range in at least one dimension, and that provides a fundamental understanding of phenomena and materials at the nanoscale and to create and use structures, devices, and systems that have novel properties and functions because of their small and/or intermediate size.*

One line of research charts the rising significance of nanotechnology in different industries and nations, by means of patent counts [53.47, 48]. *Taylor* notes that a primary purpose of patents is to encourage innovation by ensuring that inventors and the companies that support them will be able to profit, but he also points out that a poorly designed patent system can have the opposite effect of stifling innovation in nanotechnology [53.49]. Among the ways that this could happen is that nanotechnology patents could limit developments in nanoscience, unless special exemptions are made for research uses of patented techniques. While this issue affects many areas of science and technology, its is especially acute here because of the difficulty involved with drawing a conceptual line between nanotechnology and nanoscience.

The attorney *Miller* has suggested that nanotechnology might have implications for a number of legal specialties [53.50]. Criminal law might be affected through nanoscale forensics, which already employs DNA analysis and may in future exploit a very wide range of kinds of evidence collected or analyzed at the nanoscale, including surveillance information from nano-enabled microscale sensors. Health and environmental law might be complicated by the difficulty of defining nanostructured substances, the challenge of measuring their health effects, and the possibility that it could be hard to track the origins of uncontrolled releases of nanoparticles into the environment. In general, Miller argues that scientific expertise is going to play a wider role in legal cases, and she points out that it will be essential to arrive at consistent definitions of nanorelated terms when writing legislation and applying the law.

Clearly, some objectively hazardous nanostructured materials will need to be covered by government regulations, and thus they will need to be unambiguously defined so that their presence or absence can be established for any particular case. There is a danger if vague and overly broad definitions of nanotechnologies are enshrined in laws that inhibit the development and use of safe and beneficial applications. Given the uncertainties and ambiguities in this area of rapid scientific and engineering progress, it may not be feasible or desirable to establish a comprehensive set of formal regulations. The alternative is to embed appropriate values into the organizations that are doing the research, development and deployment of new nanotechnologies [53.51]. That is, the organizations themselves should be dedicated to using the technology for human benefit – which after all is what their customers want – and naturally include effective review procedures and safeguards in their standard operating procedures.

Undoubtedly, in the future many nanostructured materials will be precisely defined and covered by government health and environmental regulations, but the potential societal impacts of nanotechnology go very far beyond adding a few exotic materials to the very long list of substances that can be implicated in pollution. Thus, a very important role for government will be supporting research into the wider implications, research that is often called *science and technology studies* (STS) [53.52]. The very real but secondary effects include the way that government funding initiatives such as the National Nanotechnology Initiative influence the nature of modern universities [53.53].

One area of intense government-supported nanotechnology activity is the development of military and national security applications, such as lightweight armor, sensors, battlefield medical equipment, and a host of other possibilities [53.54–56]. Reasonable people disagree about who has the responsibility to make decisions about developing technologies that are intended to harm human beings, or to protect fighters who harm other human beings. If one has a great deal of confidence in the political system of one's nation, then it is plausible to assert that the government itself has the effective moral responsibility to make proper decisions, and individuals should not be expected to second-guess these decisions. It can also be argued, however, that individuals can never abdicate their own moral responsibilities to their government, in which case nanoscientists, nanotechnologists, and those who support them must inform themselves and consider carefully the likely uses of their innovations.

If a government does not deserve its citizens' confidence, then scientists and engineers must consider carefully before developing technologies with military applications. In addition, they should pay attention to the international context. For example, a dual-use nanotechnology may be developed by an ethically responsible nation, then diffuse to an irresponsible one.

International discussions about the societal implications of nanotechnology have begun, notably at a conference held in Arlington Virginia, from the 17 to 18 June 2004, where representatives of 25 nations contributed [53.57]. At the very least, international communication will help all of the nations inform themselves about issues and the alternative responses to them. It would seem likely that international definitions of terms will be developed. In a linguistic sense, nanotechnology was born in 1960, when the General Conference on Weights and Measures included the word *nanometer* in the internationally recognized definition of the metric system. Perhaps regulations governing international trade of some potentially hazardous nanostructured materials might be developed, and one could also imagine international partnerships to help develop especially beneficial nanotechnologies and distribute their benefits most widely around the globe.

53.4 The Cultural Context for Nanotechnology

For better or worse, our society's image of nanotechnology has largely been defined by fantasy stories, rather than by scientific research. However, the mass media are beginning to familiarize the public with the concept, and social scientists are beginning to explore public understanding and opinion about the topic. This is a crucial moment in history, when educational institutions need to examine how to incorporate nanoscience into curricula, and how to begin to produce the technically trained workforce that society will need.

53.4.1 Science Fiction

An old if minor tradition exists in science fiction where tales are set in very tiny environments. In 1919, the pulp magazine *All-Story Weekly* carried *The Girl in the Golden Atom* by Ray Cummings. Later expanded into a novel, this story concerned a scientist who was able to shrink himself down far below the nanoscale to visit an electron, which was depicted like a planet circling an atomic nucleus as its sun. This adventure begins when the scientist focuses a powerful new microscope on the wedding ring of his mother, and sees an attractive female inhabitant of the atomic planet in great danger. He has the technology to zoom down and rescue her, but if his instruments ever lose track of that particular atom, he will never find it again.

Cummings was actually well-informed about science and technology, because he was an assistant of the great electrical genius, Thomas Alva Edison. In 1911, the physicist Ernest Rutherford had discovered that atoms consisted of a nucleus surrounded by electrons, and his *solar system* model of the atom received considerable publicity. He had used gold foil as the target of his experiments, which naturally suggested a gold wedding ring for Cummings' romance. Even in 1911, it was clear that atomic structure was very different from that of our solar system, however. For example, electron orbits are not limited roughly to a plane as is true for the planets, and there can be several identical electrons in the same orbit. In 1913, Niels Bohr suggested that electron orbits are possible only at certain energies, defined by the quantum theory that developed over the next few years, and the modern scientific picture of the atom is very different from a solar system. Importantly, electrons and other fundamental constituents of matter cannot function as worlds in which intelligent creatures could live and have adventures or romances, because below a certain size (called the Planck length) or energy (called the quantum) there can be no stable structures or entities [53.58].

This is not to say that Cummings should not have written his entertaining story. It probably encouraged readers who happened to be teenage boys to feel that atoms were interesting, and when the novel came out in 1923 its science was only a decade out of date. Later writers borrowed the idea that electrons could be the settings for stories [53.59]. In the critically acclaimed story *Surface Tension, James Blish* imagined aquatic people only a millimeter in height, whose cells would of necessity be nanoscale, struggling to break through the conservatism of their water-bound culture, for which surface tension is a nice metaphor [53.60]. Perhaps the best-known novel about adventures at the nanoscale is *Fantastic Voyage*, novelized by *Isaac Asimov* from a movie with the same title starring Raquel Welch and Stephen Boyd, in which a submarine and its crew are miniaturized and injected into the blood stream of a human being to destroy a dangerous blood clot in his brain [53.61].

Quite separately, two of the contributors to *Societal Implications II*, *Whitesides* and *Alpert*, quoted one of the fundamental principles of science fiction, Arthur C. Clarke's third law: *Any sufficiently advanced technology is indistinguishable from magic* [53.62, 63]. Alpert also quoted the variant offered by science fiction writer Gregory Benford: *Any technology distinguishable from magic is insufficiently advanced* [53.64, p. 5]. The implications are two-fold: 1) to the uninitiated, nanotechnology seems like magic; 2) at the nanoscale, it is hard for nontechnical people to tell the difference between fact and fiction. Clarke's first and second laws also reveal the fact that science fiction writers wish reality were more magical than real scientists find it to be:

1. When a distinguished but elderly scientist states that something is possible, he is almost certainly right. When he states that something is impossible, he is very probably wrong.
2. The only way of discovering the limits of the possible is to venture a little way past them into the impossible [53.65, 66].

The historian of science fiction *Sam Moskowitz* explains [53.67, p. 11]:

> *Science fiction is a branch of fantasy identifiable by the fact that its eases the 'willing suspension of disbelief' on the part of its readers by utilizing an atmosphere of scientific credibility for its imaginative speculation in physical science, space, time, social science and philosophy.*

Of concern to scientists and engineers is the possibility that the general public will either adopt an excessively cautious attitude toward nanotechnology, on the basis of false assumptions derived from horror fiction, or that they will fail to appreciate real developments because they are not so wondrous as the impossible ideas in science fiction.

Fully nine of the essays in the *Societal Implications II* report refer to *Michael Crichton*'s technophobic novel, *Prey*. Crichton's introduction warns [53.68, p. X]:

> *Sometime in the twenty-first century, our self-deluded recklessness will collide with our growing technological power. One area where this will occur is in the meeting point of nanotechnology, biotechnology, and computer technology. What all three have in common is the ability to release self-replicating entities into the environment.*

Practicing chemists and engineers in the relevant fields doubt that engineered nanoscale entities could become self-reproducing in the natural environment, unless they were genetically designed on the basis of the living things that have co-evolved with the Earth, but this is a standard assumption of science fiction about nanotech. Crichton's novel postulates that a monster is created by a corporate research project gone awry, consisting of biologically launched nanoparticles that evolve rapidly and develop a swarm intelligence. In the first of two climaxes, the monster kills people, and in the second, it takes control over their bodies.

Aficionados of science fiction literature note that Crichton's novel is similar to a 1983 story by *Greg Bear*, that was later expanded into the novel, *Blood Music*, describing the fantastic transformation of all humanity by intelligent nanobots [53.69, 70]. Among the most highly acclaimed science fiction stories about nanotechnology is Neal Stephenson's novel, *The Diamond Age*, that imagines nano in terms of computer science. In a premise framed like a sociological theory, *Stephenson* explains [53.71, p. 37]:

> *Now nanotechnology had made nearly anything possible, and so the cultural role in deciding what should be done with it had become far more important that imagining what could be done with it.*

In her nanotechnology quartet of novels, *Kathleen Ann Goonan* describes a world transformed by nanotechnology, but in contrast to Stephanson's computer science model, she imagines that future nanotechnology will be more like biological genetic engineering [53.72–75].

Our civilization will benefit from artistic creativity that takes its metaphors from nanotechnology. Science fiction can inspire young people to enter technical fields, and it can communicate to a wide audience the excitement that real scientists and engineers experience in their work. Thus, we should applaud the individual writers who have begun promulgating visionary images of the future of the field, and we cannot expect them to limit their imaginations to the proven nanoscale techniques of today. However, their stories are not an appropriate basis for deciding policy, and they may misinform the public about both investment opportunities and potential hazards. The unfortunate results could be nanocrazes and nanopanics.

Future historians will need to determine how much of the current interest from people who are neither nanoscientists nor nanotechnologists was inspired by science fiction, either directly or indirectly. One reason science fiction was especially ready to embrace nano was that it had suffered great disappointment about its primary earlier obsession, spaceflight. By the end of the 1970s, when manned spaceflight lagged after the end of the Apollo Program and space probes revealed what inhospitable places the other planets are, the plausibility of colonizing other worlds had declined significantly. Small organizations that sought to advance space colonization, like the L-5 Society, thus fell on hard times, and some members were ready for something new. Among them was Eric Drexler.

Drexler is undoubtedly very important in the history of popular nanoculture, and he would be an excellent subject for an in-depth biography. In the absence of deep study, we can note just a few facts. In the mid-1970s, Drexler was an assistant to *Gerard K. O'Neill*, an engineer who wanted to build a space city at the Lagrange-5 (L-5) point in the moon's orbit [53.76, 77]. Starting in 1977, *Drexler* published in the *L-5 News* some essays about the exploitation of extraterrestrial resources, a novel way of manufacturing space structures, and the legality of space development [53.78–80]. In 1987, the L-5 Society merged with the more conservative National Space Institute to form the National Space Society. Perhaps not entirely coincidentally, that was about the time Drexler published his visionary first book about nanotechnology, *Engines of Creation* [53.81], and the leading magazine of the science fiction field, *Analog*, first mentioned Drexler's nano ideas in 1987. Soon, that magazine was regularly publishing imaginative stories that included nanotechnology.

Thus, part of the significance of nanotechnology for popular culture is that it provides a set of metaphors to sustain a sense of wondrous possibility for the future, to some extent replacing the dream of spaceflight, even as it renders actual space travel more feasible technically and economically [53.82].

53.4.2 Public Perceptions

It is likely that only a small fraction of the general public in economically advanced nations still has much idea what nanotechnology is, although the mass media are beginning to popularize it [53.83, 84]. Certainly, readers of science fiction literature are a small minority of the general population, with a range of unusual attitudes toward science and technology [53.85], although they may informally share their views with many nonreaders. In 2002, more than half of the respondents to a Eurobarometer survey could not answer an attitudinal question about how nanotechnology will impact their way of life, which was far more nonresponsive than for any of the other technologies listed [53.86]. A British survey found that only 29% of the general public claimed to have heard of nanotechnology, and only 19% were willing to try to define it [53.87, p. 6].

It turns out to be rather difficult to frame simple questions that can evaluate how much the general public knows with any accuracy without eliciting quibbles from people who are very knowledgeable. Some questionnaire items have assumed nanotechnology deals only with things that are invisible because they are the size of atoms. However, surface coatings a few nanometers thick and bulk quantities of nanoparticles certainly can be visible to the unaided human eye, and atoms are actually less than a nanometer in diameter. To measure knowledge accurately, one would need a several-item quiz of somewhat sophisticated questions, and this would be very costly in the expensive surveys of random samples of the general population; and then one would predict that almost everyone would flunk the test.

A nonrandom-sample study of members of the general public likely to be both interested and slightly knowledgeable was *Survey2001*, sponsored by the National Geographic Society [53.88]. This was an online questionnaire with an opportunistic set of respondents, and thus the percentages cannot be extrapolated to the general public. However, National Geographic readers are constantly bombarded with information about environmental degradation, and should be alert to the possible negative impacts of nano. Thus, it is interesting to see that fully 57.5% of 3909 English-speaking respondents agreed with the following statement: *Human beings will benefit greatly from nanotechnology, which works at the molecular level atom by atom to build new structures, materials, and machines.* In contrast, only 9.0% agreed that: *Our most powerful 21st-century technologies – robotics, genetic engineering, and nanotechnology – are threatening to make humans an endangered species.*

Given the sampling limitations of this study, statistical correlations within the data are probably more reliable than response frequencies [53.89, 90]. The questionnaire included 16 environmental issues, and attitudes toward nano correlated strongly with only one of them: *the introduction of genetically modified food into our food supply.* People who were worried about so-called Frankenfoods also tended to be worried about nanotechnology, an indication that nanotechnology could become publicly stigmatized if it become associated in the public mind with controversial nanoscale biotechnologies. Also less affected by the sampling limitations was an open-ended question asking respondents to comment on the two statements that named nanotechnology. Interestingly, respondents gave very few negative answers, and none mentioned a concern about the hypothetical notion of self-reproducing nanoscale robots. In contrast, they were able to mention 82 distinct possible benefits of nanotechnology.

Michael D. Cobb and *Jane Macoubrie* carried out a random-sample telephone survey of 1536 Americans that asked them to rank five benefits and five risks of nanotechnology [53.91]. By far the most highly ranked benefit was *new ways to detect and treat human diseases,* while *new ways to clean the environment* came in second. Among the possible risks, *losing personal privacy* was rated most important – presumably through nanoenabled microelectronic sensors – whereas *uncontrollable spread of nanorobots* was in last place. Interestingly, the risk that scientists currently take most seriously, *breathing in nanoparticles that accumulate in the body* was ranked in the middle of the five risks. While surveys like this are important, at the present time they probably say more about people's general concerns, rather than anything about nanotechnology itself.

53.4.3 Education

Education must be a high priority in any government nanotechnology initiative [53.92]. We can concern ourselves with two very different kinds of nanotechnology education: 1) preparing students to become professional nanoscientists, nanotechnology engineers, or technicians in nanobased industries; 2) including fundamental concepts about nanoscale phenomena in a liberal education and in the informal education that citizens receive after graduation.

To address issues in both areas, educators have begun looking at how to incorporate nano into school curricula, even in fairly early grades [53.93]. One idea of how to do so is the creation of immersive, virtual reality environments to teach students at all levels about phenomena at the nanoscale, allowing them subjectively to amble among the atoms [53.94]. For nano professionals, training in mathematics will remain crucially important [53.95].

In the case of the United States, native-born citizens have recently shown a lack of interest in technical occupations, and reliance upon foreign nanotechnologists raises both security and sustainability concerns, especially as other countries upgrade both their educational institutions and career opportunities [53.96]. Discussions have begun about whether a distinctive curriculum and degree programs in nanoengineering need to be established, as opposed to continuing to draw upon graduates from many traditional fields [53.97, 98]. For future engineers and technicians to gain experience with the real materials associated with their future work, universities and corporations may need to set up nanofabrication facilities that are shared by both academic and industrial organizations in their particular geographic areas [53.99, 100]. Whatever form their technical education takes, nanoengineers also need to learn about the social and ethical implications of the work that they will be doing [53.101].

53.5 Conclusions

Scholars and social scientists in a variety of traditions are beginning to study the societal and social implications of nanotechnology, and to develop the knowledge base necessary for proper governance. A fundamental issue is how much we can rely upon formal, government regulation, versus informal guidance of the technology by the people involved in developing it. The latter may seem more difficult, but given the fallibility of formal organizations it may be more feasible.

The challenge is to enable humanity to take advantage of nanotechnology at the most rapid rate compatible with wise choices in those (hopefully few) areas where hazards or inequities could exist. Research will be needed to identify nanostructured materials that are actually hazardous, leading to narrowly drawn regulation of these substances within the existing regulatory context for other hazardous substances. Social-scientific research will also be essential, not only to monitor nanotechnology development and identify problems promptly so they can be solved, but also to explore new avenues where nanotechnology can beneficially serve humanity. Equally important will be the public dialog and careful analysis needed to establish ethical principles and the factual knowledge with which to implement them among nanoscientists, nanoengineers, and decision-makers.

References

53.1 1. Nanoscale Science, Engineering, and Technology Subcommittee, National Science and Technology Council: *The National Nanotechnology Initiative: Research and Development Leading to a Revolution in Technology and Industry* (National Nanotechnology Coordination Office, Arlington 2005)

53.2 S. Wood, R. Jones, A. Geldart: *The Social and Economic Challenges of Nanotechnology* (Economic and Social Research Council, Swindon 2003)

53.3 W.F. Ogburn: *Social Change* (Huebsch, New York 1922)

53.4 R.K. Merton: *Science, Technology and Society in Seventeenth-Century England* (Harper Row, New York 1970)

53.5 E. Katz, P.F. Lazarsfeld: *Personal Influence: The Part Played by People in the Flow of Mass Communications* (Free Press, Glencoe 1955)

53.6 D. Crane: *Invisible Colleges: Diffusion of Knowledge in Scientific Communities* (Univ. Chicago Press, Chicago 1972)

53.7 E.M. Rogers: *Diffusion of Innovations* (Free Press, New York 1995)

53.8 J. Schmookler: *Innovation and Economic Growth* (Harvard Univ. Press, Cambridge 1966)

53.9 W.S. Bainbridge: *The Spaceflight Revolution* (Wiley-Interscience, New York 1976)

53.10 P. Freiberger, M. Swaine: *Fire in the Valley: The Making of the Personal Computer* (McGraw-Hill, New York 2000)

53.11 T.A. Ten Eyck: Communication streams and nanotechnology: The (re)interpretation of a new technology. In: *Nanotechnology: Societal Implications – Individual Perspectives*, ed. by M.C. Roco, W.S. Bainbridge (Springer, Berlin, Heidelberg 2006) pp. 280–282

53.12 W.R. Scott: *Rational, Natural, and Open Systems* (Prentice-Hall, Englewood Cliffs 1981)

53.13 L.G. Zucker, M.R. Darby: Socio-economic impact of nanoscale science: Initial results and NanoBank. In: *Nanotechnology: Societal Implications – Individual Perspectives*, ed. by M.C. Roco, W.S. Bainbridge (Springer, Berlin, Heidelberg 2006) pp. 2–23

53.14 W.R. Boulton: Managing the nanotechnology revolution: Consider the Malcolm Baldrige national quality criteria. In: *Nanotechnology: Societal Implications – Individual Perspectives*, ed. by M.C. Roco, W.S. Bainbridge (Springer, Berlin, Heidelberg 2006) pp. 24–32

53.15 J. Canton: The emerging nanoeconomy: Key drivers, challenges, and opportunities. In: *Nanotechnology: Societal Implications – Individual Perspectives*, ed. by M.C. Roco, W.S. Bainbridge (Springer, Berlin, Heidelberg 2006) pp. 32–43

53.16 S. Jurvetson: Transcending Moore's law with molecular electronics and nanotechnology. In: *Nanotechnology: Societal Implications – Individual Perspectives*, ed. by M.C. Roco, W.S. Bainbridge (Springer, Berlin, Heidelberg 2006) pp. 43–56

53.17 G. Thompson: Semiconductor scaling as a model for nanotechnology commercialization. In: *Nanotechnology: Societal Implications – Individual Perspectives*, ed. by M.C. Roco, W.S. Bainbridge (Springer, Berlin, Heidelberg 2006) pp. 56–61

53.18 L. Hornyak: Sustaining the impact of nanotechnology on productivity, sustainability, and equity. In: *Nanotechnology: Societal Implications – Individual Perspectives*, ed. by M.C. Roco, W.S. Bainbridge (Springer, Berlin, Heidelberg 2006) pp. 64–67

53.19 I. Feller: An economist's approach to analyzing the societal impacts of nanoscience and nanotechnology. In: *Societal Implications of Nanoscience and Nanotechnology*, ed. by M.C. Roco, W.S. Bainbridge (Kluwer, Dordrecht 2001) pp. 108–113

53.20 R. Freeman: Non-nano effects of nanotechnology on the economy. In: *Nanotechnology: Societal Implications – Individual Perspectives*, ed. by M.C. Roco, W.S. Bainbridge (Springer, Berlin, Heidelberg 2006) pp. 68–74

53.21 V.I. Colvin: The potential environmental impact of engineered nanomaterials, Nat. Biotechnol. **21**, 1166–1170 (2003)

53.22 C.-W. Lam, J.T. James, R. McCluskey, R.L. Hunter: Pulmonary toxicity of single-wall carbon nanotubes in mice 7 and 90 days after intratracheal instillation, Toxicol. Sci. **77**, 126–134 (2004)

53.23 D. Warheit: Nanoparticles: Health impacts?, Mater. Today **7**, 32–35 (2004)

53.24 G. Oberdörster, E. Oberdörster, J. Oberdörster: Nanotoxicology: An emerging discipline evolving from studies of ultrafine particles, Environ. Health Persp. **113**(7), 823–839 (2005)

53.25 D.M. Berube: Communicating nanotechnological risks. In: *Nanotechnology: Societal Implications – Individual Perspectives*, ed. by M.C. Roco, W.S. Bainbridge (Springer, Berlin, Heidelberg 2006) pp. 245–251

53.26 S.K. Friedlander, D.Y.H. Pui (Eds.): *Emerging Issues in Nanoparticle Aerosol Science and Technology* (Univ. California, Los Angeles 2003)

53.27 W. Luther (Ed.): *Industrial Applications of Nanomaterials – Chances and Risks* (Future Technologies Division/VDI Technologiezentrum, Düsseldorf 2004)

53.28 N. Savage: Converging technologies and their societal implications. In: *Nanotechnology: Societal Implications – Individual Perspectives*, ed. by M.C. Roco, W.S. Bainbridge (Springer, Berlin, Heidelberg 2006) pp. 164–168

53.29 M.C. Roco: Environmentally responsible development of nanotechnology, Environ. Sci. Technol. **39**, 106A–112A (2005)

53.30 L.B. Lave: Lifecycle/sustainability implications of nanotechnology. In: *Societal Implications of Nanoscience and Nanotechnology*, ed. by M.C. Roco, W.S. Bainbridge (Kluwer, Dordrecht 2001) pp. 205–212

53.31 M.C. Roco, W.S. Bainbridge: Societal implications of nanoscience and nanotechnology: Maximizing human benefit, J. Nanopart. Res. **7**, 1–1 (2005)

53.32 R. Hanson: Five nanotech social scenarios. In: *Nanotechnology: Societal Implications – Individual Perspectives*, ed. by M.C. Roco, W.S. Bainbridge (Springer, Berlin, Heidelberg 2006) pp. 109–113

53.33 B. Tonn: Co-evolution of social science and emerging technologies. In: *Managing Nano-Bio-Info-Cogno Innovations: Converging Technologies in Society*, ed. by W.S. Bainbridge, M.C. Roco (Springer, Berlin, Heidelberg 2006) pp. 309–335

53.34 I. Kant: *A Critique of Pure Reason* (Wiley, New York 1900), [1787]

53.35 G.E. Moore: *Principia Ethica* (Cambridge Univ. Press, Cambridge 1951)

53.36 J. Rawls: *A Theory of Justice* (Harvard Univ. Press, Cambridge 1971)

53.37 G.C. Homans: *Social Behavior: Its Elementary Forms* (Harcourt/Brace Jovanovich, New York 1974)

53.38 D. Gauthier: *Morals by Agreement* (Oxford Univ. Press, Oxford 1986)

53.39 J. Mosterin: Ethical implications of nanotechnology. In: *Nanotechnology: Revolutionary Opportunities and Societal Implications*, ed. by M. Roco, R. Tomellini (European Commission, Luxembourg 2002) pp. 91–94

53.40 J.R. Von Ehr: The ethics of ethics. In: *Nanotechnology: Societal Implications – Individual Perspectives*, ed. by M.C. Roco, W.S. Bainbridge (Springer, Berlin, Heidelberg 2006) pp. 195–198

53.41 R.E. McGinn: Ethical issues in nanoscience and nanotechnology: reflections and suggestions. In: *Nanotechnology: Societal Implications – Individual Perspectives*, ed. by M.C. Roco, W.S. Bainbridge (Springer, Berlin, Heidelberg 2006) pp. 169–172

53.42 J. Trumpbour: Technological revolutions and the limits of ethics in an age of commercialization. In: *Nanotechnology: Societal Implications – Individual Perspectives*, ed. by M.C. Roco, W.S. Bainbridge (Springer, Berlin, Heidelberg 2006) pp. 113–121

53.43 V. Weil: Ethical issues in nanotechnology. In: *Societal Implications of Nanoscience and Nanotechnology*, ed. by M.C. Roco, W.S. Bainbridge (Kluwer, Dordrecht 2001) pp. 245–251

53.44 V. Weil: Ethics and nano: A survey. In: *Nanotechnology: Societal Implications – Individual Perspectives*, ed. by M.C. Roco, W.S. Bainbridge (Springer, Berlin, Heidelberg 2006) pp. 172–182

53.45 R.W. Berne: Negotiations over quality of life in the nanotechnology initiative. In: *Nanotechnology: Societal Implications – Individual Perspectives*, ed. by M.C. Roco, W.S. Bainbridge (Springer, Berlin, Heidelberg 2006) pp. 198–205

53.46 US Patent Trademark Office: Nanotechnology Classification 977 (USPTO, Classification Operations, Washington 2002), http://www.uspto.gov/go/classification/uspc977/defs977.htm

53.47 Z. Huang, A. Yip, G. Ng, F. Guo, Z. Chen, M.C. Roco: Longitudinal patent analysis for nanoscale science and engineering: Country, institution and technology field, J. Nanopart. Res. **5**, 333–363 (2003)

53.48 Z. Huang, H. Chen, Z. Chen, M.C. Roco: International nanotechnology development in 2003: Country, institution, and technology field analysis based on USPTO patent database, J. Nanopart. Res. **6**, 325–354 (2004)

53.49 E.J. Taylor: An exploration of patent matters associated with nanotechnology. In: *Nanotechnology: Societal Implications – Individual Perspectives*, ed. by M.C. Roco, W.S. Bainbridge (Springer, Berlin, Heidelberg 2006) pp. 187–195

53.50 S.E. Miller: Law in a new frontier. In: *Nanotechnology: Societal Implications – Individual Perspectives*, ed. by M.C. Roco, W.S. Bainbridge (Springer, Berlin, Heidelberg 2006) pp. 182–187

53.51 F.N. Laird: Problems of governance of nanotechnology. In: *Nanotechnology: Societal Implications – Individual Perspectives*, ed. by M.C. Roco, W.S. Bainbridge (Springer, Berlin, Heidelberg 2006) pp. 207–211

53.52 B.E. Seely: Societal implications of emerging science and technologies: A research agenda for science and technology studies (STS). In: *Nanotechnology: Societal Implications – Individual Perspectives*, ed. by M.C. Roco, W.S. Bainbridge (Springer, Berlin, Heidelberg 2006) pp. 211–223

53.53 T.L. Smith: Institutional impacts of government science initiatives. In: *Nanotechnology: Societal Implications – Individual Perspectives*, ed. by M.C. Roco, W.S. Bainbridge (Springer, Berlin, Heidelberg 2006) pp. 223–232

53.54 W.M. Tolles: National security aspects of nanotechnology. In: *Societal Implications of Nanoscience and Nanotechnology*, ed. by M.C. Roco, W.S. Bainbridge (Kluwer, Dordrecht 2001) pp. 218–236

53.55 W.M. Tolles: In defense of nanotechnology in defense. In: *Nanotechnology: Societal Implications – Individual Perspectives*, ed. by M.C. Roco, W.S. Bainbridge (Springer, Berlin, Heidelberg 2006) pp. 236–240
53.56 J. Reppy: Nanotechnology for national security. In: *Nanotechnology: Societal Implications – Individual Perspectives*, ed. by M.C. Roco, W.S. Bainbridge (Springer, Berlin, Heidelberg 2006) pp. 232–236
53.57 Meridian Institute: *International Dialogue on Responsible Research and Development of Nanotechnology* (Meridian Institute, Alexandria 2004)
53.58 S. Weinberg: *The Discovery of Subatomic Particles* (Freeman, New York 1990)
53.59 F. Pragnell: *The Green Man of Graypec* (Greenberg, New York 1950)
53.60 J. Blish: *The Seedling Stars* (Gnome, New York 1957)
53.61 I. Asimov: *Fantastic Voyage* (Bantam, New York 1966)
53.62 G.M. Whitesides: Science and education for nanoscience and nanotechnology. In: *Nanotechnology: Societal Implications – Maximizing Benefit for Humanity*, ed. by M.C. Roco, W.S. Bainbridge (Springer, Berlin 2006) pp. 42–51
53.63 C.L. Alpert: Public engagement with nanoscale science and engineering. In: *Nanotechnology: Societal Implications – Individual Perspectives*, ed. by M.C. Roco, W.S. Bainbridge (Springer, Berlin, Heidelberg 2006) pp. 265–274
53.64 T. Pratchett, I. Stewart, J. Cohen: *The Science of Discworld* (Ebury, New York 1999)
53.65 A.C. Clarke: Hazards of prophecy. In: *The Futurists*, ed. by A. Toffler (Random House, New York 1972) p. 144
53.66 A.C. Clarke: *Profiles of the Future* (Bantam, New York 1963)
53.67 S. Moskowitz: *Explorers of the Infinite* (Meridian, Cleveland 1963)
53.68 M. Crichton: *Prey* (Harper Collins, New York 2002)
53.69 G. Bear: Blood music, Analog **106**(6), 12–36 (1983)
53.70 G. Bear: *Blood Music* (Simon Schuster, New York 2002)
53.71 N. Stephenson: *The Diamond Age* (Bantam, New York 1995)
53.72 K.A. Goonan: *Queen City Jazz* (Tor, New York 1994)
53.73 K.A. Goonan: *Mississippi Blues* (Tor, New York 1997)
53.74 K.A. Goonan: *Crescent City Rhapsody* (Eos, New York 2000)
53.75 K.A. Goonan: *Light Music* (Eos, New York 2002)
53.76 G.K. O'Neill: *The High Frontier: Human Colonies in Space* (Morrow, New York 1976)
53.77 M.A.G. Michaud: *Reaching for the High Frontier: The American Pro-Space Movement, 1972–84* (Praeger, New York 1986)
53.78 E. Drexler: Non-terrestrial resources, L-5 News **2**(3), 4–5 (1977)
53.79 K. Henson, E. Drexler: Vapor phase fabrication of structures in space, L-5 News **2**(3), 6–7 (1977)
53.80 E. Drexler: Space mines, space law, and the third world, L-5 News **3**(4), 7–8 (1978)
53.81 K.E. Drexler: *Engines of Creation* (Anchor Press/Doubleday, New York 1986)
53.82 S.L. Venneri: Implications of nanotechnology for space exploration. In: *Societal Implications of Nanoscience and Nanotechnology*, ed. by M.C. Roco, W.S. Bainbridge (Kluwer, Dordrecht 2001) pp. 213–218
53.83 B.V. Lewenstein, J. Radin, J. Diels: Nanotechnology in the media: A preliminary analysis. In: *Nanotechnology: Societal Implications – Individual Perspectives*, ed. by M.C. Roco, W.S. Bainbridge (Springer, Berlin, Heidelberg 2006) pp. 258–265
53.84 D.S. Hope, P.E. Petersen: A proposal to advance understanding of nanotechnology's social impacts. In: *Nanotechnology: Societal Implications – Individual Perspectives*, ed. by M.C. Roco, W.S. Bainbridge (Springer, Berlin, Heidelberg 2006) pp. 251–258
53.85 W.S. Bainbridge: *Dimensions of Science Fiction* (Harvard Univ. Press, Cambridge 1986)
53.86 G. Gaskell, N. Allum, S. Stares: *Europeans and Biotechnology in 2002: Eurobarometer 58.0* (2003) http://europa.eu.int/comm/public_opinion/archives/eb/ebs_177_en.pdf
53.87 Royal Academy of Engineering: *Nanoscience and Nanotechnologies: Opportunities and Uncertainties* (The Royal Society, London 2003)
53.88 W.S. Bainbridge: Public attitudes toward nanotechnology, J. Nanopart. Res. **4**, 561–570 (2002)
53.89 W.S. Bainbridge: Validity of web-based surveys. In: *Computing in the Social Sciences and Humanities*, ed. by O.V. Burton (Univ. Illinois Press, Urbana 2002) pp. 51–66
53.90 W.S. Bainbridge: Sociocultural meanings of nanotechnology: Research methodologies, J. Nanopart. Res. **6**, 285–299 (2004)
53.91 M.D. Cobb, J. Macoubrie: Public perceptions about nanotechnology: Risks, benefits and trust, J. Nanopart. Res. **6**, 395–405 (2004)
53.92 G.M. Whitesides, J.C. Love: Implications of nanoscience for knowledge and understanding. In: *Societal Implications of Nanoscience and Nanotechnology*, ed. by M.C. Roco, W.S. Bainbridge (Kluwer, Dordrecht 2001) pp. 129–145
53.93 K.M. Kulinowski: Incorporating nanotechnology into K-12 education. In: *Nanotechnology: Societal Implications – Individual Perspectives*, ed. by M.C. Roco, W.S. Bainbridge (Springer, Berlin, Heidelberg 2006) pp. 322–327
53.94 J. Klein-Seetharaman: Interactive, entertaining, virtual learning environments. In: *Nanotechnology: Societal Implications – Individual Perspectives*, ed. by M.C. Roco, W.S. Bainbridge (Springer, Berlin, Heidelberg 2006) pp. 152–158
53.95 M.G. Forest: Mathematical challenges in nanoscience and nanotechnology: An essay on nano-

technology implications. In: *Societal Implications of Nanoscience and Nanotechnology*, ed. by M.C. Roco, W.S. Bainbridge (Kluwer, Dordrecht 2001) pp. 146–173

53.96 G.C. Black: Human resource implications of nanotechnology on national security and space exploration. In: *Nanotechnology: Societal Implications – Individual Perspectives*, ed. by M.C. Roco, W.S. Bainbridge (Springer, Berlin, Heidelberg 2006) pp. 297–300

53.97 T. Chang: Educating undergraduate nanoengineers. In: *Nanotechnology: Societal Implications – Individual Perspectives*, ed. by M.C. Roco, W.S. Bainbridge (Springer, Berlin, Heidelberg 2006) pp. 305–317

53.98 P.E. Stephan: Human resources for nanotechnology. In: *Nanotechnology: Societal Implications – Individual Perspectives*, ed. by M.C. Roco, W.S. Bainbridge (Springer, Berlin, Heidelberg 2006) pp. 331–335

53.99 S.J. Fonash: Implications of nanotechnology for the workforce. In: *Societal Implications of Nanoscience and Nanotechnology*, ed. by M.C. Roco, W.S. Bainbridge (Kluwer, Dordrecht 2001) pp. 173–180

53.100 J.L. Merz: Technological and educational implications of nanotechnology – Infrastructural and educational needs. In: *Societal Implications of Nanoscience and Nanotechnology*, ed. by M.C. Roco, W.S. Bainbridge (Kluwer, Dordrecht 2001) pp. 186–196

53.101 B.E. Seely: Educational opportunities related to the societal implications of nanotechnology. In: *Nanotechnology: Societal Implications – Individual Perspectives*, ed. by M.C. Roco, W.S. Bainbridge (Springer, Berlin, Heidelberg 2006) pp. 327–331

Subject Index

α-Al_2O_3(0001) 650
α-actinin 1195
α-keratin 1112
μCP (microcontact printing) 316, 327
C – O vibration 1387
C=N vibration 1388
1,1′,biphenyl-4-thiol (BPT) 846
1,2-dioleoyl-sn-glycero-3-phosphocholine (DOPC) 444
1-D localization effect 141
1-D nanowire effect 138
1-butyl-3-methylimidazolium hexafluorophosphate (BMIM-PF_6)
– property 1377
1-ethyl-3-(3-dimethylaminopropyl) carbodiimide (EDC) 92, 469
1-hexadecanethiol (HDT) 440
2-D
– FKT model 930
– histogram technique 940
– wax 1423
2-DEG (two-dimensional electron gas) 695
2-pyridyldithiopropionyl group (PDP) 766
$\sqrt{3}$-Ag 645
3,3′-dimethyl bipyridinium 22
3,3′-dithio-bis(sulfosuccinimidylproprionate) (DTSSP) 778
3-D
– bulk state 681
– force 749
– force measurement 606
– force spectroscopy 749
– patterning 287
– patterning capability 303
3-aminopropyltriethoxysilane (APTES) 475

A

abrasive wear 942, 946
Abrikosov lattice 698
absorption 1538
AC dielectrophoresis 414
ac electrodeposition 124
acantha 1535
accelerated
– friction 1300
– test 1854
acceleration energy 1276
accelerometer 1666, 1767
– capacitive 365
– monolithic 367
– surface-micromachined 366
accumulation of dislocation defect 1773
acetonitrile 22, 27
acetylene (C_2H_2) 1277
acid-treated SWNT 75
acousto-optical deflector (AOD) 1208
acousto-optical modulator (AOM) 1208
actin binding protein (ABP) 1175, 1187
activation
– barrier 771
– energy 1772
actuation
– advanced 1769
– force 1719
– mechanism 554
– pneumatic 508
– voltage 1767
actuator
– nanoelectrostatic 221
adatom 642
adenine 18
adenosine triphosphate (ATP) 488
adhesion 573, 802, 890, 1128, 1134, 1173, 1277, 1309, 1323, 1331, 1390, 1532, 1592, 1611, 1683, 1708, 1719
– cell–cell 1188
– coefficient 1573, 1574, 1585
– complex 778, 1179
– design database 1581
– dry 1528
– effect of humidity 1148
– energy 1569, 1574
– enhancement 1556
– Hamaker constant 1573
– hysteresis 882, 883, 888, 910
– hysteresis, relation to friction 888, 889, 892–895, 902
– in microstructure 1722
– linkage 1195
– measurement 795, 1705
– mechanics 879, 880
– mechanics of layered system 881
– modeling 1566
– of biomolecular film 1702
– of biomolecule 1698
– of polymer 882
– on the nanometer level 1590
– primary minimum 858, 871, 873
– probability 775
– promoter 317
– rate-dependent 882, 883
– receptor-mediated 1189
– secondary minimum 871
– test 1563
– tip–surface 961
– transient 1190
– wet 1528
adhesion force 860, 861, 865, 878, 879, 882, 1352, 1496, 1505, 1506, 1509, 1554, 1559, 1569
– calculation 1581
– gecko 1564
– gecko foot-hair 1563
– gecko spatula 1576
– multiple contact 1566
– quantized 873, 900
– single contact 1567
adhesion-controlled friction 886, 887, 889, 890
adhesion-force image 779
adhesive
– bond 1530
– coating 767
– contact 880
– fibrillar 1541
– fluid 1535
– hair 1528
– interaction surface 961
– organ 1533, 1535
– pad 1527
– performance 1543
– wear 1775
adhesive force 583, 802, 939, 976, 1078, 1120, 1122, 1148, 1165, 1323, 1352, 1354, 1362, 1367, 1379, 1387, 1460, 1461, 1492, 1508, 1695, 1716, 1720, 1722
– distribution 1355
– finite element analysis 1559
– histogram 1133
– hydrophilic leave 1461

Subject Index

– influence of temperature 1361
– map 1148
– mapping 1145
– measurement 820
– scale dependent 1473
– single spatula 1562
– superhydrophobic leave 1461
adsorbed water 1720
adsorption 76, 91
– capacity 92
– sites in MWNT 77
advancing contact angle 1445
AES (Auger electron spectroscopy) measurement 1304
AFAM (atomic force acoustic microscopy) 574
affymetrix 391
AFM (atomic force microscope) 664
– adhesion 1332
– bath cryostat 668
– Binnig design 581
– calculated sensitivity 603
– cantilever 580, 598, 711, 1586, 1646
– cantilever array 1603
– carbon nanotube tip 970
– commercial 581
– contact mode 579, 1704
– control electronics 610
– design optimization 596
– dynamic mode 688, 692
– feedback loop 610
– for UHV application 927
– friction 1332
– instrumentation 595
– interferometer 668
– manufacturer 581
– mode 620
– operating mode 1069
– operational mode 737
– piezo creep 665
– probe construction 587
– static mode 688
– surface height image 1726, 1727
– surface height map 1294, 1460
– surface imaging 802
– test 1294
– thermal drift 664
– thermal noise 665
– thermomechanical recording 1604
– tip 636, 638, 641, 643, 764, 769, 772, 1078, 1325, 1355, 1585, 1705
– tip sliding 818
– variable temperature 666
– vibration isolation 666
– wear 1336
African hair 1060, 1087, 1121
– cross section 1093
– elastic modulus 1091
– surface hardness 1091
– topographical image 1116
Ag 646
– trimer 645
agarose 395
air
– damping 756, 1749
– filled hair 1413
– induced oscillation 596
– pocket 1508, 1715
air-layer thickness 1452
Al_2O_3-TiC 805
– head 1303
Al_2O_3 64, 925
alignment 130
– accuracy 324
– of nanowire 130
alkali halide 647, 648
alkaline phosphatase (AP) 522, 527
alkanethiol 654
alkanethiolate 316
– on gold 316
– on palladium 316
– on silver 316
alkylketene dimer (AKD) 1441, 1465
alkylphosphonate SAM 1330
alkylsiloxane film 1320
all-fiber interferometer 668
all-optical
– logic gate 29
– switching network 371
aluminum
– gate 41
– oxide 64, 925
amino acid (AA) 466, 467
amino silicone 1079
– conditioner 1153
aminopropyldimethylethoxysilane (APDMES) 476, 1703
aminopropyltriethoxysilane (APTES) 1699
ammonium 1377
Amontons' law 736, 882–886, 890, 892, 896, 939
amorphous carbon (a-C) 57, 212, 1271, 1697
– chemical structure 1277
– coating 1277
– coatings 1284
– phase 56
amorphous surface 858
amperometric time-based detection method 522
amphiphilic molecular building block 31
amplicon 396
amplitude modulation (AM) 638
– mode 579
– SFM (scanning force microscope) 688
amplitude-mode AFM 732
amplitude-modulation (AM) AFM 738
analog-to-digital converter (ADC) 1622
anatase $TiO_2(001)$ 650
AND
– gate 153
– operator 23
angle of
– helicity 49
– twist formula 602
angled microfiber array 1586
animal 1525
anisotropy
– of friction 933
annealing effect 1798
anodic alumina (Al_2O_3) 121
anodization 121
anodized alumina membrane (AAM) 170
anthracene 28
– channel 29
anti-adhesion performance 1720
anti-adhesive 1538
anti-adhesive coating 291
antiadhesive property 1545
antibody 390, 391, 1698
antibody–antigen complex 775
antibody-coated bead 521
anti-bovine serum albumin (a-BSA) 443
anticancer device 469
anti-fouling 1439
antimony nanowire 142
antistiction
– coating 1766
AP (alkaline phosphatase) 522
apparent contact
– angle 1445
– area 1743
application of nanowire 159
approach–retract cycle 779
APTES-coated silicon 476
aquaplaning 1543

aquatic
– bug 1528
– organism 1529
arc discharge 412, 1276
area expansion modulus 1191
areal density 1604, 1614, 1629
armchair tube 407
armchair-type SWNT 50
array 498
– cantilever 1607
– characterization 1607
– chip 1603
– of nanotube 63
artifact 665
artificial
– double helix 20
– fog 1462
– rain 1462
– surface 1426
as-deposited film 1796
Asian hair 1060, 1087, 1096, 1121
– cross section 1093
– elastic modulus 1091
– surface hardness 1091
– topographical image 1116
aspect ratio 283
asperity
– diameter 1536
– distribution 1744
– friction 1743
placement accuracy 191
assembly of CNTs 412
athermal flow stress 1772
atmospheric pressure chemical vapor deposition (APCVD) 382, 1229
atomic force acoustic microscopy (AFAM) 574, 798
atomic force microscope (AFM) 134, 258, 428, 573, 579, 635, 641, 654, 731, 764, 769, 789, 790, 860–863, 901, 904, 955, 1025, 1055, 1180, 1319, 1442, 1492, 1602, 1637, 1685, 1763, 1776
– dual-axis 1563
atomic force microscopy (AFM) 9, 462, 619, 620, 711, 763, 1055, 1057, 1179, 1229, 1309, 1347, 1406, 1496, 1498, 1536, 1663, 1746
atomic layer deposition (ALD) 238
atomic mass unit (AMU) 457
atomic resolution 578, 638, 639, 679
– image 644
– imaging 641
atomic-scale
– dissipation 947
– force measurement 595
– friction 805, 955, 978, 979
– image 575, 590, 689
– stick–slip 982
attachment
– ability 1526
– biological function 1527
– design principle 1539
– device 1525, 1526, 1537
– dry system 1535
– element 1555
– environment 1537
– force 1532, 1536
– hair density 1535
– hairy 1533, 1534
– locomotory 1533
– mechanism 1526, 1561
– organ 1530, 1543
– pad 1528
– principle of biological 1530
– smooth 1533, 1534
– time scale 1529
– wet system 1535
attachment system 1525–1527
– biological function 1527
– biologically inspired 1525
– locomotion 1527
– micropattern 1588
– single-level 1577
attraction
– long-range 865, 876
Au pattern 321
Au(111) 674
autocorrelation function (ACF) 1016
automatic nanoassembly 1648
average contact resistance 1386
averaged surface potential 1156
avidin 771
azopyridine 27

B

back-flow 511
bacmid system 458, 459
bacterium 444
BaF_2 649
ball grid array (BGA) 1837
ballistic
– phonon transport 147
– transport 136
ball-on-flat
– technique 1386
– test 1384
– tribometer 1380
– tribometer testing 1381
balloon angioplasty 318
bamboo texture 62, 72
bandgap dependence on nanowire diameter 149
bandgap of s-SWNT 406
bare surface 976
barium strontium titanate (BST) 555
barrier layer 123
barrier length scale 771
base pairing 391
basic switching operation 23
bc-MWNT (bamboo-concentric) 52
beam
– bending 1249
– clamped-clamped 381
– deflection 1188
– deflection AFM 734
– deflection FFM 925
– deflection system 621
– S-shaped 1764
– theory 1188
beam–substrate interfacial energy 1719
bearing area 1765
bee 1529, 1534
beetle 1526
behavior 1532
Bell model 773, 1190
– single energy barrier 775
Bell parameter 778
bending
– of stamps 281
– quasistatic test 1243
– stiffness 597, 1174, 1192
– strength 1231, 1239
bending-beam sensor 440
benzocyclobutene (BCB) 1776
Berkovich
– tip 1293
bh-MWNT (bamboo-herringbone) 52
BHPET chemical structure 1377
BHPT chemical structure 1377
Bi nanowire 124
$Bi_{1-x}Sb_x$ nanowire 143, 155
$Bi_{1-x}Sb_x$alloy nanowire 154
$Bi_2Sr_2CaCu_2O_{8+\delta}$ (BSCCO) 683
bias voltage 576
biased microelectrode 395
BiCMOS 364
bidirectional micropump 511
bifurcation 1750
bilayer
– resist 286

bimetallic catalyst 60, 74
bimorph
– microcantilever 445
binary compound 647
binding energy exciton 156
binding site 190, 193, 202
bio compact disk 545
bioactive module 466
bioadhesion 11, 1678, 1683, 1705
bioassay 444
biocapsule 1680
biochemical
– detection 521
– fluid 504
– reaction 513, 520
– sensor 409, 1312
biocompatibility 514
biodevice 1705
bio-electret 1159
biofilter 522
biofluidic chip 524, 1677
biofouling 1312, 1680
biohazard 441
bioinjector 224
biological
– membrane 771
– molecular machine 6
– sensing device 158
– structure 1526
biological adhesive
– reversible 1539
biological attachment
– classification 1529
biological attachment system 1544
– functional principle 1526
biological system
– guideline 1577
biologically inspired material 1543
biology 1172
biolubrication 1153
biomechanics 1526
biomedical device 318, 1227
biomembrane force probe (BFP) 768
bioMEMS (biological or biomedical microelectromechanical system) 3, 503, 513, 1228, 1245, 1309
– actuation 519
– biosensor application 1698
bioMEMS/bioNEMS 1227, 1663
biomimetic
– application 1544
– fibrillar structure modeling 1577
– gecko skin fabrication 1585
– gecko skin fabrication method 1585
– mushroom-shaped 1542
– structural basics 1427
– superhydrophobic surface generation 1428
– surface 1426, 1430
– valve 513
biomimetics 6, 1439, 1526, 1540, 1663
biomolecular component 456
biomolecule 623, 765
– synthesis 565
bionanoinjector 225
bionanotechnology 1633
bioNEMS 3, 1228, 1309
– characteristic dimension 1665
bionics 1426
bionose 496
biopolymer 1186
– model 1187
biosampling 522
– magnetic-bead-based 521
biosensing 299, 490, 491
biosensor 92, 472, 485, 490, 1679, 1698
– interface 472
bioseparator 522
biotechnology
– industry 455
biotin 92, 391
– surface roughness 1703
biotinylated AFM tip 780
biotinylated BSA 764
biphenyl 670
biphenyl-4-thiol (BPT) 1322
biphenyldimethylchlorosilane (BDCS) 846, 1326
bipyridine
– building block 20, 38
– centered LUMO 34
birefringent crystal 600
bismuth nanowire 139–142, 144, 150, 155
bistable mode 743
bit pitch (BP) 1624
bit string 1620
Bloch
– state 680
– wave 679
block copolymer 463, 465
block-like debris 1301
blood analysis 521
blood-pressure sensor chip 1668
blu-ray (BD) 272
BMIM-PF_6
– chemical structure 1377
BN nanotube 80
body mass 1536
– phylogenetic approach 1555
Bohr radius of exciton 156
Boltzmann statistics 1018
bond
– angle 79
– lifetime 773
– strength 864
bonding 1850
– adhesive 1840
– aluminum/silicon-to-glass 1847
– anodic 1839
– energy 637
– eutectic 1840
– fusion 1839
– localized 1841, 1846
– pressure 1849
– silicon 1838
– silicon–aluminum–glass 1852, 1855
– solder 1840
– technology 1530
bond-order potential 957
boron
– ion implantation 579
borosilicate tip 1461
bottom-up 389, 390, 397, 1653
– chemical strategy 18
– fabrication 1732
– paradigma 154
– synthesis approach 126
bottom-up method 1664
bouncing droplet 1468, 1485
– transition 1482, 1503
bouncing off 1509
boundary
– film 1347
– lubrication 802, 840, 884–886, 896, 899, 901
– slip 875
bovine serum albumin (BSA) 441, 443, 764, 1699
bovine serum albumine (BSA) 440
breakthrough etch 284
bridging of polymer chain 876, 877
broken coating chip 1293
Brownian motion 189, 1017
brucite-type cobalt hydroxide (BCH) 1466
brush *see* polymer brush
bubble 875
buckling 1286, 1290, 1578, 1785
– critical compressive strain 1578
– stress 1290
buffered HF (BHF) 239
buffered hydrofluoric acid 505

building block 187
bulk
– addressing 30
– atom 638
– conduction band 675
– diamond 1271, 1279, 1281
– etched silicon wafer 514
– fluid transport 512
– graphitic carbon 1302
– micromachining 1683, 1730
– phonon mode 959
– single crystal 1236
– state 681
– xenon 692
bulk material
– glass transition 724
buoyancy 1708
– force 1706
burst frequency 536
– critical 537
butterfly 1527

C

C_2 moiety 60
C_{60} 39, 49, 57, 654, 940, 1270
– film 578, 997
– island 934
– multilayered film 654
– SWNT (single wall nanotube) 83
– ultralow friction 998
C_{70} 1270
CA (constant amplitude mode) 749
CaF bilayer on Si(111) 649
CaF_2 649
– (111) 649
– (111) surface 944
– tip 944
Cajal body 1177
calibration 1015
cancer detection 397
cantilever 420, 428, 620, 637, 641, 1180, 1720
– AFM 714
– all-silicon 1606
– array 438, 444, 1606, 1607, 1609, 1667
– axis 668
– bending 736
– bending profile 440
– bending radius 431
– cell 1606
– coated 438
– coating 432
– damping in liquid 432, 433
– deflection 437, 580, 637, 639, 668, 724, 836, 924
– deflection calculation 592
– design 1609
– diode array 1608
– driving frequency 714
– dynamic mode 432
– effective mass 733, 925
– eigenfrequency 433, 665
– eigenmode 715
– elasticity 928
– electrochemistry 434
– electrode 434
– flexible 580
– flexural mode 714
– force-sensing 620, 711
– free oscillating 757
– functionalization 438
– heat mode 433
– heater 1605
– higher-harmonic vibration 714
– history 428
– in cantilever (CIC) 444
– liquid environment 434
– low-stiffness 629
– material 587, 637
– microbeam 1243
– microfabricated 428, 622, 735
– motion 579
– mount 584
– photothermal spectroscopy 434
– piezoresistive 436, 437
– Q-factor 688, 689, 746
– quality factor 432, 722
– rectangular 715, 1454
– reference and sensor 430
– resonance behavior 598
– resonance frequency 433, 1790
– response 430, 714
– sensor 429, 1605
– spring constant 432, 768
– stainless steel 1293
– static mode 430
– stiffness 586, 1606
– surface 429, 444
– surface-stress induced bending 430
– temperature 1621
– thermal noise 688
– thermomechanical noise 688
– thickness 925
– trajectory 740
– triangular 587, 597, 602
– untwisted 583
– vapor detection 439
– V-shaped 589, 926
– wear 478
– wear experiment 478
cantilever beam
– array (CBA) 1719
– deflection 435, 436
– fabrication procedure 1260
– lateral bending 1251
– torsional vibration 801
cantilever-array chip
– sensor 1613
cantilever-based probe 688
cantilever–tip assembly 798
capacitance detection 604
capacitive
– detection 609
– detector 579
– displacement sensor 1793
– RF MEMS switch 378
capacitive-type silicon 1666
capacitor
– tunable 375
capillarity 1526
capillary
– condensation 1763, 1765
– effect 1532
– electrophoresis (CE) 505
– engine 1709
– force 720, 858, 873, 878, 879, 1554, 1558, 1562, 1764, 1766
– interaction 1532
– number 558, 563
– particles assembly 204
– pressure 515
– pressure gradient 1355
– serial valve 537
– valve 536
– wave 1449, 1711
capillary assisted particle assembly (CAPA) 205
capped nanotube 972
– tip 972
capture probe down format 396
carbon 1270
– crystalline 1270
– fullerene 180
– magnetron sputtered 1277
– shell 62
– source 67
– unhydrogenated coating 1282
– vapor grown nanofiber 66
carbon nanofiber (CNF) 625
carbon nanotube (CNT) 64, 125, 382, 390, 397, 398, 624, 677, 997, 1465, 1541, 1633, 1643, 1644, 1667
– adsorption property 90

– aligned 68
– application 85
– assembly 412
– based sensor 404
– bending 629
– catalyst-free growth 71
– catalyst-supporting material 88
– chemical reactivity 79
– device fabrication 412
– diameter 75
– doubly clamped 420
– field emission 86
– formation 54
– growth mechanism 70
– helical 1643
– heterogeneity 75
– in situ filling 81
– maximum current density 86
– molten state filling 81
– oxidation 84
– production 54, 412
– property 74
– sensor 403
– structure 624
– sublimation filling 83
– synthesis condition 73
– tip 590
– transport 409
– tunable resonator 406
– wet chemistry filling 81
carbon nanotube field-effect transistor (CNFET) 403, 404, 409
carbon–carbon distance 691
carboxylates ($RCOO^-$) 654
carnivorous plant 1419, 1538
carrier
– density 140, 143
– gas 1282
– mean free path 136, 137, 139
– mobility 140
Casimir
– attractive force 215
– force 859
Cassie
– equation 1443, 1445, 1447
– state 1427, 1451
Cassie–Baxter
– branch 1453
– equation 1442, 1443, 1482
– metastable state 1481
– regime 1452, 1478, 1493, 1495, 1713, 1717
– state 1451
– state transition 1451
– transition 1452
– wetting regime 1517
casting 314
catalysis 647
catalysis-enhanced
– disproportionation 63
– thermal cracking 63
catalyst 57, 126, 131
– based SWNT 73
– free 68
– nanoparticle 69
– particle 412
– preparation 64
catalytic chemical vapor deposition (CCVD) 62, 65, 68, 71, 72, 412
catalytic decomposition 64
cathode deposit 57, 58
cathodic arc carbon 1284
cation attachment scheme 1390
Caucasian hair 1085, 1089, 1121
– chemically damaged 1084, 1125
– cortex region 1081
– cross section 1080, 1093
– elastic modulus 1091
– fatigue 1109
– surface hardness 1091
– tensile deformation 1101
– topographical image 1116
CD drive
– laser-scanning microscope 545
CD platform
– optical detection 550
CD-based microfluidic
– DNA microarray 546
– molecular diagnosis 546
CdSe nanorod 157
CE (constant excitation mode) 749
cell
– analysis 1216
– anisotropy 1184
– based assay 540
– biology 1018
– computational description 1185
– contractility 1178
– death 1706
– growth 302
– homogeneity 1184
– injection 1652
– mechanics 1171
– mechanics model 1186
– membrane 1190
– membrane complex (CMC) 1060
– microfluidic sorting 1216
– nuclei 1192
– poking 1180, 1186
– red blood cell membrane 1191
– rheological property 1183
– rheology 1219
– separation 397
– shrinking 1413
– sorting 1215
– sorting chip 1217
– strain 1194
– stretcher 1206
– surface structure 1413
– viscoelastic property 1183
– Young's modulus 1185
cell-free sample 491
cellular
– compression 1192
– force sensor 1652
– solid 1187
cement 1529
centrifugal microfluidic
– Gyrolab MALDI SP1 CD 537
– volume control 537
centrifugation 181
centrifuge theory 534
centrifuge-based fluidic
– ELISA 539
$CeO_2(111)$ 650
ceramic
– matrix composite 94
– nanobeam 1239
– slider 1302
– tip 86
CFM (chemical force microscopy) 86
chain length effect 1336
channel etching process 505
channel length 323
channel–liquid pair 512
chaotic advection 535
chaotic mixer 516
charge
– dissipation 1157
– distribution 725
– exchange interaction 859, 864, 866, 867
– fluctuation force *see* ion correlation force
– injection 1767, 1768
– mass ratio 398
– pattern 194
– regulation 871
– transfer 652, 867
– transfer interaction 866
charge density wave (CDW) 682
charge-coupled device (CCD) 626, 1745
check-valve design 510
chelate metal cation 20
chemical
– binding force 748

– bond 636, 637, 657
– bonding 638, 644, 645, 647
– bonding force 580
– characterization 1277
– contrast 285
– degradation 1309, 1323, 1338, 1369
– detection 439
– force 636, 639, 641, 657
– heterogeneity 883
– input 24
– interaction 1561
– interaction force 733
– linker method 1698
– quantum effect 137
– sensing device 158
– sensor 87, 409
– signal 27, 28
– synthesis 20
– template 203
– vapor infiltration 68
– warfare 441
chemical bath deposition (CBD) 1466
chemical force microscopy (CFM) 86
chemical mechanical polishing (CMP) 287, 1731
chemical vapor deposition (CVD) 124, 211, 212, 235, 291, 619, 626, 1271, 1465, 1589
chemically
– active ligand 764
– bonded PFPE 1721
– heterogeneous surface 1447
chemisorption 1347
chemistry route 68
chemomechanically damage 1097
– hair 1089, 1121
chemoselective conjugation 469, 471
chemotherapy 467
chiral tube 407
chirality 412
chord theorem 877
cicada 1534
circularly permuted (CP) 469
– biofunctional 469
clamp 1526, 1531
Clausius–Mossotti factor 556
clay nanoparticle 1245
cleaning process 1561
climb 1769
clinging force 1564
– of a gecko 1564
clinical diagnostic 396
clock
– field 1627
– signal 394
clogging 67
cluster
– C_{60} molecule 802
CMOS 389, 392, 399
c-MWNT (concentric multiwall nanotube) 51, 56, 71, 79
CO
– disproportionation 63
– on Cu(110) 669
Co cluster 677
coagulation 871, 874
Coanda effect 516
coated
– polymer surface adhesion 1705
– Si ball-on-flat test 1386
– silicon 1684
– tip 973
coating 1368
– cantilever 432
– continuity 1303
– damage 1301
– durability 1385
– failure 1297
– friction and wear behavior 961
– friction coefficient 975
– hardness 1296
– mass density 1279
– microstructure 1278
– substrate interface 1297
– thickness 1287, 1290
coefficient
– effective 940
– of dynamic friction 1743
– of static friction 1743
coefficient of friction (COF) 809, 824, 895, 926, 1121, 1122, 1140, 1144, 1285, 1293, 1301, 1338, 1352, 1367, 1373, 1379, 1387, 1461, 1695
– average 808
– influence of temperature 1361
– lubricant 1001
– relationship 592
– scale effect 1138
coefficient of thermal expansion (CTE) 93, 1852
coercivity 158
coherence length 675
coil inductor
– self-assembled 377
cold welding 867, 868, 896
colloidal
– dispersion 180
– force 859
– probe 862
colossal-magnetoresistive effect 697
comb drive 1790
combinatorial synthesizer 565
communicating
– between compatible molecular component 27
– molecular switch 27
compact disc (CD) 272
complementarity determining region (CDR) 470
complementary metal–oxide–semiconductor (CMOS) 273
complementary oligonucleotide 398
complex
– bond 774
– dielectric function 147, 148
– logic functions with molecular switch 27
– pattern 318
composite 93
composite interface 1448, 1475
– self-cleaning 1449
– stability 1449
compositionally modulated nanowire 126
compression 732
compressive
– forces nanotube 998
– stress 1282, 1290, 1796
– surface stress 429
computational study
– tribological process 955
computer simulation 896
– force 868, 871, 872, 875, 878, 880, 901
– friction 868, 888, 889, 895, 899–902, 906, 908, 910
computer-aided design (CAD) 223
concanavalin A (Con A) 778
concave interface 1444
concentration
– critical 943
concentric
– graphene 51
– texture 72
– type (c-MWNT) 51
conditioner gel
– physisorbed 1094
conditioner layer 1152
– conducting 1163
conditioner treated hair 1084, 1149
conductance quantization 136

Subject Index

conductive
– AFM 197
– polymer nanowire 174
confined
– nanostructure 80
confinement 858, 859, 884, 900, 909, 910
conformal molding 288
conglutination 1541
conifer 1413
constant
– amplitude mode 648
– current mode 576
– force mode 610
– height mode 576
– *NVE* 960
contact
– adhesive 880
– analysis 1577
– apparent macroscopic 885, 891, 896
– area 798, 862, 943, 944, 1539
– breakage 1539
– conductance 941
– electrification 867
– energy 1539
– force 1528
– formation 1526, 1539
– line density model 1451
– line tension 1445
– liquid–solid 1440
– mechanics 1578
– mechanics *see* adhesion mechanics
– mode 1070
– mode photolithography 324
– modeling 1720
– nonadhesive 880
– potential difference 695
– pressure 863
– printing 313, 329
– resistance 1370
– resistance image 1381
– resonance spectroscopy 798
– stiffness 800, 1287
– stress 1579
– time 320
– time scale 1529
– true molecular 861, 863, 879–881, 886, 890, 891, 910
– value theorem 890
contact angle (CA) 1309, 1383, 1385, 1388, 1420, 1438, 1462, 1482, 1562, 1712
– measurement 1476, 1485
– model 1437
– modeling 1442
contact angle hysteresis (CAH) 883, 1420, 1421, 1438, 1441, 1448, 1451, 1462, 1476, 1480, 1486, 1492, 1493, 1495, 1709, 1711, 1713, 1714, 1716
– macroscale 1482
contact line
– solid–air–liquid 1485
contacting surface 1555
contamination 581, 1538, 1776
– anti-adhesive 1424
continuous
– electrowetting (CEW) 554
– micropump 519
– stiffness measurement (CSM) 1230, 1287
continuum
– model 930
– theory 868–870, 872, 873, 884, 896
contracting cell 1179
contrast 689
– formation 647
control
– over diameter distribution 126
– system 607, 609
controllable
– adhesion 1543
– friction 1543
controlled
– bond formation 670
– desorption 670
– evaporation 519
– geometry (CG) 578
controlling nanowire position 129
convective 204
– assembly 191
Cooper pair 683
coordinate position 399
copper
– hexadecafluorophthalocyanine 326
– surface adhesion 976
– surface wear 976
– tip 945, 976
corannulene 73
core
– boundary scattering 141
– nanostructure 399
– shell nanowire 133
– shell structure 128, 135
Coriolis acceleration 367
corner frequency 1016
cortex 1062
cost of good (COG) 461
cost of ownership (CoO) 275
Couette flow 885, 909
Coulomb
– force 647
– interaction potential 958
– law of friction 936
covalent
– attachment 767
– bond 18
– scaffold 20
covalent bonding
– modeling of material 957
Cr coating 686
crack 1285
– lateral 837
– median 837
– propagation 1542
– width 834
cracking of the coating 834
crater formation
– surface 963
C-reactive protein (CRP) 443
creep 665, 1245, 1769
– displacement 1093
– on a nanoscale 838
critical
– burst condition 536
– concentration 943
– degree of bending 1791
– dimension (CD) 275, 295
– load 829, 1285
– magnetic length 142
– micelle concentration (CMC) 465
– normal load 1696
– pitch 1715
– position 929
– radius of impalement 1714
– shear stress 884, 886, 893, 901, 902
– temperature 674, 689
– velocity 902, 905, 907, 908
Crk-associated substrate (CAS) 1194
cross hair alignment 324
cross sectioning 1779
cross-correlation spectroscopy 492
crossed-I-beam structure 1075
cross-linked BPT (BPTC) 846, 1322
cross-linker 766
cross-linking 283, 847
– protein 1183
crosstalk 611, 1607
cross-track distance 1626
crustacean 1531

cryo-SEM 1535
cryostat 665
crystal
– growth 1409
– growth direction 122
– structure 581
crystalline
– carbon 1270
– growth 749
– silicon 504
– surface 858
crystallinity 1407
CSM (continuous stiffness measurement) 1287
Cu(100) 932, 933
Cu(111) 674, 932, 933, 948
– surface state 679
– tip 945
cumulative failure function 1853
cuprate 686
curing 283
– agent 318
current density effect 1798
curved surface 318
cushioning 293
cuticle 1400, 1413, 1537
– scratch 1111
cuticular
– fold 1414
– pattern 1414
cut-off distance 865, 889
CVD (chemical vapor deposition) 1271
cyclic
– fatigue 1287, 1798
– olefin copolymer (COC) 507
– voltammogram 37
cyclodextrin 198
cyclotron radius of nanowire 141
cylindrical
– pillar 1473
– roller stamp 323
cytosine 18
cytoskeleton 1019, 1020, 1174, 1189
– computational description 1185
– nonlinear response 1184

D

D_2O 676
(d, k) codes 1628
damage 868, 884, 886, 891, 892, 894, 896, 903
– mechanism 1294
damaged hair 1105
– surface 1131
damaged treated hair 1148
damped harmonic oscillator 736
damping 648
– constant 757
– force 1612
– mechanism 756
– pneumatic 668
damselfly 1528
data
– rate 1629
– rate limitation 1602
– storage 295
– storage device 1602
DC sputtering 222
DDMS (dimethyldichlorosilane) 1766
de Broglie wavelength 137
deaging 1752, 1753
Deborah number 899
debris 1301
Debye
– frequency 959
– interaction 864
– length 869, 871, 874
decylphosphonate (DP) 1329, 1695
dedicated equipment 1762
deep reactive ion etching (DRIE) 242, 504, 1652
deep-ultraviolet (DUV) lithography 274, 295
defect
– motion 665
– production 944
– site 319
defect in channel 505
deflagration 442
deflection
– measurement 637
– noise 688
deformable mirror display (DMD) 250
deformation 861, 863, 1150
– elastic 879, 897, 939
– plastic 868, 896
degradation 1348
degradation of PFPE film
– mechanical scission 1373
degrees of damage 1153
dehydration reaction 319
deionized (DI) 1454, 1699
delamination 834, 1283, 1286, 1290
delivery particle 1706
Demnum-type PFPE lubricant 1354
demolding 282, 291
dense
– magnetic memory array 159
– nanowire packing 123
density 444
– of defect 323, 324
– of fiber 1580
– of states (DOS) 405
dental wax 1488
deoxyribonucleic acid (DNA) 264, 300, 301
Department of Defense (DOD) 5
Department of Energy (DOE) 5
dependence of
– ZT on nanowire diameter 154
– $\kappa(T)$ on size diameter 146
– optical bandgap on nanowire diameter 148
– Seebeck coefficient on diameter 144
– Seebeck coefficient on temperature 144
depletion
– attraction 859, 877
– force 873
– interaction 876
– stabilization 859, 877
deposition
– rate 1276
– technique 1273
– techniques 1274
depth-sensing indentation 789, 836
Derjaguin approximation 861, 871, 873
Derjaguin–Landau–Verwey–Overbeek (DLVO)
– interaction 864, 871
Derjaguin–Muller–Toporov (DMT) 724, 1743
– model 939
– theory 1567, 1578
design
– for reliability 1761
– rule 596
designed molecule 20
destabilizing factor 1711
destructive fabrication 1647
detachment 1559
detection 1621
– biochemical 521
– system 581, 598
detection limit 439
development 1417
device
– microfluidic 503
– molecule-based 32, 35
– nanobiological 456

dewetting 1354, 1692
DFM (dynamic force microscopy) 688
DI silicon nitride 795
diagnostics 1698
diameter dependence of quantum wire superlattice 145
diamond 79, 1270
– coating 1282
– film 382, 1271
– friction 979
– nanoindentation 968
– tip 792, 944, 1293, 1297, 1367, 1685
diamond-like amorphous carbon (DLC) 212
– coating 975, 1278
diamond-like carbon (DLC) 224, 227, 816, 829, 834, 1043, 1311, 1691, 1697, 1776
diblock copolymer 123, 739
– template 122, 159
dicationic BIPY 25
dicationic ionic liquid film
– nanotribological study 1383
dichlorodiphenyltrichloroethane (DDT) 441
dielectric 1767, 1769
– charging 1766
– constant 865
dielectrophoresis (DEP) 199, 204, 397, 556
dielectrophoretic
– assembly 199
– force 129
differential scanning calorimetry (DSC) 135
diffuser micropump 511
diffusion 671
– based extractor 517
– coefficient 516
– constant 1017
– flame synthesis 68
– layer 176
– length exciton 157
– parameter 671
– thermally activated 665
diffusive
– transport 136, 137
digital
– feedback 609
– filter 1622
– fusion 565
– microfluidics 553
– micromirror device (DMD) 369, 1312, 1725, 1761, 1771, 1773
– signal in 394
– signal out 394
– signal processor (DSP) 576
– transmission between molecule 28
– versatile disc (DVD) 272
diisopropylmethylphosphonate (DIMP) 441
dilation 888, 901, 902, 910
dimension 637
dimensionless particle position 1707
dimer structure 644
dimer–adatom–stacking (DAS) fault 642
dimethyl sulfoxide (DMSO) 535
dimethylmethylphosphonate (DMMP) 442
dinitrotoluene (DNT) 442
diode laser 1217
dip coated tip 1705
dip coating 1349
dip-pen nanolithography (DPN) 199, 260, 462
direct overwriting 1620
direct write electron beam/focused ion beam lithography 315
direct-current plasma-enhanced CVD (DC-PECVD) 625
directed self-assembly 1589
directionality effect 1099, 1137
discrete track recording (DTR) 296
disease 396
disjoining pressure 1355
disk resonator 380
dislocation
– glide 1772
– motion 884
– nanoindentation 964
– stick–slip 982
dispensing 1544
dispersion
– force 657
– interaction 859, 864, 865
– mixing 517
displacement
– amplitude 1611
displacement sensor 1719
display
– paperlike 322
disposable biosensor 524
dissipated power 1621
dissipation 754
– image 757
– measurement 923
dissipative interaction 754
dissociation constant 770
distance
– cut-off 865
– distribution function 1765
distortion 608
distributed
– Bragg reflector (DBR) 328
– feedback (DFB) 328
– laser resonator 328
disulfide cross-link density 1090
dithiol
– monolayer 322
division of contact 1590
DLC (diamond like carbon)
– coating 1272, 1274, 1285, 1297, 1303
– coating microstructure 1288
– matrix 1311
– pillar 219
DLVO
– interaction 872, 873, 875
DMD (digital mirror device)
– CMOS fabrication 370
– pixel 370, 1725
– pixel array 369
DMT (Derjaguin–Muller–Toporov)
– adhesion force 1573
– theory 879, 1573
DNA 198, 390, 396, 398, 399, 442, 654
– AFM image 753
– amplification 396
– analysis 523
– analysis system 515
– hybridization 391, 397, 399
– microarray 394
– nanowire 40
– polymerase 396
– probe 395
– sensor 158
– tile 199
domain
– pattern 696
donor impurity concentration 148
doped
– anode 57
– silicon wafer 38
DOS structure 674
double
– barrier resonant tunneling 145
– emulsion 559, 564
– layer structure 176
– stranded (ds) 468
– stranded (ds)DNA 468
double-layer
– force 870, 873

– interaction 859, 869–871
double-wall nanotube (DWNT) 59, 412
doubly clamped beam 420, 1789
drag reduction 1430
dragonfly 1528
DRAM 1603
DRIE (deep reactive ion etching) 504
dried leave 1455
dried lotus leaf
– surface height map 1459
drift state 681
driving frequency 599
droplet 1453
– adhesion force 1509
– break-up 562
– capped geometry 1469
– Cassie–Baxter 1440
– dielectric constant 556
– fission 555, 562
– fusion 561
– manipulation 556
– maximum droop 1475
– microfluidic system 565
– microfluidics 553
– pinning 1484, 1504
– size 563
– sorting 563
– submicron 1472
– transport 557
– vibrational energy 1485
droplet evaporation 1468, 1714
– transition 1477
droplet system
– active 554
droplet-based EWOD 555
droplet-generating device 558
drug
– delivery 1228
– delivery device 1668
– delivery vehicle 465
– target 485
dry
– adhesion 1526, 1528, 1544
– friction 1449
– sliding friction 976
– superadhesive 1590
dry surface 886, 905, 1401
– force 864, 867
– friction 885, 886, 891, 903, 904
dual-in-line packaging (DIP) 1843
Dupré equation 883, 1442
durability 1371, 1375, 1381
dye penetrant 1778
dynamic
– AFM 637–639
– AFM mode 757
– friction 1743
– friction *see* kinetic friction
– interaction 858–860, 884, 897, 900
– mode 579
– operation mode 638
– pressure 1475
– viscosity 514
dynamic atomic force microscope (dynamic AFM) 731
dynamic force microscopy (DFM) 711, 758, 780
dynamic force spectroscopy (DFS) 771

E

e-beam lithography 1730
economy of scale 1816
ECR (electron cyclotron resonance) CVD 1273, 1292
– coating 1298
edge
– channel 695, 696
– disorder 318
EDP (ethylene diamine pyrocatechol) 504
EELS (electron resonance loss spectroscopy) 1278
effect of
– cycling 1369
– meniscus 1353
– viscosity 1357
effective
– coefficient of friction 940
– damping constant 751
– mass 689
– medium theory 147, 148
– resonant frequency 741
– shear stress 941
– spring constant 929, 938
– stiffness 1185
– viscosity 899, 909
– Young's modulus 1078
EFM 664, 696
elastic 293
– continuum 1186
– deformation 939
– energy 1539
– force 1568
– limit 834
– modulus 879, 881, 1229, 1231, 1245, 1283–1285, 1580, 1582
– modulus plot 1093
– tip–surface interaction 968
elasticity 689
elastohydrodynamic lubrication 884, 885, 897, 899
elastomer 506, 507
– precursor 506
– stamp 319
electret 194
electric
– field 398
– field transport 399
– force gradient 586
electric field gradient microscopy (EFM) 133
electrical
– resistance 1369, 1381
– stability of MEMS products 1825
electric-arc
– method 73
– reactor 56
electroactive
– fragment 22
– layer 32
– solid 30
electrochemical
– AFM 586
– cell 32
– deposition 123, 171
– detection 521
– device 1312
– immunosensor 523
– nanowire growth 129
– STM 578
electrode 284
– array 34
– with molecular layer 35
electrodeposition
– selective 128
electroforming 284
electrohydrodynamic (EHD) pumping 512
electrokinetic flow 1679
electroless deposition 173, 317
electroluminescence (EL) 155
electrolysis 395
electrolytic method 68
electromagnetic
– actuator 509
– force 864
– microvalve 509
electromagnetic trap 768
electron
– cyclotron resonance chemical vapor deposition (ECR-CVD) 1273, 1277

– energy loss spectroscopy (EELS) 1276
– interaction 681
– tunneling 575
electron beam (EB) 211
– resist 319
electron beam deposition (EBD) 619, 624
electron beam evaporation 319
electron cyclotron resonance chemical vapor deposition (ECR-CVD) 1697
electron diffraction (ED) 1407
electron energy loss spectra (EELS) 133
electron Fermi wavelength 136
electron-beam (e-beam) 462
electron-beam lithography (EBL) 278, 629
electron-beam-induced deposition (EBID) 627, 1642
electronegativity 1313
electron–electron interaction 674
electronic
– addressing 395, 396
– application 154
– hybridization 395, 396
– ink 325
– microarray 396
– newspaper 326
– noise 411, 581
– nose 496
– nose tongue 532
– stringency 395
– structure of SWNT 406
electronically lysed 397
electron-phonon interaction 681
electroosmosis (EO) 512
electroosmotic
– flow (EOF) 505
– pumping 510
electrophoresis 176
electrophoretic
– deposition 175
electroplating 284, 290, 1837
electrorheological fluid 510
electrostatic
– actuated device 508
– actuation 1719
– binding 1152
– force 636, 693, 858, 859, 864, 866, 869, 872, 1766
– force interaction 643
– force microscopy 695
– interaction 648, 866
– potential 696
electrostatic double layer (EDL) 1707
electrowetting (EW) 554, 566
electrowetting on dielectric (EWOD) 554, 555, 565
– operational principle 554
– reagent mixing 556
elongation speed 1769
embedded atom method (EAM) 958
embedding energy 958
embossing 272
– machine 292
– technique 320
emerging nanopatterning method 273
encapsulation 1836
– agent 564
end cap 55
end deflection 1787
endocuticle 1084, 1090
endofullerene 84
endurance limit 835
energy
– dispersive analysis of x-ray (EDAX) 135
– dispersive x-ray spectrometer (EDS) 132
– dissipation 648, 882, 883, 886, 888–891, 894, 903, 947
– landscape 190, 193, 774
– MD simulation 956
– resolution 673
energy-dispersive x-ray (EDX) analysis 1415
engineered nanostructure 21
enhanced solar cell efficiency 157
entangled state 884
entropic
– force 859, 878
Environmental Protection Agency (EPA) 5
environmental scanning electron microscope (ESEM) 1441
environmental study 1323
enzyme 390
enzyme-linked immunosorbant assay 539
epicuticle 1084
epicuticular wax 1439, 1453
– crystal structure 1407
epicuticular wax tubule 1709
epidermis cell 1453
– microstructure 1415
epitaxial growth 129
equilibrium
– binding enthalpy 771
– interaction *see* static interaction
– true (full) or restricted 876
equipartition theorem 1016
erasing mechanism 1618
Escherichia coli 444
etched STM tip 631
etching 284, 314
– dry (plasma) 504
– wet (chemical) 504
ethylene 278
ethylene diamine pyrocatechol (EDP) 504
ethylenediamine pyrochatechol (EDP) 239
eucalyptus tree 1418
Euler
– buckling 1578
– buckling criterion 1785
– equation 597
Euler–Bernoulli equation 714
euplantula 1533
European Union (EU) 5
evolution 1537
evolutionary superadhesive 1558
exchange
– carrier plate 928
– force interaction 652
– force microscopy 654
– interaction 652
exocuticle 1090, 1533
explosives 441
expression system 490
external
– noise 665
– vibrations 577
extreme ultraviolet (EUV) 233, 275
– lithography 274, 295
Eyring model 937

F

FAA (formaldehyde–acetic acid–ethanol) 1512
fabricated
– pillar array 1587
– silicon 1706
fabrication 221
– of CNT devices 412
– of microfluidic systems 505
– of mold masters 505
– superhydrophobic surface 1462
– technique 504
– technology 359
face-centered cubic (fcc) 261, 1270
faceted cavity 155

F-actin 1174
failure
– density function 1853
– mechanism 1303, 1761
– mode 1761
Fano resonance 677
fastener 1544
– artificial 1532
fast-on/fast-off kinetics 775
fatigue 1287, 1773
– crack 1301
– damage 1288
– failure 1798
– life 1285
– measurement 1793
– resistance 1788
– strength 1241
– test 1291, 1728
FCA (filtered cathodic arc) 1291
– coating 1284, 1289
Fe-coated tip 653
feedback
– architecture 362
– circuit 581
– loop 611, 926
– network 575
– signal 638
feline coronavirus (FIP) 444
Fermi
– energy 143
– level 40, 673
– level pinning 410
– point 681
ferrocene 70
– hydrophobic 32
ferromagnetic
– probe 696
FE-SEM imaging 157
FET
– heterojunction 475
– protein sensor 453, 472
– sensing channel 473
– structure 409
fetal bovine serum (FBS) 1705
Feynman 1634
FFM (friction force microscopy) 924
– dynamic mode 945
– on atomic scale 938
– signal 794
– tip 928, 931
fiber
– array 1541
– axial stress 1579
– buckling condition 1578
– condensation 1541
– elastic modulus 1580
– fracture 1579
– material 1580
– matrix 1186
– model 1577
– morphology 1555
– optic switching matrix 372
– optical interferometer 600
– stiffness 1578
fiber-optic sensor 1720
FIB-milled probe 590
fibrillar
– adhesive 1585
– surface construction 1590
fibrillar structure 1553, 1578
– compliant 1586
fibrous
– composite material 1537
– crystal 1465
field
– directed CNT growth 415
– effect transistor (FET) 97
– emission application 153
– emission SEM (FESEM) 1638
– emission-based display 68
– emission-based screen 86
– ion microscope (FIM) 734
– ion microscopy (FIM) 734
field effect transistor (FET) 152, 153, 158, 463, 1679
– protein sensing 472
filament
– cross-linker 1195
– elastic property 1175
– intermediate 1188
filamentous actin (F-actin) 1019
fill factor 282
filled nanotube 81
filling
– angle 1562, 1563, 1573
– efficiency 83
– factor 123
film
– C_{60} 997
– drainage 561
– hydrophilic PMMA 1472
– ionic liquid 1375
– nanoindentation thickness 971
– on substrate 1783
– substrate interface 1796
– surface interface viscosity 990
film thickness
– effect of humidity 1148
– map 1147
– mapping 1148
filtered cathodic arc (FCA) 1697
– deposition 1274, 1276
– deposition) 1289
filtration 1529
fine leak 1778
finite
– element method (FEM) 1229
– element modeling (FEM) 1767
– size effect 151
first principle
– calculation 646
– MD simulation 956
– simulation 689
FKT (Frenkel–Kontorova –Tomlinson) model 930
flap-type valve 511
flat surface 1490, 1494
– nanostructuring 1493
flat-on-flat tribometer 1143, 1165
flavonoid 1416
flexible
– cantilever 580
– electronic device 322
flexible crosslinker
– property 768
– spring constant 769
flexural
– mode 716
– resonance frequency 715
flip-chip technique (FC) 394, 1837
floating catalyst method 66
flocculation 871, 874
Flory
– radius 876
– temperature 876
flow
– cytometry 492, 498, 1216
– laminar 514
– nanoliters 1676
– rate 519
– rectifying junction 562
– resistance 516
– sensing 444
flowering plant 1416
flower-plant 1413
fluctuation
– interaction 864
fluctuation–dissipation theorem 1017
fluid 1535
– biochemical 504
– interface 560
– propulsion 533
fluidic
– assembly 189, 205
– cartridge 392
– delivery 394

– motion 520
– pattern 541
– structure 537
fluidic device
– controlled flow rate 538
– defined-volume 538
fluorescence
– activated cell sorting (FACS) 1215
– correlation spectroscopy (FCS) 492
– imaging 542
– polarization (FP) 492
– quantum yield 24
fluorescent
– label 300, 301
– probe 396
– protein 490
fluoride 647, 648
fluorinated silane monolayer 319
fluorocarbon
– hydrophobic layer 1467
fluorophore
– reporter 396
fly 1526
flyaway hair 1064
FM (frequency modulation)
– AFM image 690
FM-AFM 639–641
focal adhesion 1194
– kinase (FAK) 1194
focused ion beam (FIB) 211, 579, 590, 619, 624, 1763
– cross section 1775
focused ion beam chemical vapor deposition (FIB-CVD) 211
focused ion-beam (FIB) 212
force
– adhesion 860, 861, 865, 878, 879, 882
– adhesion, quantized 873
– advection 516
– attractive 859
– between macroscopic bodies 866, 870
– between surfaces in liquid 859, 868
– between surfaces in vacumm 859
– calibration 925
– calibration mode 795
– calibration plot (FCP) 1077, 1121, 1122, 1150, 1352, 1353
– calibration plot technique 1076
– cantilever-based 688
– capillary 858, 873, 878, 879
– Casimir 859
– charge fluctuation 871
– charge fluctuation *see* ion correlation force
– chemical force 689
– colloidal 859
– computer simulation 868, 871, 872, 875, 878, 880
– depletion 873
– detection 597
– determination 1182
– double-layer 873, 874
– double-layer (geometry dependence) 870
– dry surface 864, 867
– electromagnetic 864
– electrostatic 858, 859, 864, 866, 869, 872
– electrostatic force 695
– entropic 859, 878
– generation 1184
– gradient 746
– hydration 859, 864, 872–875, 878
– hydrodynamic 859, 863
– hydrophobic 859, 873–875
– indentation 968
– ion correlation 859, 869, 871
– law 860
– long-range 693
– MD simulation 956
– measurement 637
– measuring technique 857, 860, 861, 863
– microscopy 763
– modulation 1326
– modulation mode 798
– nonequilibrium 859
– oscillatory 858, 859, 871–875, 881–883
– osmotic 878
– peristaltic 878
– protrusion 859, 878
– repulsive 859
– resolution 688, 769
– sensing tip 606
– sensor 580, 639, 927, 1719, 1720
– short-range 858
– solvation 859, 871–875, 881, 883
– spectroscopy (FS) 610
– steric 864, 878
– structural 859, 871, 873, 881
– surface 857, 858
– thermal fluctuation 878
– transducer 1019
– undulation 859, 878
– van der Waals 858, 859, 864, 865, 868, 871, 872
– van der Waals (geometry dependence) 866
– van der Waals force 691
force field spectroscopy
– three-dimensional 694
force measurement 718
– time-varying 720
force spectroscopy (FS) 689, 692, 745, 764
– 3D-FFS 694
– site specific 693
force-clamp AFM 775
force–distance
– curve 586
force–distance curve 719, 795, 1151, 1571
force–distance profile 768, 770
force-driven dissociation 773
force-free dissociation 773
force–probe experiment 771
force–time
– behavior 764
– curve 1532
force–volume mode 1077
formate ($HCOO^-$) 654
formate-covered surface 655
foundry 394
four-fingered platform structure 1590
Fourier transform infrared spectroscopy (FTIR) 1145
Fourier-transform infrared spectroscopy (FTIR) 1378
four-quadrant photodetector 924
fracture 1578
– condition 1579
– failure 1243
– strength 1798
– stress 1231
– surface 1242
– toughness 1230, 1241, 1283, 1285, 1287, 1296
– toughness measurement 1787, 1791
free
– space wiring 217
– standing cantilever 1764
– surface energy 1359, 1442, 1446
freezing–melting transition 903, 904, 907
Frenkel–Kontorova–Tomlinson (FKT) 930
frequency
– measurement precision 606
– modulation (FM) 638, 745
– modulation (FM) mode 579

– modulation SFM (FM-SFM) 665
fresh lotus leaf
– surface height map 1459
friction 573, 790, 802, 857, 923, 955, 976, 1271, 1309, 1323, 1331, 1460, 1526, 1663, 1741
– adhesional component 825
– anisotropy 933
– coefficient 884–886, 891–893, 895, 896, 899, 907–909, 977
– computer simulation 868
– directionality dependence 1124, 1137
– directionality effect 1122
– dominant 770
– electrification 867
– experiments on atomic scale 930
– hydrophilic leave 1461
– image 932
– kinetic 885, 886, 888, 899, 901, 902, 904, 905, 907, 909
– load dependence 827
– loop 929, 930, 942
– lubrication 976
– map 899, 932
– measurement 802, 1794
– measurement method 591
– measuring technique 861, 863, 891, 899, 901
– metal surface 978
– molecular dynamics simulation 944
– of biomolecule 1698
– profile 1237
– rolling 867, 883
– SAM 990
– scale dependence 824
– sliding 867
– superhydrophobic leave 1461
– surface 976
– test 1074, 1340
– test structure 1745
friction and adhesion 1352, 1358
– influence 1358
– relative humidity 1358
– temperature effect 1358
– velocity effect 1357
friction and wear 1112, 1390, 1702
– nanoscale 1386
friction force 477, 580, 583, 821, 935, 990, 1058, 1120, 1293, 1323, 1352, 1379, 1695, 1717, 1722
– calibration 583, 594
– curve 610
– influence of temperature 1361
– magnitude 593
– map 940
– mapping 1120
– measurement 792
– microscopy (FFM) 573, 790, 791, 863, 887, 890, 923, 924, 955, 1025, 1055, 1057, 1319, 1685, 1723, 1744, 1776
– of Si(100) 1693
– profile 804
friction model
– cobblestone model 886, 887, 889
– Coulomb model 886
– creep model 904
– distance-dependent model 904
– interlocking asperity model 886
– phase transitions model 905–907
– rate- and state-dependent 907
– rough surfaces model 904
– surface topology model 904
– velocity-dependent model 904, 905
frictional
– aging 1751
– creep 1750, 1751
– system 1531
frictional property
– fullerene 997
friction-based actuator 1742
friction–torque model 1723
fringe detector 1788
fruit 1530
fruit fly
– end of the leg 1555
fullerene 49, 463, 680, 1270
– film 836
– frictional property 997
– like structure 57
fullerite 51
functional nanotool 227
functionalization
– capillary technique 438
– inkjet-spotting 439
– microfluidics 438
functionalized nanotube 84
fundamental resonant frequency 598
funnel 191
future research direction 159

G

G protein 1172
GaAs 319
– stamp 320
– wafer 322
GaAs/AlGaAs
– heterostructure 695
G-actin 1174
gain control circuit 639
GaN nanowire 131, 151, 156
gap stability 576
GaP(110) 690
gas
– phase growth 128
– sensor 406, 416
– separation 91
gaseous secondary electron detector (GSED) 1469
gas–liquid interface 200
gate control 138
gauge marker 1788
Gaussian
– height distribution 1116
– laser beam 1203, 1205
– surface 826
Gd on Nb(110) 683
gecko 1542
– attachment system 1563
– dry adhesion 1555
– toe 1556
gecko adhesion 1563, 1566
– macroscale 1586
gecko foot 1553, 1554, 1556
– adhesion mechanism 1561
– setal array 1558
– surface characteristics 1556
gecko seta
– adhesion coefficient 1573
– adhesive force 1564
– maximum adhesion coefficient 1577
– spatula 1585
– three-level model 1572
gecko skin
– fabrication 1585
– Young's modulus 1558
gecko spatula
– contact with a hydrophilic surface 1566
gel network 1148
genomic DNA 397
genotyping 396
genuine graphite 48
geometrically necessary dislocations (GND) 1027
geometry effects in nanocontacts 938
germanium
– nanowire 126
g-factor 675
gland 1416
glandular trichome 1416

Subject Index

glass 319, 1429, 1540
– bulk etching 505
– recrystallization chamber 1490
– slide 318
– transition 279, 722
– transition temperature 506, 884, 895
glassiness 909
glass-to-glass direct bonding 522
glass-transition temperature T_g 289
glucose 442
– oxidase (GOD) 525
glue 1526
gold 578
– electrode 35
– film 316
– nanowire 156
G-protein coupled receptor (GPCR) 485
– drug target 486
– ligand 487
– signaling 490
gradient force 1202, 1204
Grahame equation 869
grain
– boundary 867, 868, 1797
– boundary scattering 142
– nanoindentation boundary 964
– size 318, 1271
graphene 48, 74, 405, 406
– defect 82
graphite 665, 668, 691, 721, 930, 944, 1270
– (0001) 679
– cathode 1276
– flake 998
– pellet 55
– sheet 998
grasshopper 1527, 1534
grating 329
green
– chemistry 1376
– fluorescent protein (GFP) 490, 565
Greenwood and Williamson (GW) 1744, 1765
Griffith fracture theory 1243
gripper 1544
grooming 1528, 1529
gross leak 1778, 1779
group A-specific lectin 779
growing tubule 1410
GST (glutathione-S-transferase) 443
GTP binding 496
guanine 18
guanosine diphosphate (GDP) 488
guanosine triphosphate (GTP) 488
gyroscope 364
– dual-axis 367, 368
– vibrating mechanical element 367
– z-axis 368

H

H_2O 676
Hagen–Poiseuille equation 514
hair 1412, 1535
– alignment 1056
– care product 1056, 1058
– cellular structure 1086
– chemically damaged 1082
– creep behavior 1093
– cross section 1080
– damaging processes 1064
– effect of humidity 1086
– elastic modulus 1090
– fiber 1056
– hardness 1090
– hydrophobicity 1143
– indentation depth 1090
– nanomechanical characterization 1090
– outer layer 1148
– spectral property 1418
– type 1131
hair damage 1056, 1076
– schematic of the progress 1085
hair friction
– effect of temperature 1113
hair surface 1078, 1088, 1125
– adhesive force 1134, 1147
– adsorption of conditioner layer 1148
– binding interaction 1145, 1151
– chemically damaged 1146
– cushion protecting 1153
– nanomechanical properties 1091
– property 1115
– Young's modulus 1150
hair–hair interaction 1112
hairy
– adhesive 1542
– attachment 1553
– attachment system 1554
– pad 1536, 1537
– pad system 1535
– surface 1401
– system 1424
hairy leave
– wettability 1426
Hall resistance 695
Hamaker constant 636, 713, 748, 823, 865, 866, 868, 872, 889, 893, 1146, 1355, 1561, 1707
– gecko seta 1567
Hamilton–Jacobi Method 640
hand-held biochip analyzer 526
hard amorphous carbon coating 1273
hard coating 1697
hard disk drive (HDD) 295
hardness 836, 837, 1229, 1245, 1278, 1280, 1283, 1285, 1685, 1689, 1784
– of cuticle 1093
– scale effect 838
harmonic cantilever 716
– torsional 717
harmonic force
– component 713
harmonic oscillator 736
– model 714
harpooning
– interaction 859, 867
HDT (hexadecanethiol) 318
healthy hair 1056
heat
– curable prepolymer 314
– dissipation 1851
heater
– cantilever 1605
– temperature 1616
heavy-ion bombardment 698
height profile
– BMIM-PF_6 coating 1380
helicity 49
hemodynamic resistance 1708
hermeticity 1766, 1776, 1778
herringbone
– MWNT 52, 71, 90
– texture 72
Hertz
– theory 879, 880
Hertz-plus-offset relation 939
heterodyne interferometer 599
heterogeneous
– CCVD 63
– interface 1442
hetero-nanotube 80
hexadecane thiol (HDT) 318, 846, 1322, 1720
hexagonal close-packed (hcp) 262
hexagonal honeycomb polysilicon (HEXSIL) 254
hexagonally packed intermediate (HPI) 739

hexamethyldisilazane (HMDS) 232
hexatriacontane 1501
hierarchical
– arrangement 128
– patterned surface 1468
– roughness 1461
– spring model 1553, 1566, 1569
hierarchical spring analysis
– multilevel 1567
hierarchical structure 1424, 1494, 1496, 1516, 1715, 1716
– characterization 1711
– fabricated 1590, 1591
– fabricating 1467
– superhydrophobicity 1496
hierarchical surface
– ideal 1715
– nanowire 1467
high aspect ratio (HAR) 1463, 1534
high throughput screening (HTS) 488
high-aspect-ratio combined poly- and single-crystal silicon (HARPSS) 255
high-aspect-ratio MEMS (HARMEMS) 1731
high-cycle fatigue 1774
high-definition TV (HDTV) 295
high-end photolithography (PL) 303
higher orbital tip state 673
higher order
– resonance 713
– structure 400
higher-harmonic force 713
high-fidelity recognition 399
high-frequency force 714
highly ordered monolayer 316
highly ordered pyrolytic graphite (HOPG) 1409
highly oriented pyrolytic graphite (HOPG) 127, 265, 586, 691, 802, 1043
high-modulus elastomer 316
high-pressure phase
– silicon 967
high-resolution
– electrode 326
– FM-AFM 691
– imaging 657
– patterning 323
– printing 313, 318
– spectroscopy 673
– stamp 314
– transmission electron microscope (HRTEM) 1639
high-speed
– camera 1752
– data collection 815
high-temperature
– condition 73
– MEMS/NEMS 1729
– operation STM 667
– superconductivity (HTCS) 669, 683, 698
high-throughput screening (HTS) 485, 492, 532, 549, 550
Hill coefficient 776
hinge memory effect 1771
hinged cantilever 1794
HiPCo
– process 67
– technique 56
histidine tag 491, 492, 498
h-MWNT (herringbone) 52, 71, 90
hollow metal tubule 173
holographic optical tweezer (HOT) 1209
homodyne interferometer 598
homoepitaxial aggregation 179
homogeneous interface 1442
honeycomb
– lattice 80
honeycomb profile 1543
honeycomb-chained trimer (HCT) 645
Hooge law 411
hook 1526, 1544
Hooke's law 688
horizontal
– axis 1618
– coupling 577
horizontally arranged
– nanotube 999
hostboard 391, 397
hot embossing 271, 276, 506, 507
– lithography (HEL) 272, 275
hot-pressing (HP) 94
HtBDC (hexa-*tert*-butyl-decacyclene) on Cu(110) 671
HTCS (high-temperature superconductivity) 669, 683, 698
H-terminated
– diamond 980
Hückel equation 177
Human Genome Project 1668
human hair 1055, 1094
– African 1087, 1164
– Asian 1087, 1111, 1164
– Caucasian 1087, 1111, 1164
– chemical damage 1110
– chemical species 1059
– conditioner treatment 1110
– cortex 1058
– cross-sectional dimension 1060
– directionality friction effect 1112
– dry state 1165
– elastic modulus 1093
– electrostatic charge 1166
– fibers 1062
– fracture 1165
– nanoindentation 1055
– nanomechanical characterization 1087
– nanomechanical study 1164
– nanoscratch test 1164
– nanotribological characterization 1055
– outer surface 1122
– pyroelectric property 1159
– research 1057
– scratching 1099
– structural characterization 1080
– surface charge 1058
– surface potential 1153
– surface roughness 1126
– tensile deformation 1101
– tensile property 1106
– tensile testing 1074
human serum albumin (HSA) 444
human umbilical venous endothelial cell (HUVEC) 778
humidity 938
hybrid
– carbon nanotube 80
– continuum-atomistic thermostat 960
– mold 291
– nanotube tip 627
hybridization 49, 389, 647
– detection 546
hydration
– force 859, 864, 872–875, 878, 889
– regulation 874
hydrocarbon 63, 67, 1314
– precursor 1277
hydrodynamic
– damping 754
– force 859, 863
– radius 876
hydrofluoric acid (HF) 300, 440
hydrogel permeation layer 395
hydrogen 643, 644
– bonding 18, 875, 876
– concentration 1280
– content 1283
– flow rate 1282

– in carbon nanotube 90
– peroxide 442
– sensor 158
– storage 90
– termination 644
hydrogenated
– carbon 1278
– coating 1278
hydrogen-terminated
– diamond 972
hydrophilic 1536
– control 513
– leaf surface 1454
– leave 1441, 1462
– PMMA 1472
– SAM 994
– surface 858, 874, 1421, 1710
hydrophilic/hydrophobic nature 1509
hydrophilicity 1576
hydrophobic 1536
– break 544
– coating 1765
– control 513
– ferrocene 32
– force 859, 873–875
– interaction 464, 873–875
– leave 1441
– SAM 994
– surface 858, 874, 1711
– valve 535
– valving 536
hydrophobicity 1309, 1311, 1354, 1472, 1566, 1695, 1722
hydroxylated surface 36
hypergravity 548
hysteresis 608, 639, 665, 697
– adhesion 883
– contact angle 883
– loop 581, 611

I

IB (ion beam)
– coating 1298
IBD (ion beam deposition) 1274, 1276, 1291
ideal surface 1487
IL structure 1390
image
– processing software 612
– topography 580, 656
imaging
– atomic-scale 802
– bandwidth 587
– electronic wavefunction 679
– tools 131
imidazolium 1377
immobilized metal ion affinity chromatography (IMAC) 491
immunoassay 539
– magnetic-bead-based 522
– sandwich 521
immunoFET 472
– limitation 473
– molecular nanocomponent 480
– planar 473
– receptor 474
immunoisolation 1312
immunosensor 522
impact velocity 1482, 1484, 1503
– critical 1504
imperfectness 1452
impregnation method 64
imprint
– machine 292
– polymer 329
in situ
– displacement 1096
– growth assembly 413
– growth of SWNT 415
– growth study 132
– scratch depth 1101
– sharpening of the tips 576
in vivo 391
InAs 668
InAs(110) 675, 690
incipient wetness impregnation 88
indentation 720, 792, 836, 880, 1089, 1603, 1614, 1617, 1618
– creep 1230, 1247
– depth 836, 837, 1089, 1284, 1286
– fatigue damage 1290
– hardness 574, 836
– induced compression 1290
– rate 967
– size 1604
independent molecular operator 27
indirect transition in bismuth nanowire 150, 151
indium tin oxide (ITO) 323, 1466
induction-type EHD 512
inductor
– micromachine 376
– out-of-plane coil 377
– structure 377
inelastic regime 1228
inelastic tunneling 669, 670
inertial
– balance 420
– force 1505, 1506, 1508
– sensor assembly and packaging 1823
infrared (IR) 445
– imaging 445
injection molding 506, 507
injection-type EHD 512
ink
– electronic 325
inkjet-spotting 439
inorganic–organic solar cell 157
InP nanowire 149
InP(110) 642
in-phase signal 1625, 1626
in-plane
– actuation 1610
– mechanical strain 316
– stresses nanoindentation 965
insect 1525
instability 638, 641
insulin pump 519
intact antibody (IgGs) 474
integrated
– cantilever 1609
– circuit (IC) 231, 359
– MEMS 509
– tip 589
interaction 190
– charge exchange 859, 864, 866, 867
– charge transfer 866
– Debye 864
– depletion 876, 877
– Derjaguin–Landau–Verwey–Overbeek (DLVO) 864, 871
– dispersion 859, 864, 865
– DLVO 873, 875
– double-layer 859, 869, 871
– double-layer (geometry dependence) 870
– dynamic 858–860
– electrostatic 866, 867, 869
– energy 865, 866, 1561
– energy gradient 770
– fluctuation 864
– harpooning 859, 867
– hydrophobic 873–875
– induction 864
– Keesom 864
– length 190
– London dispersion 859, 864, 865
– nonequilibrium 883
– orientational 864
– osmotic 859, 869, 877, 878
– polymer 876, 877
– rate-dependent 883

– static 858, 859
– van der Waals 869
– van der Waals (geometry dependence) 866
interaction force 689, 741, 770
– cantilever deflection 768
Interagency Working Group on Nanoscience, Engineering, and Technology (IWGN) 4
interatomic
– attractive force 592
– force 580, 732
– force constant 638
– interaction 1797
– spring constant 580
intercellular diffusion 1090
interconnected
– operator 26
interconnection 287
interdiffusion 883
interdigitated array (IDA) 526
interdigitation 883
interface 553
– composite 1448
– heterogeneous 1442
– homogeneous 1442
– liquid–air 1447, 1453
– solid–air–liquid 1504
– solid–liquid 1503
interfacial
– defect 1291
– energy (tension) *see* surface energy (tension)
– engineering 453
– friction 886, 889, 894, 896, 901
– friction *see* also boundary friction
– stress 1296
– tension 558
interference lithography 203
interferometer 370
interferometeric detection sensitivity 601
interferometry 1763
interferon (IFN) 100
interlayer friction 1647
interleave scanning 798
interlocking
– macrocycle 34
– mechanism 1530
– of body part 1528
intermediate filament (IF) 1019
intermediate or mixed lubrication 884, 885, 898, 899
intermediate-frequency (IF) 373
intermittent contact mode 793
intermolecular communication 26
internal
– friction 1617
– stress 1785
International Technology Roadmap (ITRS) 274
intersymbol interference 1628
intertube distance 51
intraband transition 674
intracellular pathway 1194
intramolecular hydrogen bonding 1390
intramolecular junction 1648
intrinsic
– damping 756
– stress 1283
iodine 671
iodobenzene 670
ion
– channel 1196
– condensation 871
– correlation force 859, 869, 871
– implantation 1604, 1687
– plating technique 1273
– sensing 440
– source 1276
ion beam
– deposition (IBD) 1274, 1276
– sputtered carbon 1276
ionic bond 859
ionic liquid (IL) 1375
– dicationic 1390
– film 1347
– lubrication property 1377
– physical property 1377
– thermal property 1377
– tribological property 1376
ionizing radiation 1767
ion-selective optode detection 538
irreversible deformation 1769
isofunctional object 459
isoleucine–lysine–valine–alanine–valine (IKVAV) 466
isospectral structure 680
isotropic etching 318
itinerant nanotube level 677
ITO (indium tin oxide) 324
I–V characteristics 145
ivy plant 1527

J

jamming limit 193
jellium approximation 958
jet-and-flash imprint lithography (JFIL) 282
Johnson–Kendall–Roberts (JKR) 1534, 1556, 1743
– relation 939
– theory 879, 880, 1578, 1579
Joule
– dissipation 948
– heating 697
jump into contact 1708
jump-in distance 1146
jump-to-contact (JC) 638, 692, 734, 744, 962
junction 152

K

KBr 647
KBr(100) 932, 943
$KCl_{0.6}Br_{0.4}(001)$ surface 648
Keesom interaction 864
Kelvin
– equation 879
– probe force microscopy (KPFM) 657
– probe microscopy 715, 1071, 1153
– probe technique 1072
kinematics 1543
kinetic
– friction 885, 886, 888, 899, 901, 902, 904, 905, 907, 909
– friction coefficient 1589
– friction macroscale coefficient 1589
– friction ultralow 907
– rate 771
knife-edge blocking 609
Knoop hardness 805
Kondo
– effect 664
– temperature 677
Kramers diffusion model 773
Kramers–Kronig relation 148
K-shell EELS spectra 1279
kurtosis 1718

L

lab-on-a-chip 295, 520, 532, 1216, 1677
– application 521
– concept 503
– polymer 527
– system 515, 1676
– technology 505
lab-on-a-disc 549
laminar

Subject Index

– flow 514
– flow characteristics 517
– pattern 564
laminated n-channel transistor 326
Landau
– level 143, 665, 675, 696
– quantization 675
Langevin dynamics approach 959
Langmuir–Blodgett (LB) 846, 976, 1311, 1348
– methodology 42
– technique 129
– trough 205
lanthanide binding tag 491
Laplace
– equation 1713
– force 1354, 1562
– pressure 858, 878, 1475, 1558
large-area fabrication 297
large-sample AFM 793
large-scale adhesion 1590
laser
– ablation 412, 1731
– ablation method 73
– action 155
– application 155
– assisted catalytic growth 131
– assisted growth 125
– deflection technique 582
– device 54
– interference lithography 315
– scalpel 1213
– to-fiber coupler 370
– trapping 1639
– vaporization method 73
laser beam
– angular momentum 1211
– asymmetric beam profile 1212
– Bessel beam 1210
– corkscrew phase topology 1211
– counterpropagating 1206
– divergent 1206
– focused beam 1213
– Gaussian 1203
– Hermite–Gaussian laser mode 1210, 1212
– Laguerre–Gaussian laser mode 1210
laser microdissection (LMD) 1213
laser microdissection and pressure catapulting (LMPC) 1214
laser pressure catapulting (LPC) 1213, 1215
laser scanning cytometry (LSC) 1220
laser tracking microrheology (LTM) 1208
laser trap 1203
– dual-beam 1203
– two-beam 1218
laser tweezers Raman spectroscopy (LTRS) 1208
lasing action 155
lasing threshold 155
lateral
– collapse 1541
– contact stiffness 923
– deflection 1723
– force 817, 924, 936
– force calculation 602
– force microscope (LFM) 573, 791, 924
– force microscopy (LFM) 316, 654, 863
– resolution 576, 580
– resonator 1799
– scanning range 1610
– spring constant 587, 928
– stiffness 610, 941
– surface property 840
lattice imaging 580
lauric acid (LA) 1466
layer
– lubricant 985
– of resist 314
layer-by-layer (LBL) 196, 1466
layered structure
– lubricant 1000
lead zirconate titanate (PZT) 1637
leaf
– structure 1424
– surface characterization 1458
leakage rate 514
leave
– hydrophilic 1460
– superhydrophobic 1460
lectin 391
leg attachment pad 1553
Leiden MEMS tribometer 1746
Lennard-Jones (LJ) potential 691, 823, 958
leukocyte rolling 1190
level III spring 1570
levers 1614
LiF 647
– (100) surface 944
lifetime
– broadening 669, 674
lifetime–force relation 775
Lifshitz theory 865, 866, 868
lift mode 586, 796
lift scan height 798
lift-off resist (LOR) 286
lift-off technique 284
LIGA 505
– fabrication 1732
– microsample 1772
– process 1731
– technique 1227
– technology 273
ligand binding 492
light beam deflection galvanometer 601
light emission 155
light emitting diode (LED) 153, 155, 298
limit of tip radius 1579
limit-of-detection (LoD) 411
line scan 813
linear
– array 556
– recording density 1628
– variable differential transformer (LVDT) 609
linearization
– active 609
linker protein 1175
lipid
– bilayer 444, 493
– bilayer vesicle 768
– film 937
– vesicle 1191
liposome 444, 1706
Lippman's principle 554
liquid 985
– bridging 1724
– conditioner film 1146
– crystal on silicon (LCoS) 297
– front 1446
– handling system 540
– helium 665
– helium operation STM 667
– lubricant 841, 985, 1725
– lubricated surface 886, 891, 894, 905
– metering 538
– micromotor 566
– nitrogen (LN) 666
– phase growth 128
– suspension assembly 413
liquid film
– molecularly thick 1348
– spring constant 1146
– thickness 848
liquid–air contact 1711

liquid–air interface 1445, 1447, 1450, 1452, 1453, 1475, 1484, 1495, 1563, 1715
liquid-like lubricant 1392
liquid-like Z-15 1354
liquid–vapor interface 1472
Lithographie Galvanoformung Abformung (LIGA) 1729
lithography 129, 274, 462, 1428, 1463, 1664, 1680
live cell 778
lizard 1525
load
– contribution to friction 886, 889, 890
– critical 1285
– curve 1352
– dependence of friction 939
– force 1532
load-carrying capacity 1302
load-controlled friction 886, 887, 894, 896
load–displacement 1251
– characteristic 1229
– curve 838, 1089, 1231, 1286, 1287
– profile 1250
loading rate 770
local deformation 828
– characterization 797, 833
local density of states (LDOS) 673
local stiffness 610
localization effect 143–145, 154
localized heating 1841, 1846, 1848, 1852
localized surface elasticity 798
lock 1526, 1544
lock-and-key 1531
locomotion 1527
locust 1534
logic
– gate 22, 23, 152, 153
– protocol 23
London
– dispersion interaction 859, 864, 865
– penetration depth 683, 698
long-chain hydrocarbon
– lubricant 986
longitudinal
– magnetoresistance 142
– piezo-resistive effect 604
long-range
– attraction 865, 876
– force 639, 648, 693
– meniscus force 795
long-term
– measurement 665
– stability 664, 1630
– test 1854
loose association 766
lotus (*Nelumbo nucifera*) 1410, 1424
– effect 1425, 1437, 1454, 1709, 1722
– effect surface 1453
– pattern 1470
lotus leaf
– artificial 1716
– microstructure 1488
– replica 1463, 1500
– surface 1709, 1716
lotus leave 1403
– self-cleaning 1429
lotus replica
– micropattern 1712
– microstructured 1500
lotus tubule
– morphological characterization 1497
lotus wax 1501, 1506, 1508
– morphology 1491
low aspect ratio (LAR) 1463
low cost MEMS manufacturing 1807
low pressure chemical vapor deposition (LPCVD) 1229, 1235, 1730
low temperature
– AFM/STM 581
– condition 71
– SFM (LT-SFM) 668
low-cycle fatigue 1773
– resistance 1793
low-density
– lipoprotein (LDL) 443
– polyethylene 721
low-noise measurement 928
low-polarizable SAM 1322
low-pressure imprint 282
low-stress silicon nitride (LSN) 1730
low-temperature
– microscope operation 664
– scanning tunneling spectroscopy (LT-STM) 669
– SPM (LT-SPM) 664
lubricant
– chain diameter 840
– flow 1355
– fraction 1381
– liquid 840, 1691
– MEMS/NEMS application 1348
– mobile-phase 1349
– nanotribological performance 1348
– perfluoropolyether 1691
– PFPE 1348
– solidlike 1350
– Z-15 1348
– Z-DOL 1348
lubricant film 1348, 1699
– electrical property 1366
– environmental study 1373
– failure 1340, 1371
– interaction 1390
– thickness 848
lubricant layer
– hydrophobic 1725
lubricated
– Si(100) 1371
– silicon 1365
– surface 1724
lubricating
– film 1309
– thin film 985
lubrication 840, 1663, 1683
– elastohydrodynamic 885, 897, 899
– intermediate or mixed 885, 898, 899
– mobile layer 1724
– nanoscale boundary 802
lubrication property
– ceramics 978
– polymer 974
LVDT (linear variable differential transformer) 609

M

macrocyclic polyether 21, 33
macrodroplet 1472
macrohardness 838
macromolecule 454, 876
macroscale friction 812, 1137, 1684
– force 1057
macroscopic building block 18
magnetic
– actuator 509
– application 158
– bead 521
– bead microrheology 1179, 1180
– bead separator 526
– disk 1302
– disk drive 1272
– field dependent magnetoresistance 142

– field length 142
– film 296
– force 636, 638, 639
– force gradient 586
– force microscopy (MFM) 134, 297, 574, 586, 696, 752
– head 296
– immunoassay 521, 526
– length 142
– ordered nanowire 148
– phase breaking length 143
– quantum flux 698
– recording 1602
– storage device 1273
– thin-film head 1303
– tip 686
– twisting cytometry 1180
magnetic field microscopy (MFM) 134
magnetohydrodynamic (MHD) pumping 512
magnetooptical Kerr effect 148
magnetoresistance 141, 142
– temperature dependent 141, 142
magnetostatic interaction 696
magnetron sputtered carbon 1277
magnon excitation 677
MALDI MS
– CD platform 543
– sample preparation 543
– volume definition 543
manganite 686
manipulation
– individual atom 695
– individual atoms 669
– molecules 669
– of individual atom 574
– of particle 1529
manipulator 221
manmade pollutant 1560
marine organism 1527
mask, high resolution stamp 314
masking
– layer 288
– step 394
mass
– effective 689
– of cantilever 579
mat arrangement of nanowire 131
material 1533
– biologically inspired 1539
– for microfluidic devices 504
– lubricant 1000
– molecule-based 43
– structural 1788
material property
– Young's modulus 721
matrix assisted laser desorption ionization (MALDI) 537
matrix electronic paper
– active 325
matrix-method 749
Matthiessen's rule 141
maximum
– adhesive force 1579
– likelihood estimator (MLE) 1855
– thermal conductivity 146
maximum load
– nanoindentation 965
MBI (multiple-beam interferometry) 861
MCM-41 122
MD simulation 956
– constant-*NVE* 960
mean square displacement 1016
mean time to failure (MTTF) 1855
mean-field theory 870
mechanical
– coupling 904
– dissipation in nanoscopic device 923
– interlocking 1527, 1528
– micropump 510
– resonance 907
– scission 1373
– sensor 406
– stability 1607
– stability fullerene 997
– stop motion 1748
– stress 1769
– wear 514
mechanical property 836, 1228, 1243, 1783
– characterization 1798
– experimental technique 1229
– of DLC coating 1292
– of nanowire 135
– of SWNT 406
mechanics of cantilever 596
mechanism-based strain gradient (MSG) 1027
mechanochemistry 1647
mechanosensing 1196
– protein 1194
mechanosensitive channel of large conductance (MscL) 1196
mechanosensory organ 1528
mechanotransduction 1172, 1194
medulla 1062
melted SAM 1334
melting temperature 1770
membrane 778, 1173
– deflection method 1799
– deformation 1190
– protein 485
– stiffness 1190
– surface tension 1191
memory
– distance 903
MEMS 359, 520, 1179, 1273, 1783
– accelerometer 364, 367
– application 360
– capacitive switch 378
– challenge 383
– device 1669
– display 369
– fiber optic switching network 372
– nonsilicon 4
– optical device 369
– packaging 1821
– product architecture 1809
– radio-frequency 373
– resonator 379
– resonator-based oscillator 379
– sensor 364
– stiction 1826
– surface effect 1803
– surface treatment 1826
– switch 378
– technology 1795
– tribology 1746
– tunable capacitor 375
– variable capacitor 375
MEMS/NEMS
– characteristic dimension 1665
– device 1663
– lubricant 840
– lubrication 846
– mechanical property 1228
– Si-based 4
– tribological property 1683
MEMS-based
– microfluidic system 521
– storage device 1602
meniscus 200
– curvature 1355
– formation 1380
– of liquid 878, 879
meniscus force 822, 823, 1078, 1120, 1355, 1472, 1558, 1720
– time dependent 1357
mercaptopropyltrimethoxysilane (MPTMS) 320
merocyanine 26
mesoporous
– molecular sieve 122
– template 181
mesoscale osmotic actuator 519

metal 1540
– deposited Si surface 645
– electroplating 505
– evaporated (ME) tape 1271
– matrix composite 93
– nanowire 81, 123, 172
– oxide 647, 650
– particle (MP) 833
– particle (MP) tape 1303
– particle catalyst 64
– porphyrin (Cu-TBPP) 654
– tip, deformable 961
– vapor 440
metal surface
– superhydrophobic 1466
metal/insulator/metal (MIM) 326
metal–CNT junction 410
metal-evaporated (ME) 834
metallic
– microbeam 1243
– nanoparticle 397, 398
– nanotube 41
– nanowire 124, 136
metallic bonding
– modeling of material 957
metallofullerene 180
metalorganic chemical vapor deposition (MOCVD) 236
metal-organic compound 66
metal–oxide–semiconductor field-effect transistor (MOSFET) 1679
metal-strip grating 297
metamaterial 188
methylene stretching mode 1320
MFM 664, 696
– AM-MFM 696
– domain imaging 696
– FM-MFM 696
– sensitivity 586
– vortex imaging 697
MgO(001) 650
MHD (magnetohydrodynamic) pumping 512
mica 586, 933
– muscovite 943
– surface 934
micro total analysis system (mTAS) 1216
micro-/nanoelectromechanical system (MEMS/NEMS) 789, 1025, 1426, 1663
micro-/nanooptoelectromechanical systems (MOEMS/NOEMS) 1664
microarray 391
– biochip assay 485
– device 399
microasperity 1488
microbalance 432
microbuckling 1578
microbump 1456
microcantilever 429, 1667, 1674
microcantilever chemical sensor 444
microcantilever sensor 444
microcapillary array (MC) 559
microcentrifuge fluidic 534
microchannel 510, 1439
microcontact printing (μCP) 196, 272, 316, 1312
microcrystalline graphite 1281
microdevice 1227, 1785
microdispenser 517
microdissection 1213
microdroplet 1485
– condensation 1469
– contact angle hysteresis 1488
– evaporation 1469
microelectrode 389, 392
microelectromechanical
– motor 360
microelectromechanical system (MEMS) 3, 200, 295, 359, 416, 955, 1182, 1228, 1309, 1348, 1439, 1563, 1635, 1664, 1743, 1761
microelectronic array 391, 397, 399
microelectronics 647
microelectrorheological valve 510
microengine 1670
microfabricated silicon cantilever 637
microfabrication 316, 434, 1682, 1731
– technology 360
microfilament 1174
microfluidic 503, 1676
– application 505
– CD
cell lysis 547
– chamber 545
– channel 504, 519
– channels on silicon substrates 504
– control 510
– device 503
– line 508
– mixer 516
– motherboard 521
– multiplexer 517, 518
– network 517
– network (μFN) 438
– passive 513
– pumping 510
– sampling 504
– structure 506
– system 553
– unit 546
microfluidic device 520, 554
– active 507
– disadvantage 520
– materials for 504
– passive 507, 520
microfluidic dispenser
– passive 524
microfriction 808
microgear 360, 1312
microgravity 548
microgripper 1543
microhinge technology 370
microimplant 1698
microindentation 1180
microindenter 1179, 1180
microinductor 1651
microinjection molding 524
microinjector 225
micromachine 1228
micromachined
– array 1719
– notch 1792
– polysilicon beam 1719
– silicon 3, 1664
– surface 475
– test structure 1744
micromachining 361, 368
– bulk 363
– surface 363
– technology 1741
micromanipulation 1202, 1640
micromanipulator 625
micromechanical
– calorimetry 433
– sensor 439
– thermogravimetry 432
micromirror 1312
– beam-steering 373
micromirror device (DMD)
– digital 1663
micromixer
– category 516
– three-dimensional 516
micromolding in capillaries (MIMIC) 205
micromotor 1312, 1718, 1723
– lubricated 1724
micron-sized channel 516
microoptoelectromechanical system (MOEMS) 3
micro-outgrowth 1531

micropattern 846, 1441, 1450
micropatterned
– biphenyldimethylchlorosilane (BDCS) 847
– polymer surface 1471
– SAM 1325
– Si replica 1498
micropatterned surface 1468, 1473, 1511–1514, 1713
– static contact angle 1515
– with $C_{20}F_{42}$ 1513
micropillar 1477
micropipette aspiration 863, 1179–1181
microplate stretcher 1179, 1180
micropore membrane 1590
micropump 510, 1677
– bidirectional 511
– passive 519
microreflector 370, 371
microresonator 1847, 1854, 1856
microroughened surface 1463
microroughness 1441, 1450
microscale
– contact resistance 1384
– diffusion 516
– friction 805, 810, 1137, 1386, 1685, 1718
– hair 1553
– machining 360
– roughness 1439
– scratching 796, 828
– seta 1591
– silicon gate 41
– wear 796, 829, 1386, 1685
microscanner 1610, 1612
microscope eigenfrequency 596
microscratch 829, 1293
– AFM 1321
microsculpture 1722
microsensor 1683
microslip 812
microspark erosion 1731
microsphere 227
microspotting 493
microstrain gauge 1785, 1796
microstructure 1494, 1526, 1715
– biomimetic 1542
– characterization 1711
– mushroom-shaped 1543
– release 1818
– replica 1489
– surface 1501, 1717
microstructured
– surface 1585
microsucker 1530
microsyringe 1501
microsystem technology (MST) 3, 1228, 1664
microtemplate 564
microtip 820
microtiter 543
micro-total analysis system (mTAS) 507
microtransformer 327
microtriboapparatus 1719, 1720
microtribometer 1588
microtubules (MT) 1019, 1175, 1188
– filament 1196
microvalve 508
– active 508
– design 510
microwear 1293, 1297, 1696
Mie regime 1202
millimeter-scale device 1732
millipede 1667
– chip 1613
– system concept 1601, 1603, 1624
MIM capacitor 326
miniature device 1636
minimum indentation pitch 1619
Mini-Whegs 1543
misfit
– angle 930
mismatch
– of crystalline surface 881, 882, 909
mix and read type of assay 492
mix-and-match 285
mixed C-N nanotube 66
mixed lubrication *see* intermediate or mixed lubrication
mixing performance 516
Mn on Nb(110) 683
Mn on W(110) 686
mobile lubricant 1366
mobility 197
MOCVD 124
mode coupling grating 328
modeling
– of material 957
modulation codes 1628
modulus
– elastic 1283–1285
– of elasticity 574, 821
molding 272, 277, 314, 1428
– process 1715
molecular
– adsorption on SWNT 408
– AND gate 25
– assembly 33
– assembly patterning by lift-off (MAPL) 285
– building block 20, 42
– chain 1316
– cloning device 458
– cohesion 769
– conduction band 40
– conformation 1348
– dynamics (MD) calculation 931
– dynamics simulation (MDS) 923, 944
– engineering 453, 490
– force 1532
– imaging 680
– interconverting state 23
– layer 1663
– Lego 391, 398
– logic gate 23
– machine 21
– motor 21
– NOT gate 24
– precursor 68
– recognition 391
– resolution 672
– reversible transformation 23
– shape 872, 873, 903, 908, 909
– shuttle 22
– spring 1336
– spring model 846, 1325, 1694
– switch 22, 42
– weight 275, 277
molecular diagnosis
– infectious disease 546
molecular dynamics (MD) 956
molecular recognition force microscopy (MRFM) 764
molecular recognition force spectroscopy
– principle 769
molecular recognition phase (MRP) 441
molecular vapor deposition (MVD) 301
molecular-beam epitaxy (MBE) 236, 1732
molecule-based
– device 32, 35
– material 43
– switch 23
moment of inertia 1791
monocationic 1390
monodispersed 188
– droplet 558
– metal nanocluster 125
monohydride 643
monolayer

– indentation 972
– surfactant 861
– switch 35
monolayer lubrication 1746
monomer 278
monomeric bond 776
MoO_3 nanoparticle 997
$MoO_3(010)$ 650
Moore's law 275
Morpho butterfly 223
morphological characterization 1496
– tubule 1496
morphological type 1416
Morse potential 748
MoS_2 friction 933
MoSe nanowire 135
MOSFET 475
moss 1413
motherboard 397
motor
– molecular 21
– neuron 466
– protein 1178, 1179, 1196
mouthpart 1529
m-SWNT
– transmission 409
mTAS (micro-total analysis system) 507
multicellular
– hair 1416
– surface structure 1411, 1416
multi-chip module (MCM) 1838
multicomponent device 472
multifunctional
– interface 1401
– surface 1402
multilayer 286
– device 326
– soft lithography 511
– thin-film 1303
multilayered
– APTES 1701
– nanowire 124
multilevel 286
– photolithography 1590
multilevel hierarchical model
– number of spring 1573
multilevel structure
– durability 1590
multimode AFM 586, 796
multimolecular layer 1311, 1691
multiple layer 399
multiple-beam interferometry (MBI) 861
multiplexed
– analysis 396
– DNA analysis 396
multipole
– interaction potential 958
multiscale roughness 1427, 1744
multiuser MEMS process (MUMP) 1842
multiwall carbon nanotube (MWCNT) 51, 55, 57, 75, 382, 590, 998, 1465, 1589, 1644
– based catalyst-support 88
– bc (bamboo-concentric) 52
– bh (bamboo-herringbone) 52
– bunch 69
– flexural modulus 79
– MWNT-Al composite 93
– surface area 75
multiwall carbon nanotube (MWNT) 624, 794, 1589, 1682
muscovite mica 766, 943
mushroom-headed microfiber 1586
mushroom-like 1531
– contact 1588
mussel 1527

N

N acceptor 18
NaCl 647
– island on Cu(111) 947, 948
NaCl(001) on Cu(111) 647
NaCl(100) 934, 935, 940
NaF 647, 933
NAND gate 23
nano/picoindentation 798
nanoasperity 1488, 1715
nanobeam 1577
nanobiological device
– design paradigm 457
nanobiology 224
nanobiotechnology 499
nanobridge 218
nanobrush 128
nanobumps 1427
nanochannel 1312, 1439, 1697, 1730, 1731
– glass 121
nanochemistry 9, 1732
nanoclay 1245
nanocomposite
– material 36, 996
nanocrystal 1, 1706
nanocrystallite 1278
nanodeformation 1347, 1350
nanodevice 98, 456, 1309, 1324
– supramolecular 453
nanodroplet 1472, 1715
– condensation 1711
– scale effect 1450
nanoelectrode 39
nanoelectromechanic 215
nanoelectromechanical system (NEMS) 3, 359, 360, 380, 1228, 1309, 1348, 1439, 1633, 1664, 1753
– device 381
nanoelectronic 97, 390, 397, 1633
– device 39, 42
nanoelectronics 9, 1312
nanoelectrostatic actuator 221
nanofabricated surface 1585
nanofabrication 300, 316, 397, 574, 789, 796, 828, 850, 1633
– method 399
– parameter 836
– technique 504
nanofatigue 1287
nanofiber 52, 63
nanofilament 52, 71
nanofluidic
– channel 295, 300
– device 1227, 1680
– silicon array 1680
nanogap 38
Nanogen 391
nanohair 1586
– polyimide 1586
nanohardness 798, 1282
nanohelix 20
nanoimprint
– lithography (NIL) 196, 271, 272, 275, 1732
– material 288
– process 277
nanoimprint type
– application 294
nanoimprinting
– resist material 288
nanoindentation 797, 961, 1072, 1087, 1283, 1685, 1784
– MD simulation 961
nanoindentation relax
– surface atom 963
nanoindenter 838, 1072, 1229, 1234, 1296, 1788
nanoinjector 224
nano-Kelvin probe 796
nanolithography 462, 1667
nanolubrication 1378
nanomachine 1635
nanomachining 9, 574, 789, 796, 828, 836, 850

nanomagnet 296
nanomagnetism 686
nanomanipulation 750
nanomanipulator 225
nanomanufacturing 390
nanomaterial
– biologically inspired 1663
– heterologous 468
nanomechanic 1683
nanomechanical
– characterization 1072, 1728
– force 724
– switch 220, 221
nanomechanics 1171, 1663
nanometer 955
– resolution 314
nanometer-scale
– device 955
– electronic device 996
– friction 955
– indentation 955
– lubrication 955
nanometer-scale feature 1
nanometer-scale friction
– ceramics 978
nanometer-scale property
– material 961
Nanomotor 1637
nanonail 128
nanonet 227
nano-object 80
nanooptics 223
nanooptoelectromechanical system (NOEMS) 3
nanoparticle 1, 57, 412, 996
– adsorption 189
– array 188
– focusing 191, 198
– gas phase synthesis 194
– mobility 191
– printing 206
– size distribution 187
– stability 192
– supercrystal 188
– synthesis 565
– tip 997
– transfer 206
nanopattern 1712
– periodic 1463
nanopatterned
– polymer 1470
– polymer surface 1471, 1712
– surface 1468
nanophotonics 397
nanopores 1680
nanopositioning 607
nanorheology 900
nanorobotic
– assembly 1642
– manipulation 1634
– manipulator (NRM) 1637
nanorobotics 1633
nanorod
– array 170
– symmetry 129
nanorope 464
nanoroughened surface 1463
nanoroughness 1450, 1454, 1464, 1572
nanoscale 1357
– adhesion 1663, 1683
– antibiotics 467
– bending 1728
– biomolecule 18, 20
– component 397
– contact resistance 1383
– device 1697
– effect 1482
– electrode 31, 35
– FET 138
– friction 815, 819
– hair friction 1137
– indentation 836
– molecule 42
– protein interface 453
– Schottky barrier 41
– spatula 1553
– supramolecular assembly 657
– test 1319
nanoscale charging
– effect of rubbing load 1157
nanoscale dispensing (NADIS) 1469
– probe 1469
nanoscale distance
– tunable 460
Nanoscope I 576
nanoscopic device 1664
nanoscratch 831, 1072, 1073, 1087, 1096, 1097
– test 1111
nanosphere 391, 398
nanostructure 399, 1490, 1494, 1516, 1526, 1715
– bending test 1230
– characterization 1711
– hybrid 469
– hydrophobic 1467
– indentation 1229
– mechanical property 1783
– roughness 1253
– scratch 1253
– scratch test 1229
– space-filling 463
– stress and deformation analysis 1227
– three-dimensional 2
nanostructured
– material 2, 1428
– surface 1494
nanosurface 1554
nanoswitch 23
nanotechnology 389
– material 455
nanotherapeutic device
– barrier to practice 478
– construction method 461
– intellectual property 462
nanothermometer 136
nanotopography 1458
nanotractor 1742, 1747, 1748, 1754, 1755
– dynamic friction measurement 1748
– release time measurement 1752
– static friction measurement 1749
nanotransfer printing (nTP) 318
nanotransistor 39
nanotribological
– characterization 1115
– properties of silicon 1685
nanotribology 475, 1310, 1348, 1663, 1683, 1746
nanotube 11, 20, 125, 1647
– assembly 1647
– bundle 998
– chemistry 79
– defect 55
– field-effect transistor (FET) 464
– functionalization of wall 85
– functionalized 84
– growth 53
– junction 1647
– morphology 57
– nucleation 53
– oxidation 84
– probe 626, 627
– resistance 41
– transistor 41
– volume fraction 96
– yield 57, 70
nanotube production 66
– efficiency 57
– electric-arc method 56
– laser ablation 54
– solar furnace 62
– techniques 68
nanotube-based

– emitter 87
– hybrid material 84
– sensor 88, 1667
– SPM tip 86
nanotube-polymer composite 96
nanotweezer 98
nanowear map 819, 820
nanowhisker 1541
nanowire 38, 390, 397, 1667
– crystallinety 122
– diameter 125
– diameter dependent magnetoresistance 141
– diameter distribution 125
– metal 81
– morphology 126
– photodetector 156
– semiconducting 123, 472
– sensor 158
– size depent Seebeck coefficient 144
– superlattice 125, 126
– transport property 139
– waveguide 156
– Zn 144
– ZnO 129, 131, 135, 144, 149, 155
nanowire FET 472
nanowire film
– aligned 129
nanowiring 217
NaPa Library of processes (NaPa LoP) 273
naphthalene channel 29
National Institute of Health (NIH) 5
National Institute of Standards and Technology (NIST) 5
National Nanotechnology Initiative (NNI) 4, 5, 390, 1635
National Science and Technology Council (NSTC) 4
National Science Foundation (NSF) 5
navigation system 364
Nb superconductor 674
$NbSe_2$ 683, 941
NC-AFM (noncontact atomic force microscopy) 688, 750
near-field
– scanning optical microscopy (NSOM) 148, 620
– technique 689
negative
– contact force 992
– differential resistance 145
– force gradient 734
– logic convention 34
Nelumbo nucifera 1425
NEMS 359, 382, 1783
– application 360
– carbon nanotube 382
– fabrication technique 381
– first generation 381
– material 381
– packaging 383
– technology 383
net
– displacement 1567
– fluid transport 512
– free-energy profile 1452
network shear modulus 1188
neuroactive agent 466
new product risk 1815
Newtonian
– flow 279, 884, 896, 897, 899
– viscosity 1187
next-generation lithography (NGL) 273
N-formyl peptide receptor (FPR) 498
NH_2 group
– tip 766
N-hydroxysuccinimidyl (NHS) 766
Ni(001) tip on Cu(100) 946
Ni(111) tip on Cu(110) 946
Ni, Co catalyst 55
Ni/Y catalyst 58
nicked strand 396
nickel iron permalloy 509
NiFe valve membrane 509
NiO 668
NiO(001) 647, 650
– surface 652, 653
nitrilotriacetate (NTA) 491, 766
nitrogenated carbon 1278
noble-metal surfaces 676
n-octadecyltrichlorosilane (OTS) 1320
noise 637, 641
– electronic 581
– external 665
– performance 641
– source 603
Nomarski interferometer 600
nonacosanol 1410
nonadhesive contact 880
nonbranched attachment system 1566
nonbuckling condition 1579
nonconducting film 1276
nonconductive
– material 654
– sample 577
noncontact
– AFM 635, 636, 641–647
– AFM image 645
– atomic force microscopy (NC-AFM) 642, 647, 746, 947
– dynamic force microscopy (NC-AFM) 947
– friction 948
– imaging 579
– mode 643
noncontant mode 638
nondestructive contact mode measurement 945
nonequilibrium interaction *see* dynamic interaction
nonlinear
– *I–V* behavior 145
– optics of nanowire 149
nonliquidlike behavior 900
nonmagnetic Zn 683
non-Newtonian flow 884, 885, 897, 899
nonpolar end group 1691
nonsilicon MEMS 1665
nonspherical tip 940
nonsticking criterion 1580
nonsymmetrical passive valve 519
nonwetting 874
NOR gate 23
normal friction 884, 892, 896
normal load 812, 1542
normalized frequency shift 747
Nosonovsky–Bhushan equation 1447
Nosé–Hoover thermostat 959
NOT gate 23
nozzle structure 225
NSOM as a defect probe 149
nTP (nanotransfer printing) 326
nuclear
– mass resonance (NMR) 1407
– radiation 1767
– stiffness 1193
nucleation of nanowire 130
nucleic
– acid (NA) 390
nucleus
– mechanical property 1193
numerical aperture (NA) 1013
numerical modeling 1357

O

O acceptor 18
octadecyldimethyl(dimethylamino) silane (ODDMS) 1329

octadecylmethyl(dimethylamino) silane (ODDMS) 1694
octadecylphosphonate (ODP) 1330, 1695
octadecyltrichlorosilane (OTS) 263, 1325, 1463, 1746, 1753
octyldimethyl (dimethylamino)silane (ODMS) 1329, 1694
oil droplet 1509, 1514
oil–air interface 1509, 1511
oil–water interface 1511
oleophilic surface 1450, 1509
oleophobic
– fabrication 1509
oleophobic surface 1509, 1518
– fabrication 1511
oleophobicity 1512, 1513
oligonucleotide 391
– discriminator 396
– primer 396
on-board control 399
on-chip
– actuator 1792
– microvalve 510
one-dimensional (1-D) 463
– quantum effect 136
– system 119
one-level
– hierarchical model 1569
– hierarchy elastic spring 1580
one-level structure 1463
– hydrophobic 1466
open platform 396
operation 637
optical
– absorption 151
– cavity 155
– cell manipulation 1201
– cell stretcher 1206
– chromatography 1217
– deformability 1218, 1221
– detector 579
– fiber 318
– force sorting 1216
– head 584
– head mount 585
– lever 601
– line tweezer 1212
– MEMS assembly and packaging 1822
– microswitch 1674
– output 24
– penetration depth 150
– profiler 1468
– property of nanowire 148
– property of SWNT 408
– ridge waveguide 318
– rotation 1211
– spanner 1211
– stretcher 1218
– switch 155
– transmission 148
– trapping 1013
optical beam deflection
– read out 437
– readout 435
optical lever
– angular sensitivity 603
– deflection method 610
– optimal sensitivity 603
optical microscope (OM) 1639
optical system
– self-aligning 370
optical trap 1179–1181, 1202
– dual-beam 1204
– single-beam 1217
optical tweezer (OT) 768, 864, 1013, 1181, 1204
– history 1207
– holographic 1209
– principle 1207
– Raman spectroscopy (OTRS) 1208
optical-electronic-optical (O-E-O) switch 371
optical-lever signal 715
optimal beam waist 603
optimum sampling phase 1627
optode detection 538
optoelectronic
– component 327
– module 370
OR
– gate 153
– operation 30
– operator 23
order
– in-plane 882, 900, 901, 909
– long-range 884
– out-of-plane 872, 873, 882, 884, 900, 901, 908, 909
– parameter 884, 902
ordered molecular assembly 1317
ordering nanowire 128
organic
– compound 1314
– droplet 557
– electronics 295
– inverter circuit 326
– light-emitting device (OLED) 295
– monolayer film 806
– transistor 323, 326
organized molecular array 31
organofunctional bond 1348
organometallic vapor-phase epitaxy (OMVPE) 1732
organosilsesquioxane 321
organs of plants 1400
orientational
– interaction 864
oscillating
– cantilever 725, 792
– tip 583
oscillation
– amplitude 639
– cycle 756
oscillatory
– force 858, 859, 871–875, 881–883
osmosis response time 519
osmotic
– force 878
– interaction 859, 869, 877, 878
– pressure 864, 870, 877, 878, 891
– stress technique 863
OTS (octadecyltrichlorosilane) 1766
outer epidermis 1412
out-of-plane coil 377
overwriting mechanism 1619
ovipositor 1532
oxide
– coating 128
– enhanced growth 126
– nanorod array 174
– sharpening 623
oxidized nanotube 84
oxygen
– content 1304
– sensor 525

P

packaging 383
– hermetic 1841
– IC 1836
– MEMS 1835
– polymer-MEMS 1849
– polymer-to-glass 1849
– polymer-to-polymer 1849
– polymer-to-silicon 1849
– reliability 1853
– vacuum 1842
– wafer-level 1849
packed system
– friction 991
PAH (polycyclic aromatic hydrocarbon) 100

palladium nanowire 158
p-aminophenol (PAP) 522
p-aminophenyl phosphate (PAPP) 522
paperlike display 322
papilla
– cell 1424
– density 1454
papillose epidermal cell 1439, 1454
parasitic charge 1767
parasitism 1528
partial erasing 1622
particle
– control 1820
– manipulation 1528
– removal 1501
– track-etched mica film 123
particle assembly
– classification 194
– rate 192
– setup 205
– statistics 193
– template 202
– yield 192
passive
– check valve 513
– droplet control 557
– droplet generation 558
– linearization 608
– microfluidic 513
– microfluidic device 507, 520
– microfluidic dispenser 524
– micromixer 515
– micropump 519
– microvalve 507, 513
– structure 1785
– valve geometry 515
pathogen microorganism 1423
pathophysiology 1172
patient-monitoring system 526
pattern
– density 282
– generating system 214
– of ink 314
– transfer 277, 284, 319
patterned
– adhesive 1534
– magnetic media 295
– polymer 1541
– self-assembled monolayer (SAM) 317, 318
– silicon chip 33
– structure 1441
– substrate 130
patterned surface 1714
– surface topography 1473
patterning 129
paxillin (Pax) 1194
Pb on Ge(111) 682
PbI_2-filled SWNT 82
PCR amplicon 396
– down format 396
PCR hybridization 547
PDMS (polydimethylsiloxane) 282, 319, 505
– blend silicone 1079
– silicone conditioner 1163
peak indentation load 1287
peak-to-peak load 1230
peak-to-valley (P–V) 1239, 1456
peapod 84, 464
pearl-shaped nanowire 132
PECVD (plasma enhanced chemical vapor deposition) 235, 1311, 1607
– carbon sample 1278
peel number 1765
peeling action 1558
Peierls instability 682
pentaerythritol tetranitrate (PETN) 441
peptide toroid 467
perfluorinated SAM 1331
perfluorodecanoic acid (PFDA) 1312
perfluorodecylphosphonate (PFDP) 1329, 1695
perfluorodecyltricholorosilane (PFTS) 1329, 1694
perfluorodecyltriethoxysilane (PFDTES) 1463, 1471, 1697, 1705, 1712
perfluorooctanesulfonate (PFOS) 1465
perfluoropolyether (PFPE) 840, 1347, 1348, 1376, 1691
– nanotribological property 1347
perfluoropolyether (Z-DOL) 812
periodic
– boundary condition (PBC) 958
– potential 929
peristaltic
– force 878
– micropump 511
permalloy 509
– microarray 526
permanent dipole moment 657
perpendicular
– magnetization 296
– scan 593
persistence length 300
perturbation approach 640
perylene 654
PES (photoemission spectroscopy)
– generation 1625
petal 1414
PFPE lubricant film 1350
– chemical degradation 1366
– surface topography 1354
pH sensor 158
pharmacophore 464
pharmacy 462
phase
– angle 1070
– imaging 755
phase change
– medium 1602
phase transformation
– nanoindentation 967
phase-breaking length 141
phenanthrene 227
phenoxy-centered HOMO 34
phonon
– assisted optical transition 151
– boundary scattering 146
– confinement effect 151
– excitation 677
– phonon scattering 146
– quantum confinement effect 146
phosphate-buffered saline (PBS) 1699
phosphazene lubricant 1366
phospholipase C (PLC) 491
phosphosilicate glass (PSG) 235, 249
photoacid 27
photoactive
– fragment 22
– solid 30
– stopper 22
photoemission spectroscopy (PES) 674
photogenerated hole 38
photoinduced
– electron transfer 24, 25
– proton transfer 27
photolithographic fabrication 1729
photolithography (PL) 271, 273, 274, 300, 389, 1712
– near-field conformal 316
– pattern 314
– proximity mode 315
photoluminescence (PL) 147, 148, 154, 155
photonic
– crystal (PhC) 188, 295
– force microscope (PFM) 1015

photoresist 317, 319, 375
– master 1260
– nanorod 1590
photosynthesis 1418
photothermal
– effect 750
– spectroscopy 434
physical properties of nanowire 130
physical vapor deposition (PVD) 1465
physical wear 1153, 1369
physisorption 1347
phytolith 1415
pick and place 188, 189, 389, 391, 399
– technique 188
pick-up tip 627
picoindentation 836
– system 798
picomotor 1637
piezo
– ceramic material 608
– effect 607
– excitation 736
– hysteresis 665
– position plot 1150
– relaxation 665
– stack 609
– tube 607
piezoelectric
– actuation 509
– crystal sensing 492, 496
– drive 575
– leg 928
– positioning elements 664
– scanner 621, 665
piezoresistive
– cantilever 436, 442, 445, 604
– coefficient 604
– deflection 726
– detection 603, 637
– property of SWNT 407
– sensor 1563
piezoresistor 361
piezoscanner 611
piezotranslator 611
piezotube calibration 577
pigment particle 325
pile-up
– characteristic 1617
– nanoindentation 966
– surface 968
– surface atom 963
pillar 1545
– array 1541
– array fabrication 1587
pilot signal 1627
pinning 698
– of a meniscus 200
pin-on-disk tribotester 1319
pipe flow 536
piriform gland 1527
PL NSOM imaging 155
placement accuracy 191
planar immunoFET 453, 473
planarization layer 286
plane-to-plane separation 20
plant 1525, 1537
– hair 1413
– surface structure 1411
plant surface 1399, 1400, 1536
– AFM 1408
– aquaplaning 1419
– architecture 1402
– hydrophobic structure 1403
– multifunctional 1417
– reflection and absorption 1417
– self-cleaning 1420
– slippery 1418
– wettability 1420
– wetting 1421
plant wax 1441, 1490, 1715
– crystallinity 1407
– morphology 1403
– self-assembly 1407
– spectral property 1418
plasma 284
– assisted growth 125
– deposited coating 1765
– deposition 1765
– enhanced chemical vapor deposition (PECVD) 326, 588, 1235, 1273, 1277, 1465, 1697
– etching 1730
– frequency 148
– resonance 147
plasmon waveguide spectroscopy 492
plasmon-waveguide resonance (PWR) 495
plastic
– circuit 323
– deformation 1290
– electronics 327
– fluidic chip 524
– large area circuit 324
plastic deformation 985
– surface 964
plastic/elastic model 1765
plastically deformed 1078
platelet nanofiber 71
plate-like 1531
platform 1610
– fabrication process 1611
platinum microelectrode 393
platinum-iridium 578
– tip 36
plug polysilicon 1731
PMMA (poly(methyl methacrylate)) 505, 507, 722, 1604, 1615, 1618
– physical property 1245
p-n junction 153, 155
– diode 152
– nanowire 137
pneumatic
– actuation 508
– damping 668
p-nitrophenolate 27
pocket formation 1513
point-of-care testing (POCT) 526
Poisson distribution 193
polarization effect 151
polarizer 297–299
polishing 1271
pollen grain 1529
poly(dimethylsiloxane) (PDMS) 505, 1227, 1229, 1260, 1580, 1732
poly(ethyleneterephthalate) (PET) 321
poly(methacrylic acid) (PMAA) 442
poly(methyl methacrylate) (PMMA) 252, 275, 278, 289, 505, 507, 627, 1227, 1229, 1244, 1470, 1604, 1615, 1618, 1712, 1732
poly(propyl methacrylate) (PPMA) 1229, 1244
poly(tetrafluoroethylene) (PTFE) 1463
polyacrylamide 395
polyaromatic shell 74
polycaprolactone (PCL) 302
polycarbonate (PC) 272, 507, 1465
– membrane 170
polycrystalline graphite 1279
polydimethylsiloxane (PDMS) 204, 293, 316, 540, 548, 1079, 1312, 1471
polyetheretherketone (PEEK) 436
polyethylene (PE) 507
polyethylene glycol (PEG) 441, 765
polyethylene naphthalate (PEN) 1214
polyethylene terephthalate (PET) 834
polyethyleneimine (PEI) 442
polymer 278, 505

– adsorption 876
– biocompatible 1227
– bioMEMS 1229, 1679
– blend film 722
– brush 877, 878, 889, 897
– *brushes* 989
– cantilever 1261
– chain relaxation 882
– chemistry 463
– end-adsorbed 877, 878
– fibril 1578
– filament 1187
– film 476
– grafted 877
– in solution 876
– interaction 876
– liquid (melt) 876, 897, 899
– matrix composite 95
– medium 1614
– microbeam 1260
– microfabrication technique 505
– microstructure 507
– mushroom 878
– nanojunction 40
– nanopatterned 1470
– nonadsorbing 877
– optics 299
– physisorbed 877
– property 514
– transparent microcapsule 325
polymer beam
– mechanical property 1248
polymer fiber orientation
– process step 1587
polymerase chain reaction (PCR) 542, 565
polymeric
– domain 463
– magnetic tape 836
– microbeam 1244
– unit self-assembling 466
polymerization 283
polymethylmethacrylate (PMMA) 381, 1465
polypeptide 18
polypropylene (PP) 654, 757
polypyrrole (PPy) 1465
polysilicon 360
– cantilever beam 1719
– deposition 1731
– doped 1689
– fatigue 1798
– film 1235, 1689, 1787
– fracture strength 1797
– fracture toughness 1797
– friction 1798
– layer 514
– mechanical property 1795
– microstructure 1797
– residual stress 1796
– undoped 1689
– Young's modulus 1797
polystyrene (PS) 275, 507, 1229, 1244, 1465, 1470, 1618, 1712
polystyrene/nanoclay composite (PS/clay) 1244
polysulfone 1618
polytetrafluoroethylene (PTFE) 293, 932
polyurethane (PUR) 757
– film 1115
polyvinyl alcohol (PVA) 1261
polyvinyledene fluoride (PVDF) 1637
polyvinylidene chloride (PVDC) 1849
polyvinylsiloxane (PVS) 1588
pore size 1466
poroelastic solid 1186
porous anodic alumina (PAA) 1590
position
– accuracy 611
– error signal (PES) 1625
position control
– closed-loop 373
positioning 130
position-sensitive detector (PSD) 434
positive
– logic convention 26, 34
– replica 1489
potential 957
power
– dissipation 756, 947
– spectrum 1016
power spectra
– MD simulation 983
power-spectral density (PSD) 1016
p-phenylenediamine (PPD) 1068
PPMA
– physical property 1245
precipitation hardening 1773
pre-crack length 1787
precursor 66, 81, 82
preferential flow direction 511
preferred crystal orientation 122, 123
preferred growth direction 131
preparation improvement 64
prepolymer 318
press 293
pressure
– and flow sensing 444
– contact 863
– homogeneity 294
– injection technique 122
– Laplace 858, 878
– osmotic 864, 870, 877, 878
– sensing 363
pressure sensor 361
– environmental resistance 1822
– feedback architecture 362
– interface 362
– monolithic 364
– packaging 1821
– schematic 361
– touch-mode 362
pressure-sensitive adhesive 1542
prestress 1188
prey capture 1528
primary creep 1772
printed
– circuit board (PCB) 287
– coil 327
– DFB resonator 328
probabilistic fastener 1531
probability force distribution 773
probe
– colloidal 862
– FIB-milled 590
– surface 766
probe tip
– performance 622
– standard 621
probe–sample distance 740
process
– gas 1277
– sequence 276
processing aid
– condition 52
production efficiency 55
production of MEMS products 1803
projection mode photolithography 315
promyelocytic leukemia body (PML) 1177
properties of a coating 1273
prostatespecific antigen (PSA) 443, 444
protective nitride layer 1731
protein 485
– biologically produced 471
– circular permuted 471
– coating 1698
– database 456
– engineering 490

– expression 489
– kinase (PKA) 443
– kinase inhibitor (PKI) 443
– macromolecule 390
– protein interaction 1179
– sensing interface 475
– sensor 453
– unfolding 1179
protrusion
– force 859, 878
PS (polystyrene) 722
– physical property 1245
Pt alloy tip 579
PTFE (polytetrafluoroethylene)
– coated Si-tip 932
Pt-Ir
– tip 578
pull-in
– voltage 1768
pull-off 1588
– cycle 1567
– force *see* ahesion force879, 927, 1350, 1353, 1534, 1565, 1566
pull-off force *see* adhesion force
pull-out voltage 1768
pulsed
– clamping 1748
– electrodeposition 124
– force mode 798
– laser deposition (PLD) 238
pumping chamber 511
P–V (peak-to-valley distance) 1699
pyramidal tip 624
pyrazoline 24
pyrolysis of hydrocarbon 70
pyrolytic method 68
PZT (lead zirconate titanate)
– scanner 792
– tube scanner 576, 582
Péclet number 561

Q

Q-control system 743
Q-factor 300, 637, 638, 692
qPlus sensor 726
quad photodetector 582, 792
quadrant
– detector 602
– photodiode (QPD) 1015, 1208
quadrature signal 1625, 1626
quality factor *Q* 587, 596, 637, 639, 736, 757, 1749, 1854
quality system for MEMS manufacturing 1824
quantized
– conductance 136
– thermal conductance 147
quantum
– box (QB) 9, 1667
– contact breakage experiment 136
– corral 679
– dot 390, 397, 398, 1732
– dot transistor 1665
– Hall regime 695, 696
– limit for the thermal conductance 147
– limit of thermal conductance 147
– size effect 139, 154
– size regime 137
– subband 137
– tunneling 671
– wire (QWR) 9, 1667
– wire superlattice 145
– yield 24
quantum confinement 138
– effect 137, 148, 156, 159
quartz crystal microbalance (QCM) 956
quasi-continuum model 967
quasi-optical experiment 672
quasistatic
– bending test 1240, 1243
– mode 638
quasistructure 223

R

radial cracking 1286
radiation pressure 1207
radiofrequency (RF) 373
– identification (RFID) 189
– MEMS/NEMS 1664
radioligand binding 492
radius
– hydrodynamic 876
– of gyration 278, 876
Raleigh's method 596
Raman
– spectra 1280
– spectra of nanowire 151
– spectroscopy 1208, 1278
Rame–Hart model 1454
random
– access memory (RAM) 35
– nanoroughness 1589
– rough surface 1593
– sequential adsorption (RSA) 193
randomly oriented polysilicon 1797
ranking of various SAMs 1324
rapid thermal processing (RTP) 1844, 1852
ratchet mechanism 807, 808, 979
rate of deformation
– simulation 967
rate-and-state friction 1747
Rayleigh
– regime 1202
RbBr 647
reaction
– biochemical 513, 520
reaction force 1723
reactive
– ion etching (RIE) 241, 284, 285, 504
– oxygen species (ROS) 100
– spreading 318
read
– channel 1621
readback
– rate 1614
– signal 1622
readout
– capacitive 437
– electronics 602
– interferometric 437
– optical beam deflection 437
– signal 1625
reagent mixing 560
real contact area 1743
rebinding 771
receding contact angle 1481
receptor
– immobilization 766
– transmembrane 1194
receptor binding
– Bell model 1190
receptor–ligand
– bond 763
– complex 764, 1189
– interaction 859
– pair 765
receptor–ligand recognition
– bond formation 764
recognition
– force spectroscopy 771
– imaging 763, 779
recombinant protein 488, 490
rectangular
– cantilever 587, 948
– Si 1454
reduced Young's modulus 721
reduction of average wire diameter 125
reduviid bug 1528
reflection interference contrast microscopy (RICM) 864
refractive index 1206

– effective 1205
regime
– Cassie–Baxter 1452, 1478
– Wenzel 1452, 1478
relative humidity (RH) 1358, 1360, 1454
– influence 1358
relative stiffness 800
relaxation time 278, 884, 885, 889, 899, 903, 909, 910
releasable attachment 1533
release
– layer degradation 290
– time 1752
reliability 1761
– packaging 1851
relief on a surface 314
remanence 158
remote detection system 602
repetitive mechanical stress 1773
replication 291
reporter dye 396
reptile 1537
repulsive
– tip–substrate 967
residual
– film stress 1783
– layer 281
– layer height 280
– stress 1282, 1285, 1290
resin 271
resist 275, 276
– material 288
– pattern 273
resisting
– pushing force 992
resolution
– vertical 580
resonance
– curve detection 587
– mechanical 907
– Raman effect 151
– spectroscopy 495
resonant frequency 1749
restoring force 1612
rest-time effect 1355, 1356
retardation effect 865, 868
retract
– curve 768
– trace approach 768
reusable engineering 1816
reversal NIL 287
reversible attachment 1544
RF magnetron sputtering 1273
RF MEMS 1767, 1769, 1774
– switch 1767
rheology 277
rigid stamp 280
rigidity 1188
RLL (run-length-limited) code 1628
RNA 390, 399, 442
robotics 1633
roll embossing 292, 294
rolling
– friction 867, 883
roll-to-roll 294
– fabrication 207
room temperature (RT) 1334
root mean square (RMS) 476, 688, 1492, 1498, 1569
– amplitude 1581
– height 1492
rotaxane 21
rotor–hub interface 1717, 1719
rotor–stator interface 1669
rough surface 1443, 1447
roughness 910, 1058, 1255, 1536, 1765
– amplitude 1594
– angle 826
– energy scale 778
– factor 1462, 1472
– hierarchical 1454, 1461
– image 801
– induced hierarchical surface 1729
– of surface 872, 873, 881, 883
– size 1447
– structure 1585
roughness factor 1444, 1461, 1710
– calculated 1455
Ru(II)-trisbipyridine 21
rubbing/sliding surface 1775
run-length-limited (RLL) 1628
rupture force 770
rutile $TiO_2(100)$ 650

S

(*S*-acetylthio)propionate (SATP) 766
sacrificial
– layer 1680
– layer lithography 1731
– layer lithography (SLL) 1730
– LIGA (SLIGA) 254
– oxide 1731
Salvinia 1416
– hair 1417
SAM (self-assembled monolayer) 1473, 1765
– chemically adsorbed 846
– coated surface 1706
– coating 1765
– deposition 1323
– edge resolution 316
– friction 846
– nanotribological property 1348
– on aluminum 1321
– on copper 1321
– stiffness 846
– sudden failure 1696
sample holder 667
sampling 1528
– rate converter (SRC) 1622
sandwich immunoassay 521
sandwiched molecule 34
sandwich-type 396
saturated calomel electrode 32
scaling effect 1536
scan
– area 584
– direction 593
– frequency 584
– head 667
– range 607, 608
– rate 577, 795
– size 585, 1116
– speed 608
scanner 370
– piezo 692
scanner-displacement amplitude 1613
scanning
– acoustic microscopy (SAM) 1778
– capacitance microscopy (SCM) 574
– chemical potential microscopy (SCPM) 574
– electrochemical microscopy (SEcM) 574
– electron microscope (SEM) 130, 131, 215, 223, 257, 830, 1057, 1453, 1490, 1493, 1497, 1556
– electron microscopy (SEM) 621, 1400, 1609, 1763
– electrostatic force microscopy (SEFM) 574
– head 577
– ion conductance microscopy (SICM) 574
– ion microscope (SIM) 218
– Kelvin probe microscopy (SKPM) 574
– lateral range 1610
– magnetic microscopy (SMM) 574
– near field optical microscopy (SNOM) 574

Subject Index

– near-field OM (SNOM) 1639
– probe lithography 315
– probe microscope 620
– probe microscope (SPM) 574, 636, 1637
– probe microscopy (SPM) 258, 619, 620
– speed 584
– spreading resistance microscopy (SSRM) 1369
– system 607
– thermal microscope (SThM) 134, 574
– tunneling microscope (STM) 258, 635, 636, 646, 732, 746, 790, 1602, 1635
– tunneling microscopy (STM) 9, 573, 619, 620, 630, 647, 963
– tunneling probe 133
scanning force
– acoustic microscopy (SFAM) 574
– microscopy (SFM) 428, 574, 688
– spectroscopy (SFS) 688
scanning tunneling microscopy (STM) 133
scanning tunneling spectroscopy (STS) 134
scattering force 1202, 1204
SCC (stress corrosion cracking) 1774
Schottky barrier (SB) 410
scratch
– critical load 1302
– damage mechanism 1296, 1297
– depth 977
– depth profile 1096, 1097
– drive actuator 1792
– drive actuator (SDA) 1742
– induced damage 1293
– resistance 1230, 1237, 1248, 1327
– test 1101, 1235, 1293
scratching 828, 1099
– force 977
– measurement 574
screening 874, 875
S-curve 1773
sealing 1836
second
– anodization 121
– harmonic generation 149
secondary alcohol 1406
Seebeck coefficient 144, 154
– temperature depent 144
seed 1530
selected area electron diffraction (SAED) 131, 135
selected growth direction 132
selectin 775
selective glue 398
selectively modified 399
self
– assembly 1746
– cleaning 1708
self-assembled
– growth 671
– nanotube bunch 69
– patterned monolayer (SAM) 317, 318
– structure 990
self-assembled monolayer (SAM) 35, 193, 251, 316, 414, 440, 497, 971, 1309, 1311, 1316, 1348, 1463, 1470, 1694
– nanotribological property 1309, 1319
– tribological property 846, 1694
self-assembly 152, 153, 188, 389–391, 397, 463, 671, 1407–1409, 1441, 1500
– chemical 1428
– directed 1589
– of organic layer 31
self-cleaning 1399, 1421, 1424, 1425, 1437, 1560, 1663
– efficiency 1501, 1717
– mechanism 1722
– surface 1437, 1439, 1495, 1709, 1710
self-cleaning plant leave 1453
– microstructure 1453
self-excitation mode 745, 758
self-lubrication 945, 946
self-repairing process 1426
self-replicating assembly 1648
self-similar scaling 1558
SEM monitoring 215
semiconductor
– nanowire 173
– quantum dot 680
– surface 642
– SWNT 87, 98
semimetal–semiconductor transition 139, 143
semisynthetic nanodevice 469
sensing device 471
– biological 157
– chemical 157
sensing element 361
sensitivity 411, 604, 605, 641, 925
sensitizer 38
sensor
– application 138
– biochemical 409, 1312
– biological 380
– capacitive 361
– characteristic 411
– characterization 411
– chemical 380
– concept 403
– force 380
– inertial 364
– mechanical 428
– micromachine 367
– neuron 466
– piezoresistive 361
– pressure 361
– self-cleaning 1429
– surface, cantilever 429
separation of bacteria 397
sequential logic circuit 28
sequential microfluidic manipulation 519
servo
– demodulation 1626, 1627
– field 1624
– loop 1624
– self-writing 1627
seta 1535
setal orientation 1559
setpoint amplitude 714
SFM (scanning force microscope) 664
– bath cryostat 668
– dynamic mode 688, 692
– interferometer 668
– piezo creep 665
– static mode 688
– thermal drift 664
– thermal noise 665
– variable temperature 666
– vibration isolation 666
SGS-SWNT (small-gap semiconducting SWNT) 406
shape memory alloy (SMA) 510
shark
– skin 1426, 1439, 1512
– skin replica 1513, 1516, 1517
shear 1769
– flow 1179, 1182
– focusing 559
– force *see* kinetic friction, static friction
– melting 884, 902
– mode 800
– modulus 925, 1183, 1185, 1615
– plane 862
– strength 939
– thinning 897, 899, 900, 904

– wave transducer 815
shear flow
– detachment (SFD) 767
shear stress 1180
– critical 886, 893, 901, 902
– effective 941
shearing design 558
shearing system 560
SH-group 765
SHiMMeR system 1762
shoot 1413
short-cut carbon nanotube 680
short-range
– chemical force 693
– electrostatic attraction 649
– electrostatic interaction 648
– magnetic interaction 647, 652, 653
shot noise 603
shrinkage 283
Shubnikov–de Haas quantum oscillator 143
Si 319, 580, 604, 638
– adatom 643
– based mold masters 506
– beam 1239
– crystalline 504
– dioxide (SiO_2) 504
– dioxide layer 38, 40
– grain 1796
– membrane 514
– micromachining 361
– nanobeam 1259
– nanomechanical beam 381
– nanowire 126, 132
– nitride (Si_3N_4) 504
– nitride (Si_3N_4) 580
tip 927
– on-insulator (SOI) 375, 1652
– polycrystalline 360
– stamp 319
– terminated tip 689
– tip 642, 644, 689, 927
– trimer 645
– uncoated 1388
– wafer 32, 35, 314, 318, 1611
Si cantilever 637, 925, 1454
– gold-coated 765
Si replica
– hierarchical 1716
– micropatterned 1500
Si surface
– micropatterned 1473, 1712
Si_3N_4 tip 589, 621, 624, 794, 795, 802, 819, 824, 827, 842, 1319, 1352, 1365, 1705
Si(001)(2×1) 644
Si(001)(2×1):H surface 644
Si(100) 1296, 1298
– coefficient of friction 1360
– substrate surface 1349
Si(111) surface 654
Si(111)-($\sqrt{3}\times\sqrt{3}$)-Ag 646
Si(111)-7×7 932
Si–Ag covalent bond 646, 647
Si-based surface 1698
Sic film
– doped 1689
– undoped 1689
signal
– transduction 28
signal-to-noise ratio (SNR) 411
silane 285, 291, 292
– bubbler 1319
– film 476
– polymer 476
– polymer film 1701
silanization process 1699
silanol group (SiOH) 1321
silicon 1540, 1683
– cantilever 795
– friction force 1333
– microimplant 1698
– micromotor 1352
– nitride (Si_3N_4) cantilever 1469
– nitride layer 1731
– oxide 1413
– pillar 1467
– surface 1318, 1390
– tribological performance 1684
silicon tip 622, 627, 1146
– single-crystal 795
silicon-on-insulator (SOI) 243
silicon-on-sapphire (SoS) 236
siloxane polymer lattice 1702
single
– cell 1411
– CNT probe 628
– gecko seta 1585
– magnetic flux quantum 143
– mode optical fiber 318
single asperity 821, 939
single atomic bond
– measurement 732
single crystal
– nanorod 179
– nanowire 122
– silicon 1784
– silicon cantilever 589
single molecule
– packing density 846
– recognition force detection 767
– spring constant 846
single nanowire
– TEM image 124
single spatula
– contact angle 1576
single-chain fragment variable (SCFv) 474
single-crystal
– reactive etching and metallization (SCREAM) 247
– silicon cantilever 799
single-crystal Si(100) 1375, 1378
single-crystal silicon 1712
– hardness 837
single-domain magnetic nanowire 159
single-level attachment system 1577
single-molecule force spectroscopy
– dissociation rate 772
single-particle wavefunction 673
single-spring contact 1567
single-step NIL 293, 294
single-wall carbon nanotube (SWCNT) 2, 463, 464
single-wall carbon nanotube (SWNT) 40, 48, 403, 590, 691, 1644, 1665, 1682
single-wall nanocapsule 57
single-wall nanotube (SWNT) 624
sintering 864, 867
SiO_2 64
– beam 1239
– film surface 1318
– MEMS semiconductor 1561
– nanobeam 1259
Sixth Framework Program (FP6) 5
size of lamella 1594
size of seta 1593
skewness 1718
skin
– synthetic 1135
sliding
– contact 944
– direction 901
– distance 905
– friction 814, 867
– induced chemistry 978
– of tip 924
– velocity 819, 889, 901, 937, 1323, 1332, 1722
– work 984
s-like tip state 673
slip 1769
– plane 176, 862

slippery surface
– wax induced 1419
slipping and sliding regime 902
SMA (shape memory alloy) 510
– actuation scheme 510
– driven micropump 511
small specimen handling 1785
smart
– adhesion 1553, 1554, 1591
– material 188, 1399
– plastic biochip 524
– system 1706
smooth
– nanobeam 1255
– pad 1537
– sliding 901
– surface 1443
S–N (stress–life) diagram 835
snap 1526, 1531
snap-in 1353
– distance 1146, 1470
$SnO_2(110)$ 650
soft
– lithography (SL) 272, 282, 506, 512, 1712, 1732
– polymer stamp 319
– stiction 1727, 1728
– stuck micromirror 1727, 1728
– substrate 975
– surface 755, 793
solar furnace
– device 62
– method 73
sol–gel 291
– foam 1466
solid
– boundary lubricated surface 885, 886, 891, 893, 894, 905, 906
– lubricant 1725
– phase Z-DOL 1354
– solution catalyst 65
– surface 1710
– thin film 1000
– xenon 692
solid–air interface 1453, 1509
solid–air–liquid
– contact line 1485
– interface 1482, 1503, 1504
solid–air–oil interface 1509
solidification 895, 899
solidlike
– behavior 909
– lubricant 1725
solid–liquid interface 1440, 1475, 1503, 1508
solid–liquid–air interface 1711
– schematic 1444
solid–liquid–vapor system 1472
solid–lubricant interaction 985
solid–oil interface 1509, 1513
solid–oil–air system 1450
solid–oil–water system 1450
solid–solid contact 1153
solid-state switch 373
solid–water interface 1513
solid–water–oil interface 1509, 1511, 1516
solution
– hardening 1773
– phase growth 127
solution-hardened 1772
solvation
– force 859, 871–875, 881, 883
soot 54
sp^3
– bonded carbon 1292
– bonded tip 968
– bonding 1284
spacer 1526, 1531
– chain 1316, 1317
spark-plasma sintering (SPS) 94
spatial
– frequency doubling 298
– light modulator (SLM) 1209, 1212
spatula 1534, 1542
spatula–substrate bifurcation 1560
specific
– adhesion energy (SAE) 1539
– discriminator 396
spectroscopic resolution in STS 669
spherical
– particle 1706
– tip 1582
spider 1525, 1528
– dry adhesion 1555
Spiderman toy 1586
spin
– casting 321
– density wave 685
– excitation 677
– quantization 675
spiropyran 26
split droplet 563
split-diode photodetector 801
SPM (scanning probe microscopy) 664
spore 444
spring
– force–distance curve 1569
– level 1566
– sheet cantilever 590
– system 575, 1610
spring constant 430, 637, 639, 640, 643, 688, 732, 733, 925, 1791
– calculation 595
– change 665
– effective 929, 938
– lateral 587, 928
– measurement 594
– vertical 587
spring-damper-mass model 1613
spring-shaped nanowire 132
SP-STM 686
sputtered coatings
– physical property 1282
sputtering
– deposition 1277
– power 1283
square-pyramidal Si 1454
squeeze flow 279–281
SrF_2 649
$SrTiO_3(100)$ 650, 651
S-shaped 1764
S-shaped beam 1764
s-SWNT
– bandgap 406
– Schottky barrier 409
– work function 409
stability
– mechanical 1607
STA-biotin
– wear 1704
stable production of MEMS products 1808
stacked-cone nanotube 625
stamp 291
– Au coated PDMS 322
– Au/Ti coated 319
– composite 316
– copy 290
– depth 319
– fabrication 290, 314
– geometry 280, 281, 283
– material 290
– mechanical property 316
– positioning 324
– protrusion 280
– rigid 319
standard temperature and pressure (STP) 1408
static
– AFM 637–639, 732
– deflection AFM 732
– friction 886, 892, 895, 899, 901, 902, 904, 905, 907, 909, 1794
– friction force 1724

– friction force (stiction) 1723
– friction torque 1724
– indentation 839
– interaction 858, 900
– mode 637, 688
– mode AFM 602
– random access memory (SRAM) 369
static contact angle 1322, 1330, 1424, 1468, 1492, 1493, 1495, 1508, 1695
– calculated 1474
statistical method 1016
statistical optical reconstruction microscopy (STORM) 1201
statistically stored dislocation (SSD) 1027
steady-state
– sliding 989
– sliding *see* kinetic friction
Stefan
– adhesion 1532
– equation 280
step and flash imprint lithography (SFIL) 282
step edge 127
step-and-repeat 292
– imprint 291
– NIL 294
step-and-stamp (SSIL) 294
stepper motor 928, 1742
stereoelectronic property 20
steric
– force 864, 878
– repulsion 859, 869
Stern layer 176
sticking
– condition 1580
– phenomenon 1467
– regime 901
stick–slip 867, 886, 892, 895, 899–901, 905, 906, 908, 1361
– behavior 804
– friction 977
– mechanism 928
– movement 804
– phenomenon 805
stick–slip testing 1754
– nanotractor 1754
Stickybot 1543
stiction 839, 895, 1309, 1663, 1763, 1767
– due to electrostatic attraction 1767
– phenomena 1725
stiff hair surface 1153
stiffness 1297, 1325
– continuous measurement (CSM) 1287
– nuclear 1193
– torsional 597
Stillinger–Weber potential 957, 967
stimulated emission depletion (STED) 1201
STM (scanning tunneling microscope) 573, 664
– bath cryostat 668
– cantilever 578
– cantilever material 588
– light emission 672
– piezo creep 665
– principle 575
– probe construction 578
– spectroscopy 673
– spin-polarized 686
– thermal drift 664
– thermal noise 665
– tip 574, 667
– variable temperature 666
– vibration isolation 666
Stokes's equation 1615
Stone–Wales defect 74, 80
Stoney
– equation 1784
– formula 431
stop valve 515
storage
– capacity 1628
– field 1624
– surface topography 1623
strain 407, 1181, 1769
– energy difference 1287
strain gauge
– rotating 1785
strand displacement amplification 396
stray capacitance 605
strength 1783
– of adhesion 1556
streptavidin (STA) 391, 395, 771
– adsorbed 476
– adsorbed to silicon 475
– binding 476
– coating 1699
– protein binding 1698
stress 1611, 1783
– distribution 1255
– fiber 1175, 1196
– field 1784
– gradient 1786
– induced crystalline 128
– intensity 1233
– maximum 1787
– measurement 1785
– relaxation test 1771
stress–strain
– behavior 1789
– curve 1110
– ratio 1187
– relationship 1578
Stribeck curve 885, 900
strip domain 697
structural
– characterization 130, 132, 1070
– force 859, 871, 873, 881
– integrity 1188
– material 1788
structure
– active 1788
– of a SWNT 405
STS 673
– energy resolution 674
– inelastic tunneling 676
Stuck comb drive 1670
stuck micromirror 1726
SU-8 resist 1610
subangstrom deflection 621
subangstrom positioning 620
subcellular operation 227
subcellular structure
– mechanics 1188
subcuticular insert 1415
subcuticular layer 1415
sublimation 62
submicron droplet 1472
submicron particle 1706
subnanometer precision 18
substrate 128
– curvature measurement 1772
– curvature technique 1784
– particle interaction 1561
– strain 1179
– surface energy 1562
subwavelength 297
– metal wire grating 298
– metal wire/strip grating 295, 297
– metal-strip grating 297
sucker 1530
suction cup 1528, 1530, 1544
sulfonium 1377
sulfur-gold bond 321
summit 1492
– density 1496, 1498
superadhesive tape 1553
supercapacitor 98
superconducting
– gap 683
– magnetic levitation 575

Subject Index

– matrix 698
– nanowire 123
superconductivity 669, 683, 685, 697
– vortex 683, 697
– vortices 697
superconductor 683
– type-I 683
– type-II 683
superhydrophobic
– contact angle 1438
– natural 1439
superhydrophobic surface 1399, 1401, 1421, 1425, 1427, 1429, 1456, 1518
– how to make 1462
– self-cleaning 1429
– wetting 1446
superhydrophobicity 1437, 1439, 1492, 1493, 1496
– hierarchical structure 1496
– roughness-induced 1440, 1663, 1708
– technical use 1426
– underwater use 1429
superlattice
– nanowire 123, 133
– structure 128
superlubric state 986
superlubricity 933
superparamagnetic 158
– limit 296
supersaturated nanocrystal 125
superstructures of nanowire 128
supramolecular
– assembly 25
– chemistry 463
– force 42
– interaction 198
surface 1501, 1539
– acoustic wave (SAW) 295
– adsorption 514
– air-retaining 1426
– amorphous 858
– atom layer 638
– attachment 491
– band 675
– biomimetic 1426
– charge 871
– charge density 869–871
– chemistry 318
– crystalline 858, 881, 882
– diffusion 318
– elasticity 798, 838
– energy 1270, 1322, 1445, 1556
– energy (tension) 861, 865–867, 879, 880, 883, 888, 889, 893
– free energy 1330
– friction 976
– functionalization 414
– functionalized 429
– heterogeneity 883
– hierarchical 1428
– hierarchical structured 1501
– hydrophilic 858, 874, 1421
– hydrophobic 858, 874, 1423
– imaging 802
– immobilization 491
– inhomogeneity 1711
– insect-inspired 1541
– interaction 790
– interaction energy 1764
– lubricated 1362
– material 1557
– micromachining 360, 363, 1730
– modification 138
– nanometer-scale mechanical property 961
– natural water-repellent 1462
– oxide 132
– parameter 1593, 1594
– pattern 1470
– patterned 1441
– phonon mode 959
– plasmon resonance (SPR) 492, 494, 497
– profiler 1719
– property 1402
– protection 1310
– sculpture 1412
– self-cleaning 1437, 1439, 1495
– self-cleaning efficiency 1501
– separation 860
– slippery 1419
– slope 1449
– state lifetime 674
– stress 429, 430
– superhydrophilic 1421
– superhydrophobic 1423, 1429, 1456, 1462
– topography 924
– transfer chemistry 319
– ultrasmooth 823
– unlubricated 1362
– wetting 1420
surface characterization 792, 1458, 1459
– 2-D profile 1459
surface film
– nanotribology 1310
surface force 858, 1763
– apparatus (SFA) 767, 790, 860–862, 865, 866, 891, 901, 904, 956, 1025, 1146, 1719, 1746
– optically induced 1206
surface height 809, 1368, 1369
– image 1378
– profile 1593
surface height map 1456, 1713
– contact mode 1458
– tapping mode 1458
surface potential 796, 869–871, 1072, 1153, 1154, 1276, 1381
– effect of external voltage 1155
– Fermi energy level 1369
– mapping 796
– study 1071
surface potential map 1368, 1385
– chemically damaged hair 1161
surface roughness 792, 810, 872, 873, 881, 883, 1122, 1323, 1443, 1448, 1500, 1561, 1623, 1717, 1744
– map 806
– plot 1134
surface structure 131, 858, 872, 873, 881, 884, 903, 908
– multicellular 1411, 1416
surface temperature
– simulation 967
surface tension 515, 1191, 1192, 1322, 1348, 1509, 1558
– force 1562, 1563
surface topography 833, 1354
– gray-scale plot 848
surface with hair 1413
surface-directed CNT growth 415
surface-emitting laser (VCSEL) 435
surface-mounted biofilter 523
surfactant 122, 123, 127
– monolayer 861, 878, 882, 884, 885, 892–894, 896, 902, 905, 906
suspended
– beam 1248
– membrane 1790
sustained drug delivery 519
switch
– metal-to-metal contact 379
– molecule-based 23
– monolayer 35
– nanomechanical 220
– two-state 27
switching
– device 22
– energy 155

– network 371
– speed 155
SWNT (single-walled nanotube) 49, 62, 691
– adsorption property 75
– assembly 413
– based FET 98
– based material 75
– biosensor 1682
– catalyst-based 73
– conductance 87
– direct gap material 408
– doubly clamped 419
– electronic property 406, 407
– electronic structure 406
– epoxy composite 95
– field-effect transistor (FET) 409
– flexural modulus 79
– gauge factor (GF) 419
– in situ growth 415
– matrix interaction 95
– mechanical property 406
– metallic 406
– optical property 408
– piezoresistive property 407
– production 412
– production technique 53
– property 405
– resonant inertial balance 406
– semiconducting 406
– sensor fabrication 412
– structure 49, 405
– surface area 75
– synthesis 412, 413
– tensile strength 79
– thermal conductivity 79
SWNT (single-walled nanotube) sensor
– biochemical 416, 417
– chemical 416
– design consideration 404
– piezoresistive 418
– resonator 420
sync pattern 1627
synchronization 1627
syndrome associated coronavirus (SARS-CoV) 444
synthesis 121
– of SWNT 412
– reactor 55
– yield 55
synthetic vesicle 1706
syrphid fly
– end of the leg 1555
system integration of MEMS components 1816

T

Tabor parameter 1744
talin 1194
tape 1545
– lubricated 1386
– unlubricated 1386
tapping amplitude 720
tapping force 714
tapping mode (TM) 580, 732, 800, 801, 829, 1070, 1454, 1704
– AFM 712, 715, 738, 757
– etched silicon probe (TESP) 589
Taylor dispersion pattern 517
Teflon layer 945
TEM environmental chamber 133
temperature 900
– critical 674, 689
– dependence of friction 938
– Flory 876
– sensing 393
– theta 876
temperature-dependent resistance 138
template 123, 129, 202
– approach 170
– assisted self-assembly (TASA) 262
– assisted synthesis 121, 122
– effect 1410
– filling 180
– material 170
– nanowire growth 124
– synthesis 121
templated particle assembly 188
templating 1428
– technique 67
tenent hair 1527
tensegrity structure 1186, 1188
tensile
– load 1787
– stress 429, 1231, 1243, 1785, 1796
– stress relaxation 1772
– test 1797
– testing 797
tensile deformation 1073
– effect of ethnicity 1106
– effect of soaking 1106
terminal element 1538, 1555
test environment 1302
test specimen 1771
tether
– length 772
– nanomechanical 382
tetraarylmethane 21
tetracene 28
– channel 29
tetracysteine motif (TCM) 491
tetrahedral 399
tetrahydroperfluorodecyltrichlorosilane (PF_3) 1463
tetramethyl ammonium hydroxide (TMAH) 239, 504
tetramethylsilane (TMS) 1465
Texas Instruments (TI) 250
T-filter 517
therapeutic device 457
therapeutic nanodevice 453
– assembly approach 461
– utility 461
therapeutics 1698
– commercial 461
thermal
– conductivity 146
– conductivity of nanowire 145
– CVD 626
– desorption 1338
– drift 664, 667
– effect 935
– expansion 445
– expansion coefficient 638
– fluctuation 925, 1187
– fluctuation force 878
– frequency noise 665
– NIL 271, 275, 276, 302
– noise imaging 1018
– processing 1229
– sensing 1605
– stability 135
– stress 1851
thermal conductivity
– temperature dependent 146
thermionic emission 410
thermocoefficient 143
thermocouple 143
thermoelectric
– application 154
– figure of merit 145
– figure of merit (ZT) 154
– property 143
thermogravimetric analysis (TGA) 84
thermomechanical
– noise 688
– write/read 1604
thermoplastic material 272
thermoplastic molding 275
thermoplastic resist 277
thermoplastics 507
thermopneumatic actuation 508
thermostat 958

theta
– condition 876
– temperature 876
thick film lubrication *see* elastohydrodynamic lubrication
thin film 279, 970
– deposition 393
thiol terminated SAM 320
third harmonic generation 149
third-body molecule 982
– frictional force 983
three-axis sensing 368
three-body deformation
– contact region 826
three-digit input/output string 28
three-dimensional (3-D) 212, 272, 273, 473, 1015, 1558
– force field spectroscopy 694
– wax tubule 1500
three-input NOR gate 30
three-level hierarchical model 1569
three-level model
– adhesion coefficient 1573
three-state molecular switch 26
three-terminal device 38
three-way microvalve system 508
through-thickness
– cracking 1286
thymine 18
Ti atom 655
TiC grain 805
tilt angle (TA) 1420, 1424, 1467, 1476, 1492, 1711, 1714, 1716
tilt configuration 1328
tilted surface 1445
time-resolved studies of emission dynamics 156
time-varying force detection 711
time-varying nanomechanical force 714
timing
– field 1624
– recovery 1627
TiO_2 676
– (110) surface 650
– (110) surface simultaneously obtained with STM and NC-AFM 651
– substrate 655
TiO_2(110) 656
tip
– apex 689, 948, 1607
– artifact 643
– atom 665
– bound antibody 779
– bound antigen 772
– cantilever assembly 624
– carbon nanotube 624
– conductive 630
– displacement 1623
– electron-beam deposition 624
– fabrication 621
– Fe-coated 653
– focused ion beam 624
– geometry 624
– immobilized ligand 766
– induced atomic relaxation 690
– jumping 982
– load 1615
– mount 577
– multiwalled carbon nanotube 795
– oscillation 740
– oscillation amplitude 947
– oscillation trajectory 747
– oxide-sharpened 623
– polymer interface 1617
– preparation in UHV 927
– preparation method 578
– radius 1363
– radius effect 587, 1293, 1362
– sharpness 1610
– Young's modulus 713
tip sample force 638
tipless cantilever 627
tip–particle
– distance 36
tip–sample
– adhesion 1379
– distance 638, 647, 691, 732, 744, 749, 798, 823
– electric field 948
– energy dissipation 639
– interaction 801, 947
– interface 1333
– potential 640
– separation 754
– separation distance 1350
tip–sample force 636, 638–641, 688, 692, 712, 719, 735, 747
– frequency spectrum 713
– gradient 640, 641
– time-varying 716
tip–sample interaction 641, 688, 712, 713, 737
– chemical 689
– electrostatic 695
– hysteresis 720
– van der Waals 691
tip–surface
– distance 780
– interaction 944, 979
– interface 963
– potential 928
tissue engineering 301
titanium oxide (TiO_x) 319
Ti–O–Si bond 319
Tokay gecko 1556
tolerance to contamination 1542
Tomanek–Zhong–Thomas model 805
Tomlinson model 818, 923, 928
– finite temperature 935
– one-dimensional 928
– two-dimensional 929
top-down 389, 1653
– approach 18, 154
– method 1664
topographical
– asymmetry 653
– defect 149
– image 580
topography
– and recognition image (TREC) 780
– induced effect 808
– measurement 583
– of human hair 1068
– scan 796
topped pyramidal asperity 1710
toroid structure 467
torsion mirror 372
torsional
– dissipation 948
– harmonics 722
– mode 717
– resonance (TR) 800, 1069, 1070
– vibration 716
torus model calculation 607
tosylate 1375
total adhesion force 1575
total internal reflection (TIR)
– fluorescence (TIRF) 493
– microscopy (TIRM) 864
toxicity monitoring 526
toxin 441
TR
– amplitude 801
– mode 800, 801, 812
trace and retrace 809
track
– centerline 1625, 1626
– etched polymer 121
– following 1625
– pitch (TP) 1624
tracking servo loop 1625
traction force microscopy 1182
transcription factor 1184
transducer 427

transistor 152, 324
transition criteria
– Cassie–Baxter 1474
– Wenzel 1474
transition criteria range 1476
transition metal
– complex 38
– oxide 652
transitions between smooth and stick–slip sliding 901
transmembrane transport 779
transmission electron microscope (TEM) 131, 132, 212, 257, 622, 830, 1057
transmittance 298
transport
– active 394
– in carbon nanotubes 409
transverse
– magnetoresistance 142
transversel piezo-resistive effect 604
trap-flower 1419
trapping
– constant 1016
– site 1767
traveling direction of the sample 594
TREC imaging 781
triangular cantilever 587, 597, 602, 795
triaryl 21
tribo-apparatus 1776
tribocharging 1766
tribochemical reaction 841, 1358
– mechanism 1721
tribochemistry 984
triboelectric contact 1159
triboelectrical emission 1338
triboelectrification 867
tribological
– C_{60} 997
– characterization of coating 1283
– performance of coating 1298
– problem 1669
triboluminescence 867
tribometer 1684
tribotest apparatus 1347, 1369
trichome
– water absorbing 1423
triethanolamine 22
triethoxysilane 937
triflamide 1375
trimer tip 734
trinitrotoluene (TNT) 441
triple line 1449
tropical plants 1423
true atomic resolution 640–642, 689
– images of insulators 647
truth table 25, 29
tubelike carbon nanotube 625
tubular nanostructure 1491
tubule-forming nanostructure 1496
tungsten 578
– sphere 925
– tip 578
tuning fork force sensor 750
tunneling 410
– current 575, 635, 636, 638, 639, 650, 676
– detector 579
– junction 32
– tip 686
turbulence 535
turn-on voltage 137
twirling 558, 560
two-digit input string 28
two-dimensional (2-D) 212, 272, 1190
– electron gas (2-DEG) 695
– electron system (2-DES) 681
– surface 1450
two-level
– hierarchical model 1569
– model 1568
– structure 1467
two-photon excitation
– cross-correlation spectroscopy (TPE-FCCS) 492
– fluorescence cross-correlation spectroscopy (TPE-FCCS) 492
two-state switch 27

U

ultrahigh vacuum (UHV) 630, 641, 668, 731, 745, 927, 961
– environment 931
– UHV-AVM 927
ultrananocrystalline diamond (UNCD) 1776
ultrasmooth surface 823
ultrathin DLC coating 1273
ultraviolet (UV) 1418
Umklapp process 146
unbinding
– force 765, 770, 771
– force-driven 775
unbonded Z-DOL film 1355
uncoated Si 1379
underwater
– adhesion 1542, 1543
– superhydrophobicity 1430
– use 1429
undoped polysilicon 1689
undulation
– force 859, 878
unfilled polymer 1249
unidirectional electron transfer 36
unified modeling language (UML) 458, 459
unimolecular level 30
unit
– cell 1186
– gravity 548
universal reporter 396
unloading curve 965
unlubricated
– motor 1724
– Si(100) 1364
– surfaces *see* dry surface
unnatural AA (UAA) 471
unperturbed motion 640
useful fiber radius 1579
user-defined microarray 396
UV exposure 283
UV-curable resist 276, 289, 290
UV-LIGA 505
– lithography 517
UV-NIL 271, 276, 282, 283, 302
UV-transparent stamp 271

V

vacancy
– diffusion 1769
van der Waals (vdW) 636, 639, 654, 657, 1639
– adhesion 1445, 1565
– adhesive force 1585
– attraction 820, 1354
– attractive force 1708
– force 648, 720, 733, 795, 858, 859, 864–866, 868, 871, 872, 1147, 1530, 1554, 1706
– interaction 69, 279, 1707
– interaction energy 866
– molecular force 1763
– surface 691
van Hove singularity 408
vapor barrier 537
vapor grown
– carbon fiber 63
– carbon nanofiber 66
vapor phase cocondensation 320
vapor–liquid–solid (VLS)
– mechanism 71, 124

Subject Index

vapor-phase epitaxy (VPE) 236
vapor-phase etching (VPE) 242
variable capacitor
– surface-micromachined 374
variable force mode 610
variable heavy–heavy (VHH) 474
variable temperature STM setup 666
vehicle stability 367
velocity
– critical 902, 905, 907, 908
– rescaling 959
velocity dependence
– of friction 937
vertical
– coupling 577
– nanotube 1000
– noise 641
– RMS-noise 689
– torsion mirror 372
vertical-cavity surface-emitting laser (VCSEL) 1222
vertically aligned MWNT 70
Verwey transition temperature 696
very large-scale integration (VLSI) 1602
vibrating droplet 1468, 1505
– model 1505
vibration 747
– external 577
– model 1505
vibrational energy 1485
vibrational spectroscopy 676
vibrometry 1763
vibromotor 371
Vickers
– indentation 1235
– indenter 1230
vinculin binding site (VBS) 1195
vinylidene fluoride 654
virgin Caucasian hair 1081
virgin hair
– charge deposition 1159
virus 444
viscoelastic 278
– mapping 792, 800
– model 1616
– property 800
– solid 1186
viscoelasticity 1219, 1221, 1534
– mapping 798, 838
viscosity 278, 279, 444, 862, 1526
– local 1017
viscous
– damping 736, 769
– dissipation 804
– drag 1015
– force 558, 885, 896, 900, 1532
VLS (vapor–liquid–solid)
– growth 128
– mechanism 71
VLSI (very large-scale integration) 1602
volatile organic compound (VOC) 440
voltage/current sourcing 393
voltage-controlled oscillator (VCO) 373, 1628
Volterra series 1623
volume definition 543
vortex in superconductor 697
V-shaped cantilever 589, 926

W

W tip 579
wafer 293
– bonding 1818
– curvature technique 1796
– singulation 1820
walking machine 1543
wall-climbing robot 1553, 1585
wall-walking robot 1543
water
– contact angle 1589
– film thickness 1573
– repellent plant 1427
– roll-off (tilt) angle 1438
– roll-off angle 1448
– vapor 439
– vapor content 1302
water–air interface 1484, 1509
waterproofing of clothes 1429
water-repellent plant 1709
wax 1400
– chemistry 1404
– compound 1404
– crystal 1405, 1406, 1410, 1419
– film 1416
– morphology 1404, 1405
– platelet 1410, 1493
– recrystallization 1408
– structure 1403, 1423
– terminology 1405
– three-dimensional (3-D) 1406
– tubule 1406, 1496
– two-dimensional (2-D) 1406
waxy surface 1538
wear 790, 955, 1271, 1285, 1309, 1663, 1741
– contribution to friction 943
– damage 1302
– damage mechanism 1301
– debris 831
– depth 477, 829, 1365
– fullerene 997
– initiation 1366
– map 1700
– mapping 819
– mark 1727
– measurement 574
– mechanical 514
– mechanism map 819
– nanoscale 828
– of biomolecule 1698
– process 1300
– profile 1365
– region 833
– resistance 1298, 1301, 1327
– test 477, 1115, 1300, 1382, 1696
– testing 1755
– tip 982
– track 977
wear *see* damage
wear and stiction
– mechanism 1727
wear detection 1367, 1381
– electrical resistance measurement 1369
wear mark
– AFM image 830
wearless friction 985
web spider 1528
Weibull
– distribution function 1853
– statistics 1797
weight function 641
well aligned nanowire 131
Wenzel
– equation 1442, 1513, 1712
– model 1471
– regime 1452, 1478, 1495, 1713, 1717
– state 1441, 1451, 1476
– transition 1452
Wenzel–Cassie transition 1452
wet adhesion 1526, 1528, 1535, 1544
wet environment 820
wettability 720, 841, 1375, 1420, 1496, 1692, 1708
wettable leaf surface 1455
wetting 883, 1446, 1472
– behavior 1421, 1516
– contrast 196, 202
– dewetting 1480
– dewetting cycle 1485
– hysteresis 1451

– phenomena 1544
– regime 1451, 1493
– surface 962
winding channel 561
window opening 277, 284
wire
– boundary scattering 142
– cantilever 587
– grid polarizer 299
work
– of adhesion 883, 1556
write/read scheme 1613
writing mechanism 1614

X

xenon 668
xerography 194
XNOR 25
XPS spectra 1388
x-ray
– analysis 131
– diffraction 134, 135
– diffraction pattern 122
– energy dispersion analysis (EDAX) 134
– imaging 1779
– lithography 1730
– photoelectron spectroscopy (XPS) 1321, 1378
– photon spectroscopy (XPS) 1145
– powder diffraction (XRD) 1407
XRD characterization 122

Y

yield point 884, 885, 901
yield strength
– simulation 967
Young equation 1442, 1446
Young's modulus 93, 215, 594, 610, 665, 939, 1174, 1557, 1643, 1719, 1783, 1784
– effect of humidity 1094
– mapping 1150
– measurement 1788
– of elasticity 789, 798, 836, 838
Young's modulus of hair
– effective 1094

Z

Z-15 841, 1348
– coefficient of friction 1357
– fully bonded 1357
– Z-DOL 1357
Z-15 film 841, 1392
– molecularly thick 1362
– wear depth 1364
Z-15 lubricant 840
Z-DOL 1348
– diffusion coefficient 1351
– durability 1373, 1374
– film 849
– molecule 1349
– schematic 1366
– unbonded 1350
Z-DOL (fully bonded) 841, 1349, 1392, 1693
– durability 1365
– hydrophobicity 1364
– wear depth 1364
zeolite 64
zero-deflection (flat) line 1353
zero-shear viscosity 275
zeta potential 176, 512
zigzag tube 407
zigzag-type SWNT 50
zinc
– nanowire 141
Zisman plot 1322, 1330
Zn nanowire 144
ZnO nanowire 129, 131, 135, 149, 150, 155
zone folding 405, 406
Z-TETRAOL 1366, 1371, 1382
– durability 1373, 1374
– property 1377
– schematic 1366